BRIDGING RESEARCH AND GOOD PRACTICES TOWARDS PATIENT WELFARE

PROCEEDINGS OF THE 4TH INTERNATIONAL CONFERENCE ON HEALTHCARE SYSTEMS ERGONOMICS AND PATIENT SAFETY, TAIPEI, TAIWAN, 23–26 JUNE 2014

# Bridging Research and Good Practices towards Patient Welfare

## Healthcare Systems Ergonomics and Patient Safety 2014

*Editors*

Yuh-Chuan Shih

*Department of Logistics Management, National Defense University, Taipei, Taiwan*

Sheau-Farn Max Liang

*Industrial Engineering and Management, National Taipei University of Technology, Taipei, Taiwan*

CRC Press
Taylor & Francis Group
Boca Raton London New York Leiden

CRC Press is an imprint of the
Taylor & Francis Group, an **informa** business
A BALKEMA BOOK

Published by:
CRC Press/Balkema
P.O. Box 447, 2300 AK Leiden, The Netherlands
e-mail: Pub.NL@taylorandfrancis.com
www.crcpress.com – www.taylorandfrancis.com



ISBN 13: 978-1-138-02716-9 (hbk)
ISBN 13: 978-1-315-73628-0 (ebk)

Typeset by MPS Limited, Chennai, India

*Bridging Research and Good Practices towards Patient Welfare – Shih & Liang (Eds)*
*© 2015 Taylor & Francis Group, London, ISBN 978-1-138-02716-9*

# Table of contents

*Preface* IX
*Preface by Dr. Mao-Jiun J. Wang* XI
*Preface by Wen-Ko Chiou* XIII

## *Part A: Aging and healthcare system*

The "Art of Aging" from an ergonomics point of view 3
*H. Strasser*

Attention performance in elderly people in complex visual environment 17
*Y.-Y. Huang & M. Menozzi*

The incident reporting systems and organizational learning in Indonesian public hospitals 27
*I. Dhamanti, S. Leggat & S. Barraclough*

What can Indonesia learn from Taiwan's successful patient-safety reporting system? 37
*I. Dhamanti, S. Leggat, S. Barraclough, A. Utarini & S. Liao*

The influence of age and object form on the motivation of haptic 47
*S.-J. Chen, C.-W. Chien & C.-L. Lin*

Comparison of public health care system between Taiwan and Japan – Focusing on long-term oxygen therapy 55
*S. Tatsukawa, N. Shintani & A. Kajimura*

Five Qualities (5Qs) and ergonomic approach – "Towards improved patient safety, medical and healthcare quality" 61
*M. Zineldin, J. Zineldin & V. Vasicheva*

## *Part B: Healthcare, mobile application and usability*

How do you support breast cancer patients on work-treatment balance? 71
*N. Shintani, H. Nagaoka, E. Koike, S. Tatsukawa & A. Kajimura*

Patient safety in the home: Examining the relationship between therapeutic self-care and adverse events for home care clients in Canada 79
*W. Sun & D. Doran*

Designing user interfaces of a handheld device for physical therapists 87
*S.-F. Max Liang, L.-W. Cheng, P.-F. Jennifer Tsai & C.-L. Rau*

A scoliosis survey in children with overweight at age of 6 to 13 years old 93
*C. Yin Wu & M.-J. Wang*

A review of the effect on postural stability while using mobile phone 101
*N.R. Nurwulan, H. Iridiastadi & B.C. Jiang*

A heuristic evaluation on the usability of health information websites 109
*Y.N. Chen & S.L. Hwang*

Investigation on lifestyle factors that influence obesity of adult workers 117
*N. Takayama, T. Amemiya, K. Nishikawa, Y. Suzaki & H. Ariyoshi*

Evaluation of the accuracy of mobile healthcare applications 123
*C.-J. Su & S.-F. Huang*

## *Part C: Safety, hazards and MSDs*

Importance of occupational hazards in nurses MSD symptoms 133
*F. Serranheira, A. Sousa-Uva & M. Sousa-Uva*

The survey of sharps injuries by unsealing the neck of ampoule and safety tool design 141
*C.-W. Lien, B.-S. Liu & J.-T. Lee*

Effect of different wrist positions on median nerve cross-sectional area at proximal carpal tunnel 149
*P. Yeap Loh, H. Nakashima & S. Muraki*

Relationships of musculoskeletal disorder symptoms and perceived workload among hospital workers 155
*T. Ryu, S. Kwon, S.-C. Jee, M.-D. Oh & M. Hwan Yun*

A survey study on musculoskeletal pain: Prevalence, regional distribution and risk factors among Indian physiotherapists 163
*D. Sharan, M. Mohandoss, R. Ranganathan & J.S. Rajkumar*

Risk factors for musculoskeletal disorders in healthcare professionals: A systematic review 167
*D. Sharan, M. Mohandoss, R. Ranganathan & J.S. Rajkumar*

Factors affecting the perceived slipperiness rating 175
*W.-R. Chang, M.F. Lesch, S. Matz & C.-C. Chang*

An integrated programme to promote safe surgery best practices in Italian hospitals 181
*R. Onofrio, P. Trucco, C. De Capitani, F. Lorenzi, M. Montinari & E. Pernazza*

Consideration of occupational stressors that influence senses of satisfaction at work in Japan 191
*A. Kajimura, N. Shintani & S. Tatsukawa*

## *Part D: Simulation, modeling and decision making*

Integration of biophysiological information with a point-of care decision support system to promote patient safety at home 199
*W. Sun & D. Doran*

Simulation study on the impact of working time systems on the work-life balance of employees in a hospital 207
*G. Zülch, P. Stock, D. Schmidt & M. Leupold*

Application of a modified SIR-based model-case of H1N1 in Taiwan 213
*K.H. Yang, C.Y. Juan, J.Y. Hsu & Y. Chen*

A computer-aided system to categorize characteristics of children with special needs 223
*C.Y. Juan, J.W. Wu & K.H. Yang*

Simulation-based optimization method for operation scheduling using combined heuristics rules 229
*Y. Guo, F. Liang, Q. Wang & J. Sui*

Finding significant factors for readiness of ISO 9001 implementation in a health care system 239
*S.I. Chang, K. Douglas-Mankin, L.A. Ang & B. Tavakkol*

Optimizing surgery schedule with PICU nursing constraints 251
*D. Ben-Arieh*

New mathematical models for vector-borne disease: Transmission of tomato spotted wilt virus 259
*J. Wu, S. Zhao, Y. Kuang, D. Ben-Arieh, J. Nechols & D. Margolies*

Medical device training classification system 269
*M. Rys, M. Prieb & C. Taylor*

## *Part E: Environment and system design*

An experience and evaluation of objective structured clinical examination in nursing education at Kaohsiung Medical University 281
*S.-Y. Jian & F.-H. Chou*

Retained Surgical Instruments: Using technology for prevention and detection 289
*Y.-C. Teng, J. Ward, T. Horberry, P.J. Clarkson & V. Patil*

Holistic enhancement in hip surgery by design 299
*A. Albayrak, R.H.M. Goossens, S.B.W. Vehmeijer, H. Bodewes, T. Sunderman & J. Lancaster*

Establishment of a checklist for triaging STEMI patients in the emergency department 309
*R.F. Lin, R. Pang, K.-C. Tsai & A.-H. Li*

Examined the CCT 4000K lighting environment on visual fatigue during working 319
*C. Chou, T.-Y. Chen & C.-Y. Chen*

Facilitating teamwork in orthopaedic surgery by design 329
*M. Melles, G. Anastasiadis, C.C.M. Moes, R.H.M. Goossens, J. Lancaster & S.B.W. Vehmeijer*

Classification of soothing music using Fuzzy C-Means clustering algorithm 337
*Y.-W. Hsu, H.-P. Tsai, M.-C. Chiu, S.-L. Hwang, H.-L. Shih, F.-T. Huang & C.-T. Lee*

A tangible user interface of map for hospital wayfinding 347
*C.-W. Chen & Y.-F. Kao*

## *Part F: Human factors and product design*

A user centered design approach for minimal invasive surgery 357
*A. Albayrak, M. Melles & R.H.M. Goossens*

The study of anthropometry of Indonesian infant age 0–2 years in Jakarta 363
*Yanto, C.-W. Lu & D.C. Lioenita*

Effects of virtual scene parameters on shoulder joint motor performance during throwing movement 371
*C.-H. Huang & T.-L. Sun*

Fitting pressure garments analysis for burn victim's scars 379
*P.-W. Huang & C.-W. Lu*

Assessment of range of joint motion using Kinect 387
*J.-L. Lin, H.-C. Chuan, P.-C. Kuan & S.-M. Ni*

Synchronizing the arm-speed of toothbrushing with sounds of differing tempos 393
*T.Y. Wu & C.-T. Lin*

The anthropometric data for Taiwanese female military personnel 399
*Y.-C. Shih, C.-C. Pao, C.-L. Lin & L.-W. Liu*

A study of five "F" in product design 409
*C.-W. Chien, C.-L. Lin & R.-T. Lin*

Author index 419

*Bridging Research and Good Practices towards Patient Welfare – Shih & Liang (Eds)*
*© 2015 Taylor & Francis Group, London, ISBN 978-1-138-02716-9*

# Preface

The theme of this book is "Bridging research and good practices towards patient welfare", which is one way to improve healthcare systems and patient safety. This volume contains forty-nine selected papers contributed to the fourth HEPS (Health care Ergonomics and Patient Safety) conference, which was held from June 23 to 26, 2014, at the Grand Hotel in Taipei, Taiwan. These papers address the latest research and applications, covering the areas of Aging and Healthcare Systems; Healthcare, Mobile Applications, and Usability; Safety, Hazards, and MSDs; Simulation, Modeling, and Decision Making; Environment and System Design; and Human Factors and Product Design.

This collection is intended for people who are working in healthcare systems, those who are interested in products/procedure design/improvement of the healthcare systems and patient safety, and ergonomists.

The success of the conference was due in great part to the support of the sponsors, especially the Ministry of Science and Technology, Ministry of Economic Affairs, Ministry of Labor, Ministry of Foreign Affairs, Ministry of Health and Welfare, Department of Information and Tourism, Taipei City Government, National Defense Industrial Development Foundation, Taiwan Joint Commission on Hospital Accreditation and Quality Improvement, Ergonomics Society of Taiwan, National Tsing Hua University, Chang Gung University, and Chung Yuan University. We are immensely grateful for the contributions of these organizations to the conference.

Editors
Yuh-Chuan Shih
Professor of National Defense University, Taipei, Taiwan

Sheau-Farn Max Liang
Associate Professor of National Taipei University of Technology

*Bridging Research and Good Practices towards Patient Welfare – Shih & Liang (Eds)*
*© 2015 Taylor & Francis Group, London, ISBN 978-1-138-02716-9*

# Preface by Dr. Mao-Jiun J. Wang

Health care ergonomics and patient safety (HEPS) conference is a triennial conference. The first HEPS conference starts in Florence, Italy in 2005. This is the fourth HEPS conference. This year we have more than 300 delegates from about 20 countries to participate the event. The theme of this year's conference is "Bridging research and good practices towards patient welfare".

Like many other countries, Taiwan is already an aging society. The government is promoting active aging and wellness programs to facilitate the well-being of the society. The Ministry of Science and Technology launched special research programs in the development of innovative assistive technology from APP to ICT devices. The Ministry of Health and Welfare also initiated long-term care service act and long term care 10 year's strategic plan to enhance the health care service quality and effectiveness. This timely trends and growing demands provide unique opportunities for ergonomics and health care professionals to have important contribution to the society through research and development, training and education. Certainly, this conference can serve as a platform for exchanging ideas, sharing research findings, bridging researchers and practitioners, and building collaboration networks.

After the conference, the organizing committee of HEPS 2014 decided to select about 50 papers presented at the conference to publish a special edited book entitled "Bridging Research and Good Practices towards Patient Welfare". The topics covered in this special book include: smart healthcare, aging care, patient safety, MSD prevention, anthropometry, and etc. The book can serve as a very useful reference in Health Care Ergonomics related areas.

We would like to thank Vice President of Taiwan ROC, Mr. Den-Yih Wu, and Deputy Minister of Ministry of Health and Welfare, Dr. Tzou-Yien Lin and Commissioner of Department of Health, Taipei City, Dr. Chih-Hung Lin to show their great support to this conference and to deliver opening remarks. I would like to take this opportunity to thank President of Ergonomics Society of Taiwan, Prof. Chin-Jung Chao and his team who have done a remarkable job in organizing this conference. Indeed, this conference was a great success.

Conference Chair
Dr. Mao-Jiun J. Wang
Chair Professor of National Tsing Hua University

*Bridging Research and Good Practices towards Patient Welfare – Shih & Liang (Eds)*
*© 2015 Taylor & Francis Group, London, ISBN 978-1-138-02716-9*

# Preface by Wen-Ko Chiou

Patient safety is a global concern. In 2004, the World Health Organisation (WHO) initiated the World Alliance for Patient Safety to promote infection control, fall prevention and medication safety. These areas were highlighted by the US Institute of Medicine (IOM) as requiring urgent attention. Patient safety is impacted by a wide range of issues including improper prescription of medication, physician fatigue, and poor working conditions for paramedics. Increasingly, efforts to improve patient safety have included the systematic implementation of health care policies, certification, review and professional censure of medical professionals and institutions; however, there is still much room for improvement.

In 2000, the International Ergonomics Association (IEA) defined Human Factors and Ergonomics (HF/E) as including three domains: physical ergonomics, cognitive ergonomics and organizational ergonomics. The broad application of these three domains can facilitate research into patient safety, medical malpractice and adverse medical outcomes events; it can also help to improve patient safety in terms of medication, falls, surgery, blood transfusions, medical treatment, public safety, injuries, pipeline flow planning, sudden cardiac arrest, narcosis and medical inspections.

Taiwan's medical system is dominated by two institutions: the Hospital Accreditation System (HAS) and the National Health Insurance (NHI) system. Launched in 1978, Taiwan's HAS program was the first in Asia and the fourth globally, and aims to ensure high quality patient-centred treatment. The NHI system was established in 1994 and offers patients full freedom in choosing doctors and hospitals. High public satisfaction and accessibility rates have made Taiwan's public health system an example to which other countries now turn for insight and experience. To further promote patient safety, the Taiwan Patient-safety Reporting (TPR) System has recorded more than 100,000 medical incidents, with many being selected as case study material.

The Taiwan Joint Commission on Hospital Accreditation and Quality Improvement has collaborated with the Ergonomics Society of Taiwan to establish the Good Ergonomics Practices Database, which focuses on healthcare and patient safety, in connection with the IEA Good Ergonomics Practices Database. Past efforts to promote TPR and Root Case Analysis (RCA) in Taiwan have been very successful in preventing medical malpractice and adverse medical outcomes. RCA is used to improve project planning, execution and modification in the PDCA (Plan, Do, Check, Action) process and a feedback learning loop. Based on lessons learned from in a series of successful HEPS conferences, one important criteria for determining the host country was to evaluate the bids—for the 2014 HEPS conference, preference will be given to the proposals that are outside of Europe. Taiwan's healthcare systems feature many aspects of good ergonomic practices and, in hosting HEPS 2014, we hope to share these experiences.

"Linking Research and Good Practices to Promote Patient Welfare" is the focus of HEPS 2014. With its human-centred philosophy and scientific solutions, HF/E plays an indispensable role in ensuring effective healthcare and patient safety. In addition, to benefit patients, research results must be applied in practice. I hope that this focal themeand living document can help us move in the right direction. This premise has been inspired by Taiwan's notable achievements and unique experiences in healthcare management and patient safety, and hosting HEPS 2014 gives us a platform to share success stories and insight from Taiwan and our colleagues from around the world.

This compilation would not have been possible without considerable effort and assistance of many individuals. I am especially indebted to Chin-Jung Chao, Hsieh-Ching Chen, Sheau-Farn Liang, and Yuh-Chuan Shih, who devoted many hours to this publication. Also, I wish to thank the staff of the HEPS Executive and Publication Committee who created this final electronic version.

Finally, I am most grateful to each of the present contributors who helped to translate this work from a dream to a reality.

Wen-Ko Chiou
Chang Gung University, Taiwan
August 2014

# *Part A: Aging and healthcare system*

*Bridging Research and Good Practices towards Patient Welfare – Shih & Liang (Eds)*
*© 2015 Taylor & Francis Group, London, ISBN 978-1-138-02716-9*

# The "Art of Aging" from an ergonomics point of view

H. Strasser
*Ergonomics Division, University of Siegen, Siegen, Germany*

ABSTRACT: After some introductory comments on unavoidable biological changes as a consequence of advanced age, the role of the human operator in the control loop "work" is addressed, whereby it may not be faded out that the human performance potential is substantially impacted by age. The thereupon following survey on age-related changes from an ergonomics point of view is limited to rough knowledge on physical and sensorimotor performance, the course of aging on the vision and the hearing and on central nervous and cognitive age-related changes. The not always pleasant facts that will emerge are garnished with aphorisms and qualifying "wisdoms on age". The latter convey the main ideas on the "art of aging".

## 1 INTRODUCTION

When reporting on the topic "The Art of Aging" from an ergonomics point of view, special attention has to be given to the aspect of work potential in dependence on age. The survey of the course of aging of some performance functions from an ergonomics point of view does not have the same scientific rigor as a paper of technical sessions. Instead, it is limited to some fundamental, unavoidable biological changes, whereby the not always pleasant facts that will emerge are garnished with some "aphorisms" (see amongst others N.N. 1986) that will hopefully encourage people to say "yes" to the course of the working life and to help accept that some things will be different at an advanced age than people would like them to be. Even if positive thinking is practiced – and who would be opposed to that? – and even if nowadays to apply the classical "deficit model" is refused, changes that will come with old age cannot be ignored. Two statements that are polar opposites, although both contain a kernel of truth, shall represent the starting point of the survey. The first is the popular challenge to "*Stay young, even as one gets older.*" The second one is a poem by Eugen Roth, one of the most read German lyricists of the 20th century, that contains much self-awareness.

> *"Wet at birth – a son or daughter – An infant's made of mainly water, Sucking, drinking, they abound, 'til baby dries and runs around. And dry he gets – behind the ears – so does his humor, over the years. Alas, the human degrades, yet still lives on for decades."*

This is not exactly an enjoyable outlook since – despite the admittedly humorous exaggeration – changes manifest themselves which will invariably catch up with us at some point in the future. After the initially desirable water losses, it is expected, of course, that the not-so-pleasant age-related water losses will not occur for some time. When one thinks about the problems with our skeletal structure and our motor function, it seems unfortunately to be true that, among other things, the intervertebral discs as a sort of liquid cushion between the vertebrae of the spinal column become less elastic. Also, the skeletal bones become more brittle and the body's height, due to the compressed discs and the humpback, will shrink by a few centimeters. As will be shown later, it is partially this age-related dehydration that causes the decreased ability of the eye's lens to adjust its focus between far and near, which occurs at a relatively young age. While the eye's vision may diminish quickly, there also is a positive side, according to Jean Paul, to the vision of the spiritual eye: "*The more sand has run through the hour glass, the clearer is the view through the empty glass*". Also it does not take a scientist to determine that the formation of wrinkles on the

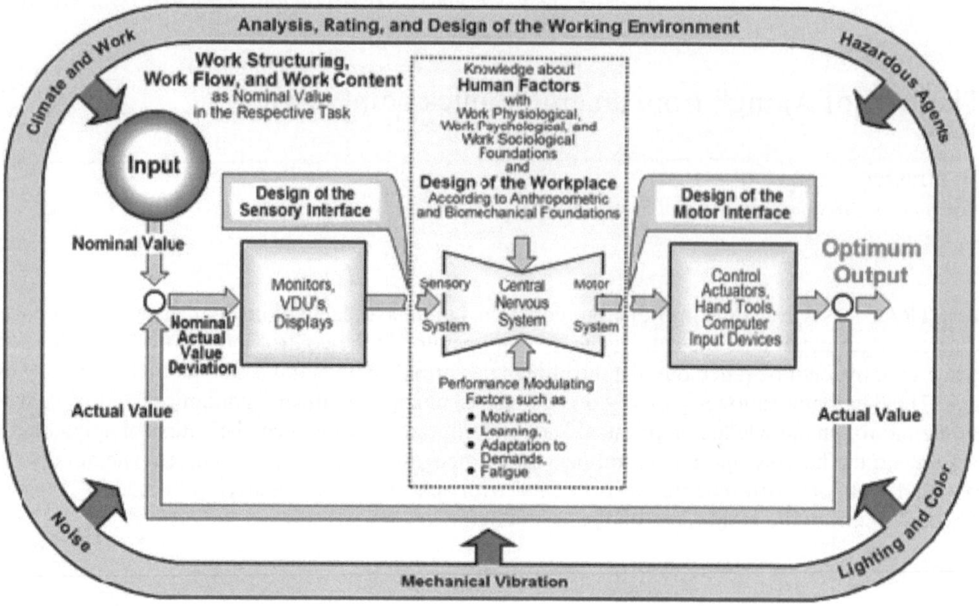


Figure 1. Control system "work".

aging skin is due to a loss of water. While we enjoy drinking for the most part, even large amounts will not prevent such wrinkle formation. After this unappealing perspective on drying out and its potential results – if Eugen Roth is indeed correct – more work-relevant insights will be presented. At first, how and under which conditions human work occurs will be examined.

## 2 THE HUMAN BODY IN THE CONTROL SYSTEM "WORK" AS IMPACTED BY AGE

The human operator as a cybernetic entity (cp. center part of Fig. 1) of sensory and motor systems controlled by the central nervous system (CNS) acts as a controller who communicates via his sensory and motor systems with the outer world. Hereby he continuously receives information, e.g. about the work's task and the work flow (as the nominal value) as an input of the control loop, either via direct observation or via displays, monitors or other information-providing instruments. Via his motor system, he operates controls, actuators, tools, and input devices. The result should be an optimal output. This will only be possible, however, if the current output, the actual value, is compared to the target, the nominal value, and any discrepancies (nominal/actual value deviations) are iteratively corrected. Of course, the output can only be optimal if the work design applies the knowledge of "human factors", if specific ergonomic design takes place in the sensory interface, and if the equation "suitable for the human body = suitable for the hand" holds in the motor interface, i.e., if the dimensions of manual tools are based on the characteristics of the human hand-arm system. The guiding principle of work design must be compatibility, in particular stimulus-stimulus compatibility, response-response compatibility, and finally stimulus-response compatibility.

An additional task of ergonomics is of course the optimization of the working environment. It must be ensured that "climate and work" harmonize, that individuals are protected from detrimental acoustic stress (i.e., noise) and "mechanical vibrations" or "hazardous agents," and that "lighting and color in the workplace" are appropriate. As will be demonstrated shortly, age-adequate work design must be discussed. Knowledge about the effect of aging on various performance variables should be incorporated into work design with respect to the sensory system, the CNS, and the motor system.

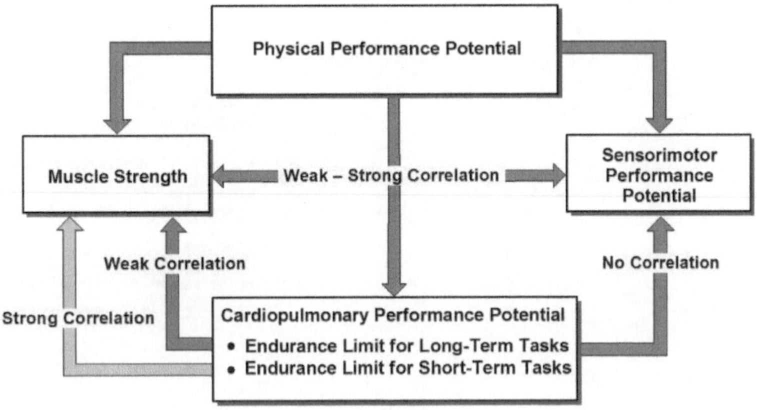


Figure 2. Outline of physical performance potential.

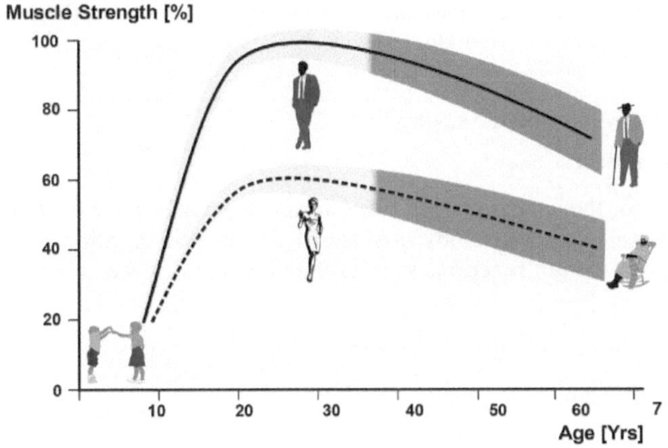


Figure 3. Muscle strength in dependence on gender and age.

## 3 PHYSICAL AGE-RELATED CHANGES IN PERFORMANCE POTENTIAL

Beginning with the functions of the physical performance potential and the existing interdependencies as shown in Fig. 2 and as described by Rutenfranz (1983), "muscle strength" and the "sensorimotor performance potential" have to be considered.

The course of aging of muscle strength in dependence on gender and age that is shown in Fig. 3, probably is well-known (details see, e.g., Hettinger 1960, Rohmert et al. 1992 or Rühmann and Schmidtke 1992). This is less of a concern nowadays with the diminishing role of physical work than in a time when it was the human body's physical strength that was of the highest importance. Many tasks which involve carrying heavy loads can be taken over by robots. In health care, however, it may be of high relevance when, e.g., a heavy patient has to be transported from his bed to a wheel-chair and vice versa.

To maintain a 20- to 30-year old's maximum muscle force, except to health care, will also not be requested over time. With respect to the age-dependent muscle strength it would be foolish, however, to ignore that

> *"While physical strength cannot increase beyond a certain mid-day peak of life, there is no such limit for intellectual strength. This is the true consolation to an aging person".*

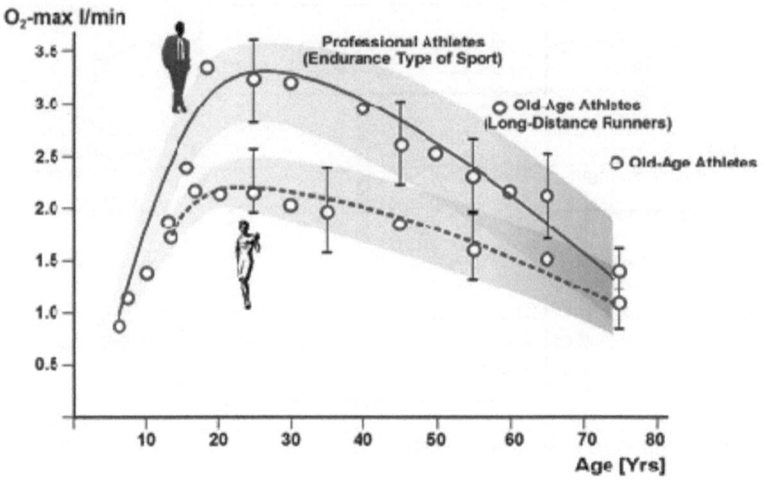


Figure 4. Maximum oxygen uptake of men and women in dependence on age (Means from 2334 subjects, according Hollmann, from Müller-Limmroth, 1984/85).

And there is a positive side to the saying that

*"A rolling stone gathers no moss".*

The same is true for the saying that *"Resting means rusting"*, since even at an advanced age, physical work and exercise keep a body in shape. The range of variation for, e.g., the course of aging of some physical output functions such as the maximum oxygen uptake as shown in Fig. 4 is substantial.

## 4 SENSORIMOTOR COURSE OF AGING

With respect to motor function and swiftness (cp. Fig. 5) the picture, of course, is less rosy.

The required time for gripping and transposing an item is substantially longer for 60-year olds than for men and women around the age of 30.

According to Fig. 6, the same is true for the number of assembled units per time unit. Already 40-year olds and especially 50-year olds cannot keep pace with 30-year olds. Appropriately repetitive tasks, however, should nowadays be left to robots and operation systems. Aging becomes a truly serious issue when it comes to vision and hearing, the two most important sense organs.

## 5 ON THE COURSE OF AGING OF THE VISION

According to well-known early and fundamental research in physiology (e.g., shown in Keidel, 1967) which is also found in serious ergonomics textbooks or encyclopedias (e.g., Grandjean 1979, Kroemer, Kroemer and Kroemer-Elbert 1994, Karwowski 2006) a dramatic decrease in accommodation of the eye lens is associated with aging. Fig. 7 exhibits that twenty-year olds have an average accommodation of more than 10 diopters, i.e., that they are capable of adjusting their eye lens such that an object that is at a distance of approximately 0.1 m from the eye can be focused onto the retina. This minimum feasible distance at which the eye can still focus already increases to 0.25 m for 40-year olds because the refraction power has decreased from 10 to 4 diopters, i.e., less than half the initial value. 50-year olds, having experienced a further decrease to 2 diopters are at the limit of their maximum accommodation when they work at a distance of 0.5 m to a computer

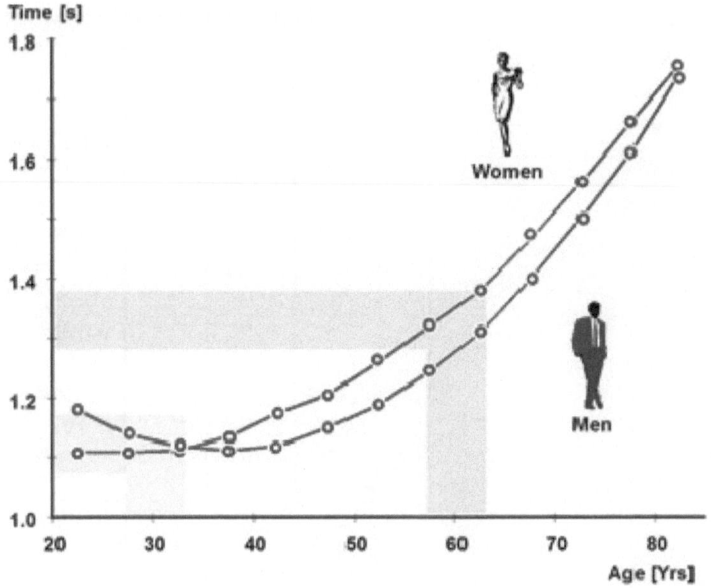


Figure 5. Performance of the right hand during gripping and transposing in dependence on age and gender (according Rohmert 1980, from Müller-Limmroth 1984/85).

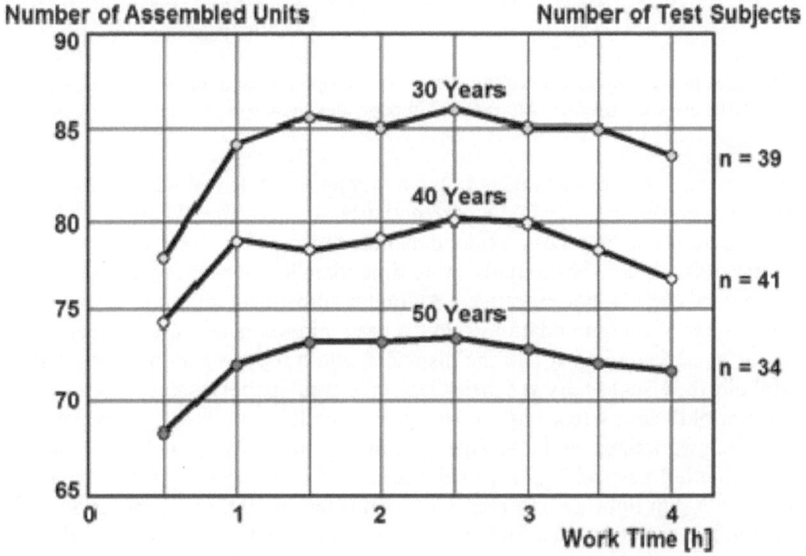


Figure 6. Assembly performance in dependence on age (from Müller-Limmroth 1984/85).

monitor and 60-year olds, with an average value of 1 diopter, are only able to focus on items that are at a distance of at least 1 m from the eye unless they use a type of vision aid. This explains the origins of the saying that

*"The eyes still work fine, but the arms are too short,"*

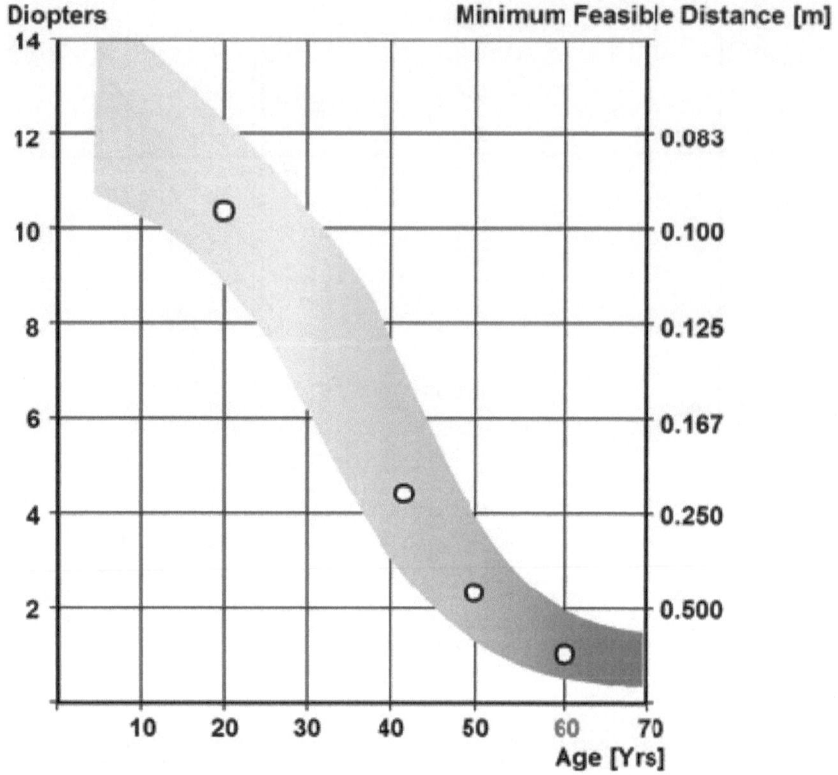


Figure 7. Decrease in accommodation of the eye lens (in diopters) and increase in the (minimum feasible) distance at which the eye can still focus (in meters) in dependence on age (from Strasser 1993b).

when the arms are stretched out while reading a newspaper. While vision aids may compensate for a lack in refraction power, the speed of accommodation, i.e., how quickly the lens can focus on an item that is close by, is also to a good extent dependent on age (cp. Fig. 8).

Even 40-year olds require substantially more time than 30-year olds to adjust their vision from far-away to close-by objects. For example, a 4-diopter adjustment takes almost 1 s. While 40-year olds reach their limit of accommodation at 25 cm after approximately 2 s, 30-year olds are able of accommodation at 12.5 cm, i.e., at half the distance, within the same amount of time.

In a 60-year old, the visual acuity as a measure for the eyes' spatial resolution is still approximately 75% of a 20-year old's acuity (cp. Fig. 9). However, such limitations are still cause for concern if the demands in the workplace are high. This is shown in the following Figures 10 and 11.

As shown in the left part of Fig. 10, visual acuity of 100% (i.e. 1.0), can be attained by 20-year olds even with dim lighting or a marginal luminance of 100, i.e., 1 candela/m$^2$. In contrast, 60-year olds require a 1,000 times higher luminance to achieve the same visual acuity. Hence, older individuals will still have limitations even at high luminance. Younger individuals, e.g., 20-year olds, on the other hand, can double their performance even with less advantageous lighting, and hence experience a substantially better spatial resolution during testing and inspection tasks. Ensuring a visual acuity of 1.0 for 50- or even 60-year olds, which is essential for many industrial inspection tasks, would require a substantial investment in terms of lighting.

In order to achieve a visual acuity of 1.0, due to Fig. 11 even 40-year olds already require an increase in the field of vision's luminance relative to 20-year olds from approximately 7 to 70 candela/m$^2$, and thus a tenfold increase. Alternatively, if the luminance is held constant, the contrast C must be increased from 0.5 to 0.9.

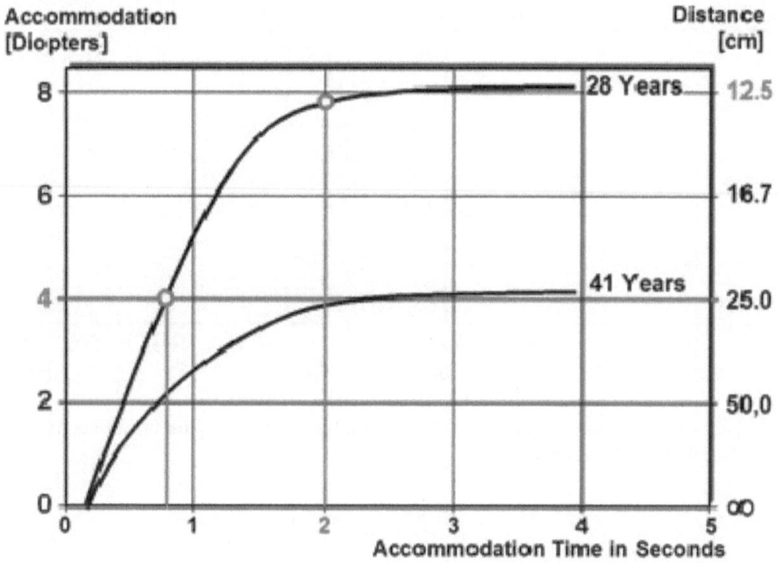


Figure 8. Approximation curves for accommodation time of 40- and 30-year olds during vision adjustment from far-away to close-by objects.

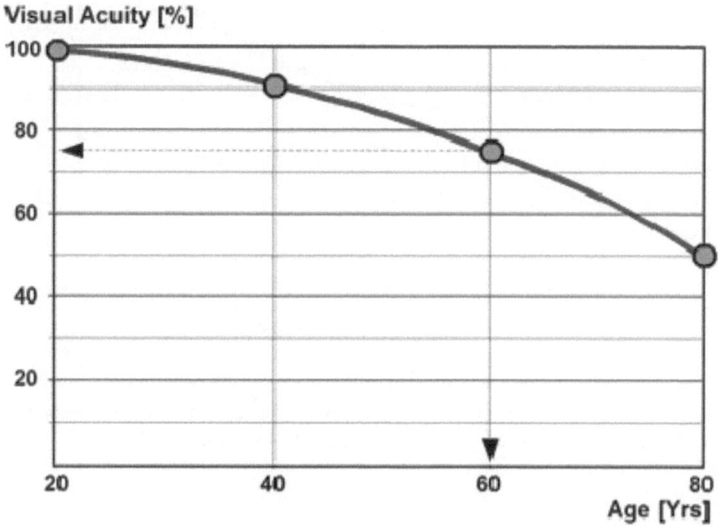


Figure 9. Course of aging of the visual acuity.

In summary, Fig. 12 drastically shows how the vision gets worse with increasing age unless an investment in additional lighting is made. Assuming an average age of 40 for the working population, a 50-year old requires double and a 60-year old five times the amount of light in order to achieve the same visual performance. Relative to a 20-year old who only requires half as much light as a 40-year old, there is a difference of a factor of ten in the need for light between 20- and 60-year olds to provide age-adequate lighting. Thus, the range is on the same magnitude as for the accommodation.

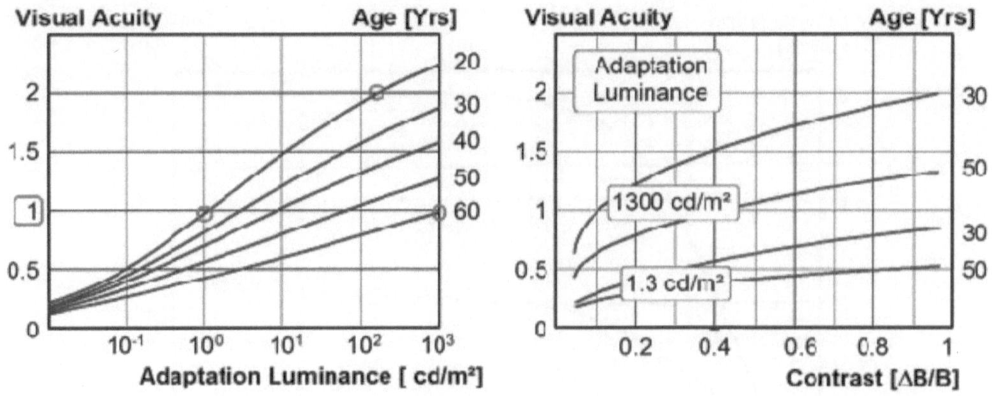


Figure 10. Impact of adaptation luminance (left) and contrast (right) on the visual acuity of different age groups (from N.N. 1964 and Strasser 1993b).

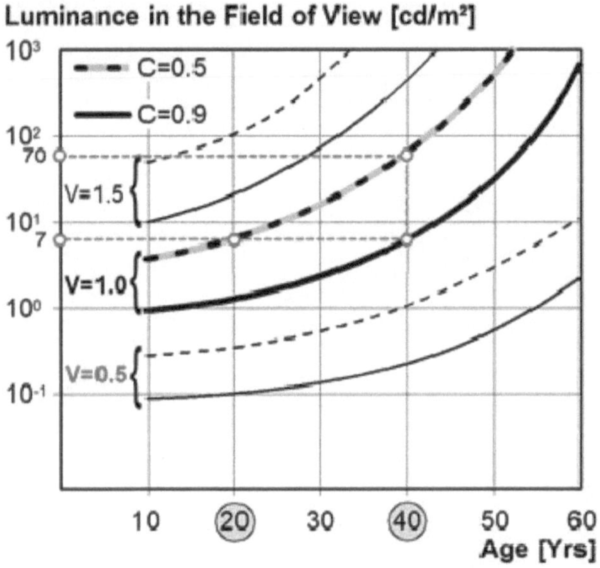


Figure 11. Luminance and contrast requirements with the goal of ensuring that different age groups achieve a certain visual acuity (which is required, e.g., for inspection tasks) (source N.N. 1964).

## 6 ON THE COURSE OF AGING OF THE HEARING

Not only the eye as the most important sense organ, which plays the crucial intermediary role with the environment, with its channel capacity of 80-90%, is affected by drastic age-related performance losses. Unfortunately also substantial age-related performance losses in the hearing must be expected. Especially the acoustics in the upper frequency range are adversely affected (cp. Fig. 13).

A 20-year old's hearing experiences a wonderfully broad acoustic world of frequencies of approximately 10 octaves that are in the audible range from 16 Hz to 16,000 Hz. With every decade, approximately 2 kHz are lost at the upper end of the hearing range, i.e., at the age of 30, the maximum value is 14 kHz. At 40, this value has declined to 12 kHz and at 50, frequencies up to 10 kHz represent the limit that is detectable by the hearing. And once the age of 60 has been reached, the

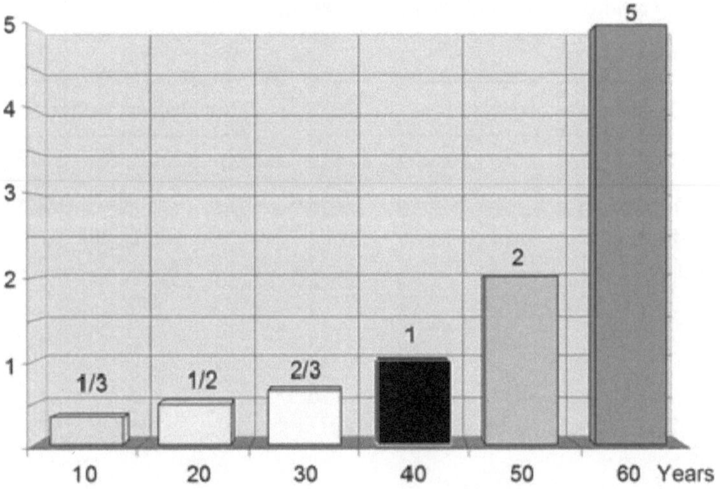


Figure 12. Age and lighting needs as multiples of a 40-year old's needs for constant visual performance (according to Dall, from Strasser 1993b).

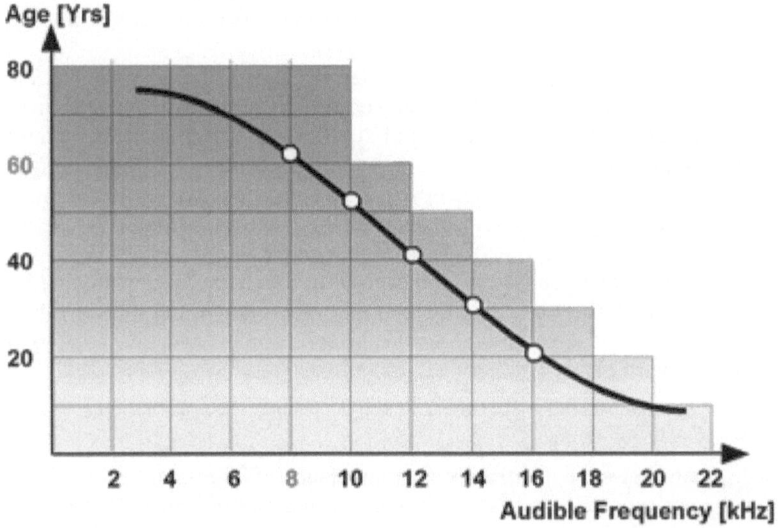


Figure 13. Upper hearing limit in dependence on age.

purchase of expensive stereo equipment with a "valuable" bandwidth of 20 kHz does not make sense because of a hearing range with an upper limit of only 8 kHz. Such a purchase only benefits the seller, but not the aged consumer. The age-related hearing loss as an elevation of the hearing threshold can also be measured audiometrically at different frequencies (cp. Table 1).

Table 1 illustrates the age-related hearing loss in dB between 1 and 6 kHz in steps of 5 years. Young men up to the age of 25 typically do not experience any limitations of their hearing and the following years usually pass without any substantial losses, as measured in dB. As the years go by, however, substantial hearing losses are almost impossible to avoid. Approximate hearing losses from 10 to 30 dB and more mean that people above the age of 60 experience an acoustic world that is considerably attenuated. By the way, what was just established for men and their age-related

Table 1. Age-related hearing loss [dB] (from Strasser 1993a).

| | Men | | | | | Women | | | | |
|---|---|---|---|---|---|---|---|---|---|---|
| | Test Frequency [kHz] | | | | | Test Frequency [kHz] | | | | |
| Age [Years] | 1 | 2 | 3 | 4 | 6 | 1 | 2 | 3 | 4 | 6 |
| <25 | 0 | 0 | 0 | 0 | 0 | 0 | 0 | 0 | 0 | 0 |
| 25–30 | 0 | 0 | 1 | 1 | 2 | 0 | 0 | 1 | 1 | 1 |
| 30–35 | 0 | 1 | 3 | 4 | 5 | 0 | 1 | 2 | 2 | 3 |
| 35–40 | 1 | 3 | 5 | 7 | 9 | 1 | 2 | 3 | 4 | 6 |
| 40–45 | 2 | 5 | 8 | 11 | 14 | 3 | 4 | 5 | 7 | 9 |
| 45–50 | 4 | 7 | 12 | 16 | 18 | 4 | 6 | 8 | 10 | 13 |
| 50–55 | 5 | 9 | 17 | 20 | 24 | 6 | 8 | 11 | 13 | 18 |
| 55–60 | 7 | 12 | 21 | 26 | 30 | 7 | 11 | 14 | 17 | 22 |
| 60–65 | 9 | 16 | 26 | 31 | 37 | 10 | 14 | 18 | 22 | 28 |

hearing losses is slightly different for women (cp. right part of Table 1). Women are the stronger sex as they age. Their age-related hearing losses are less severe by approximately 5 to 10 dB. Probably biological reasons are responsible for the gender difference. The more finely branched arteries in women are less prone to arteriosclerosis than those in men. This, of course, also applies to the delicate structures of the inner ear.

As previously mentioned, presbyacusis is characterized by limitations of the hearing, especially at higher frequencies. A reduction of 10 dB already corresponds approximately with a halving of the subjectively experienced loudness, a reduction of 20 dB already decreases the dynamics of sound to $^1/_4$, and a reduction of 30 dB means a further dramatic decrease to 1/8. Considering that normal communication takes place at levels around 60 dB, it becomes clear that an elderly individual whose hearing has diminished by 30 dB is rather isolated from his acoustic surroundings. As is well known, increasing the volume by the missing "dB" is a double-edged sword.

Again, the table shows average values, and each of us can hope, of course, that he personally, statistically speaking, is in the lower end of the range of variation and thus in a better position. Should one be among the "losers," and there must be some, he may find comfort in the following bit of wisdom by Bernard Shaw:

> *"There are two great advantages to old age: One's teeth no longer hurt, and one no longer hears all the nonsense that people say."*

Fig. 14 once again shows the not exactly pleasant results of large-scale epidemiological studies of men who were not exposed to any explicit noise during their lifetime. These somewhat dated, yet still representative, data show that drastic losses on the magnitude of 20 to 50 dB and more can be expected between 2 and 6 kHz when the population in the western world and the northern hemisphere is in the range of 40 to 60 of age. Interestingly, the age-related hearing threshold shifts of the Mabaan are less severe. The hearing of the hunters and gatherers of this tribe of the African outback who have neither drums nor firearms, even that of 70- to 80-year olds, exhibits few limitations and can almost be described as excellent. Typically, this is no longer the case for the population in the western world once they reach the age of 30. Presumably, this is due not only to the typical high-tech environment at work, but also because our everyday life is much too noisy.

## 7 CENTRAL NERVOUS AND COGNITIVE AGE-RELATED CHANGES

As far as age-related changes in the central nervous system and in mental-cognitive performance areas are concerned (cp. upper part of Table 2), the information that has been presented on hearing

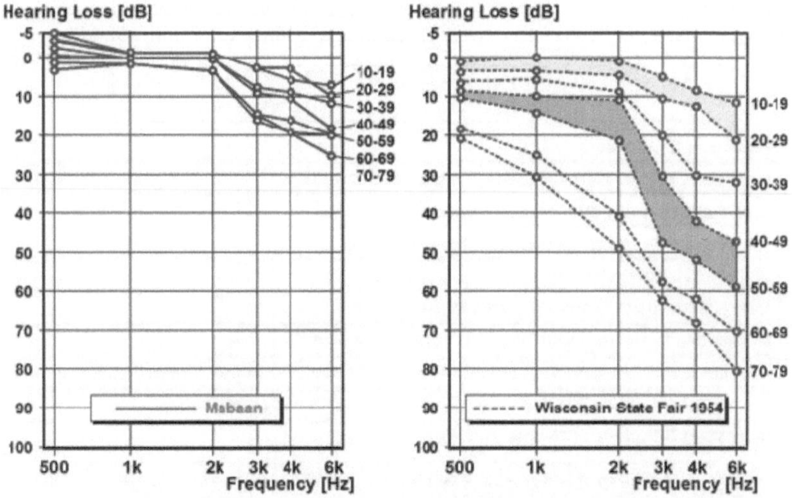


Figure 14. Range of audibility in several decades of life (in men).

and vision leaves no doubt that a certain "fading of sensations" occurs. This is confirmed by the saying that

*"All senses get weaker in old age, but the stubbornness thrives."*

Hence, the "limitation of the mental capacity to the well-known" should be added to the list in the upper part of Table 2. The "slowing down of psychomotor processes" has already been mentioned in a preceding chapter. A person's true age (in biological rather than in calendaric terms), however, according to an aphorism can allegedly be seen by observing whether he takes two steps or two pills at a time. To continue the list of the deficit hypothesis, a "slowing down of the mental reaction potential" and an "increase in egocentricity" can also be expected.

This can possibly also be applied to the "decrease of both inhibitions and self-criticism" which allegedly occur with age. "The difficulties to grasp multi-step thought processes" and the "problems with the sense of orientation with respect to space, time, and personal issues" as well as wavering and hesitating may have contributed to the following wisdom that was already quoted by Plato, one of the most important ancient Greek philosophers

*"Old age measures and weighs. Youth wants no delays."*

Two final hypotheses of alleged "increased affective penetrability with frequent overreacting" and the hardly disputed increased difficulty of "name and word retrieval" conclude the list of the deficit theory. About 30 years ago, Müller-Limmroth drew the following – possibly slightly exaggerated – conclusions with respect to work potential:

Elderly workers above the age of 50 fare worse than younger workers if formal education and experience are of less importance in the workplace, if the current situation must be processed as quickly as possible in the context of finding a solution, if new experiences are relevant to the job and a quick adjustment and consistent processing of the new information are required, and when rudimentary material must be processed mentally.

## 8 OUTLOOK

Despite all the listed "changes in behavior" that not necessarily do occur but become more likely in old age, it would be absolutely inappropriate to equate aging with throwing someone to the "scrap heap" when "experience," "prudence," and "strategic and cross-linked thinking" are needed in the

Table 2. Age-related changes in the central nervous system and in mental performance areas as well as "demanding" requirements for "older" employees (according to Müller-Limmroth 1984/85).

| Age-Related Changes in the Central Nervous System and in Mental-Cognitive Performance Areas |
|---|
| Fading of Sensations |
| Limitation of the Mental Capacity to the Well-Known |
| Slowing Down of Psychomotor Processes |
| Slowing Down of the Mental Reaction Potential |
| Increase in Egocentricity |
| Decrease of Both Inhibitions and Self-Criticism |
| Difficulties to Grasp Multi-Step Thought Processes |
| Problems with the Sense of Orientation with Respect to Space, Time, and Personal Issues |
| Increased Affective Penetrability with Frequent Overreacting |
| Increased Difficulty of "Name and Word Retrieval" |
| **Elderly Workers Above the Age of 50 Fare Worse than Younger Workers,** |
| If Formal Education and Experience are of Less Importance in the Workplace |
| If the Current Situation Must be Processed as Quickly as Possible in the Context of Finding a Solution |
| If New Experiences are Relevant to the Job and a Quick Adjustment and Consistent Processing of the New Information are Required |
| When Rudimentary Material Must be Processed Mentally |

workplace. According to Landau and Weißert-Horn (2007), qualities that tend to improve with age are the "ability to plan ahead," a "sense of responsibility" and "reasoning power", i.e., the ability to draw inferences or conclusions through the use of reason. Furthermore, "independence" as well as "diligence in the workplace" are qualities that can increase with age. A leading former German politician, who was not always correct in his decisions when he was young, was probably right when he said that:

> *"Pushing out older workers would mean that society would ignore experience, remembrance, and history."*

and

> *"Even a grandmother's advice possibly contains more wisdom than the science of a twenty-year old expert in social matters."*

With regard to the latter, nowadays one may question the "wisdom" of a young computer scientist who is too fixated on the digital world.

A few more optimistic aphorisms may conclude the outlook:

> *"There is no way to avoid getting old, but it is possible to fight getting antiquated."*

Goethe's rich experiences go well with the topic "The Art of Aging", when he found out

> *"It is not hard to become old, but bearing it is".*

or

> *"It is never too early to find out how expendable one is in this world".*

Hence, we should be guided by the following metaphors:

> *"Aging is like an ocean wave. Let the wave take over, and you will float on top. Try to fight it, and you will drown".*

and

> *"You're old, if the past gives you more pleasure than the future".*

or

*"What delight, if it is said: Old man, you may be rich in years, but your intellect is flourishing."*

Age-differentiated work design, fortunately, nowadays again is an important work-scientific research topic. All the time, besides gerontechnology (cp. N. N. 1994), ergonomics research also included "Aging, Work and Health". In this context, amongst others "Work and Aging: A European Perspective (Snel & Cremer 1995) or "The Paths to Productive Aging" (Kumashiro 1995) may be mentioned. But the first comprehensive studies on "Aging and Work" and "The Role of Ergonomics for Maintaining Work Ability during Aging" were carried out by Ilmarinen in Finland already in the early 90th of the last century (e.g., see Ilmarinen, 1995). This Finn-age action program included altogether 28 research projects and was carried out with international collaborating parties such as ICOH, ILO, NIOSH and WHO. Finally, the results allowed to develop the internationally well-known "Work-Ability Index". Nevertheless, a lot of open questions remained and, therefore, another comprehensive research project (a six-year priority research program, see Frieling 2006) was funded by the German Research Foundation (DFG). Some results of the meanwhile published final report (see Schlick, Frieling & Wegge, 2013) to which more than a dozen university research groups contributed provide means for the hypothesis that the "strength" of the various stages of life should be purposefully utilized. Thus, also in the work environment one should follow the following advice.

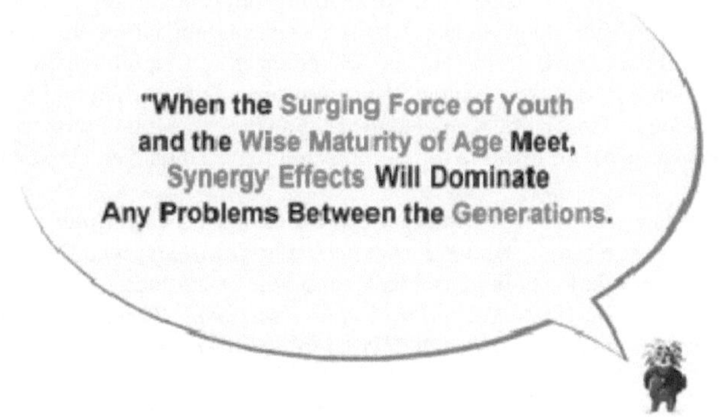


This survey of the course of aging of the performance potential shall be concluded with the following request to all of us, being still "young" or already in an advanced age with the verses of Freidank, a lyric poet of the early 13th century:

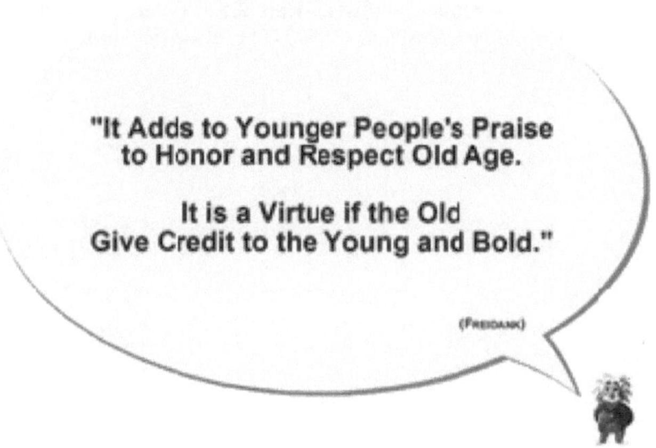

REFERENCES


Frieling, E. 2006. Editorial "Altersdifferenzierte Arbeitssysteme". *Z.Arb.wiss.* 60 (3): 149–150.

Grandjean, E. 1979. *Physiologische Arbeitsgestaltung – Leitfaden der Ergonomie.* Third Edition. Thun: Ott Verlag.

Hettinger, Th. 1960. Muskelkraft bei Männern und Frauen. *Zbl.Arb.Wiss.* 14: 79–84.

Ilmarinen, J. 1995. Aging and Work: The Role of Ergonomics for Maintaining Work Ability during Aging. In A. Bittner & Champney, P. C. (eds.), *Advances in Industrial Ergonomics and Safety VII*: 3–17. London: Taylor & Francis.

Karwowski, W. 2006. *International Encyclopedia of Ergonomics and Human Factors.* Second Edition. London: Taylor & Francis.

Keidel, W. D. 1967. *Kurzgefasstes Lehrbuch der Physiologie.* First Edition. Stuttgart: Thieme Verlag.

Kroemer, K., Kroemer, H. & Kroemer-Elbert, K. 1994. *Ergonomics – How to Design for Ease & Efficiency. Prentice Hall.* NJ, USA: Englewood Cliffs.

Kumashiro, M. (ed.) 1995. *The Paths of Productive Aging.* London: Taylor & Francis.

Landau, K. & Weißert-Horn, M. 2007. Ältere Arbeitnehmer. In K. Landau (ed.), *Lexikon Arbeitsgestaltung*: 36 – 39. Gentner & Ergonomia Verlag, Stuttgart.

Müller-Limmroth, W. 1982/83. *Jung bleiben, auch wenn man älter wird.* Manuskript eines Vortrages anlässlich des Personalforums der Süddeutschen Zeitung.

Müller-Limmroth, W. 1984/1985. Lebensalter und Arbeit (I-VIII). Arbeitsmedizinische Tafeln. Sonderdruck der Zeitschrift *Arbeitsmedizin, Sozialmedizin, Präventivmedizin*, 19. und 20. Jahrgang, Hefte 9–12 (1984) und Hefte 5, 6, 9, 10 (1985).

N.N. 1964. Sehen im Betrieb. Grundlagen und Anwendung physiologischer und beleuchtungstechnischer Daten. Arbeitstechnische Reihe. Angewandte Arbeitswissenschaften. Nürnberg: Siemens AG.

N.N. 1986. Ältere Mitarbeiter im Betrieb. Fakten – Tendenzen – Empfehlungen. München: Arbeitswissenschaftliche Schriftenreihe des Bayerischen Staatsministeriums für Arbeit und Sozialordnung.

N.N. 1994. Gerontechnology – Translating Knowledge of Aging Processes into Products and Services to Benefit the Aging: 1–20. Brochure of the Institute for Gerontechnology, Eindhoven University of Technology, The Netherlands.

Rohmert, W. 1980. Arbeitswissenschaft 1. Umdruck zur Vorlesung, TU Darmstadt.

Rohmert, W., Rückert, A. & Schaub, K. 1992. *Körperkräfte des Menschen.* Darmstadt: Selbstverlag des Instituts für Arbeitswissenschaft der Technischen Hochschule Darmstadt.

Rühmann, H. & Schmidtke, H. 1992. *Körperkräfte des Menschen – Perzentilierung isometrischer Maximalkräfte sowie Ausdauer und Beanspruchung bei konzentrischer und exzentrischer Muskelarbeit.* Köln: O. Schmidt Verlag.

Rutenfranz, J. 1983. Leistungsfähigkeit und Leistungsbereitschaft. In W. Rohmert & Rutenfranz, J. (eds.), *Praktische Arbeitsphysiologie*: 99–104. Stuttgart: Thieme-Verlag.

Schlick, Ch. M., Frieling, E. & Wegge, J. (eds.) 2013. *Age-Differentiated Work Systems*: 448 pp. Berlin, Heidelberg: Springer.

Snel, J. & Cremer, R. (eds.) 1995. *Work and Aging: A European Perspective.* London: Taylor & Francis.

Strasser; H. 1993a. Ergonomie – Umgebungseinflüsse. Kap. 2.5.1. Lärm. In Th. Hettinger & Wobbe, G. (eds.), *Kompendium der Arbeitswissenschaft*: 243–274. Ludwigshafen: Kiehl-Verlag.

Strasser, H. 1993b. Ergonomie – Umgebungseinflüsse. Kap. 2.5.3. Beleuchtung. In Th. Hettinger & Wobbe, G. (eds.), *Kompendium der Arbeitswissenschaft*: 288–318. Ludwigshafen: Kiehl-Verlag.

*Bridging Research and Good Practices towards Patient Welfare – Shih & Liang (Eds)*
*© 2015 Taylor & Francis Group, London, ISBN 978-1-138-02716-9*

# Attention performance in elderly people in complex visual environment

Y.-Y. Huang & M. Menozzi
*Human Factors Engineering, ETH Zurich, Zurich, Switzerland*

ABSTRACT: The aging workforce has become a trend as well as a challenge in our modern lives. With such changes we may expect some unavoidably reduced performance in elderly people. To minimize the impact on the outcome also to keep the working life quality, it is important to provide a proper and comfortable working environment for the aging workforce considering the actual status of the performance ability. In this study we aim to investigate the potential effects on the attention performance in elderly people when being exposed to a complex visual environment. Attention performance has been examined in a video-based task that required participants to attend to visual information under various background complexity levels. Experimental results have shown that elder participants performed the attention task with a reduced performance than the younger ones. Besides, the elder group seemed to shift their attention resources to process information located in the central visual field when a more complex environment was presented. Elderly people not only reduce their attention performance in general, but also change their strategies for acquiring information under heavy mental load. As a result, a simpler visual environment should be considered for the elderly workforce especially when the task is more crucial.

## 1 INTRODUCTION

The changing demographics have transformed workforce structures and extended working ages, which also have an impact on the healthcare system. The aging process does not only cause various sorts of physiological changes, but also affect the cognitive performance. High-quality performance in attention is one major factor in many safety concerns. From our daily lives, to our profession, our performance quality and decision making mechanism rely on what input information we have perceived and how we have proceeded with it. As a result, for certain crucial tasks, we must ensure the attention performance in order to avoid any neglected key information and/or unwanted errors.

Many factors may cause reduced attention performance, such as alcohol consumption, illness, fatigue, pressure, outer environment, and so on. Besides, the aging factor might affect the attention ability as well, and may further result in risky events. With the increasing aging population in our household and working environment, it has become a major topic for everyone to learn about how the aging process would affect the perception of various information inputs and how the perceived information would be processed. In the aging workforce, we might expect a lower quality of the perceived information due to some physiological impairment in visual or acoustic functions, for instance. With such facts, how can we enhance the outcome quality based on the limited information perceived to an acceptable level?

Moreover, the medical working environment has become more complex in modern days, for example the increased amount of examination devices with complicated user interface, or the operating room with various equipments and multi-monitoring media. People working at these positions would require some time getting acquainted with all the changes of newly introduced devices and user interfaces. An even extremer yet common scene would be the emergency room with the tension involving over-crowded people, long queue at the reception, newly-coming patients, ringing phones, shouting noise, and all instrumental operation. Medical staffs are required to make immediate decisions and work efficiently in such complex environments. In addition, the elderly people might have less flexibility and control over the allocation of processing resources. Since our attention resource for information processing is limited (Wickens et al. 2005), and with some degeneration issues, it is likely that elderly people are more disturbed by physical environment complexity and worse in dealing with complex multitasking.

In a previous study (Huang et al. in press), we were able to show that the peripheral visual attention has been impaired with the presence of some disturbing visual stimuli. The irritating and distracting sensation of the unwanted visual information has affected our visual perception and shared parts of our attention resources. As a result, when being exposed to certain unnecessary information, we have less available resources for detecting the targets in our peripheral visual field. Nowadays we have to access huge amount of information presented in a wide visual field in our daily tasks. In a hospital it is common for the healthcare personnel to monitor a multi-screen display which requires them attending visual information subtending a wide visual field at one time and very carefully. In our daily driving a car driver also needs to collect various visual information coming from a wide visual field to ensure the traffic safety. As for the elderly, the visual task consisting of a wide visual field might be even more stressful since both the physiological function and the cognitive performance are not at the top levels anymore. For the safety concerns from various aspects in our daily lives, we should take good care of the information being provided to the users. The right amount and non-irritating information for each individual could avoid some risky events also reduce the mental load. When we are less frustrated, a higher life quality is achieved.

In this study, we aim to investigate the potential impairment in the visual attention of elderly people, especially when elderly people are exposed to complex visual environment varying in degree of complexity. Besides, we aim to examine the attention performance for detecting visual targets located in different visual fields, i.e. in the central visual field and in the peripheral visual field. Furthermore, we aim to make suggestions for the working/living environment for elderly people to improve the working/living quality and to keep them (and people around them) in a safe condition.

## 2 METHODS

A video-based visual attention test has been carried out in a validated instrument. Participants performed the attention test by watching a video consisting of different complexity levels of the visual information background. Random visual stimuli superimposed on the background video were presented in three different locations in the visual field.

### 2.1 *Participants*

A total of 151 participants, 58 females and 93 males, were included in this study. The total participants ranged in age from 15y to 71y, with the mean age of 40.34y (SD = 12.56) and a median of 42y. All participants had normal or corrected-to-normal vision and wore their habitual optical corrections for far vision during the experiment, if applied. The present study has been approved by the Research Ethics Committee of ETH Zurich. All participants were given a full explanation about the experimental procedures orally and in a written form. A consent form was obtained with the right to withdraw from the study at any time without giving reasons and without any negative consequences.

### 2.2 *Instrumental setting*

The test video was displayed on a 24″ TFT screen (Samsung 245B Plus 2400) with a resolution of 1920 by 1200 pixels at a distance of 75 centimeters to the participant. The viewing distance considered some slight uncorrected myopia in the general population and also enabled presbyops to perform the test without their corrections for near vision. Participants were seated in a quiet room

Figure 1. A participant performed the attention test with a fixed viewing distance. A computer mouse was used to give the feedbacks.

and performed the visual tasks with a fixed head position by putting the head on a forehead and chin rest. No auditory effect was provided during the test. To give their responses to the attention tasks, they pressed two corresponding buttons on a computer mouse. Figure 1 demonstrates a participant taking part of the attention test. Recorded data were sent to an output log file simultaneously for later analysis.

### 2.3 *Experimental design & procedure*

During the test, participants were instructed to keep their eye fixation on a fixation cross which was centrally superimposed on a video of a recorded car drive took in Zurich city. The fixation cross was aligned with the sagittal and Frankfurt horizontal plane of participants. Immediately after the fixation cross disappeared, the visual stimulus consisting of a 6-digit number was presented either in the central visual field (0°), in the left periphery (−15.6°), or in the right periphery (15.6°). Each 6-digit number stimulus was presented for the duration of 300 ms. Participants were guided to perform a two-alternative forced choice (2AFC) task, in which they were required to detect if the integer "3" was shown in each trial of the 6-digit number presentation. Participants were asked to report their answers by pressing two specified buttons on a computer mouse, one for the integer "3" detected, and the other one for the integer "3" not detected. The video background was varying in complexity levels from a simpler visual environment, i.e. the uncrowded residential area with low traffic as shown in Figure 2, to a more complex visual environment, i.e. the city center with much traffic and crossing pedestrians as shown in Figure 3.

In the main test, a set of 72 trials with the 6-digit number stimulus was presented to each participant. In 50% of the total trials the integer "3" was included and in the rest 50% trials the integer "3" was not included. A subset of 24 trials was presented at each presentation location, i.e. center, left periphery, and right periphery. The integer "3" was shown two times in each digit position of the 6-digit number. Other digits of the 6-digit number were filled with random integers different from the integer "3." A pre-created set of 72 6-digit stimuli was generated before conducting the attention test. All participants were given the same set of the visual stimuli but with random presentation sequences.

The duration of the whole test was about six minutes. Before running the main test, participants were explained about the procedure of performing the task in a practice run with 10 trials. Participants were asked to give their answers as fast and accurate as possible. The main test consisted of three continuous sections – 1. training section, 2. lower complexity level section, and 3. higher complexity level section.

Figure 2. A template scene of the car drive background in a residential area with low traffic and simple traffic guides which corresponds to a lower visual complexity level.

Figure 3. A template scene of the car drive background in a city zone with heavy traffic, pedestrians, and traffic signs which corresponds to a higher visual complexity level.

### 2.4 *Data analysis*

The theory of signal detection (SDT) was applied to evaluate the experimental results (Gescheider 1985). And the detectability d' was used to estimate the attention performance in the detection task. The theory of signal detection, a method for quantifying the ability to distinguish a signal from a noise, introduces four types of the outcome from our decision making system while detecting if a signal is present – 1. "Hit" means that one perceives the signal when a signal is present; 2. "False alarm, FA" means that one perceives a signal when no signal is present; 3. "Miss" means that one perceives no signal when a signal is present; 4. "Correct rejection, CR" means that one perceives no signal when no signal is present. In this study, the presence of a signal means that the integer "3" was included in the 6-digit number stimulus.

## 3 RESULTS

### 3.1 *Detectability d' – Effect of age*

In order to evaluate the potential effects of the age factor, the total 151 participants were divided into two age groups by a median split. Age group A (15y to 41y) had 23 females and 50 males with the mean age of 29.56y (SD = 6.98) and a median of 29y. Age group B (42y to 71y) had 35 females and 43 males with the mean age of 50.44y (SD = 6.95) and a median of 49y. In Age

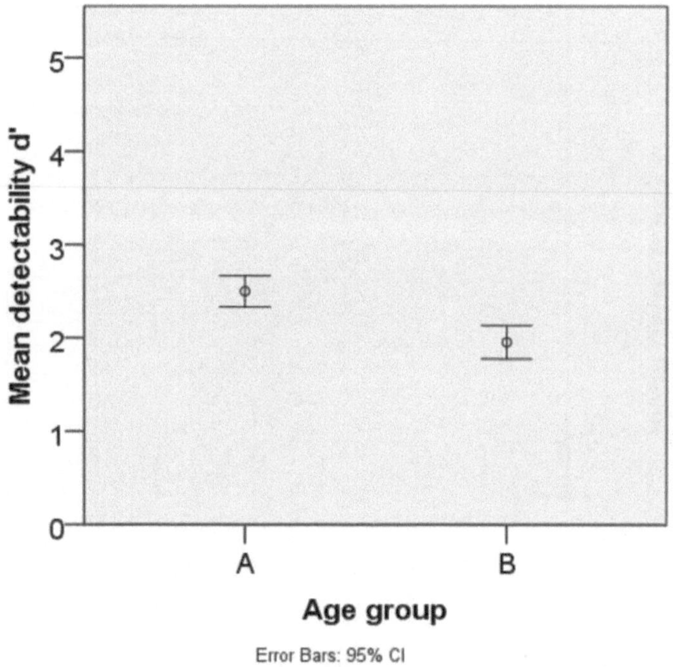


Figure 4. Error bar diagram of the mean detectability d' in Age group A (15–41y) and in Age group B (42–71y).

group A, the mean detectability d' among all trials was 2.50 (SD = 0.72). In Age group B, the mean detectability d' among all trials was 1.96 (SD = 0.79). The mean detectability d' in two age groups are reported graphically in Figure 4. Statistical results reveal that the overall performance in Age group A, i.e. the younger participants, is better than Age group B, i.e. the elder participants (Univariate ANOVA, one factor of age groups, F(1,149) = 19.665, p = 0.000, effect size partial $\eta^2 = 0.117$, observed power = 0.993).

### 3.2 *Detectability d' – Effect of complexity level vs. age*

A two-factorial ANOVA was run considering the two-level within-subjects factor of the complexity level of the visual environment (lower complexity level and higher complexity level) and the between-subjects factor of the age group. Results show a significant effect of the complexity level (F(1,149) = 4.536, p = 0.035, effect size partial $\eta^2 = 0.030$, observed power = 0.562). Figure 5 shows the mean detectability d' of the three sections (training section and two complexity levels) in both age groups graphically. With the increasing complexity of the visual background, the attention performance was decreased in both age groups.

### 3.3 *Detectability d' – Effect of stimulus location vs. age*

In order to compare the attention performance in centrally presented visual tasks and in peripherally presented visual tasks, we combined the trials of visual stimuli located in the left periphery (−15.6°) and in the right periphery (+15.6°) as a group of peripheral targets. A two-factorial ANOVA was run considering the two-level within-subjects factor of the visual stimulus location (center 0° and periphery +/−15.6°) and the between-subjects factor of the age group. Statistical results show a significant effect of the stimulus location (F(1,149) = 240.118, p = 0.000, effect size partial

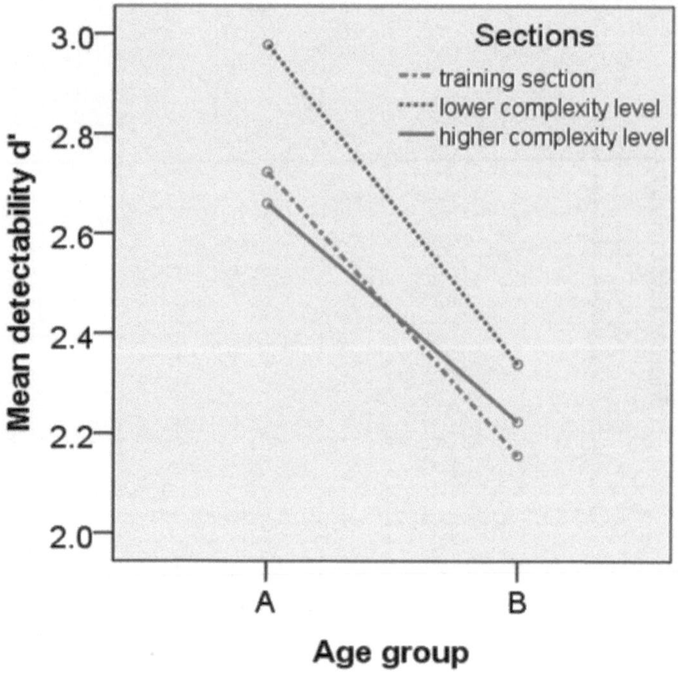


Figure 5. Graphical representation of the mean detectability d' in Age group A and B under different complexity levels of the background information.

$\eta^2 = 0.617$, observed power = 1.000). Both age groups performed the attention test better when the visual stimulus was presented in the central visual field as shown in Figure 6.

### 3.4 *Detectability d' – Effect of complexity level vs. stimulus location vs. age*

A three-factorial ANOVA was run considering the two-level within-subjects factors of the complexity level (lower and higher complexity) and the visual stimulus location (center and periphery) and the between-subjects factor of the age group. Results show no significant effect of the complexity level ($F(1,144) = 0.030$, $p = 0.863$, effect size partial $\eta^2 = 0.000$, observed power = 0.053) and a significant effect of the visual stimulus location ($F(1,144) = 148.736$, $p = 0.000$, effect size partial $\eta^2 = 0.508$, observed power = 1.000). Figure 7 and Figure 8 represent the mean detectability d' for the central visual targets and the peripheral visual targets under two complexity levels of the video background in Age group A and B respectively.

## 4 DISCUSSION

Based on our experimental results, we can see the overall performance in the attention test was significantly worse of the elder participants than the younger ones. This finding is not too surprising in our video-based detection tasks since the aging process might affect the visual function and the attention ability. Elderly people might be less sensitive to the visual stimuli and have lower capacity in processing the total input information across the visual field. As we must rely on as well as take good care of the aging population and workforce, an appropriate living/working environment which is not over-challenging the user's ability should be well-defined and provided. When being exposed to a complex environment, which is frequently encountered nowadays, it is important

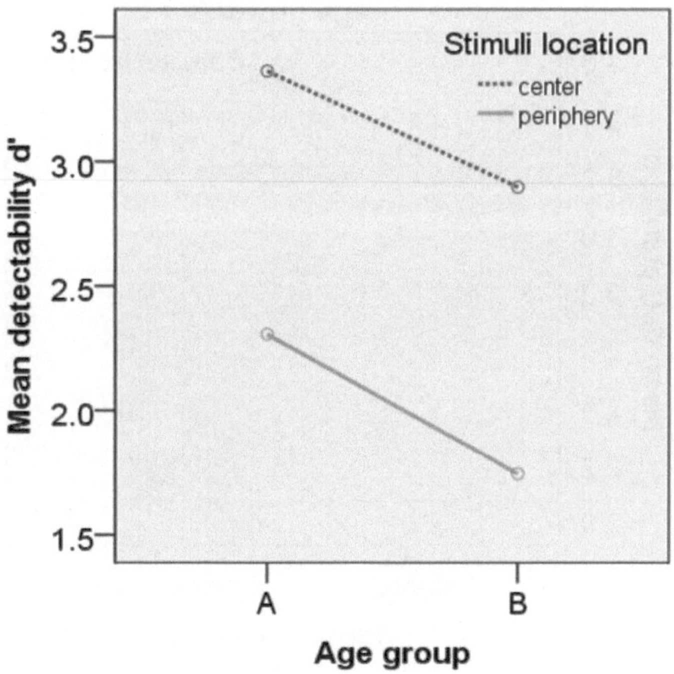


Figure 6. The mean detectability d' for attending centrally (0°) and peripherally (+15.6° & −15.6°) presented visual stimuli in both age groups.

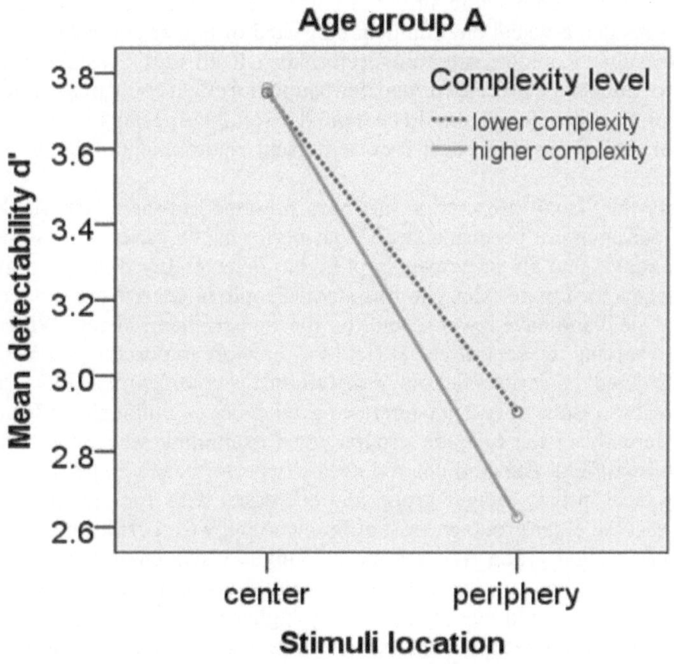


Figure 7. The mean detectability d' in the two complexity levels for the visual stimuli located in the center or in the periphery of Age group A (15–41y).

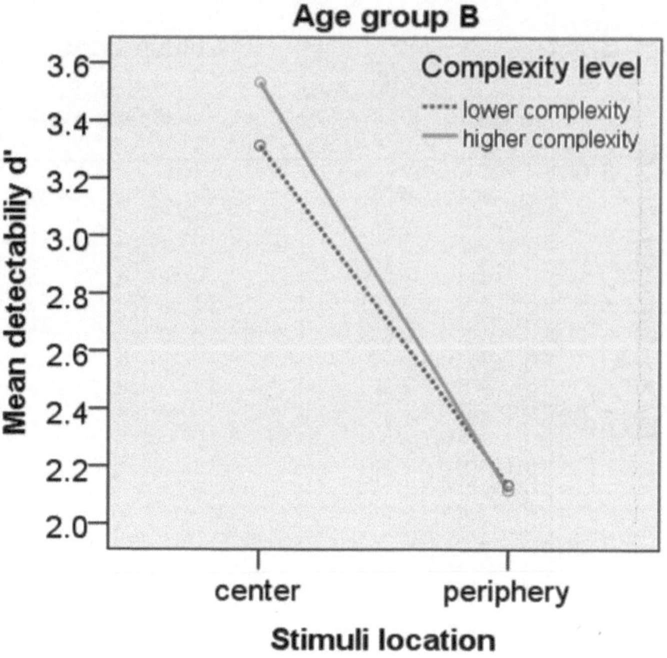


Figure 8. The mean detectability d' in the two complexity levels for the visual stimuli located in the center or in the periphery of Age group B (42–71y).

to examine if we could keep a stable and high performance in our everyday tasks. As shown in the experimental results, a visual environment consisted of higher complexity level of presented information has resulted in a lower attention performance in all ages. Possible reasons could be the over-loaded information to be processed, also the distracting effect could play a role. The significant reduction in attention ability for all should be treated carefully especially when a more serious task is conducted. For critical cases, an over-crowded visual environment and any distracting factors should be avoided.

Besides, where the visual information has been presented in the visual field also influences the outcome of our attention performance. Visual performance varies across the visual field. It is quite understandable that all participants got higher detectability d' in the trials with centrally located visual targets than in periphery in this study. People in general are more concentrating on the central visual field and have lower resolution for the peripheral vision. With such finding we would suggest preserving the central visual field for the more important information.

When we considered the various factors in this study, i.e. complexity level vs. stimulus location vs. age, parts of the results reveal an interesting tendency. Complexity affected the attention performance differently in the two age groups when examining where the visual targets were presented in the visual field. The gradient of decrease in performance for peripheral tasks increased with complexity level in the younger group and decreased with complexity level in the elderly ones. It seemed that the elder group chose a different strategy for performing the attention task in comparison to the younger group. When a more complex visual environment was presented, the elder group intended to focus on the tasks in the central vision in which important information is presented usually. This could be the reason why higher detectability d' has been found in the elder group when the stimuli were located in the central visual field. The elderly group seemed to alter their strategies for acquiring information under different mental load. When performing heavy mental load tasks, the elderly people might shift their attention resources towards the central visual tasks in order to handle a more critical event. As for a relatively normal case, i.e. not so heavy mental load, they might preserve some efforts (less resources are used) and keep the overall performance at a certain level based on their experiences in life. In the younger group, however, a different strategy which could be more consistent has been chosen. The younger people probably would try to keep top performance and take control of the overall information under different situations and did not shift their focuses among all tasks. This could explain the stable and higher performance in the younger group for the central targets in both complexity levels. Since the attention resource is limited, for processing the heavier load of information in the higher complexity condition, the capacity for maintaining the performance for peripheral vision dropped down unavoidably.

While the overall attention performance is better in the young people, the elderly people tend to better distribute limited resources across the visual field when being exposed to a complex environment and/or under heavy mental load. The elderly, the more-experienced workforce, are likely more reliable when dealing with complex situations, and are probably better in supervisory tasks.

## 5 CONCLUSIONS

In this study we have conducted the attention test with participants in different ages. We were able to show that aging has certain impacts on the attention ability. Despite better experiences and practices of the elderly people which might help in better performance and decisions in certain tasks, the significantly reduced ability in attention may be still influential. In particular, for tasks in a complex environment, e.g. control-room monitoring (multi-display & real-time notification), vehicle driving (driving behavior adaption & on road objects), complex visual inspection (examination screen), emergency room assistance (crowded area & pressure), elderly people may behave differently from the younger ones. To improve the attention performance of elderly people, better designs in ergonomics should be considered, e.g. to reserve the central visual field for high importance information. For critical situations, we should prevent any risky uncertainty from putting the elderly people in such an environment, as well as other involved people. We therefore suggest a simpler environment for elderly people when crucial tasks are expected, by minimizing the background and periphery complexity in the visual environment, avoiding multi tasks at the same time, and decreasing the mental load for the elderly people.

## ACKNOWLEDGEMENTS

The authors thank all the volunteers for their participations in this study.

## REFERENCES

Gescheider, G. A. 1985. *Psychophysics: Method, Theory, and Application.* 2nd ed. Houndmills: Lawrence Earlbaum Associated.

Huang, Y.-Y. and M. Menozzi. (in press) "Effects of Discomfort Glare on Performance in Attending Peripheral Visual Information in Displays." *Displays.*

Wickens, C. D., J. S. McCarley, A. L. Alexander, L. C. Thomas, M. Ambinder, and S. Zheng. 2005. *Attention-Situation Awareness (A-SA) Model of Pilot Error (Technical Report AHFD-05-15/NASA-04-5).* Savoy, IL: University of Illinois, Aviation Research Lab.

*Bridging Research and Good Practices towards Patient Welfare – Shih & Liang (Eds)*
 *ISBN 978-1-138-02716-9*

# The incident reporting systems and organizational learning in Indonesian public hospitals

I. Dhamanti
*Airlangga University, Surabaya, East Java, Indonesia*

S. Leggat & S. Barraclough
*La Trobe University, Melbourne, Victoria, Australia*

ABSTRACT: Hospital incident reporting has become one of the most important tools to capture errors, and within incident reporting, learning from errors can be applied and potentially disseminated across staff at all levels. The aim of this study is to investigate the implementation of patient safety incident reporting in Indonesia and to see how public hospitals have implemented the organizational learning arising from incidents. Three public hospitals were purposively selected from three regions in East Java. Hospital directors or managers, patient safety teams, and ward supervisors were interviewed. We applied thematic analysis to analyze the transcripts. Issues found included delays in reporting, unwritten or incomplete reports, lack of staff training, lack of leadership, absence of feedback, and inadequate management supports. Accordingly, we recommend that hospitals conduct and allocate resources for Root Cause Analysis, follow up the corrective actions for an agreed time, conduct further analysis on their data, standardize feedback and dissemination mechanisms, and improve the leadership and organizational factors preventing satisfactory incident reporting.

*Keywords*: patient safety, incident reporting system, organizational learning, public hospital

## 1 INTRODUCTION

Almost a decade ago, Indonesia began its patient safety movement. Since 2011, the implementation of patient safety programs, including patient safety incident reporting, has become compulsory for all Indonesian hospitals, for accreditation and quality improvement purposes. The patient safety incident reporting system is a mechanism provided by the Government for reporting clinical incidents that can cause any harm for patients. Patient safety incidents could be categorized as near miss (a serious error that fails to cause adverse events because of chance), adverse event (an event that resulting in harm or injury for patients), or sentinel event (an event that resulting in death or serious injury for patients). Ideally, all Indonesian hospitals should report their incidents to the Indonesian Hospital Patient Safety Committee (IHPSC). However, according to the IHPSC's Annual Report for 2011, only 132 incidents were reported by 132 hospitals. Since there are 2,267 hospitals in Indonesia (MOH, 2013), fewer than 6% of the total number participated in reporting.

The main purpose of a reporting system is to collect patient safety information from health care providers in order to report errors, to identify hazards and risks for patients to enable organizations to change their practices in ways that reduce adverse event rates (Harper & Helmreich, 2005; WHO, 2005), and to detect preventable adverse events (Evans et al., 2006). When used effectively, incident reporting systems enable organizations and their staff to learn from the mistakes of others (Chuang et al., 2007). Learning from incidents, however, can be difficult, and it is challenging to make such learning a part of organizational learning. The incident learning system consists of several steps in a cyclical process, such as identification and response, reporting, investigation, corrective actions, and learning (Cooke, 2003; Cooke et al., 2003) or before the incident, immediate response, prepare for analysis, analysis process, follow through and close the loop steps (Incident Analysis Collaborating Parties, 2012). Any incident happens in hospital is an opportunity for system improvement, there should be a learning process for individual staff, teams, work units, hospitals, and even at the national level.

As described in the Indonesian Patient Safety Incident Reporting Guidelines, the incident reporting system in Indonesia consists of internal and external reporting. Internal reporting is conducted at the hospital level and after it is completed, reports are sent to IHPSC, as external reporting. Internal reporting is paper-based, and anyone who has observed an incident submits a report to the supervisor or head of unit, who then assesses the incident to determine its grade. This then determines the type of investigation needed: simple or comprehensive. Adverse or sentinel events need comprehensive investigation using Root Cause Analysis, and are estimated to take 14 to 45 days. In order to analyze the causes of an incident, staff use Root Cause Analysis (RCA) to identify the contributing factors. In the external reporting form, the contributing factors are placed into eight categories: external, organization and management, work environment, teamwork, staff, patient, communication, and work (IHPSC, 2008). After the internal form is completed at the work unit level, for adverse events, the hospital patient safety team (HPST) must conduct an investigation using RCA, and complete the external form before sending it to the IHPSC. Most hospitals lack the awareness of the requirements for identifying and investigating the incidents and for learning effectively at multiple levels (Benn et al., 2012). Often, the learning is kept local, or may not exist.

The present study was carried out in three accredited Indonesian public hospitals in East Java Province. Hospital A and C have full accreditation status for 16 services while hospital B has obtained full accreditation status for 5 services. The aim of this study is to investigate the implementation of patient safety incident reporting and find out how the hospitals implemented their organizational learning from reported incidents. Even though learning occurs at different levels, including individual staff, groups and organizations, this study focuses mainly on how learning takes place at the organizational level.

## 2 METHODS

### 2.1 *Setting and participants*

Three public hospitals were purposively selected from three regions in East Java Province, Indonesia. The inclusion criteria were public hospitals with B classification that had obtained accreditation status. Since implementing patient safety programs has been a part of accreditation requirements, it is assumed that the hospitals that have been accredited would have implemented their patient safety programs, including patient safety incident reporting. In each hospital, we interviewed senior management such as hospital directors, vice-directors, or managers, patient safety teams, and one randomly selected ward supervisor in the Inpatient Unit. We interviewed a total of nine staff members in three hospitals. Ethics approval was obtained from Faculty Human Ethics Committee La Trobe University.

### 2.2 *Data collection methods*

Semi-structured interviews and observations were conducted for data collection. The audio-recorded interviews of 20–45 minutes each took place in participants' offices, with interview questions given out in advance, to familiarize participants with the questions. The researcher asked questions related to the implementation of patient safety incident reporting in the hospital, the procedures for reporting, the learning process and procedures for providing feedback and dissemination of reports, as well as the level of management support for reporting. The first author also observed completed incident forms and the form's location in each hospital.

## 2.3 *Data analysis*

All data collected from the interviews were transcribed and translated by the principal researcher. Thematic analysis steps, as proposed by Fielden et al. (2011), were applied to the transcript: these steps are built on the notes and ideas generated through transcription, coding, searching for themes, further coding, defining and naming the themes, up to the last step, report production. The four major themes identified are "Reporting Process", "Data analysis", "Feedback" and "Management Support".

# 3 RESULTS

## 3.1 *Reporting process*

Hospital A has implemented the internal patient safety incident reporting procedure as prescribed in the Indonesian Patient Safety Incident Reporting Guidelines. Every incident is reported to the HPST by completing a form, though there are often delays in submission, thus delaying the process as mentioned by one of the participants:

> We have made reports on patient safety incident. Yes, we, the Hospital Patient Safety Team (HPST), distributed the form to each of our wards. Later these forms are collected and sent to HPST. So, we have the form and already distributed.. eee... but sometimes asking them to return the form has its own problems.(A, senior management)

Reporting without completing a form often delays investigation and analysis since not enough information is available about the event, its location, and possible causes, and it could even be forgotten because it is not given priority, or ignored because of lack of attention from the unit staff and the team. Previously, in hospital A, a book was used in the wards to document incidents, and for a period this book was submitted at intervals to the hospital patient safety team. The team then completes the form and holds an investigation to uncover possible causes and make appropriate recommendations. However, the team could not complete the form and conduct an investigation on time, especially if a new incident was reported that caused frequent delays in reporting. Recently, therefore, the team had taken to distributing the forms in the wards with the expectation that the staff would start to complete them by themselves. The reports are then collected monthly, and an evaluation focusing on any post-reporting action or improvements, are conducted every six months, though the team could not provide any incident reporting data for the period of this study.

The internal reporting procedure in hospital B is similar to that in hospital A. As the ward supervisor commented:

> Um... for the reporting system, we received socialization about this system. The reporting starts from an incident location, continues directly to HPST. Later, management will take action... (E, ward supervisor)

The incident forms are distributed to the wards, so that whenever an incident occurs, ward staff have access to it. Four incidents were reported during 2012, and between January–August 2013, there was only one incident, reported by a nurse. A similar problem of unwritten reports also occurs, as the hospital patient safety team frequently receives oral rather than written reports. Even though solutions are provided for incidents, and some actions for improvement carried out without documentation the number of incidents is under-reported.

In hospital C, internal reporting is effectively implemented, thus the incidents data is accurately recorded every month. According to the Hospital Incident Report, there were 33 incidents reported March–September 2013, four of them sentinel events. The reporting procedure is similar to those of the two other hospitals, though for emergencies, where an incident causes a serious risk, actions following a reported incident are taken immediately by the team and hospital management, without waiting for the form to be completed by the reporter.

> ... because some of cases are urgent and need to be followed up immediately and some of them can be postponed. If the case is urgent, it should be followed up immediately. So, they [the reporters] don't have to make any written report. They could directly call head of the ward. Head of the ward contact

HPST directly. And if the team could not make any solution, the case will go straight to management for follow up. So, at first, reports don't have to be in written. If a case needs immediate action, it can be reported through a telephone call or another informal way. (H, senior management)

With urgent cases, the team will conduct a meeting with the hospital director immediately then decide an interim recommendation. The secretary of the hospital patient safety team then coordinates actions with doctors, nurses and the related work unit to carry out the root cause analysis. Some incomplete reports have been found in those submitted, reports mentioning only the incident's date, patient's name, reporter, and incident location, with further information related to incident description, type of incident, possible consequence, actions taken are left blank. The incomplete reports may happen because staff postpone the form completion, or staff lack the necessary knowledge and training for completing the form.

### 3.2 *Data analysis*

In Hospital A, if incidents are categorized as adverse or sentinel events, the supervisor or head of the unit, the hospital patient safety team, and the management, will hold a meeting. In this case the hospital has not used Root Cause Analysis (RCA) to investigate the incident, even though most of heads of units have attended the RCA training.

So, actually, they (most heads of units) understand the RCA. But, it's because they are busy, this RCA has not been done. (L, hospital patient safety team)

Every time an incident occurred an interim recommendation to the incident was decided at a meeting. Any result or recommendation was fed back to the reporter by the head of unit or supervisor. This participant also emphasized the lack of staff, especially in the team, since there is only one active member, who runs and handles everything – for example, the monthly walk-around, and also manages and analyzes the incident reports. Lack of staff was also the reason why the hospital never submitted its external reporting to the Indonesian Committee on Hospital Patient Safety (ICHPS). Further analysis of data has not been implemented yet, and incidents are reported to the hospital management only every six months as part of work unit performance.

In hospital B, RCA also had not been implemented by the hospital patient safety team, which decided on recommendations and solutions for an incident at a meeting, without using it. As described by the participant:

But eh… good data management the detail analysis has not been done yet. To be honest, I haven't done it. Because it should be HPST's thing. But, the other thing is that because in our HPST, there are not many doctors in it, and they have many patients.in the end the analysis is not that detail. (E, senior management)

As this participant also commented, to investigate using RCA is the patient safety team's responsibility, where management may be involved, and after completing it, reporting to the director. The head of this particular hospital patient safety team is a busy surgeon, who works with other medical specialists and ward supervisors in the team. In this case, despite lack of training for the team members, lack of leadership may also be an important factor in failure to use RCA.

In hospital C, the team has implemented RCA to investigate direct and non-direct causes of an incident, the investigation involving doctors, nurses, patients, the hospital patient safety team, and hospital managers, assisted by the secretary of the hospital patient safety team. In the report, however, the reporter had not completed some parts of the form, including specifying the type of incident, an omission which may result from confusion in categorizing the incident. Thus it becomes the responsibility of the supervisor or head of unit to undertake any correction, though the participants admitted lack of time as one of the barriers to validating the content of the form. Lack of time was also the reason for delay in the investigation, because a new incident appeared to be more urgent.

Limited time ya…, because work overload. As secretary of HPST, my work is overload. While other people still do not realize that this is our common interest. So they put it all on me… that become a problem. (W, hospital patient safety team)

In hospital C, the secretary of the team was in the nursing division and transferred to her new position two years ago. She has been initiating some improvements in the patient safety program and reporting, though delegating some of her work could possibly improve the team performance.

### 3.3 *Feedback*

In hospital A, the reporter is least likely to attend the meeting of the patient safety team, because he/she is already represented by the head of units or ward supervisors. At the meeting, the team gives feedback about incidents directly to those senior staff, instead of to the reporter, giving feedback to the reporter at a morning meeting in the ward.

> From meeting's results, in the morning before they do their activities, every ward supervisors make morning reports. That is where feedback will be communicated. (L, hospital patient safety team)

The reporter may not always be the person making errors, but a reporter's presence at the meeting is important to provide accurate information. In addition, direct specific feedback may be influential in changing the condition or the behaviour of the person concerned in the incident.

In hospital B, the hospital patient safety team never provides direct feedback to the reporter or to the work unit, but collects all the internal reporting forms from nurses and reports the incidents orally to management.

> That is the thing. The report stops at… eee...the secretary… Mrs. S. The reports go to the directors, so far, only in spoken form not in written. If the report is in written, we will give a feedback. But because it is spoken reports, ya… when we receive it [the report], we just follow it up. So, [the reporting] stops at her because she only reports in oral form. (E, head of division)

The hospital patient safety team and management then discuss the incident and determine the recommendations or solution. However, neither the team nor the management has provided feedback to the reporter. As described by the ward supervisor:

> Since 2012, I have made (reports)… but until now I have never received any feedback. (E, ward supervisor)

It is evident that this hospital has not fully understood the importance of providing feedback. The team has not made effort to disseminate feedback to reporters, or related work units, and provides information on solutions regarding an incident only to the work unit.

Hospital C in contrast, had implemented better feedback methods for reporters, providing it either to some reporters or to related work units in written form, *via* a meeting or informal conversation.

> We will report, management will then follow up directly to the ward. During this time, Mrs. W will investigate to the ward, to find the causes from related work unit. After that she will give guidance. It is personal guidance to each of related-work unit. Usually, is she comes to the ward, we hold a meeting straight away. So it is not with one staff… but with all staff in ward, so all of them will know. (D, ward supervisor)

The team also avoids giving individual feedback in a meeting, most often, the individual feedback related to performance or behavior will be given on a different occasion. The hospital patient safety team emphasizes the solution and how it could prevent the incident from recurring, despite the effective implementation of the internal reporting system, the dissemination of the solution is still limited to only certain people, and not across work units that involve all staff.

### 3.4 *Management support*

To encourage reporting, once in a month in hospital A, the HPST conducts a walk around the wards in order to raise questions about incidents that have happened in the unit. The team also asks staff to complete the internal reporting form if they find incident, and submit it to the team. Hospital A also performs patient safety training every year.

> Yes, we have patient safety training every year in hospital. The participants are all nurse, doctors. Some of them were sent to Jakarta, usually members of the HPST. Every year the hospital conducts patient

> safety training for adverse events in the hospital. All the nurses attend, also doctors and some also go to Jakarta for training, but usually that is only the hospital patient safety team member. (T, ward supervisor)

Morning reports involving doctors and nurses are routinely held weekly. At the meeting, nurses and doctors discuss patients' conditions and incidents that have occurred. Nurses or doctors reporting an incident are asked to complete and submit the form to the hospital patient safety team.

During accreditation preparation in 2011, hospital B conducted staff socialization related to patient safety incident reporting, inviting heads of units and other hospital staff. Following this socialization, nurses reported four incidents, but the implementation did not proceed well: between January–August 2013, there was only one incident reported by a nurse. – As participant reported feeling a lack of management support:

> We don't get active encouragement… it seemed none. It's our own initiative from the ward. The first thing, maybe a clearer socialization on perhaps the reporting system. Socialization should be repeated, problems on reporting to be refreshed. Besides… supports… every time we make a report it needs to be followed up shortly. (E, ward supervisor)

In contrast with the conditions in hospital A, where there is only one active member in the hospital patient safety team, in hospital B, the team has 32 members, but none of them has actively encouraged reporting. The socialization program promoting reporting incident, has been held only once, and did not include all staff, which indicates that incident reporting may not be a high priority of the hospital, just as providing feedback and disseminating solutions widely are not yet promoted.

In hospital C, the patient safety team always actively encourages hospital staff to report incidents comprehensively. For example, morning reports are conducted four times a week with the purpose of finding new incidents. Most importantly, the secretary of the hospital patient safety team always attends, and actively follows up any incidents. The hospital had also conducted training for incident reporting:

> We attended the four days training in incident reporting, and some units held exercises for incident reporting. Ward supervisors, heads of units and doctors also joined in. (D, ward supervisor)

In addition, the interview revealed that some staff have attended four days patient safety incident reporting, to raise awareness and understanding about it. Often, if reporters feel unsure, for example, when determining the type of incident, they can discuss it, and will get assistance from the hospital patient safety team.

## 4 DISCUSSION

The incident learning system is described as a way of improving an organization's capability for benefiting from incidents by extracting valuable information leading to improved organizational performance (Cooke, 2003). It is expected that, with patient safety incident reporting, this learning could be applied over time, and potentially disseminated across levels. The themes in the interviews are reporting process, data analysis, feedback and management support, were resembled the framework of incident learning system suggested by Cooke (2003); Cooke et al. (2003) and the Incident Analysis Collaborating Parties (2012).

This research found many problems in the reporting process within the three hospitals – for example, incomplete or unwritten reports and validation problems. These hospitals have provided various direct or indirect responses, depending on the category and severity of each incident. However, most of the learning process ends at reporting. Investigation, the third step, involving examination of the site, interviewing witnesses, evaluating, and gathering data to establish the sequence and description of events (Cooke, 2003) is not carried out in all hospitals. Only one hospital implemented RCA, but had not applied it to all adverse events. Some adverse events have not been followed by investigation, for various reasons. Therefore, RCA training, as well as allocating resources for conducting that training, are crucial.

In the other two hospitals, instead of conducting an investigation, the staff moved straight to corrective action and determined a recommendation by only looking at the problem in general. Since root cause analysis is an important component of quality improvement, this is not consistent with the principles of quality improvement, with recognition of the need to identify the root cases (Connelly, 2012). It is a risk taking action without implementing a correct investigation thus fixing the problem only at a specific time without understanding the contributing factors and effectively disseminating information about them, and about subsequent corrective action. This omission could thus lead to failure to prevent a similar incident from recurring. Furthermore, not all reporters were included in the investigation. As the literature suggests the incident reporter is required to be involved in an investigation for recalling the incident and visualizing possible failures that have not yet occurred (Cooke, 2003). More importantly, effective corrective actions derived from an investigation ideally should follow, to ensure fix the problem (the Incident Analysis Collaborating Parties, 2012).

Not all these hospitals have developed an effective organizational learning mechanism – in two, learning does not exist. Those hospitals have evidently not understood the principal aim of reporting, which is to learn from incidents. As suggested by Blike et al. (2006), a wide range of learning mechanisms, such as after action reviews, audits, investigations, performance appraisals, simulation and benchmarking, are available, but these hospitals focus only on investigation, and meetings for disseminating the solutions to a limited extent. By leaving out this learning stage, the hospital puts patient safety, as well as the hospital's wider image and reputation, at risk. Most hospitals also have not utilized incident data for further analysis, which means they lose the opportunity to acquire valuable information conducive to improving quality.

When hospital staff report incidents, but do not receive any feedback, there was no subsequent learning experience. We found in the two hospitals, A and B that despite its importance, effective feedback of learning and corrective actions to prevent the same incidents recurring was not implemented. Often the feedback from the team or hospital management remained isolated at individual staff or work unit level, and there were no mechanisms in pace to spread the learning across all work units. Chuang et al. (2007) have emphasized the organization's responsibility to translate and transfer knowledge and experience acquired from incidents, and to take corrective action to prevent the occurrence of similar events. Unfortunately, most of the three hospitals lack a clear mechanism for conducting all those stages – each has its own non-standard methods, especially when it comes to deciding what action should be taken, as well as what is the best method to provide feedback and dissemination of learning.

As suggested by Cooke et al. (2007), the success of the learning system will depend on the degree of organizational encouragement for reporting, the emphasis on incident reporting as part of the job, and commitment to providing resources for investigation and corrective actions. Local leaders must value safety, set safety as a priority, hold group staff accountable for actions in patient safety, and create structural arrangements that reinforce learning and the transfer of knowledge (Chuang et al., 2007). Ideally, leaders should be committed to creating norms, procedures and routines that support learning, and develop structures to facilitate learning that supports patient safety. As we can see in hospital C, leadership that values safety influences all staff and affects the success of incident reporting. This finding is consistent with Chuang et al. (2007) referring to a positive relation between leadership in patient safety, and the effective assessment of patient safety, and of subsequent organizational improvement.

In addition, creating and developing learning systems mostly depends on organizational and leadership factors. Most problems found in hospitals, for example, delays in reporting, unwritten or incomplete reports, lack of training for staff, lack of leadership, absence of feedback and low levels of management supporting the implementation of incident reporting and organizational learning, are rooted in organizational rather than individual factors. In summary, leadership and organizational factors are essential preconditions for implementing effective patient safety incident reporting, and learning (Blike et al., 2006).

## 5 CONCLUSION

This study revealed there are problems associated with the implementation and learning from incident reporting system in Indonesian hospitals. A systematic learning has not yet developed within the three hospitals. Accordingly, in order to improve learning, it is essential that hospitals conduct RCA training, as well as allocating resources for RCA, follow-up with corrective actions – to ensure it has fixed the problem, conduct further analysis of data, standardize the feedback and learning dissemination mechanisms, and improve the performance of leadership and other organizational factors. However, leaving all those efforts, the hospitals will only putting the patients at a risk.

## ACKNOWLEDGEMENT

The first author would like to thank the Indonesian Directorate General of Higher Degree for providing financial support during her doctoral study in La Trobe University, Australia.

## REFERENCES

Benn, J., Koutantji, M., Wallace, L., Spurgeon, P., Rejman, M., Healey, A., & Vincent, C. 2009. Feedback from incident reporting: information and action to improve patient safety. *Quality and Safety in Health Care*, 18(1), 11–21.

Blike, G., Campion, P., & Schlosser, E. 2006. Implementation of a patient safety collaborative forum facilitates organizational learning from medical error. *Proceedings of the Human Factors and Ergonomics Society Annual Meeting*, 50(10): 944–948.

Connelly, L.M. 2012, "Root Cause Analysis", *Medsurg Nursing*, vol. 21, no. 5, pp. 316–316, 313.

Incident Analysis Collaborating Parties. 2012. Canadian Incident Analysis Framework. Edmonton, AB: Canadian Patient Safety Institute.

Chuang, Y. T., Ginsburg, L., & Berta, W. B. 2007. Learning from preventable adverse events in health care organizations: Development of a multilevel model of learning and propositions. *Health care management review*, 32(4): 330–340.

Cooke, D. L. 2003. *Learning from incidents. In 21st System Dynamics Conference*, New York City.

Cooke, D. L., Dunscombe, P. B., & Lee, R. C. 2007. Using a survey of incident reporting and learning practices to improve organisational learning at a cancer care centre. *Quality and Safety in Health Care*, 16(5): 342–348.

Edmondson, A. C. 2004. Learning from failure in health care: frequent opportunities, pervasive barriers. *Quality and Safety in Health Care*, 13(suppl 2): ii3–ii9.

Evans, S. M., Berry, J. G., Smith, B. J., Esterman, A., Selim, P., O'Shaughnessy, J., &DeWit, M. 2006. Attitude and barriers to incident reporting: a collaborative hospital study. *Quality Safety in Health Care*, 15: 39–43

Fielden, A.L., Sillence, E., & Little, L. 2011. Children's understandings of obesity: a thematic analysis. Int *J Qual Stud Health Well-being*, 6(3): 10.

Harper, M.L., &Helmreich, R.L. 2005. Identifying barriers to the success of a reporting system. *Advances in Patient Safety*. Vol.3. Retrieved from http://www.ahrq.gov/qual/advances/

IHPSC. 2008. *National Guidelines for Hospital Patient Safety*. Jakarta: IHPSC

IHPSC. 2011. *Annual report on Hospital Patient Safety Incident in Indonesia*. Unpublished Report.

Jeffs, L., Berta, W., Lingard, L., & Baker, G. R. 2012. Learning from near misses: from quick fixes to closing off the Swiss-cheese holes. *BMJ quality & safety*, 21(4): 287–294.

Ministry of Health. 2013. *Online hospitals data: hospitals recapitulation by category*. Retrieved from http://202.70.136.52/rsonline/report/report_by_catrs.php.

Stirling, J.A. 2001. Thematic networks: an analytical tool for qualitative research. *Qualitative research*, 1: 385.

Tamuz, M., Franchois, K. E., & Thomas, E. J. 2011. What's past is prologue: Organizational learning from a serious patient injury. *Safety Science*, 49(1): 75–82.

Tafvelin, S. 2013. Comparing employees and managers perceptions of organizational learning, health, and work performance. *Advances in developing human resources* (1523–4223), 15(2): p. 163.

WHO. 2005. *WHO Draft Guidelines for Adverse Events Reporting and Learning Systems*. Geneva: WHO Press.

*Bridging Research and Good Practices towards Patient Welfare – Shih & Liang (Eds)*
 *ISBN 978-1-138-02716-9*

# What can Indonesia learn from Taiwan's successful patient-safety reporting system?

I. Dhamanti
*Airlangga University, Surabaya, East Java, Indonesia*

S. Leggat & S. Barraclough
*La Trobe University, Melbourne, Victoria, Australia*

A. Utarini
*Gadjah Mada University, Jogjakarta, Indonesia*

S. Liao
*Taiwan Joint Commission on Hospital Accreditation, Taipei, Taiwan*

ABSTRACT: Indonesia and Taiwan began patient safety reporting at almost the same time. To date, there are substantial differences between the numbers of cases reported annually in each country. This study compares the Indonesian and Taiwan patient-safety reporting systems in order to learn from Taiwan's successful patient-safety reporting system. We searched the patient safety agencies websites, guidelines, and reports, published and unpublished articles and performed descriptive comparative analysis on relevant documents. We found some similarities on who can report, content of report, the procedure, confidentiality and the nature of reporting. Supported by its advanced data management technology, Taiwan is more comprehensive on methods of reporting, incident analysis; providing feedback, learning and system evaluation. Indonesia has established a framework for the reporting system but some improvements can be made, specifically on the utilization of data to enhance learning, providing various feedback mechanisms, and the evaluation of the system.

*Keywords*: patient safety, incident reporting, Indonesia, Taiwan, comparative analysis

## 1 INTRODUCTION

Patient safety has been a national priority in many countries for more than a decade. Patient safety reporting systems have been developed to allow learning from experience, in order to prevent the same mistakes from recurring. Through such reports, senior staff acquire evidence essential for investigating an incident, identifying and analyzing the hazards and risks that cause system failure, and establishing ways to adjust or re-design the system, to avoid similar errors in the future (WHO, 2005).

Indonesia is a unitary state with a population of 244.5 million with US$867.5 billion GDP and U$3,499 GDP per capita (DFAT, 2013). Indonesia has 2,262 hospitals of which 1,273 hospitals have been accredited. There are 471,136 health workers (IMOH, 2013). Taiwan's population in 2013 reached 23.3 million with US$484.7 billion GDP in current prices, and estimated per capita GDP of US$20,706 (DFAT, 2013). According to Taiwan's Ministry of Foreign Affairs (2013), in 2009, there were 514 hospitals and 19,792 clinics operating in Taiwan.

Indonesia began its patient safety movement in June 2005 by forming the Indonesian Hospital Patient Safety Committee (IHPSC). In the following year, the Committee collaborated with the Ministry of Health, the Indonesian Commission on Hospital Accreditation, and the Indonesian

Table 1. Population, participating hospitals and incidents reported in Indonesia and Taiwan.

| | 2008 | 2009 | 2010 | 2011 |
|---|---|---|---|---|
| **Indonesia** | | | | |
| Population | 237,939,000 | 240,715,000 | 243,423,000 | 246,065,000 |
| Participating hospitals | 61 | 114 | 103 | 132 |
| Number of incidents reported | 61 | 114 | 103 | 132 |
| **Taiwan** | | | | |
| Population | 22,987,000 | 23,065,000 | 23,127,000 | 23,174,000 |
| Participating hospitals | 248 | 293 | 553 | 631 |
| Number of incidents reported | 25,283 | 32,988 | 50,310 | 55,020 |

Sources: TJCHA (2013), IHPSC (2011), US Census Bureau (2013).

Hospital Association to set the National Guidelines for Hospital Patient Safety, and developed the Hospital Patient Safety Standards. The purpose of reporting is to reduce the number of adverse events and 'near miss' cases, as well as to improve patient safety and thus the quality of care (IHPSC, 2008). During the early system development phase, the IHPSC conducted a campaign to raise awareness by visiting twelve major cities to promote patient safety (WHO, 2007). The Committee also encouraged hospitals to implement patient safety programs, including safety reporting. The first method of reporting an incident was by post or courier, but in 2008, the Committee developed e-reporting. During 2006–2007, 145 hospitals reported 145 cases (IHPSC, 2012). In comparison to other countries with reporting systems in place, IHPSC's Annual Reports reveal a low number of incidents being reported to the Committee from year to year, with only 132 cases reported in 2011.

The Taiwan Department of Health introduced the Taiwan Patient-safety Reporting System (TPR) in December 2003, although the service began only in January 2005 (Chiu et al., 2008). The Department established the Patient Safety Committee in 2004 (Chou et al., 2011), and in the first stage of system development, the Taiwan Joint Commission on Hospital Accreditation assisted healthcare institutions to install software and databases, and to manage the ensuing work. According to Wu et al. (2009), the purpose of TPR is to integrate the national patient safety reporting data, analyze trends, and provide advice for staff about high frequency and severe incidents. In the first year, 77 institutions were enrolled and 1,760 incidents reported (Chiu et al., 2008). As described by Chiu et al. the reporting methods used are online reporting, software upload, and data-mapping, with 3,910 facilities participating and 189,210 events entered to the system database in 2011.

If we compare the two systems in the period 2008 to 2011, the number of incidents reported in Taiwan far exceeds that of in Indonesia (Table 1). Since both systems began operations at almost the same time, the difference in the number of incidents reported is significant, suggesting that Indonesia can draw on Taiwan's successful experience to improve its own reporting system. The aim of this paper is to compare the patient safety reporting systems in Indonesia and Taiwan, in terms of the system, process, classification, analysis and evaluation of each, with the aim of learning from Taiwan's TPR.

## 2 METHOD

We searched the official websites and homepages of the responsible leading agencies or organizations, or the patient safety agencies of the two countries, and reviewed the publicly available report information – for example, the website of the Taiwan Joint Commission on Hospital Accreditation and the Taiwan Patient-safety Reporting System, also IHPSC's website. We also looked at guidelines, regulatory documents, and government reports, as well as published and unpublished articles concerned with patient safety reporting in Indonesia and Taiwan, written in English, Indonesian or Chinese. An Indonesian professor and a staff member from the Taiwan Joint Commission on Hospital Accreditation assisted reviewing the documents in Indonesian and Chinese respectively.

We used the WHO's components of patient safety reporting that consists of type of system, process, classification, and analysis as the framework for comparison (WHO, 2005). We also added evaluation as one of the components of any system, and used descriptive comparative analysis on relevant documents.

## 3 RESULTS

### 3.1 *Type of system*

Indonesia's patient-safety reporting system is voluntary, covering all hospitals in Indonesia, and consists of both internal and external reporting. Internal reports are compiled by staff who were either involved in, or who observed the incident, or by a staff member who established the occurrence of the incident for the first time. The internal reports are submitted to the supervisor or head of the unit; external reports go from the head of a hospital patient-safety team to the IHPSC either by post or online. The name of the reporter should be recorded in the internal reporting form, for the purpose of gathering more information related to the incident, but in external reports, the names of the patient, reporter and hospital are not disclosed.

The reporting system itself is a learning system, because it emphasizes the learning processes, to prevent similar incidents from recurring. There are two levels of learning, at the hospital and national levels. Within the hospital, learning occurs when the patient-safety team investigates the incident and analyzes the causes. Based on the investigation, the patient-safety team sends recommendations designed to improve performance, and 'feedback' for the staff of the work unit where the incident happened. At the national level, the experiences of the hospitals are shared among them, through IHPSC's website and through seminars attended by personnel from some hospitals. Access to the website is limited, since hospitals need a code and password to gain entry and read the reports. The IHPSC Annual Report in 2011 shows that only 220 hospitals had access, that is, about 10% of hospitals in Indonesia.

TPR also has a voluntary nationwide patient safety reporting system which aims to observe the principles of anonymity and confidentiality; it is voluntary and non-punitive, and foster mutual learning (TJCHA, 2013; Chiu et al., 2008; Wu et al., 2009). It also practices internal and external reporting (Lin et al., 2012). For internal reporting, staff can submit a report online and it will be audited. For confidentiality, any identification of patient or hospital is removed before the data are stored. In the next process, external reporting, hospitals report to a national reporting system (TPR) own by the government and maintain by Taiwan Joint Commission on Hospital Accreditation, where the data are analyzed and integrated.

As in the Indonesian reporting system, the learning process in TPR starts from the hospital investigating an incident, to find the cause and solve the problems. Similarly, hospitals in Taiwan also submit accounts of incidents, what has been learned, and proposed improvements to the system, with 'signal alert' and instructive cases developed from the database quarterly, by using incidents of high severity, high frequency, and which TPR reviewers judge as valuable for learning (Wu et al., 2009). Unlike the Indonesian reporting system, the TPR-developed 'alerts' and 'valuable learning' case details are fed back to the hospitals through two different websites, and through an individual email to each hospital. The websites have no restrictions on access, so anyone can read or download the learning material. In addition, the teaching materials developed after consultation with experts, while the cases are also used in doctors' education workshops (Lin et al., 2012; Wu et al., 2009).

### 3.2 *Process*

The Indonesian 'Guidelines for Patient Safety Incident Reporting' (IHPSC, 2008) includes reporting adverse events and 'near misses'. Actually, the *potential* hazards are reported at the hospital level, but not to the IHPSC. The internal reporting system is paper-based, so that anyone, including patients and their relatives, or hospital visitors, are able to report an incident to staff, and later those staff complete an internal reporting form. The process of reporting starts when staff submits a report to the supervisor or head of the unit (Figure 1). The supervisor or head of the unit then conducts an assessment to determine the 'grade' or category of the incident, which determines the type of investigation, which can take from seven to 45 days. The incident report, with what has been learned from it, and recommendations, are then reported to the hospital director, before going on to the IHPSC, while the originating unit gets some 'feedback' from the hospital patient-safety team. For internal reporting, an incident must be reported within 48 hours, though there is no specific time limit on reporting externally to the IHPSC.

External reports are available in paper-based or online form. This form is a mix of a structured questionnaire form and a narrative text. 'Narrative' covers points such as the incident chronology, the action taken after the incident, the outcome, and whether a similar incident has occurred in other work units, so the reporter can cover all the necessary information. When the investigation is complete, the team sends the paper-based form by post or submits the online version through the website.

In TPR, anyone in the hospital can report a patient safety incident that they observed or were involved in (Lin et al., 2012). Types of events reported are adverse events and 'near misses', and both internal and external reports are online. In internal reporting, staff can submit reports by paper or online and the report will be audited, while statistical and root cause analysis is conducted before submission of the report to the national database (Figure 2). The data are analysed and integrated after validation processes de-identifying any information related to a patient or hospital, as well as checking the correctness of categorization of an event and its severity (Lin et al., 2009). Data are then published in the quarterly or yearly reports distributed to all hospitals or to health-care quality journals, publication's section on the website, added to 'frequently asked questions' and download sections on the TPR website (MOH, 2012).

### 3.3 *Classification*

The Indonesian reporting system has 13 incident types, as used in WHO international classification for patient safety (WHO, 2009), except that the Indonesian system has not included incidents caused by staff behaviour in their classification, adding instead 'laboratory incident'. TPR has the same number of incident types as does Indonesia but its classification system is different (Table 2) and does not include categories related to staff behavior or organizational and management factors. The classification was mainly designed by user friendly service domain or process to make it easier both for incident reporting and analysis.

### 3.4 *Analysis*

Effective analysis is the critical component of a patient safety reporting system (Clarke, 2006). Indonesian patient safety reporting has undertaken data analysis at national level, using SPSS, and presents only the frequency and distribution of an incident and its ranking by order of frequency – for instance, the number of incidents based on the type of hospital ownership, the province where the hospital is located, the sex and age of patients, incident location, and the reporter. Instead of merely focusing on frequency distribution, the analysis of national data in TPR also seeks the relationship between two variables, but also the possible reasons for frequently occurring incidents (Lin et al, 2012). For example, it was found that the most common type of medical incident was drug-related, and the most frequent cause was the human factor. In addition, TPR analyzes the time of incident occurrence in order to understand the time periods or the certain times when incidents occur most frequently and their relation to the reasons for their occurrence.

### 3.5 *Evaluation*

Evaluation includes ongoing monitoring of programs, as well as one-shot studies of program processes or program impact (Wholey et al., 2010). The evaluation of the whole system has never

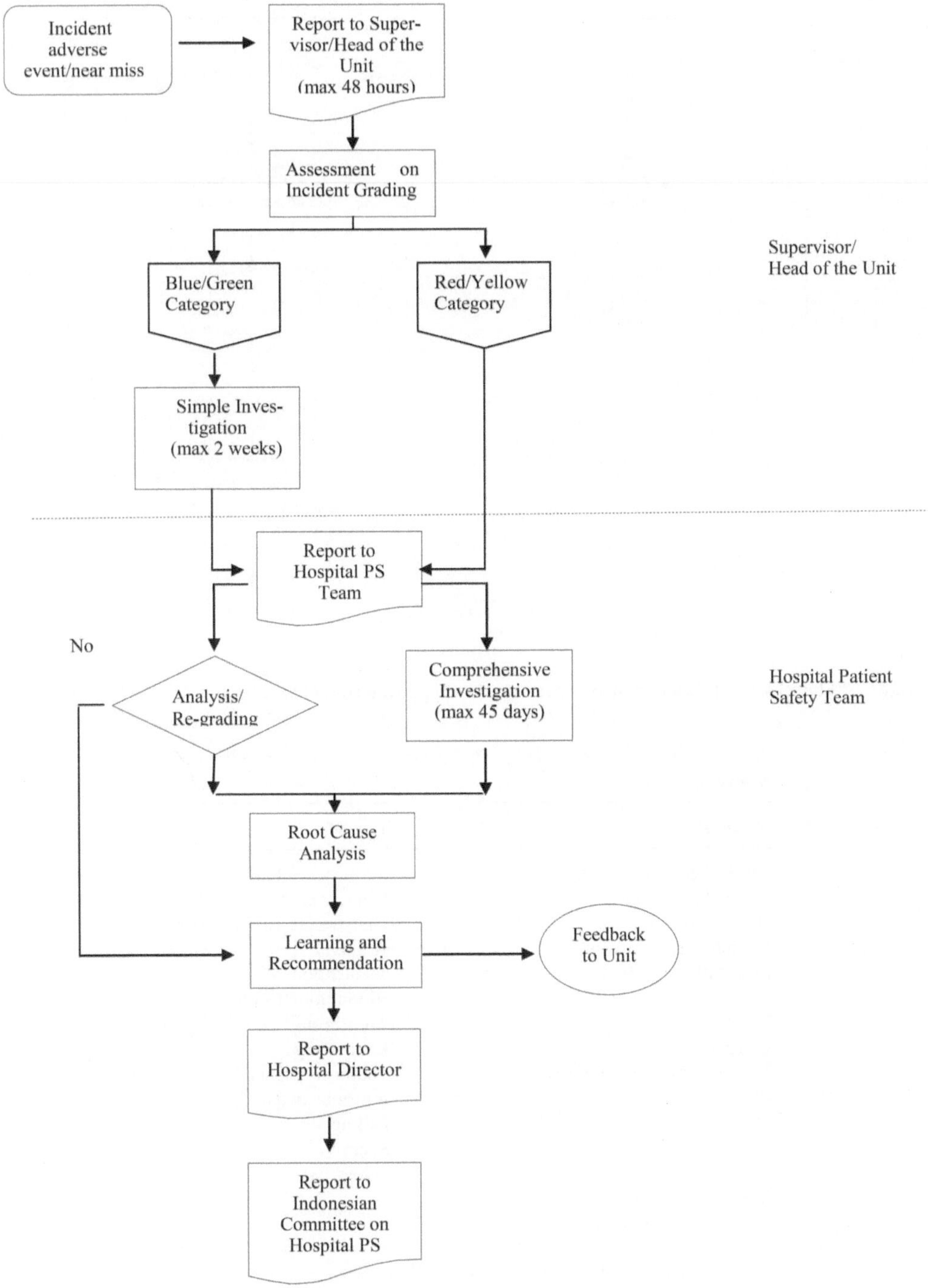


Figure 1. Flowchart of Indonesian patient-safety incident reporting (IHPSC, 2008).

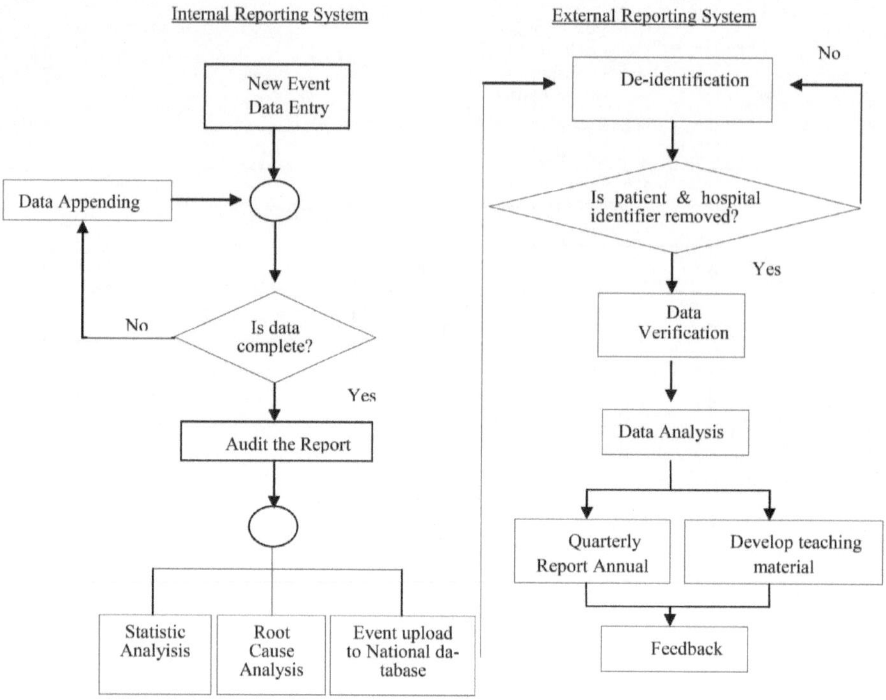


Figure 2. Flowchart of Taiwan's patient-safety incident reporting (Lin et al., 2012).

Table 2. Incident types used in Indonesian and Taiwan patient-safety reporting.

| Indonesia | Taiwan |
|---|---|
| Clinical administration | Surgery-related |
| Clinical procedure | Clinical care |
| Documentation | Unexpected cardiac arrest |
| Hospital associated infection | Tubing |
| Medication/IV fluids | Medication |
| Blood/blood product | Blood transfusion |
| Nutrition | Anesthesia |
| Oxygen/gas | Laboratory |
| Medical device/equipment | Public accidents |
| Behaviour | Injurious behavior |
| Falls | Falling |
| Patient accidents | Security |
| Infrastructure/building/fixtures | Others |
| Resource/organizational management | |
| Pathology/laboratory | |

Sources: IHPSC (2008) and Lin et al. (2012).

been conducted in the Indonesian reporting system, though some improvements in the content of the patient safety incident form were made on 2007 and 2008 (IHPSC, 2011). In contrast, as described by TJHA (2006) on its website, Taiwan's evaluations of its patient safety goals, including the reporting procedures, take place every year and their improvement is reviewed annually.

## 4 DISCUSSION

This paper has compared the patient safety reporting systems in Indonesia and in Taiwan, to learn from the latter, the Taiwan Patient Safety Reporting. Even though both systems began almost at the same time, there are many differences between the two. On the whole, however, TPR is a more comprehensive and responsive system in comparison to the Indonesian reporting system.

The development of the TPR was carried out in three different stages, including a pilot phase that lasted for two years, in which healthcare institutions were assisted to install software and databases which were vital to the system infrastructure. During this pilot phase, the method of reporting was improved and changed, based on healthcare institution needs, so it is not surprising that the number of institution enrolled in the system increased significantly (Chiu et al., 2008). This contrasted with Indonesia, where in the development phase, the Committee only 'encouraged' the hospitals to implement the reporting system, without, actually supporting them, by providing resources such as internet access for rural or remote hospitals, or free training for relevant hospital staff.

Regarding confidentiality, both systems protect the privacy of staff as reporters, and patients involved in an incident, while at national level, hospital identification is also removed, to protect the privacy of the organization. This is consistent with Doupi's (2009) view that the management of confidentiality in patient safety reporting should include confidentiality for the hospital, the patient, and lastly, the reporter.

Learning from and within a patient safety system should occur on different levels: at the local unit or clinic, the hospital, and in the wider environment (Pronovost et al., 2011). Both countries have conducted their learning activities at a number of levels. However, TPR is better at the national level; especially because it developed the system of 'alert' and learning cases, with the learning widely disseminated using several channels with has no restriction on access to reading or downloading on the website, and even notifying hospitals by email about a report. In the Indonesian reporting system, for anonymity reasons, the IHPSC does not know the names and the identities of the hospitals concerned; thus it cannot provide them with direct feedback or notifications. As Leape (2002) recommends, results that are aggregated and analyzed by an external body should be disseminated broadly. This limitation on access to the website's learning section is a drawback, since it limits other hospitals who could learn from others' mistakes, and prevents sharing information about the many methods used in investigations, and what is learned, which would be important in improving the Indonesian reporting system. Doupi (2009) advocated dissemination as the mechanism that provides direct feedback to the reporting organisation, and recommended the widest possible dissemination of such information among healthcare professionals and providers. Consistent with this, we can see, in TPR that a measure like email notification looks simple yet is important, as the acknowledgment and timely feedback that motivates hospitals to report its incidents consistently.

In both systems, processes are similar in the type of incident reported and the existence of internal and external reporting. However, in the Indonesia system, the validation has not taken place either in hospital or national level. Within TPR, the validation process is conducted at the national level to ensure the quality of data reported, but in Indonesia, no further validation is conducted, the collected reports from the hospitals are tabulated and entered into SPSS. In addition, TPR has many more methods for giving feedback – through reports, publications, frequently asked questions, and a download section on the TPR website.

The TPR reporting system includes more comprehensive analysis and evaluation than the Indonesian system. For analysis, TPR has sought to establish the surrounding relationships, or the possible reasons for incidents, to identify a set of frequent contributing factors. Leape (2002) recommends this feature, mentioning the advantages of a national voluntary reporting system, in which analyses from multiple hospitals can lead to the identification of common contributing factors. Moreover, TPR also has set an annual assessment to review its patient safety goals, including assessing patient safety. This act as a mechanism for monitoring whether any system changes instituted have improved the system, or have instead caused new risks (Kaplan & Fastman, 2003).

## 5 CONCLUSION

There are some similarities between the Indonesian and Taiwan patient-safety reporting systems, such as who can report, the content of reporting, reporting procedures, confidentiality, and the nature of reporting. Supported by its advanced data management technology, the Taiwan Patient Safety Reporting System is more comprehensive in providing open channels for reporting, in conducting incident analysis, in providing feedback and details of learning from incidents to the hospitals, and regularly evaluating the system. Some improvements of the Indonesian system that can be made in hospital and national levels include using data to enhance learning, provide various feedback mechanisms, conducting data verification and setting requirements for evaluation of the system at regular times.


## ACKNOWLEDGEMENT

The principal author would like to thank the Indonesian Directorate General of Higher Degree for providing financial support during her doctoral study in La Trobe University, Australia.


## REFERENCES


Chiu, C.T., Huang, S.Y., Liao, H.H., Wung, H.Y., Wung, H, Y. & Shih., C.L., et al. 2008. *Enhancing TPR participation with IT.* TJCHA Publication. http://www.tjcha.org.tw/.

Chou, H.L, Cheng, C.H., Chen, F.G. & Chen., J.S. 2011. *A study on patient safety and accident prevention measures with modern information technology.* CSEE 2011, Part V, CCIS 218: 190–195.

Clarke, J.R. 2006. How a system for reporting medical errors can and cannot improve patient safety. *The American Surgeon*, 72: 1088–1091.

Department of Foreign Affairs and Trade. 2013. *Countries and Regions.* https://www.dfat.gov.au/geo/

Doupi, P. 2009. National reporting systems for patient safety incidents : A review of the situation in Europe. *Report.* Persephone Doupi and National Institute for Health and Welfare.

IHPSC. 2011. *Annual report on Hospital Patient Safety Incident in Indonesia.* Unpublished Report.

IHPSC. 2008. *Guidelines fo Hospital Patient Safety Incident Reporting.* Jakarta: IHPSC.

Indonesian Ministry of Health. 2008. *National Guidelines for Hospital Patient Safety.* Jakarta: IHPSC.

Indonesian Ministry of Health. 2013. *Online hospital data.* http://202.70.136.52/rsonline/report/report_by_catrs.php

Kaplan, H.S. & Fastman, B.R. 2003. Organization of event reporting data for sense making and system improvement. *Qual Saf Health Care,* 12(Suppl II): ii68–ii72.

Leape, L.L. 2002. Health Policy Report: Reporting of adverse events. *N Engl J Med*, 347(20): 1633-1638.

Lin, C.C., Shih, C.L., Liao, H.H. & Wung, C.H. 2012. Learning from Taiwan patient-safety reporting system, *Int. J. Med. Inform.* 81(12): 834–841.

Lin, P.J, Liao, H.H. & Shih, C.L. 2009. *Assuring patient safety report data quality through a validation process.* TJCHA Publication. http://www.tjcha.org.tw/.

Ministry of Foreign Affairs, Republic of China (Taiwan). 2013. *Medical care system.* http://www.taiwan.gov.tw/

Pronovost, P. J., Thompson, D. A., Holzmueller, C. G., Lubomski, L. H., Dorman, T., Dickman, F. & Morlock, L. L. (2006). Toward learning from patient safety reporting systems. *Journal of critical care*, 21(4): 305–315.

Pronovost, P. J., Morlock, L. L., Sexton, J. B., Miller, M. R., Holzmueller, C. G., Thompson, D. A. & Wu, A. W. 2011. Improving the value of patient safety reporting systems. *Advances in patient safety: new directions and alternative approaches*, 1.

Taiwan Joint Commission on Hospital Accreditation. 2006. *Patient Safety Annual Goals.* http://www.patientsafety.mohw.gov.tw/en/Content.asp?cid=4

Taiwan Joint Commission on Hospital Accreditation, 2013. *Taiwan Patient-safety Reporting System.* http://www.tpr.org.tw/english.htm.

Wholey, J. S., Hatry, H. P. & Newcomer, K. E. (2010). *Handbook of practical program evaluation* (Vol. 19). Wiley. com.

WHO. 2005. *WHO Draft Guidelines for Adverse Events Reporting and Learning Systems.* Geneva: WHO Press.

WHO. 2007. *First Regional Workshop on Patient Safety: A Report.* New Delhi: WHO Regional Office for South-East Asia New Delhi.
WHO. 2009. *Conceptual Framework for the International Classification for Patient Safety.* http://www.who.int/patientsafety/taxonomy/icps_full_report.pdf
Wung, C.H., Yu, T.S., Shih, C.L., Lin, C.C., Liao, H.H. & Chung, K.P. 2011. Is it enough to set national patient safety goals? An empirical evaluation in Taiwan. *International Journal for Quality in Health Care*, 23(4): 420–428.
Wu, T.Y., You, Y.L., Chang, F.L., Chen, I., Wung, C. & Shih, C.L. 2009. *Developing Alert and Learning Cases from Taiwan Patient-safety Reporting System*. TJCHA Publication. http://www.tjcha.org.tw/.

*Bridging Research and Good Practices towards Patient Welfare – Shih & Liang (Eds)*
*© 2015 Taylor & Francis Group, London, ISBN 978-1-138-02716-9*

# The influence of age and object form on the motivation of haptic

Si-Jing Chen & Chiu-Wei Chien
*Graduate School of Creative Industry Design, National Taiwan University of Arts, Taiwan*

Chih-Long Lin
*Crafts and Design Department, National Taiwan University of Arts, Taiwan*

ABSTRACT: The main purpose of this research is to investigate the effect of user characteristic – the nature of need for touch (high group and low group), age (18–30 years old, 31–50 years old, greater than 51 years old), and object form (spheroid, cube, regular tetrahedron, cone and cylinder) on subject's motivation of touch, preference and sense. The dependent variables including the willing of touch, preference and 7 adjectives of sense were measured by questionnaire interview. The ANOVA results showed that the willing of touch and subjective preference were mainly affected by the nature of need for touch and object form. The results of regression equations showed that the willing of touch was mainly affected by subjective preference. Moreover, the subjective preference was mainly affected by the rating of beauty and the rating of funny for an object. Therefore, the subjective preference increased for an object was followed the rating of beauty and then the willing of touch was increased. The findings of this study can give an insight into the motivation of touch, and further provide some guidelines and recommendations about the product design and selling method to increase the competitive advantage of product.

## 1 INTRODUCTION

In spite of the "Do No Touch" notices or warning signs commonly seen in front of artworks, luxury goods or dangerous devices at various occasions such as art galleries, boutique stores and public facilities, we can still frequently observe many rebellious souls touching the items against the advice. What are the reasons behind the desire to touch or the motives so strong that people are tempted to ignore the warning signs and reach out their hands? Peck and Childers (2003b) suggested that from the perspectives of the consumers, touching a product allows them to assess it with more confidence and thus increase purchase intention and determination. Due to the diversification of shopping channels in recent years, consumers can now purchase goods via catalogs, the Internet, television and other media rather than buying them at physical stores. However, these channels are inherently limited to images or sounds in communicating visual or auditory properties of products while the delivery of texture, hardness, weight or other tactile information of products are even more difficult. Therefore, the potential compensations for the loss of tactile experiences of products have become a dominant area of study in the fields of design and marketing in recent years. This study will discuss researches related to the influences of product features and personal characteristics on the motivation to touch.

### 1.1 *Influences of product features on the motivation to touch*

The behavior of touching a product by the consumers prior to purchasing depends on the different types of products. Peck and Childers (2005) conducted a survey in which the hand gestures and verbal descriptions of the subjects were filmed as they assessed a number of products. The results suggested a high degree of correlation between their verbal expressions and hand gestures.

Furthermore, the durations of touch were the longest for products requiring assessments on the material properties (such as sweaters); followed by products with less needs for assessments such as mobile phones or calculators; and the products with the shortest durations of touch were those with the lowest needs for assessment such as food and toothbrushes.

Are there other methods to compensate for the needs to touch if the tactile experiences are not available to the consumers during product assessments? Holbrook (1983) used a sweater as the sample for an experiment and discovered that the subjects relied heavily on tactile information to assess the product. Hence it was recommended to provide actual products for assessments rather than presenting only the visual information with images. McCabe and Nowlis (2003) further investigated the influences of product material properties and the methods of assessment (visual and tactile, visual and no tactile, or visual only) on the purchasing desires. The results indicated that for product assessments requiring tactile senses such as towels and rugs, the subjects displayed higher purchasing desires with tactile assessments compared to the lack of tactile assessments. However, for products with similar material properties such as audio or video cassette tapes, the availabilities of tactile assessments did not affect the purchasing desires because visual assessment was the primary factor affecting purchasing decisions for this type of products as explained by the authors. An interesting finding was that the influence of touch on the purchasing desires reduced once supplementary information on the characteristics of product materials were provided to the subjects. The research of Peck and Childers (2003b) also indicated the lack of tactile assessment could be compensated with text descriptions.

### 1.2 *Influences of personal characteristics on the motivation to touch*

In addition to the investigation on the potential compensations for the needs to touch, the research of Peck and Childers (2003b) also examined the influences of material properties and individual differences on product preferences. The study indicated that different reactions were created in response to different material properties, for example the plasticity of a material generated more delightful sensations for the subjects compared to its weight. Furthermore, Peck and Childers believed the presence of individual differences in the preferences for tactile senses and developed a Need for Touch Questionnaire from a validation process involving seven experiments. The Need for Touch (NFT) is defined as the inclination to collect the required information through touch, which can be further classified as Instrumental Dimension and Autotelic Dimension. Instrumental Dimension is the behavior of touch driven by the desire to collect the information required for a specific purchasing task, in other words, it is the action of objective-oriented searching by the consumers to find the required information for a purchase until the final decision has been resolved. In contrast, Autotelic Dimension is the behavior of touch driven by the desire to enjoy or an impulse or irresistible temptation to explore, which can be interpreted simply as a consumer behavior based on the needs for fun, sensory stimulation or joy rather than for purchasing a product (Peck and Childers 2003a).

Peck and Childers (2003a) discovered from their research that the subjects with high NFT were more familiar in dealing with tactile information and would consider the material properties with priority in their product assessments. Consequently, the lack of direct contact with a product would lead to more struggles and less confidence in the decisions of these subjects. In comparison, hints on tactile information could be collected through visual inspections for those with lower dependencies on tactile assessments (low NFT), who could then proceed to decide on a purchase. Field observations at the stores suggested that subjects with high NFT paid more attention to the sampling information within the stores and proceeded to the sampling stalls. In contrast, subjects with low NFT completed purchases with confidence even in the absence of direct contacts with the products. Last of all, the authors indicated that the compensations for the loss of touch exerted more influences on the subjects with high NFT, in particular, the needs for touch could be compensated with text descriptions for functional tactile information (such as weight). However, text description was insufficient to compensate for the absence of experiential tactile information (such as plasticity). Peck and Childers (2005) analyzed the hand gestures of the subjects during product assessments and noticed that the time spent assessing the material properties by the subjects with high NFT was shorter than those with low NFT, except for the assessment of surface texture of products. The reason behind this phenomenon was that subjects with high NFT extracted information with more efficiencies during the tactile process (Peck and Childers 2003a); but the comforts and delights offered by the surface textures of products attracted them to spend more time touching and feeling the products.

### 1.3 *Research purpose*

Tactile stimulation may enhance the positive assessments of products by consumers in many contexts. Millar & Tesser (1986) proposed that elevated emotional values may be derived from the experiences of direct contacts with the products. Breckler & Wiggins (1991) also observed that the degree of consistency between the attitudes and behaviors of consumers increased with more exposures to direct contact experiences. Hence it is evident that tactile sensation and product assessment are highly correlated, but only a number of researches discussed the triggers driving the motivation to touch. The impulse to "Just Touch It!" does not intend to damage, but it is a simple desire to experience the sensation of touching an object or to know if it feels as soft and smooth or solid and coarse as suggested by the visual information. Previous studies have discovered that different motivations to touch are induced by various product categories, and that the various levels of NFT of an individual may develop different motivations to touch. Therefore, this study further investigates the influences of the object shapes on the motivation to touch and analyzes the differences between various age groups of the subjects in-depth. The findings from this research will not only contribute to the understanding of the underlying mechanisms of the motivation to touch, but can also be applied to product design and marketing strategy to transform the motivation to touch into motivation to buy.

## 2 METHODS

### 2.1 *Subjects*

This study recruited a total of 60 subjects, arranged according to age into 3 groups of 18~30 years of age, 31~50 years of age, 51 years of age and above with 10 males and females in each age group, to participate in the experiment.

### 2.2 *Experimental design and procedure*

This study utilized multifactorial experiment design in which the between-subject factors were the levels of NFT and age groups and the within-subject factors were the object shapes. The questionnaire developed by Peck and Childers, with a total of 12 questions, was used as the primary reference for determining the NFT levels of subjects for this study. The available answers were divided into 7 grades ranking from $-3$ to $+3$, in which a grade of $-3$ corresponded with "strongly disagree"; a grade of 0 corresponded with "no comment"; and a grade of $+3$ corresponded with "strongly agree" (Peck and Childers 2003a). The total scores for each the 60 subjects were calculated and the participants were assigned into high NFT and low NFT groups based on the median score of 12. The subjects were also arranged according to age into 3 groups of 18~30 years of age, 31~50 years of age, 51 years of age and above. 5 sample models, each with a dimension of 9 cm, in the shapes of sphere, cube, tetrahedron, cone and cylinder were used as factors of object shapes. The samples used in this experiment are illustrated in Figure 1.

The dependent variables of the experiment were the levels of motivation to touch a specific sample (1 being "very undesirable"; 3 being "average desirability"; 5 being "very desirable"); the levels of preferences for a specific sample (1 being "very unappealing"; 3 being "average appeal"; 5 being "very appealing"); and 7 sets of sensory descriptions (light, warm, delicate, smooth, comfortable, beautiful and delightful) each ranked on a scale of five grades (1 being "strongly disagree"; 3 being "agree"; 5 being "strongly agree").

Figure 1. The sample models used in this experiment: sphere, cube, cone cylinder and tetrahedron ordered from left to right.

The procedures for the experiment began with the preparation session for the researchers to introduce the purpose and process of the experiment and the subjects to complete the NFT questionnaires. This was followed by the official experiment in which the 5 sample models were presented in a random order to the subjects for visual inspections and their subjective assessments recorded on the questionnaires with no time limitations.

### 2.3 *Data analysis*

This study used the repeated measures analysis of variance (ANOVA) to explore the presence of statistically significant influences of between-subject factors and within-subject factors on the motivation to touch, preferences and sensory descriptions. The Pearson correlation analysis was then conducted to investigate the levels of correlation between the motivation to touch, preferences and sensory descriptions. Thirdly, using the scores on the motivation to touch as dependent variable and the gender, scores on NFT, preferences and sensory descriptions as independent variables, the regression model for the motivation to touch was constructed with multiple stepwise regression analysis. Lastly, multiple stepwise regression analysis was utilized again in the construction of the regression model for preferences. The series of data analysis conducted in this study could facilitate the understanding of the correlations between the different factors and the underlying mechanisms for the motivation to touch.

## 3 RELUTS

### 3.1 *Results of the average and variance analysis*

The results of the various measured variables in the average and variance analysis of the levels of each factor in this study are illustrated in Table 1. The table indicates that the levels of NFT of individual subjects present statistically significant influences on the motivation to touch ($p < 0.01$), preferences ($p < 0.05$) and the sensory descriptions of delicate ($p < 0.05$), beautiful ($p < 0.001$) and delightful ($p < 0.001$). In terms of the average values, the motivation to touch and preferences for the group with high NFT ranked higher compared to the group with low NFT and the sample models were perceived as delicate, beautiful and delightful. Age, as a factor, did not present significance on all of the measured variables. However, the average values of the preferences and motivation to touch were the highest for subjects in 31~50 years of age and lowest for subjects in 51 years of age and above. The object shapes presented influences of statistical significance ($p < 0.05$ and above) on all of the measured variables, in which the motivation to touch was the strongest for the sphere

Table 1. The results of the various measured variables in the average and variance analysis of the levels of each factor.

| Factor | Level | Motivation to touch | Preference | Sensory description | | | | | | |
|---|---|---|---|---|---|---|---|---|---|---|
| | | | | Light | Warm | Delicate | Smooth | Comfortable | Beautiful | Delightful |
| Individual | Low | **3.43** | **3.25** | 2.75 | 3.19 | **3.45** | 3.32 | 3.43 | **2.96** | **3.09** |
| NFT | High | **3.81** | **3.51** | 2.87 | 3.39 | **3.70** | 3.46 | 3.62 | **3.50** | **3.57** |
| Age | 18~30 | 3.60 | 3.39 | 2.76 | 3.18 | 3.56 | 3.25 | 3.47 | 3.33 | 3.48 |
| Group | 31~50 | 3.80 | 3.41 | 2.88 | 3.36 | 3.58 | 3.42 | 3.52 | 3.28 | 3.25 |
| | 50 and above | 3.44 | 3.32 | 2.79 | 3.33 | 3.57 | 3.49 | 3.57 | 3.06 | 3.24 |
| Object | Sphere | **4.00** | **3.70** | **3.00** | **3.67** | **3.82** | **3.92** | **3.88** | **3.47** | **3.62** |
| Shape | Cube | **3.37** | **3.25** | **2.33** | **3.13** | **3.38** | **2.93** | **3.38** | **3.13** | **3.28** |
| | Cone | **3.50** | **3.27** | **3.00** | **3.15** | **3.58** | **3.52** | **3.38** | **3.22** | **3.23** |
| | Cylinder | **3.70** | **3.40** | **2.43** | **3.48** | **3.58** | **3.48** | **3.78** | **3.22** | **3.27** |
| | Tetrahedron | **3.50** | **3.25** | **3.28** | **3.02** | **3.48** | **3.08** | **3.17** | **3.08** | **3.22** |

Results in bold indicate that the measured variable showed significance ($p < .05$ and above) within the level of the factor.

Table 2. The results of the pearson correlation analysis between all of the measured variables in experiment 2.

| | Motivation to touch | Preference | Sensory descriptions | | | | | | |
|---|---|---|---|---|---|---|---|---|---|
| | | | Light | Warm | Delicate | Smooth | Comfortable | Beautiful | Delightful |
| Motivation to Touch | 1 | | | | | | | | |
| Preference | 0.54 | 1 | | | | | | | |
| Light | 0.18 | 0.23 | 1 | | | | | | |
| Warm | 0.29 | 0.34 | 0.19 | 1 | | | | | |
| Delicate | 0.31 | 0.26 | 0.36 | 0.38 | 1 | | | | |
| Smooth | 0.27 | 0.31 | 0.33 | 0.38 | 0.56 | 1 | | | |
| Comfortable | 0.34 | 0.36 | 0.26 | 0.39 | 0.46 | 0.48 | 1 | | |
| Beautiful | 0.48 | 0.53 | 0.28 | 0.21 | 0.38 | 0.38 | 0.40 | 1 | |
| Delightful | 0.46 | 0.50 | 0.28 | 0.17 | 0.36 | 0.33 | 0.37 | 0.68 | 1 |

All correlation coefficients of the variables achieved significance ($p < .05$ and above).

(score of 4.00) and the lowest for the cube (score of 3.37). Similarly, the level of preferences was the strongest for the sphere (score of 3.70) and lowest for the cube and tetrahedron (score of 3.25). In analyzing the sensory descriptions, the scores for the sphere was the highest for all descriptive terms; the scores for the cube were the lowest in the descriptions of light, delicate and smooth; the scores for the tetrahedron were the lowest in the descriptions of warm, comfortable, beautiful and delightful. The interactions between the levels of NFT and age groups as well as the levels of NFT and object shapes did not show any statistical significance.

### 3.2 *Pearson correlation analysis*

The results of the Pearson correlation analysis is illustrated in Table 2. The values of the correlation coefficients ranged between 0.17~0.68 and all of the coefficients achieved statistical significance ($p < 0.05$ and above). The correlation coefficient between the motivation to touch and preferences was the highest ($r = 0.54$); and the preferences and sensory descriptions of beautiful ($r = 0.63$) and delightful ($r = 0.68$) showed the highest correlations.

### 3.3 *Multiple regression analysis for the motivation to touch*

In this analysis, the regression models for the motivation to touch for various groups consisting of all subjects, the age groups of 18~30 years of age, 31~50 years of age and 51 years of age and above were constructed. The results are presented as follows:

(1) For the sample consisting of all subjects

$$\text{Motivation to Touch}_{(\text{all subjects})} = 0.92 + 0.39(\text{Preference}) + 0.23(\text{Beautiful}) + 0.12(\text{Delicate})$$

$$\text{Adj. } R^2 = 0.36,\ p < 0.001$$

(2) For the sample group of 18~30 years of age

$$\text{Motivation to Touch}_{(18\sim30 \text{ years of age})} = 0.54 + 0.54(\text{Preference}) + 0.27(\text{Delicate})$$

$$\text{Adj. } R^2 = 0.40,\ p < 0.001$$

(3) For the sample group of 31~50 years of age

$$\text{Motivation to Touch}_{(31\sim50 \text{ years of age})} = 1.67 + 0.40(\text{Preference}) + 0.26(\text{Delightful})$$

$$\text{Adj. } R^2 = 0.32,\ p < 0.001$$

(4) For the sample group of 51 years of age and above

$$\text{Motivation to Touch}_{(51 \text{ years of age and above})} = 0.93 + 0.39(\text{Preference}) + 0.29(\text{Beautiful})$$

$$\text{Adj. } R^2 = 0.38,\ p < 0.001$$

From the models, it is observed that regardless of the age groups, the preferences always play a factor in affecting the motivation to touch. In particular, the group of 18~30 years of age was influenced by delicate sensations; the group of 31~50 years of age was influenced by delightful sensations; and the group of 51 years of age and above was influenced by beautiful sensations. As the preferences toward the sample models increased or as the beautiful or delicate sensations of the sample models increased, the motivation to touch by the subjects also increased.

### 3.4 *Multiple regression analysis for preferences*

In this analysis, the regression models for the preferences for various groups consisting of all subjects, the age groups of 18~30 years of age, 31~50 years of age and 51 years of age and above were constructed. The results are presented as follows:

(1) For the sample consisting of all subjects

$$\text{Preferences}_{(\text{all subjects})} = 1.20 + 0.29(\text{Beautiful}) + 0.25(\text{Warm}) + 0.27(\text{Delightful})$$

$$\text{Adj. } R^2 = 0.37,\ p < 0.001$$

(2) For the sample group of 18~30 years of age

$$\text{Preferences}_{(18\sim30 \text{ years of age})} = 1.05 + 0.50(\text{Delightful}) + 0.26(\text{Warm})$$

$$\text{Adj. } R^2 = 0.36,\ p < 0.001$$

(3) For the sample group of 31~50 years of age

$$\text{Preferences}_{(31\sim50 \text{ years of age})} = 1.29 + 0.40(\text{Beautiful}) + 0.28(\text{Comfortable})$$

$$\text{Adj. } R^2 = 0.35,\ p < 0.001$$

(4) For the sample group of 51 years of age and above

$$\text{Preferences}_{(51 \text{ years of age and above})} = 1.21 + 0.31(\text{Delightful}) + 0.37(\text{Beautiful}) + 0.21(\text{Warm})$$

$$\text{Adj. } R^2 = 0.49,\ p < 0.001$$

From the models, it is observed that the preferences of the groups 18~30 years of age and 51 years of age and above were influenced by the delightful and warm sensations while the preferences of the group of 31~50 years of age were primarily influenced by beautiful and comfortable sensations. Overall, as the beautiful, warm, delightful sensations of the sample models increased, the preferences also increased.

## 4 CONCLUSIONS

This study has observed that the levels of NFT for individuals and the object shapes are both important factors with significant influences on the motivation to touch and preferences. However, the factor of age group did not show influences of significance. Subjects with high NFT displayed higher preferences and perceived the sample models as more delicate, beautiful, delightful, and hence their higher motivation to touch. The motivation to touch and preferences were the highest for the sphere and lowest for the cube. The regression models constructed indicated that the sample models suggesting more beautiful and delicate sensations would lead to higher motivation to touch while beautiful, warm and delightful sensations would increase the preferences of the subjects. It can be concluded that in addition to imparting a perception of beauty, a product also needs to provide a sense of delight to increase its preferences and thus the motivation to touch. The results of this research not only contributed to the construction of models for the motivation to touch, but can also be used to enhance the motivation to touch and thus the motivation to buy. As presented in this study, the results can provide useful guidance on the key factors of product design and marketing strategy in the future.

## REFERENCES

Breckler, S. J. and Wiggins, E. C., 1991. Cognitive responses in persuasion: Affective and evaluative determinants. Journal of Experimental Social Psychology, 27 (2), 180–200.

Holbrook, M.B., 1983. On the importance of using real products in research on marketing strategy. Journal of Retailing, 59 (1), 4–23.

Mccabe, D.B. & Nowlis, S.M., 2003. The effect of examining actual products or product descriptions on consumer preference. Journal of Consumer Psychology, 13 (4), 431–439.

Millar, M. and Tesser, A., 1986. Thought-induced attitude change: The effects of schema structure and commitment. Journal of Personality and Social Psychology, 51, 259–269.

Peck, J. & Childers, T.L., 2003a. Individual differences in haptic information processing: The "need for touch" scale. Journal of Consumer Research, 30 (3), 430–442.

Peck, J. & Childers, T.L., 2003b. To have and to hold: The influence of haptic information on product judgments. The Journal of Marketing, 67 (2), 35–48.

Peck, J. & Childers, T.L., 2005. Self-report and behavioral measures in product evaluation and haptic information: Is what i say how i feel? *Advances in Consumer Research,* 32 (1), 247.

*Bridging Research and Good Practices towards Patient Welfare – Shih & Liang (Eds)*
*© 2015 Taylor & Francis Group, London, ISBN 978-1-138-02716-9*

# Comparison of public health care system between Taiwan and Japan – Focusing on long-term oxygen therapy

Shigeki Tatsukawa, Nanae Shintani & Ayako Kajimura
*School of Health Sciences, Department of Nursing, Kansai University of International Studies, Japan*

ABSTRACT: This study aims to find potential issues of Long-Term Oxygen Therapy between Taiwan and Japan by comparison of Public Health Care System, and to make a proposition of solution approach for COPD patients bringing better quality of life. The Method is to collect health care systems data for long-term oxygen therapy from appropriate parties by interviews. In conclusion, I recommend; Japan to implement an education program about oxygen therapy service for appropriate parties to increase the number of eligible patients using the home-visit nursing care service for COPD as covered by Long-Term Care Insurance. And to introduce a qualification of "Respiratory Therapist" is suggested. Taiwan to discuss an item "Oxygen concentrator" of treatment for Long-Term Oxygen Therapy, which should be covered by future public health care system "Long-Term Care Insurance" planned in Taiwan. And it is suggested to introduce the system of "Prescription of oxygen inhalation".

## 1 BACKGROUND

### 1.1 *Chronic Obstructive Pulmonary Disease (COPD)*

Chronic Obstructive Pulmonary Disease (COPD) is a major cause of morbidity and mortality worldwide, and is among the top 10 global contributors to the global burden of disease as measured by disability-adjusted life years (Lopez AD, Mathers CD, et al. 2006). In 2004, COPD was the forth-leading cause of death. According to the World Health Organization (WHO), the total number of deaths in the world from COPD is projected to increase by 30% in the next 10 years. And by 2030, COPD would become the third-leading cause of death worldwide (World Health Organization. 2002, World Health Organization. 2008).

Among all major chronic diseases, COPD is the only disease that shows a rising mortality. This trend is predicted to rise further in the next decades with the escalating increase in the associated risk factors for COPD and an aging population in many parts of the world (World Health Organization. 2002, World Health Organization. 2008). The burden of COPD in Asia is currently greater than that in developed Western countries, both in terms of the total number of deaths and the burden of disease, as measured in years of life lost and the number of years spent living with disability, and is mainly linked to the epidemic of tobacco exposure and indoor and outdoor air pollution in Asian countries (Lopez AD, et al. 2006)

According to Global Initiative for Chronic Obstructive Lung Disease: GOLD, there are 4 important treatments for managing stable COPD.

### 1.2 *Four treatments of global initiative for chronic obstructive lung disease: GOLD*

1. Education:
   Patient education alone does not improve exercise performance or lung function. But it can play a role in improving skills, ability to cope with illness, and health status. Patient education regarding smoking cessation has the greatest capacity to influence the natural history of COPD. Education

also improves patient response to exacerbations. Prospective end-of-life discussions can lead to understanding of advance directives and effective therapeutic decisions at the end of life.

Education program include: smoking cessation; basic information about COPD and pathophysiology of the disease; general approach to therapy and specific aspects of medical treatment; self-management skills; strategies to help minimize dyspnea; advice about when to seek help; self-management and decision-making during exacerbations; and advance directives and end-of-life issues are suggested (NHLBI/WHO workshop report, 2000, NHLBI/WHO workshop report, 2007).

2. Pharmacologic Treatments:
   Pharmacologic therapy is used to prevent and control symptoms, reduce the frequency and severity of exacerbations, improve health status, and improve exercise tolerance. None of the existing medications for COPD have been shown to modify the long-term decline in lung function that is the hallmark of this disease. However, this should not preclude efforts to use medications to control symptoms (NHLBI/WHO workshop report, 2000, NHLBI/WHO workshop report, 2007).
3. Rehabilitation:
   The principal goals of pulmonary rehabilitation are to reduce symptoms, improve quality of life, and increase physical and emotional participation in everyday activities (NHLBI/WHO workshop report, 2000, NHLBI/WHO workshop report, 2007).
4. Oxygen Therapy:
   The long-term administration of oxygen (>15 hours per day) to patients with chronic respiratory failure has been shown to increase survival. It can also have a beneficial impact on hemodynamics, hematologic characteristics, exercise capacity, lung mechanics, and mental state (NHLBI/WHO workshop report, 2000, NHLBI/WHO workshop report, 2007).

In addition, there is increasing interest in home care for end-stage COPD patients, although the exact criteria for this approach as opposed to hospital treatment remain uncertain and will vary by health care setting (NHLBI/WHO workshop report, 2000, NHLBI/WHO workshop report, 2007).

The Japanese experience suggests that the challenges in the delivery of home oxygen are awareness in the public and health-care giver coordination and cost considerations, which are similar to those in the West (Takahashi Y, and Fukuda T. 2003).

Although prevalence estimates for COPD are being published for many areas of the world, high-quality estimates are lacking for key regions, and differences in measurement methodology hinder meaningful comparison of published study (R. J. Halbert, J. L. Natoli, et al. 2006). Moreover, there is almost no previous study for health care system comparison of Long-Term oxygen therapy among countries.

## 2 PURPOSE

This study aims to find potential issues of Long-Term Oxygen Therapy between Taiwan and Japan, by Comparison of Public Health Care System, and to make a proposition of solution approach for COPD patients bringing better quality of life.

## 3 METHOD

The Method is to collect following four categories of health care systems' information for Long-Term Oxygen Therapy from appropriate parties as interviews including Pulmonologists, Home Visiting Nurses, Respiratory Medicine Out-Patient Clinic Nurses, Respiratory Therapists and Oxygen Concentrators and Potable Oxygen Tanks Providers in each country (Taiwan and Japan).

1. Health care systems for Long-Term Oxygen Therapy
2. Prescription of oxygen inhalation
3. Frequency of routine out-patient visit for Long-Term Oxygen Therapy
4. How to obtain portable oxygen tanks and oxygen concentrators

## 4 RESULTS

### 4.1 *Japan*

In Japan, I interviewed a Pulmonologist, a Home Visiting Nurse, a Respiratory Medicine Out-Patient Clinic Nurse and an Oxygen Concentrators and Potable Oxygen Tanks Provider. According to these appropriate parties, Long-Term Oxygen Therapy is covered by National Health Insurance, Welfare Service, and Long-Term Care Insurance. And mainly out-patients with Long-Term Oxygen Therapy use National Health Insurance for continuous treatment. However Home Visiting Nurses' service is covered by National Health Insurance, Welfare Service, and Long-Term Care Insurance, they do not have much experience of actual Nursing visit for Long-Term Oxygen Therapy. There is an existing system of Prescription of oxygen inhalation.

Out-patient visit is continuously set once per month under cover of National Health Insurance. Rental and delivery systems of portable oxygen tanks and oxygen concentrators have been developed under cover of National Health Insurance as a treatment of COPD. And the same company provides both portable oxygen tanks and oxygen concentrators.

### 4.2 *Taiwan*

In Taiwan, I interviewed a Pulmonologist, a Home Visiting Nurse, two Respiratory Therapists and two Oxygen Concentrators and Potable Oxygen Tanks Providers. According to these appropriate parties, Long-Term Oxygen Therapy is covered by National Health Insurance and Welfare Service. And mainly out-patients with Long-Term Oxygen Therapy use National Health Insurance for continuous treatment. Despite the limitation for using Home Visiting Nurses' service as twice per month, there is a qualification of"Respiratory Therapist" and these therapists also visit out-patients for maintaining their condition. There is no existing system of Prescription of oxygen inhalation, although Pulmonologists decide the amount of oxygen inhalation individually. Out-patient visit is set once per month under cover of National Health Insurance. However out-patient visit only provide approximately three months period after Long-Term Oxygen Therapy is introduced. Welfare service is only the health care system to provide portable oxygen tanks and oxygen concentrators. Different companies provide portable oxygen tanks and oxygen concentrators. And out-patient is able to choose the provider for portable oxygen tanks and oxygen concentrators to rental or buy. Moreover, there are many welfare equipment stores around each hospital in Taiwan, and they sell those portable oxygen tanks and oxygen concentrators. So that, any out-patients are easily able to buy those portable oxygen tanks and oxygen concentrators.

Table 1. Health care systems for COPD.

| Country service | Japan | Taiwan |
|---|---|---|
| Long-term oxygen therapy treatment | National Health Insurance<br>Welfare Service<br>Long-Term Care Insurance | National Health Insurance<br>Welfare Service |
| Home Visiting Nurses' service | National Health Insurance<br>Welfare Service<br>Long-Term Care Insurance (No limitation) | National Health Insurance<br>Welfare Service<br>(limitation; 2 per month)<br>(Respiratory Therapist) |
| Prescription of oxygen inhalation | o | × |
| Frequency of routine out-patient visit | 1 per month | 1 per month<br>(only provide 3 months period after introduced) |
| Portable oxygen tanks and oxygen concentrations | National Health Insurance<br>Welfare Service | Welfare Service easily able to buy in many welfare equipment stores |

Table 2. COPD Diagnosis and indication of treatment.

| Country | Japan | Taiwan |
|---|---|---|
| Criterion for COPD Diagnosis | GOLD | GOLD |
| Indication for Long-term Oxygen thrapy treatment | GOLD | GOLD |

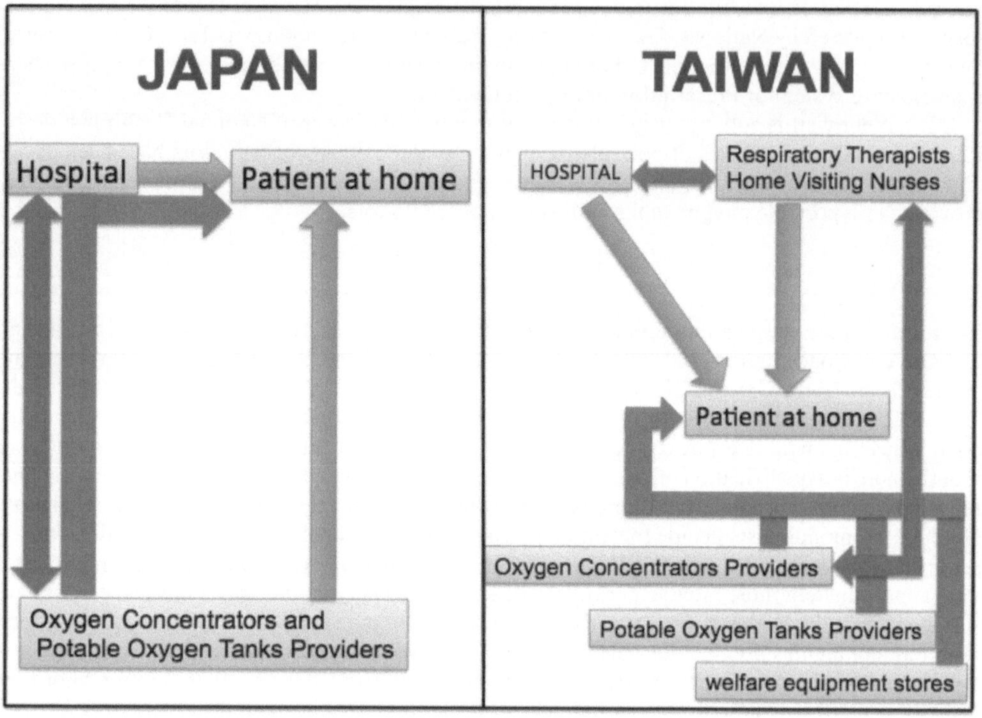


Figure 1. Comparison about Health Care Systems for COPD between Taiwan and Japan.

## 5 DISCUSSIONS

Both Taiwan and Japan offer Long-Term Oxygen therapy for COPD under cover of Public Health care systems mainly using National Health Insurance.

In Japan, it is well known that rental and delivery systems of portable oxygen tanks and oxygen concentrators have been developed under cover of National Health Insurance as a treatment of COPD. In addition, Long-Term Care Insurance covers home-visit nursing care service, although only 27% of COPD patients in Japan used this service in 2004 (在宅呼吸ケア白書 2005).

In Taiwan, both portable oxygen tanks and oxygen concentrators are not covered under National Health Insurance. The rental systems for both items are not even prevalently well developed.

Japan to implement an education program about oxygen therapy service for appropriate parties to increase the number of eligible patients using the home-visit nursing care service for COPD as covered by Long-Term Care Insurance. And to introduce a qualification of "Respiratory Therapist" is suggested.

Taiwan to discuss an item "Oxygen concentrator" of treatment for Long-Term Oxygen Therapy, which should be covered by future public health care system "Long-Term Care Insurance", planned in Taiwan. And it is suggested to introduce the system of "Prescription of oxygen inhalation".

## 6 CONCLUSION

This study aims to find potential issues of Long-Term Oxygen Therapy between Taiwan and Japan by Comparison of Public Health Care System, and to make a proposition of solution approach for COPD patients bringing better quality of life. The Method is to collect health care systems data for long-term oxygen therapy from appropriate parties by interviews. Results are; In Japan, Rental and delivery systems of portable oxygen tanks and oxygen concentrators have been developed under cover of National Health Insurance as a treatment of COPD. In addition, Long-Term Care Insurance covers home-visit nursing care service although few COPD patients use this service. In Taiwan, both portable oxygen tanks and oxygen concentrators are not covered under National Health Insurance, and are not prevalently well developed. Because these items provide under thoughts of Welfareness.

Japan to implement an education program about oxygen therapy service for appropriate parties to increase the number of eligible patients using the home-visit nursing care service for COPD as covered by Long-Term Care Insurance. And to introduce a qualification of "Respiratory Therapist" is suggested.

Taiwan to discuss an item "Oxygen concentrator" of treatment for Long-Term Oxygen Therapy, which should be covered by future public health care system " Long-Term Care Insurance" planned in Taiwan. And it is suggested to introduce the system of "Prescription of oxygen inhalation".

## REFERENCES


Lopez AD, Mathers CD, Ezzati M, et al. 2006. Global and regional burden of disease and risk factors, 2001: systematic analysis of population health data. Lancet 2006; 367:1747–1757.

World Health Organization. 2002. World health report 2002: reducing risks, pro-moting healthy life. Available at: www.who.int/whr/2002. Accessed December 21, 2008.

World Health Organization. 2008. World Health Statistics 2008. Available at: http://www.who.int/whosis/whostat/EN_WHS08_Full.pdf. Accessed July 15, 2009.

NHLBI/WHO workshop report, 2000. Global Initiative for Chronic Obstructive Lung Disease. Global strategy for the diagnosis, management, and prevention of chronic obstructive pulmonary disease: Available at: http://www.goldcopd.com/. Accessed December 21, 2008.

NHLBI/WHO workshop report, 2007.Global Initiative for Chronic Obstructive Lung Disease. Executive Summary: Global strategy for the diagnosis, management, and prevention of chronic obstructive pulmonary disease updated 2007. Available at: http://www.goldcopd.com/. Accessed April 15, 2008.

Takahashi Y, and Fukuda T. 2003. The indication and practice of home oxygen therapy. Nippon Rinsho 2003; 61:2193–2199.

R. J. Halbert, J. L. Natoli, A. Gano, E. Badamgarav, A. S. Buist and D. M. Mannino. 2006. Global burden of COPD: systematic review and meta-analysis. European Respiratory Journal. 2006 Sep;28(3):523-32. Epub 2006 Apr 12.

在宅呼吸ケア白書　第1版第1刷発行 2005.

*Bridging Research and Good Practices towards Patient Welfare – Shih & Liang (Eds)*
*© 2015 Taylor & Francis Group, London, ISBN 978-1-138-02716-9*

# Five Qualities (5Qs) and ergonomic approach – "Towards improved patient safety, medical and healthcare quality"

Mosad Zineldin
*Linnaeus University, Sweden*

Jonas Zineldin
*Karoliniska Institutet: A Medical University and Växjö Hospital, Sweden*

Valentina Vasicheva
*Linnaeus University, Sweden*

ABSTRACT: According to the Institute of Medicine (IOM) report "To Err is Human" the injuries from medical errors are immense and the majority of the errors do not result from individual recklessness but from the flaws in the way the healthcare is delivered. This study is about a five qualities (5Qs) model which is a system including Human Factors Approach to quality and Patient Safety. The purpose of this paper is to demonstrate the integrated 5Qs and ergonomic as a new approach to build an effective and safe patient care environment and provide the guard of reducing medical and non-medical errors and increasing patient safety. The new 5 Qs-Ergonomic approach can be used as diagnostic instrument to identify and understanding of how good ergonomics makes a service (medical or health service) easy to use and contribute to patient safety. This study provides an empirical base to understand the complex and multidimensional nature of the 5Q's, ergonomics and medical and healthcare safety. Quantitative research strategy was utilized and a relevant sampling technique was used to collect quantitative data from 223 inpatients of 8 Chinese hospitals. Hypotheses were tested by using multiple regression analysis to the obtained data in SPSS. The study implies that although there is high human and non-human competences at the Chinese hospitals, the Ergonomics factors related to patient's safety still needs to be improved to avoid any possibility of medical errors and other harm during the process of the care.

*Keywords*: Atmosphere, Process, Interaction. Infrastructure, Safety, Medical, Ergonomic, 5Qs, Healthcare, Hospitals

## 1 INTRODUCTION

More people die and injure as a result of MEs than as a result of AIDS, breast cancer or car accidents. 23 percent of Europeans argue that they have been directly affected by lack of medical and health care safety. 18 percent indicate that they or their family members have experienced a serious ME in a hospital whereas 11 percent announce having been prescribed wrong medication (European Commission, 2006).

Many of the biological responses of the individual are related to the social environment acting upon him or her. The responses are often influenced by previous experience, relationships and interactions as well as by genetic make-up, but there is sufficient regularity of the responses to suggest that the more efficient and right perspective and approach is to understand how the social and other non-biological and non-medical related factors impact biology to cause disease and how patient safety is strongly influenced by other factors such as interactions and atmosphere between healthcare professionals and the designers and equipment manufacturers. Although, they should work together in an effective and ergonomics manner, much evidence reveals that these professionals do not collaborate well together.

A social structure view emphasizes on prevention rather than cure. An integrated approach of Quality and ergonomic is vital when measuring the quality of medical and healthcare, patient safety and satisfaction (Williamson et al., 2012).

Systematic methods of patient safety (PS) and healthcare quality are still evolving. Patient safety and good quality of care are considered to be the right of all patients and the responsibility of all staff within the hospital (Zineldin et al., 2014). The medical and healthcare sector have to cope with social and environmental pressures and the emergence of new treatments and technologies in order to remain competitive (Andaleeb, 1998; Lim and Tang, 2000). Not surprisingly, health-care quality, patient safety and satisfaction constructs are of vital concern for health-care organizations (Zineldin, 2006).

PS is the absence of the harm during the care process. PS is defined by The Agency for Healthcare Research and Quality (AHRQ, 2003) as freedom from injuries, illnesses or harm to patients resulting from the process of care. Safety and prevention of accidents have long been a high priority in ergonomics. This includes consideration to the error risks built into complex systems including humans. Human error is a major cause of safety problems as well as of quality deficiencies. Poorly delivered care can cause infections, injuries, and even death. Thus, safety of patients should be assured to avoid any kind of harm as a result of patients' encounters with the health care system. Healthcare institutions must attend to ensure that their personnel and patients feel safe and secur; without reliable safety, there can be no quality.

Li et al. (2011) argued that sound physical environments or sufficient equipment, for example, can match peoples' expectations but not sure that it can offset dissatisfaction with long waiting times. Patients may be satisfied with core services (i.e. quality of the object such as doctor' diagnoses or biological health outcomes), but the law and poor quality of supplemental services or the process of the services (e.g., inconvenience and slow processing speed at the payment counter or pharmacy) can give patients a bad impression.

Patient trust and satisfaction with care and outcomes, are not just about determining what care is preferred (object) and why, but also about how it is provided (process) e.g. frequency and duration, and valued endpoints of care. It could also include measures of the way information is exchanged (interaction) with patients to assist them to make care decisions, the way their choices are incorporated into treatment decisions, efficacy of treatment options presented, and the quality of both the interaction and the information exchanged (Brien et al., 2009; Zineldin, 2006)

Trust and satisfaction enable patients to provide feedback that organizations can subsequently use to improve existing performance. Patients' satisfaction with how their personal issues were addressed, as well as with tests and treatments also had strong and significant relationships with trust of the caregiver. Meeting and exceeding the exceptions of the patients are one of the main goals of any health care organization to improve the levels of patient satisfaction which is very critical to their long term success (Torres and Guo, 2004).

Davila (2002) believes that defining or measuring the healthcare quality is a dilemma. He states the following:

> *"For many, quality healthcare is like beauty or pornograph—they know it when they see it but they just can't define it. Yet, a widely accepted and specific definition of quality healthcare is required for its assessment and promotion, and a lack of this definition makes these impossible. The sum and substance, then, is 'What is an acceptable and specific definition of quality healthcare?"* (p. 84).

Much evidence reveals that these professionals such as physicians and equipment designers do not collaborate well together (Scott et al., 2009). Ergonomically designed medical equipment and products may improve performance. One example from electric connector assembly shows that factors such as visibility, position, layout, sequence, size, shape, weight, colour and surface properties are all important clues for the operator in his or her strive to select and connect the right connectors. Designs not considering this have been identified as potential causes of quality deficiencies and medical errors. That constitute examples of how insufficient interaction between technology and humans may be the origin of such quality deficiencies and medical error.

5Qs approach highlights the role of quality and the impact of the internal and external environment and professional cooperation on health care rules and performance, on relationships and networks, on communications and interactions with different internal and external actors (Zineldin et al., 2011).

Shaw and Miller (2000) argue that it is essential to develop multifactorial domains for disease outcomes (clinical, humanistic, and economic). The clinical domain was divided into healthcare systems, healthcare needs, and individuals. Thus, Zineldin (2006) developed the multifactorial measures of quality: object (technical quality), process (functional quality), infrastructure (basic resources needed to perform healthcare services), interaction (quality of information exchange), and atmosphere within a hospital or wherever the medical and healthcare is provided. Patient care technology has become increasingly complex, transforming the way nursing care is conceptualized and delivered. Before extensive application of technology, nurses relied heavily on their senses of sight, touch, smell, and hearing to monitor patient status and to detect changes. Over time, the nurses' unaided senses were replaced with technology designed to detect physical changes in patient conditions. A brief presentation of the OPIIA is:

Q1 Object: Quality of object (the Outcome) – the technical quality (what patient receives). It measures the treatment itself; the main reason of why a patient is visiting a hospital.

Q2 Process: the functional quality (how the health-care provider provides the core service (the functional). Examples include waiting times, how and where, waiting places, and speed of performing the health-care activities. Process of clinical care includes primary quality; treatment process and its outcome; reliability; understanding of illness; diagnosis, dental services; freestanding medical imaging facilities; emergency room services; managed care programs; outpatient psychotherapy practices.

Q3 Infrastructure: A pleasant environment can significantly improve the patient's mood, satisfaction and perceived quality of the healthcare experience. Q3 measures the basic resources which are needed to perform the health-care services: the quality of the internal competence and skills, experience, know-how, technology, motivation, internal resources and activities.

Q4 Interaction: How people interact with the equipment in healthy, comfortable, and efficient manner are also part of the Q4.

Q5 Atmosphere: The atmosphere indicators should be considered very critical and important because of the belief that lack of frankly and friendly atmosphere explains poor quality of care in developing countries. Light levels, colour rendering, luminance and reflections.

## 2 RESEARCH GAP, RESEARCH MODEL AND HYPOTHESES

The existing quality instruments are suffering from the lack of the critical structure, process, and outcomes measures. 5Qs' OPIIA closes this gap as it also measures the structure, the process, the outcomes and beyond. Structure measures cover the organizational context of care. Process measures how care is provided. Outcome or the object measures complications in any secondary diagnosis fields, morbidity and mortality. To measure the 5Qs system success, a 5Qs system success model is proposed as a framework shown in Figure 1. According to the OPIIA model, the hypotheses that follow directly from the proposed model are summarized below in short.

H1. Patient Safety of the OPIIA system is positively associated with Q1
H2. Patient Safety of the OPIIA system is positively associated with Q2
H3. Patient Safety of the OPIIA system is positively associated with Q3
H4. Patient Safety of the OPIIA system is positively associated with Q4
H5. Perceived patient safety is positively associated with Q5.

Table 1. Reliability coefficient and descriptive statistics of patients' perceived hospital OPIIA and safety (SAF).

| Scale | No. of items | Alpha | M | SD |
|---|---|---|---|---|
| Q1 | 19 | .94 | 3.62 | .71 |
| Q2 | 12 | .94 | 3.54 | .81 |
| Q3 | 12 | .93 | 3.79 | .72 |
| Q4 | 12 | .93 | 3.75 | .82 |
| Q5 | 12 | .92 | 3.55 | .89 |
| SAF | 13 | .93 | 3.54 | .70 |

n = 223.

## 3 RESEARCH METHODOLOGY

Data was collected through a web survey administered to senior executives of different size private and public hospitals in China. The diversity of sizes and ages of the hospitals and the regions was designed to assure the possibility of generalization of the research results and outcomes. The questionnaire was designed as a web-survey with a link to the survey platform or portal on line and as an e-mail survey. The hospital inpatients were encouraged to login with the specific password to conduct online reply or to return the answered survey to the researchers e-mail.

### 3.1 *Sample*

Data had been collected form 223 hospital inpatients from different Chinese cities and regions. 41% of inpatients were male and 59% were females. 89% were public hospitals and 11% were private.

### 3.2 *Measurements*

The research questionnaire was adapted from previous research by Zineldin (2006) and Zineldin et al. (2011). Our questionnaire contains a total of 67 items of the 5Qs OPIIA dimensions which were identified to be the most relevant attributes for hospitals and 20 demographic questions/items such as age, sex, occupation, income, etc. A 5-point Likert scale was used.

We draw upon human and non-human scales as well as clinical/medical and non-clinical or behavioral scales to further the process of validation for established scales. Majority of the scales had been tested on previous occasions and had been proven to be reliable (Hussain and Urehman, 2012; Byrd, 2009; Zineldin, 2006; Zineldin et al., 2011) with Cronbach alpha values generally in excess of 0.85 and apparent simple structure as indicated from confirmatory factor analysis.

## 4 RESULTS

The Descriptive statistics and the Reliability coefficients of the studied variables were tested by internal consistency. The impact of each item was evaluated on the basis of the reliability of the combined items designed to measure each of the 5Qs OPIIA dimension. As shown in Table I, the reliability analysis of all the constructs used in this study generated a Cronbach alpha score of 0.94, which was found to be above the standard set by Nunnally (1978), which is 0.50–0.60. Mean scores had been computed by equally weighing the mean scores of the all items of each construct.

On a five-point scale, the mean scores of patient's perceived medical and healthcare quality of the private and public hospitals range from 3.54–3.79 indicating that patients perceive that medical and healthcare quality of the OPIIA being offered by the Chinese hospitals is relatively high. The mean score of patient's SAF (safety) is 3.54(sd = 0.70) which is less than the mean score of the other

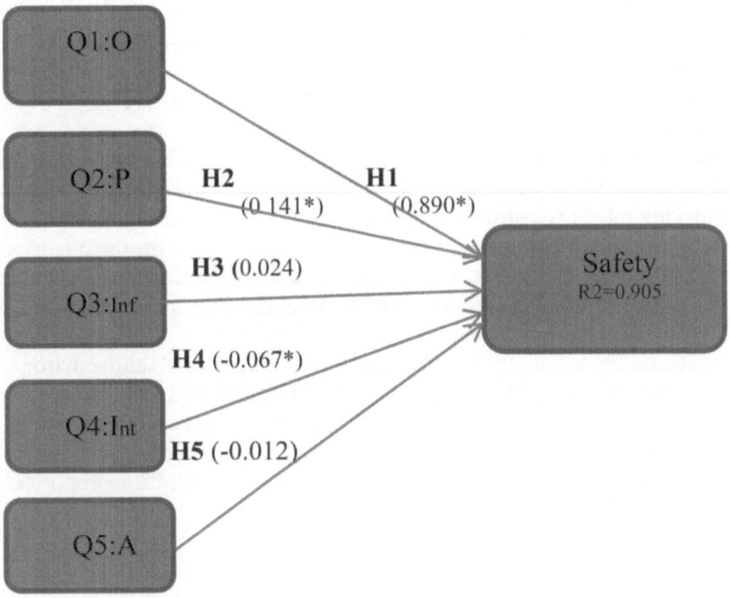


Figure 1. Structural 5Qs OPIIA Model Fit.

qualities implying that although there is high human and non-human competences at the Chinese hospitals, the patient's safety still needs to be improved to avoid any possibility of medical errors and other harm during the process of the care.

The theoretical model was tested using regression analysis, developing a composite measure for the dependent and independent variables by taking an average of different items on a scale. We examined the estimated coefficients of the causal relationships between constructs, which would validate the hypothesized effects. Figure I illustrates the estimated coefficients and their significance for the structural model. The coefficients of determination (R2) for each endogenous construct are also shown.

Hypotheses 1–5 examined the impact of OPIIA on patient safety. Each of Q1O (Object), Q2P (process) had a significant impact on safety ($\beta = 0.890$, $\beta = 0.141$, $P < 0.01$). The OPIIA Q3Iinf (Infrastructure) and Q5 (Atmosphere) had no significant impact on safety. The Q4 (Interaction) had a very minor negative significant on safety $\beta = -0.067$, $P < 0.10$) The regression model explained 90.5% of the variance in patient safety in the hospitals ($R2 = 0.905$). Thus, hypotheses 1.2 and to same extent 4 are accepted while hypotheses 3 (infrastructure) and 5 (atmosphere) are rejected.

## 5 DISUSSION AND CONCLUSION

The findings from this study led to the development of a novel evidence-based framework to measure OPIIA quality based on the 5Qs and ergonomic approach.

This study shows that the notion of maintaining high levels of both technical (Q1: object) and functional (Q2: process) quality are critical and relevant to the medical and healthcare sector to assure patient safety. Technical quality such as efficacy, effectiveness, and efficiency are related to the quality of object and the core quality of treatment (e.g., infection, mortality rates, and length of stay). Clinical outcomes and patient satisfaction are also part of the quality of the object. Functional quality refers to service process (Q2) and delivery issues (e.g., acceptability, legitimacy, and equity), and also includes patient trust and satisfaction (Kohli et al., 1999; Campbell et al., 2000). Clinical (Q1), process (Q2) as well as interpersonal care impacting the quality of interaction (Q4) and the overall atmosphere (Q5), infrastructure human and non-human competences and resources (Q3) are the basic or the minimum requirements which should exist to legitimate the activities of any health care institution. The mean score of the Q3 was the highest among all OPPIA ($m = 3.79$) which provide reasonable base for patient safety. High competence level (Q3) explain the strong impact of the clinical Q1 ($\beta = 0.890$).

The most unexpected surprising result was that while the infrastructure (Q3) and atmosphere (Q5) had good mean scores ($m = 3.79$ and $m = 3.55$), they were not significant in relation to patient safety. A study using same 5Qs instrument conducted by Hussain and Rehman (2012) of a Swedish hospital also showed that the Swedish patients' highest scores were for the quality of object (Q1) and interaction (Q5). our study shows that the Chinese patients ranked Q3 and Q4 as highest. Could it be due to the fact that Swedish Patients are more clinical and healthcare knowledgeable than the Chinese patients?

Patients usually perceive quality on the basis of interpersonal and environmental factors (Zineldin, 2006; Zineldin et al., 2011; Piligrimene and Bucieoniene, 2008; Yesilada and Direktör, 2010). Finally, through an evaluation of patients' safety, a reliability of 0.94 showed that this OPIIA based ergonomic system is able to effectively construct the integrated patient-safety centric environment. Thus, determining the Ergonomic factors associated with PS is important topic for the health care provider.

## REFERENCES


Agency for Healthcare Research and Quality (AHRQ). (2003). *Quality indicators: Guide to patient safety indicators* (AHRQ Publication 03-R203). Rockville, MD.

Andaleeb, S.S. (1998). Determinants of customer satisfaction with hospitals: a managerial model. *International Journal of Health Care Quality Assurance.* Vol. 11(6), pp. 181–190.

Atinga, R.A., Abekah-Nkrumah, G. & Domfeh, K.A. (2011). Managing healthcare quality in Ghana: a necessity of patient satisfaction. *International Journal of Health Care Quality Assurance.* Vol. 24(7), pp. 548–563.

Brien, S.E., Dixon, E. & Ghali, W.A. (2009). Measuring and reporting on quality in healthcare: a framework and road map for improving care. *Journal of Surgical Oncol*ogy. Vol. 99, pp. 462–466.

Byrd, L. (2009). *An examination of information technology and it perceived quality issues in single system hospitals in the United States*, Doctoral dissertation Auburn University, Alabama USA, December 18.

Campbell, S.M., Roland, M.O. & Buetow, S.A. (2000). Defining quality of care. *Soc Sci Med*, Vol. 51, pp. 1611–1625.

Davila, F. (2002). What is an acceptable and specific definition of quality healthcare? *Baylor Univ Med Centre Proc*, Vol. 15, pp. 84–85.

Duggirala, D., Rajendran, C. & Anantharaman, R.N. (2008), "Patient perceived dimensions of total quality service in healthcare", *Benchmarking: An International Journal*, Vol. 15 No. 5, pp. 560–83.

European Commission. Special Eurobarometer 241 (2006*), Medical Errors: Fieldwork*: September–October 2005, European Commission.

Hussain, S.N. & Ur Rehman, S. (2012). *Patient Satisfaction Regarding Hospital Services: A Study of Umea Hospital.* Master Thesis defended at Umeå School of Business, Umeå University, Sweden.

Kohli, R., Tan, J.K., Piontek, F.A., Ziege, D.E. & Groot, H. (1999). Integrating cost information with health management support system: an enhanced methodology to assess healthcare quality drivers. *Top Health Inform Manage*, Vol. 20, pp. 80–95.

Li, S., Huang, Y. & Yang, M.M. (2011). How satisfaction modifies the strength of the influence of perceived service quality on behavioral intentions. *Leadership Health Serv* 24(2), pp. 91–105.

Lim, P.C. & Tang, K.H. (2000), "A study of patients' expectations and satisfaction in Singapore hospitals", *International Journal of Health Care Quality Assurance*, Vol. 13 No. 7, pp. 290–99.

National Quality Forum. (2002). *Better healthcare through safe practices*. Washington, DC.

Nunnally, J.C. (1978). *Psychometric Theory*. McGraw-Hill: New York.

Piligrimene, Z. & Bucieoniene, I. (2008), Different Perspective of Health Care Quality: Is the Consensus Possible? *Engineering Economics*. No. 1 (56), pp. 104–108.

Rose, R.C., Uli, J., Abdul, M. & Ng, K.L. (2004), "Hospital service quality: a managerial challenge", International Journal of Health Care Quality Assurance, Vol. 17 No. 3, pp. 146–59.

Scott, S.D., Hirschinger, L.E., Cox, K.R., McCoig, M., Brandt, J. & Hall, L.W. (2009), "The natural history of recovery for the healthcare provider 'second victim' after adverse patient events", *Qual Saf Health Care*, Vol. 18, pp. 325–330.

Shaw, L.J. & Miller, D.D. (2000). Defining quality healthcare with outcomes assessment while achieving economic value. *Top Health Inform Manage*. Vol. 20, pp. 44–54.

Thom, D.H., Hall, M.A. & Pawlson, L.G. (2004). Measuring patients trust in physician when assessing quality of care. *Health affairs*, Vol. 23 (4), pp. 124–132.

Torres, E.J. & Guo, K.L. (2004). Quality improvement techniques to improve patient satisfaction. *International Journal of Health Care Quality Assurance*, Vol. 17, No. 6, pp. 334–338.

Williamson, P., Altman, D., Blazeby, J., Clarke, M., & Gargon, E. (2012). Driving up the quality and relevance of research through the use of agreed core outcomes. *Journal of Health Serv Res Policy*, Vol. 17, Iss.1.

Yesilada, F. & Direktör, E. (2010). Health care service quality: A comparison of public and private.

Zineldin, M. (2006). The quality of healthcare and patient satisfaction: an exploratory investigation of the 5Qs model at some Egyptian and Jordanian medical clinics. *International Journal of Healthcare Quality Assurance*, Vol. 19, pp. 60–92.

Zineldin, Akdag, H. & Vasicheva, V. (2011). Measuring, evaluating and improving hospital quality parameters/dimensions – an integrated healthcare quality approach, *International Journal of Health Care Quality Assurance,* Vol. 24, Iss. 8, pp. 654–662.

Hospitals, *African Journal of Business Management*, Vol. 4(6), pp. 962–971.

Zineldin, M., Zineldin, J. & Vasicheva, V. (2014). Approaches for reducing medical errors and increasing patient safety: TRM, quality and 5Qs method, *The TQM Journal*, Vol. 26, No. 1, pp. 63–74.

# *Part B: Healthcare, mobile application and usability*

*Bridging Research and Good Practices towards Patient Welfare – Shih & Liang (Eds)*
*© 2015 Taylor & Francis Group, London, ISBN 978-1-138-02716-9*

# How do you support breast cancer patients on work-treatment balance?

Nanae Shintani
*Faculty of Health Sciences, Department of Nursing Kansai University of International Studies, Japan*

Hiroyasu Nagaoka
*The Graduate School of Information, Production and Systems, Waseda University, Japan*

Eriko Koike
*Gihoku Kousei Hospital, Japan*

Shigeki Tatsukawa & Ayako Kajimura
*Faculty of Health Sciences, Department of Nursing Kansai University of International Studies, Japan*

ABSTRACT: The number of patients who contract or die of breast cancer in Japan is increasing year by year. A peak age of onset of breast cancer is late 40s'. So some patients have to work and receive chemical treatments at the same time in their productive years. The purpose of this study is to clarify difficulties of continuing work while getting chemotherapy for breast cancer outpatients, to clarify patients' thoughts and problems behind them, and to contribute to improving the Quality Of Life of breast cancer patients. This study targets 11 patients who were diagnosed with breast cancer in a center hospital in A prefecture, and they continue to work while receiving chemotherapy. A semi-structured interview was conducted for them, its results were analyzed by means of text-mining approach. The software used in the study for text-mining approach was IBM SPSS Text Analytics for Survey Ver.4.0, and characteristic sentences were summarized to review the results, which were gained to categorize what the patients said as "In Difficulty" by use of emotion analyzing function. This study gained approval by ethical committee at the institution the author once belonged to. The results suggest the need to encourage and support the patients by various institutions to which the patients belong to work in consideration of their disease characteristics.

## 1 BACKGROUND

In Japan, the breast cancer has become the top disease in female cancers and developed about 35 thousands people a year. The age-adjusted incidence rate of breast cancer was 21.7 persons per 100 thousands persons but it becomes 43.6 persons per 100 thousands persons in 1998. It increased to almost double in 23 years and the peak age is between 45 and 49 years old. Thus, 40's is a peak age of onset in Japan (Ministry of Health, Labor and Welfare, 2011). In this background, it is said that there is an westernized way of living such as a dietary life. In addition, an increase in body and mind burden caused by women's aggressive social advancement and tendency to marry later are also regard as the cause of the increase in breast cancer.

In the past, removal of the breast by surgery was a standard way of medical treatment of breast cancer but in these years, a direction of improving probability of survival and possible maintaining of QOL (Quality Of Life) by the multimodality therapy such as partial removal of the breast, medical treatment by anticancer/hormone drug, and radiotherapy are regarded as important.

However, most of these treatments are done by outpatient visit and only the surgery such as removal of the breast requires hospitalization. Even if the outpatient visit treatment, the damage on the body by the treatment is significant and it is not easy to cope with side effects especially from the chemotherapy. There are a lot of side effects from chemotherapy. Those are the systemic symptoms such as a risk of infection caused by the low number of white blood cells and general malaise, the local symptoms such as vomiting, alteration in taste and angialgia, and things which influence women's identity involving hair loss, dye fixation and swelling. There are cases where a lot of patients are confusing with emerging unexperienced side effects attacked in waves. The peak onset age of breast cancer is the same period as taking a lot of roles in social life such as household chores, raising children, nursing care, and so on. Under this circumstances, patients have not only breast cancer treatment but also a lot of burdens which make them difficult to continue the treatment.

In addition to this, patients who are working have more physical and mental burden than ones who are not working.

According to the research conducted by Health, Labor and Welfare Ministry in 2003, there is a data that 31% out of about 2600 patients, who had been working when they were informed of the cancer, retired voluntarily and 4% of them are laid off. It is heard that there were cases where patients were transferred to a different department even though they were not laid off. Based on the Cancer Control Act legislated in 2006, "The Basic Plans to Promote Anti-Cancer Measures" was released in June, 2007 and anti-cancer measures are promoted comprehensively but working support measures for patients in working generation were untouched at this point. However in the revised version of the Basic Plans to Promote Anti-Cancer Measures released in March, 2012, it is called into question about the influences that affected working generation leave apart. It is acknowledged that the cancer is a big problem for working generation and it is indicated that a measure for working of cancer patients is an urgent issue. However the reality is that there will be a lot of patients who have difficulties in dealing with their job and treatment at the same time because of the delay in the support measures.

## 2 PURPOSE

The purpose of this study is to clarify difficulties of continuing work while getting chemotherapy for breast cancer outpatients, to clarify patients' thoughts and problems behind them, and to contribute to improving the Quality Of Life of breast cancer patients.

## 3 METHOD

### 3.1 *Survey target*

Eleven breast cancer patients who were agreed by themselves and their primary doctors and continued working were selected from 32 patients who were diagnosed as breast cancer at the hub hospitals of cancer in "A" prefecture and getting chemotherapy. The average age of the patients was 49.1+-1.8. They were new-onset breast cancer patients and finished the second chemotherapy after the operative treatment.

### 3.2 *Collecting data*

Eleven patients were selected and a semi-structured interview about "the difficulty in continuing job and treatment" was conducted. An interview guide was made for clarifying difficulties to continue both treatment and job, physical, mental and social influences on roles such as professional, woman, wife, mother and daughter, and thoughts of patients.

### 3.3 *Analysis method*

After converting all the contents of interview results to text data, codes classified as "have a difficulty" that exactly matched for this survey propose, were extracted by using an emotion analysis feature of the text mining software, IBM SPSS Text Analytics for Survey Ver.4.0. The extracted codes were categorized with discussing by three researchers and certified cancer nurse.

## 4 ETHICAL CONSIDERATION

We explained the purpose, method, complying of confidentiality, respecting of voluntary study participation and the right to drop out anytime by explaining with an explanatory document, and got agreement from administrators and patients in the hub hospitals of cancer in A prefecture. In addition, this study was approved by the ethical committee of the organization to which the researchers belong

## 5 RESULTS

From the extracted codes, we got 14 categories as follows:

1. "Psychological conflict by continuous treatment"
2. "Psychological depression from anxious thought of prognosis"
3. "Difficult judgment when one can resume work as before"
4. "Lack of understanding that there are no cancer experiencer in one's boss"
5. "Always think about only disease when there are no jobs"
6. "Irregular mealtime due to the side effects"
7. "Being seen by others when side effects occur"
8. "Being noticed by colleagues about the disease"
9. "Seriousness of the hair loss damage"
10. "Seriousness of the economic burden"
11. "Not inform children of the disease"
12. "Lack of support from family"
13. "Distance to a medical institution to visit"
14. "Difficulty of the treatment plan to take into account of life events"

## 6 DISCUSSIONS

The categories described previously became clear as the results of interview about the difficulties in continuing both treatment and job, and influences on roles such as professional, woman, wife, mother and daughter when continuing.

About the category 1. "Psychological conflict by continuous treatment", various kind of treatments such as mastectomy, chemotherapy, radiation therapy, endocrine therapy start after getting diagnostic confirmation of breast cancer. On this occasion, a treatment plan giving the highest priority to prognosis is explained by the primary doctor but it is up to patients' decision finally. During patients are at the mercy of diagnosis of breast cancer and in a chaotic condition, they have to select as treatment. Furthermore, the decision is very important because it will change the prognosis. In many cases, treatments are done continuously and decision making points come accordingly. To take into account of treatment effects, decision making time is short and there are not enough time to decide. Patients result in a psychological confusion due to the unexperienced change in lifetime event caused by anxiousness about unexpected prognosis side effects. In the research conducted by Kokufu group in 2012 about the factors exert influence on conflict of treatments selection by breast cancer patients, the lack of advice or support at the decision making point influence on conflict such as satisfaction or fight. On the other hand, Nishio group reported in 2011 that understanding of explanation of contents and psychological support were done relatively favorable but information providing including "Daily Life" was not enough at the decision making point. Thus the advice and support are important at the point of decision making. After considering personality and living status of patients, it is important to provide information to make the decision making easy and environment to support patients after making decision. Furthermore, physiological viewpoint focused on personality and ability to advice to take into account of difficulties of impaired daily life by body invasive.

According to the Category 2. "Psychological depression from anxious thought of prognosis", it is said that 15 to 25 percent of cancer patients are in depressive state (Lloyd-Williams M, 2011). It becomes clear that because of the depressive state, patients get into a feeling of inability, it influences martially communication in the family relations, and induces problems by influencing emotional and behavioral aspects for children (Miaskowski C, 2004)As for this measure, there are some medical institutions which have medical examination by psychiatrist and counseling by clinical psychotherapist for the patients in the course of a treatment. Many of the patients are anxious about reoccurrence and it is said that the increase in anxiousness is influenced by relatives and celebrities who become worse in disease condition. For these kind of psychological support, it is necessary to open consultation services by professionals is required. There are medical institutions which perform spiritual intervention to aim patient's peace of mind.

About the Category 3. "Difficult judgment when one can resume work as before", due to the emergence of an oral medicine named dexamethasone, the side effects at the chemotherapy have been reduced significantly and the QOL of breast cancer patients has been increased. However, there are cases that some disorders such as lymphedema and decrease in a range of upper arm movement are remained and force to get pain at the daily life. For these bad effect of treatment, there were a lot of patients who have anxiousness about "When they can resume their work as before?" depending on the type of a job. Furthermore, there are various kinds of cases if there have side effects or not and the degree of side effects even if they get same treatment, and each working condition is different. It is necessary to have precise support to cope with each condition and it is necessary to inform the estimated timing of returning job as before with considering working condition once the treatment plan is decided.

About the Category 4. "Lack of understanding that there are no cancer experiencer in one's boss", the research conducted by Sakai group in 2009 about the changes in job of 224 (67.1%) female cancer patients, around the half of them answered "work at the same company" but 20 people (6.0%) answered "quit voluntarily" and 10 people (3.0%) answered "fired". Furthermore, 61.6% of the employed person answered "There are concerns about continuing work in the future" and there are a lot of people who answered the necessary things for continuing work is "to be understood the disease condition by colleagues and superiors at work". At the treatment of breast cancer, there are side effects such as breast removal, hair loss and dye fixation which damage the body image, but it is difficult to distinguish the damages that patients are receiving just from the outer appearance because there are order-made hair wigs available for hair loss and advanced breast reconstruction surgery for the after surgery damages. However, there are cases of developing infection coming from low number of white blood cells and patients are forced to visit other treatment department besides the regular breast cancer treatment. Furthermore, since the working conditions are various such as food manufacture, hospitality industry, transportation industry and office worker, it is desirable to support including those working conditions. Currently, Japanese top cause of death is cancer and there are quite a lot of cases of onset at the working ages of 40's and 50's. (Center for Cancer Control and Information Services, 2012) From these points, it is imaginable that there are a lot of cases of cancer experiencers working in the office, and it is desirable to utilize the knowledge of cancer experiencers for caring patients and institutional reform.

About the category 5. "Always think about only disease when there are no jobs", it is heard from the patients that "It was a big psychological burden to get treatments without having a job "It was good to have a job". After becoming breast cancer, the patients loose several things such as breast, hair on the head, healthy body and money even if a certain period of time. There are some cases to loose forever. We could find out how things really are, that patients were helped other than economy if there are working such as devoting their jobs by forgetting about disease during working and get encouragement from colleagues and superiors.

About the Category 6. "Irregular mealtime due to the side effects", mealtime and taste are changed if there are vomiting and taste abnormality due to the side effects of chemotherapy. It is desirable to establish a system to take mealtime and breaks appropriately only during the chemotherapy treatment as the measure in the Labor Standards Article 32. At the time to introduce the system it is required to describe what part of labor and how long should it be covered according to the emerging symptoms for personnel staffs to understand easily as the instruction items communication card of maternity health management.

About the Category 7. "Being seen by others when side effects occur", the location of the locker room for changing clothes should be considered as much as possible to take into account the feeling of the patients themselves.

The Category 8. "Being noticed by colleagues about the disease", it is necessary for superiors to confirm the patient's intention when and what area the information can be opened in terms of personal information protection. Furthermore, regarding the sensitive information such as disease, it is necessary to establish a system in the organization such as a line composition which makes report easily to superior from patient and not to leak the information.

From the Category 9. "Seriousness of the hair loss damage", this is a category to ooze identity as a woman. A lot of patients told that "They got damages more than they expected" in the interview survey about the hair loss. For this reason, it is desired to discuss support by communicating with industrial medical staff, doctor and nurse by making use of knowledge of experiencer.

From the Category 10. "Seriousness of the economic burden", the magnitude of burden on medical expenses for cancer treatment was reflected. "Consciousness survey of 1,600 cancer patients" conducted by Health and Global Policy Institute in 2009, 70 percent of the patients answered the burden on treatment fee is significant. The high cost illness insurance is adapted when the copayment of medical expense in a month exceed a specific amount of money but the problem is the payment at the outpatient visit. According to the outpatient visit chemotherapy becomes popular, new anticancer drug and molecularly-targeted drug are expensive, so the number of people who have to pay expensive copayment of medical expense are increasing. In this case, patients' copayment will be reduced to several tens of thousands yen by the high cost illness insurance. (in case of average income) However, in case of not using the Eligibility Certificate for Ceiling-Amount Application, it is necessary for patients to pay the amount in full at the window. The expenses such as travel expense, hair wig for hair loss and an underwear after removing breast are required other than the medical expense. In terms of disease rate by age of breast cancer, since there are quite a few patients need funds for education expenses at the same period of life stage, there are patients who abandon treatment due to the medical expense presses the family budget. Furthermore for other than regular employment, there are patients who cannot take sick leave due to the possibility to get pressure to take early retirement. It is necessary to give advice to patients by connecting to social workers working at a family support consultation room in a hospital so that patients have the opportunity to get consultation for not to abandon treatment and not to have pressure during treatment. Furthermore as the institutional enhancement, we want to propose to improve inadequacy of the high cost illness insurance that the Eligibility Certificate for Ceiling-Amount Application cannot be used at outpatient visit so that the burden on treatment fee won't be a further stress on patients.

From the Category 11. "Not inform children of the disease", there were cases that they didn't tell about their mothers' breast cancer, "Since patients have children, they don't want their children to watch their hair loss head because of the emotional trauma of their children." or since their children are in school age, they might have any influences on daily life if they hear about the disease and get shocked.

From the Category 12. "Lack of support from family", there were statements that patients anxious about the burden on the specific family such as husband and biological mother. As the results of semi-structured interview to mothers and their children to clarify the situation of notifying mothers' breast cancer to their children and reaction of children, two things that the anxiety of children who received notify cannot become independent and cannot continue their life and schoolwork were clarified. (Nemoto, 2005) On the other hand, since there were a gap of feeling between parents and children who were not notified, it is desirable to notify the truth to their children after taking into

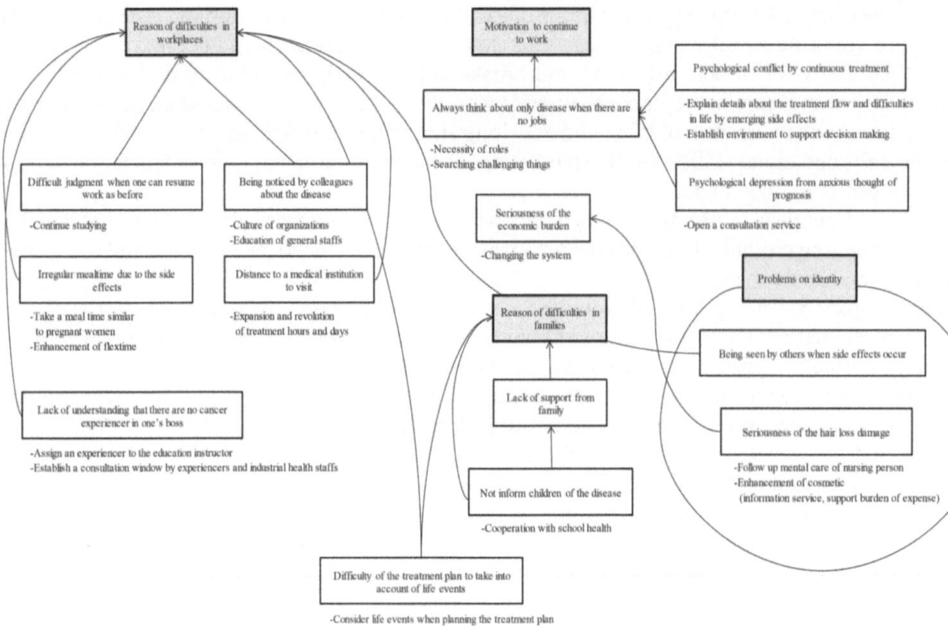


Figure 1. The relationship between obtained categories into the form of a diagram by adding lacking words.

account of the maturity level at their age. From this point, it is considered necessary for children to know the reality and face mothers' disease as a member of family but not concealing mothers' disease. Also, it is considered that the power to overcome helps shape children as a human being such as the abilities of coping with own stress and problem solving. It is necessary for industrial health staffs to exchange information with medical staffs not only in the facility but also the nursing teacher and counselor in school and watch development of children. Furthermore, industrial health staffs need to reduce patients burden and establish environment where patients confront the disease with power of their family in response to this.

From the Category 13. "Distance to a medical institution to visit", in case of visiting hospital every day for such as radiation treatment, it is found that patients showed constraint to others because of far distance to the medical institute to visit. Also, the reason why patients selected the far distant hospital was "Patients want to be treated by skillful doctor." From the situation that there are a lot of accesses to the web site listed good doctors and famous hospitals, it is found that many patients want to get treatment by experienced and popular doctors even if there is some distance. It is required that company and hospital pay flexible attention to patients to get treatment at their preferred hospital.

From the Category 14. "Difficulty of the treatment plan to take into account of life events", although treatment will start after getting diagnostic confirmation, at that time, it is natural that treatment effect is given priority over job and family. However to increase treatment effectiveness, support by family and economy must have been important factors. At the time patient start treatment, it is considered that patients QOL will be increased by explaining treatment details and possible side effects that might be emerged, and proposing treatment plan to take into account of patents life events and sense of value.

Also, we put the relationship between obtained categories into the form of a diagram by adding lacking words. (Figure 1) As the result, it was clarified that there are a lot of factors such as social system and corporate efforts are mixed for dealing with treatment and job at the same time. Even when we get sick, it is necessary to exchange information and cooperation crossover between various areas such as personnel staffs, social workers, school health, beauty care, as well as medical institution and industrial medical staffs to continue opportunity to utilize ability by working in the society. For securing women work forces to connect to the future development, it is desirable to have reformation of social system same as the liability relief of medical expense and maternity care.

## REFERENCES


Center for Cancer Control and Information Services, 2012.

Health and Global Policy Institute, 2009.

Hidemi, N. 2005: Confession and Response from Patients under advanced breast cancer to children, Japanese Society For Clinical Thanatology, 10(1), 18–27.

Japan Labour Health and Welfare Organization 2010: Mutual Support Between Cancer and Work; Support for the balancing of Work and Cancer.

Kokufu, H. 2012: Conflict accompanying the choice of initial treatment in breast cancer patients, Japan Journal of Nursing Science, 9, 2, Page 177–184.

Lloyd-Williams, M. & Friedman, T. 2001: Depression in palliative care patients–a prospective study, Eur J Cancer Care (Engl) 10 (4): 270–4.

Miaskowski, C. 2004: Gender differences in pain, fatigue, and depression in patients with cancer, J Natl Cancer Inst Monogr (32): 139–43.

Ministry of Health, Labor and Welfare, 2011.

Nishio, A. & Fujii, T. 2011: Consideration between ward nursing staff Support for decision-making of cancer patients' treatment options and influence factors, Japan Journal of Nursing Science, 31, 1, Page 14–24.

Sakurai, N., Yamamoto, N. & Suguri, M. et al. 2009: Recommendation ofThe current status of working cancer patients and support of them continue, Japan Medical Journal Right Reserved, 4442, Page 89–93.

Seiichiro, T., Nobuhito, T. & Akiji, M. 2012: The Survey about supporting job opportunity for working cancer patients through interviews by industrial physicians, The Institute for Science, 88, 4, Page 148–153.

*Bridging Research and Good Practices towards Patient Welfare – Shih & Liang (Eds)*
*© 2015 Taylor & Francis Group, London, ISBN 978-1-138-02716-9*

# Patient safety in the home: Examining the relationship between therapeutic self-care and adverse events for home care clients in Canada

Winnie Sun & Diane Doran
*University of Toronto, Toronto, Ontario, Canada*

ABSTRACT: This research study is a retrospective cohort design and utilized secondary data to test the relationship between therapeutic self-care ability, and the frequency and types of adverse events experienced by home care clients using secondary data sources housed at the Institute of Clinical and Evaluative Sciences in Ontario, Canada. Using logistic regression analysis, the relationship between home care clients' therapeutic self-care ability was examined in relation to two types of adverse events: (1) use of health care resources, including emergency room visits and hospital readmissions; (2) safety outcomes, including falls; unintended weight loss; urinary tract infection; decline in activities of daily living; pressure ulcer; compliance with medication; and decline in caregiver conditions. This study provides implications to health professionals about the need to focus on improving client self-care functioning, as well as providing evidence for policy formulation related to the importance of home care services in supporting chronic disease management.

## 1 INTRODUCTION

Home care forms an important bridge between acute care received in hospital and a patient's ultimate recovery or ability to return to a more independent level of functioning (Hollander, 2008). A study by Hollander and Tessaro (2001) found that home care service is of critical importance in the management of chronic diseases by avoiding unnecessary visits to emergency room and unnecessary hospitalization, as well as delaying admissions to long-term care homes. In 2006, there were one million hospital discharges in Ontario (The Change Foundation, 2006). Among these hospital discharges, home care potentially plays a critical role in maintaining the independent function of the growing older population, as well as restoring and encouraging self-care in clients with chronic conditions (Suter et al., 2008). McCormack (2003) suggests that the issue of self-care is especially important for those with chronic diseases because the client and the family members are responsible for his/her day-to-day care over the length of the illness.

Self-care encompasses the ability to take medications as prescribed, to recognize and manage symptoms that may be experienced such as pain, to perform and adjust regular activities of daily living, and to manage changes in condition (Sidani, 2003). In home care, self-care relates to the client's ability to manage their condition at home, such as after discharge from the hospital. Dai, Chang, Hsieh & Tai (2003) describe how the discharge of a client with a chronic disease is not only a medical problem, but it is also a condition with social implications because both clients and their families may face tremendous challenge related to self-care. Often, clients and their family caregivers have to cope with psychosocial problems generated by chronic disease, and must manage daily living according to their financial and social conditions (Lorig, & Holman, 2003). Barlow, Wright, Sheasby, Turner and Hainsworth (2002) propose that successful self-care of chronic conditions requires sufficient knowledge of the condition and its treatment, performance of management activities and application of the necessary skills to maintain adequate psychosocial functioning.

Despite the increasing demand for home care services to support the management of chronic diseases, there is a growing body of evidence that suggests at least one-third of all home care clients and caregivers were found unprepared for self-care in the next care setting (Coleman, 2006). A Canadian study by Charles et al. (1994) assessed hospital discharge planning, and found that 20% of clients were not told about important side effects; 39% were not told what signs and symptoms to watch for at home; 32% were not told when they could resume normal activities; and 29% were not told what activities they could or could not do at home. This study suggests that hospitals often discharge clients with insufficient planning on self-care skills, poor instruction and inadequate information on self-care knowledge.

Naylor et al. (2003) reviewed 94 studies, and found that the transition of older adults from hospital to home is associated with high rates of preventable poor post-discharge outcomes. In Ontario, a study by Forster et al. (2004) found that an estimated 23% of patients discharged from hospital to home experienced at least 1 adverse event. Among these adverse events, 12% were preventable; 12% lead to hospital readmission; and 72% were due to medication errors. In particular, home care clients who were older adults coping with multiple chronic conditions and complex health care regimens were significantly more vulnerable during the transition from acute care to home (Naylor et al., 2004).

As pressures have mounted to discharge hospital patients sooner and sicker, home care will play an important role in maintaining the independent function of the growing older population, as well as restoring and encouraging self-care in clients with chronic conditions. However, the current trends in home care suggest that a greater proportion of home care resources have been directed toward post-hospitalized acute patients, with fewer resources available to support the long-stay clients with chronic health care needs (Williams et al., 2010). As a result, many home care clients, especially older adults are at increased risk of losing independence in self-care, and this situation may put clients in unsafe situations, leading to safety problems or adverse outcomes (OACCAC, 2009).

Underlying all of this is a recognized concern about the quality and safety of current home care services to support chronic disease management. Along with prevention, one of the best ways to reduce the impact of chronic conditions on people's lives and on their need for expensive health care services is by supporting clients in the development of self-care skills. In the context of chronic disease management, providing self-care support to help clients manage their diseases is an important focus of home care services. However, the current literature reveals gaps in care that raise concern about the safety of Canadian home care clients, and highlight the need for increased self-care support in home care. While patient safety outcomes are well documented in acute care settings, only limited data exist about adverse events experienced by home care clients (Doran et al, 2009). In particular, there is a lack of research that explores the relationship between the client's self-care ability and safety problems in home care settings. In an era of a rapidly increasing number of older people who require home care services, clients must possess or develop self-care ability in order to manage their health safety in their homes.

To date, little is known about the concept of therapeutic self-care in the home care setting. The current study attempts to address the existing research gap by using an exploratory research approach to examine the relationship between home care clients' therapeutic self-care level, and their likelihood of experiencing adverse outcomes. An exploratory research approach is chosen for this study because this research topic is relatively new and unstudied. In this research study, therapeutic self-care is defined as the ability to take medications as prescribed, to recognize and manage symptoms that may be experienced such as pain, to perform and adjust regular activities of daily living, and to manage symptoms, performing and adjusting regular activities, and managing changes in condition (Sidani, 2008).

## 2 STUDY PURPOSE

The purpose of this study was to evaluate the relationship between therapeutic self-care, adverse events, and health system utilization; while controlling for client demographics and clinical characteristics. The aim of this research was to test the hypotheses about the causal relationship between therapeutic self-care ability, and the frequency and types of adverse events experienced by home care clients. This study used the secondary data sources housed at the Institute of Clinical Evaluative Sciences in Ontario, Canada to answer the following question:

What is the relationship between home care clients' therapeutic self-care scores and adverse events measured using the following data sources: 1) HOBIC (Health Outcomes for Better Information and Care); 2) Home Care Reporting System RAI-HC (Resident Assessment Instrument-Home Care); 3) NACRS (National Ambulatory Care Reporting System) for emergency admissions; 4) DAD (Discharge Abstract Database) for hospital admissions.

In this study, the relationship between home care clients' therapeutic self-care ability was investigated in relation to two types of adverse events: (1) use of health care resources, including new emergency room visits and unplanned hospital readmissions; (2) safety outcomes, including client falls; unintended weight loss; new urinary tract infection; ADL decline; new pressure ulcer or ulcer deterioration; compliance/adherence with medication; and new caregiver decline. This study suggested that clients with low measured therapeutic self-care would be less likely to become activated to engage in self-care, and therefore may be linked to poor health outcomes, including safety problems and adverse events. The research proposed that therapeutic self-care may play an important role in contributing to more activated clients who have the knowledge, skill and confidence to manage their condition and maintain their health, and thus leading to better health outcomes.

## 3 METHOD

The quantitative study employed a retrospective cohort design. The focus of this study was the HOBIC therapeutic self-care measured in the home care sector. This study used a cross-sectional design where each home care client with a HOBIC and RAI-HC assessment was selected as the index assessment during the accrual window period (April 1, 2010 to September 30, 2011). The cohort of home care clients with data on HOBIC and RAI-HC assessments were followed forward for one-year period to investigate the occurrence of adverse events. The index assessment and follow-up period were used to examine whether therapeutic self-care scores predicted the frequency and types of adverse events experienced by home care clients. Specifically, HOBIC assessment on therapeutic self-care measure were linked prospectively with data from: (1) DAD for new acute care admissions; (2) NACRS for new ER visits; (3) RAI-HC assessment for falls, unintended weight loss, urinary tract infection; ADL decline; new pressure ulcer or ulcer deterioration; compliance/adherence with medication; and new caregiver decline.

In this study, therapeutic self-care was the independent variable that was posited to influence adverse events in home care. Specifically, the Therapeutic Self-Care Scale that was developed by Sidani and Doran (Doran et al., 2002), and collected as part of the HOBIC assessment used to obtain data on clients' therapeutic self-care measures. The Therapeutic Self-Care scale is a 12-item instrument that captures the domains of self-care: taking prescribed medications; recognizing and managing symptoms; performing and adjusting regular activities; and managing changes in condition (Sidani, 2008). The items measure the clients' perceived ability to perform these self-care activities with a 5-point numeric rating scale anchored with "not at all" and "very much", with higher total scores indicating high levels of self-care ability.

To prepare for the independent variable, the researcher has dichotomized the 1470 HOBIC therapeutic self-care scores into low self-care group (score 0 to 4), and high self-care group (score 5) where low self-care was coded as 0 and high self-care was coded as 1 in the logistic regression model. This grouping has created a final sample of 820 individuals in the low self-care group versus 650 individuals in the high self-care group. Adverse events were the dependent variables that were the outcomes of interest in this study. The World Health Organization (WHO) (2008) defines adverse events as the incidents that result in harm to a client. The final logistic regression model was developed using backward stepwise selection process.

Backward stepwise selection method involved dropping out the statistically non-significant risk factors for the indicator outcome. This method was appropriate for this study because stepwise

Table 1. Characteristics of study cohort and follow-up from 2010 to 2011.

| Cohort Characteristics | N = 1470 | % |
|---|---|---|
| Age over 65 | 1026 | 69.8 |
| Female | 832 | 56.6 |
| With Subsequent RAI-HC Assessments | 615 | 41.8 |
| Long-Term Care Admissions in one year follow-up | 81 | 5.5 |
| Death in one year follow-up | 230 | 15.6 |
| New Hospital Visits | 717 | 48.8 |
| New ER Visits | 836 | 56.9 |

selection dealt directly with redundancy and helped identify those client risk factors that strongly affected the probability of the outcome. The final model resulting from the stepwise selection was reviewed for the statistical analyses, including regression coefficients, Wald Chi-Square, odds ratios, and 95% confidence intervals for odds ratios for each of the predictors. To interpret the output of regression analysis, the significance criteria for regression coefficient, $p \leq .05$ was used in the logistic regression model. The Wald Chi-Square was used to evaluate the contribution of an individual predictor to a model. The odds ratios were computed to examine the ratio of the probability of an adverse event occurring where the independent variables and the outcome were not related when the value of an odds ratio was close to 1.

## 4 RESULTS

### 4.1 *Characteristics of study cohort*

The quantitative study of this research employed a retrospective cohort design with a study population consisted of long stay home care clients who were adults with the age of 18 years or older. Long-stay home care clients are defined as those who are expected to receive home care services for 60 days or longer, and are qualified to receive the RAI-HC assessment. The cohort was constructed by linking HOBIC-HC assessment where a baseline RAI-HC assessment was completed 90 days prior to or 30 days after the index HOBIC-HC assessment. The accrual of study cohort started from April 1, 2010 and ended on September 30, 2011, which resulted in a final sample of 1470 linked HOBIC-HC and baseline RAI-HC assessments.

The results presented in Table 1 describe the characteristics of study cohort. Among the 1470 assessments, the majority of home care clients were female who were in the age range of 65+ with the average age of 71.9 years (standard deviation 14.9). After the creation of study cohort, the 1470 linked assessments were being followed up to investigate the occurrence of adverse outcomes. The purpose of the cohort follow-up was to examine whether therapeutic self-care scores predict the frequency and types of adverse events experienced by home care clients. During the one year follow-up period, 48.8% of home care clients had a new hospital visit that was urgent or non-elective admission with an overnight stay, while 56.9% of home care clients had a new emergency room visit. Among the 1470 assessments, there were 607 individuals who had a subsequent RAI-HC assessment that allowed for follow-up on the occurrence of safety outcomes. For those individuals who did not have a subsequent RAI-HC assessment, 81 home care clients were admitted to long-term care facility and 230 individuals died during the one-year follow-up period.

### 4.2 *Therapeutic self-care: high self-care vs. low self-care*

Among the 1470 home care clients, 820 individuals were in the low self-care group and 650 individuals were in the high self-care group. Among 820 individuals in the low self-care group, the

Table 2. Prevalence rates of adverse events identified in RAI-HC, NACRS and DAD for home care clients from 2010 to 2011.

| Adverse Event | N = Number of home care clients follow-up from 2010 to 2011 | n = Number of home care clients with adverse events | Prevalence Rates % (n/N) |
|---|---|---|---|
| New ER Visit | 1470 | 836 | 56.9% |
| New ADL Decline | 615 | 318 | 51.7% |
| New Hospital Visit | 1470 | 717 | 48.8% |
| New Fall | 615 | 215 | 35.0% |
| New Caregiver Distress | 615 | 166 | 27.0% |
| Unintended Weight Loss | 615 | 83 | 13.5% |
| Non-Compliance/Adherence with Medications | 615 | 68 | 11.0% |
| Newly Detected Urinary Tract Infection | 615 | 40 | 6.5% |
| New Pressure Ulcer/Ulcer Deterioration | 615 | 34 | 6.0% |

majority of them were female who were over the age of 65 and living with their informal caregivers at home. The low self-care individuals were characterized as having more complexity in clinical status such as having recent hospitalizations; multiple chronic diseases, polypharmacy (taking more than 9 medications); higher CHESS scores (increasing health instability); and physical symptoms such as edema. They also demonstrated poor functional status with impaired self-reliance and difficulties with ADL activities such as mobility issues, as well as difficulties with IADL activities such as medication management. Further, the low self-care individuals were found to be more cognitively impaired with depressive symptoms, and behavioral symptoms including wandering, verbally abusive and resistive to care.

### 4.3 *Prevalence of adverse events*

The prevalence rates of adverse events identified in RAI-HC, NACRS, and DAD for the home care clients are presented in Table 2 for 2010 to 2011. During the one year follow-up period, new ER visit; new ADL decline; new hospital visit (any urgent/non-elective admission to hospital with an overnight stay); new fall; and new caregiver distress were ranked among the top five most frequently occurring adverse events. Unintended weight loss; Non-compliance or adherence with medications; newly detected urinary tract infection; and new pressure ulcer/ulcer deterioration were less frequently identified events.

### 4.4 *Analysis of association between therapeutic self-care and adverse events*

Based on the results of the logistic regression analyses, (1) new ER visit; (2) new ADL decline; (3) new fall; (4) unintended weight loss; and (5) compliance/adherence with medication were the five outcomes that were found to have association with therapeutic self-care scores. In particular, high therapeutic self-care group was associated with increased odds of new ER visit whereas low therapeutic self-care group was associated with increased odds of adverse events including new ADL decline, new fall, unintended weight loss and non-compliance/adherence with medication. Table 3 provides a summary of the results of the logistic regression analyses with adjusted odds ratio estimates for therapeutic self-care scores in relation to each adverse event.

## 5 CONCLUSION

In summary, the relationship between home care clients' therapeutic self-care ability were investigated in relation to two types of adverse events: (1) use of health care resources, including new

Table 3. Adjusted odds ratio estimates for therapeutic self-care in relation to adverse events.

| Adverse Events | Adjusted Odds Ratio for Therapeutic Self-Care | 95% Confidence Interval | p-value |
|---|---|---|---|
| New ER Visit | 1.26 | 1.02, 1.57 | 0.04 |
| New ADL Decline | 0.58 | 0.40, 0.84 | 0.04 |
| New Fall | 0.64 | 0.42, 0.99 | 0.05 |
| Unintended Weight Loss | 0.58 | 0.34, 0.99 | 0.05 |
| Non-Compliance/Adherence with Medication | 0.45 | 0.25, 0.78 | <=0.00 |
| Newly Detected Urinary Tract Infection | 1.74 | 0.90, 3.36 | 0.09 |
| New Pressure Ulcer/Ulcer Deterioration | 3.15 | 0.72, 13.87 | 0.13 |
| New Caregiver Distress | 1.34 | 0.89, 2.00 | 0.16 |
| New Hospital Visit | 0.85 | 0.66, 1.09 | 0.19 |

emergency room visits and unplanned hospital readmissions; (2) safety outcomes, including client falls; unintended weight loss; new urinary tract infection; ADL decline; new pressure ulcer or ulcer deterioration; non-compliance/adherence with medication; and new caregiver decline. Based on the results of the logistic regression analyses, (1) new ER visit; (2) new ADL decline; (3) new fall; (4) unintended weight loss; and (5) compliance/adherence with medication were the five outcomes that were found to have association with therapeutic self-care scores. In particular, high therapeutic self-care group was associated with increased odds of new ER visit whereas low therapeutic self-care group was associated with increased odds of adverse events including new ADL decline, new fall, unintended weight loss and non-compliance/adherence with medication.

This study explored the extent to which clients' therapeutic self-care ability was associated with safety problems. In particular, the study results provided a better understanding of the nature of relationships between therapeutic self-care ability, and the prevalence and incidence of safety outcomes for the older home care clients. This study added valuable contribution to the field of health services research by generating evidence about the role of therapeutic self-care in explaining variation in adverse events for home care clients. The knowledge gained from this study provided insight about the need to support risk mitigation and promote effective disease management for the clients and their caregivers. Clients and caregivers are central to the success of home care. Therefore, it is important that the decisions regarding client services in home care take into account of what the client and caregiver need to maintain and manage their own health and well-being. In the context of chronic disease management, shortened hospital stays, and increased complexity of home care clients, home care professionals will continue to have an important role in supporting clients in the development of self-care skills. A clear understanding of the nature of relationships between therapeutic self-care ability and safety outcomes will help focus health care delivery on improving client health and functioning, and to provide evidence for policy formulation related to the role of home care services in supporting therapeutic self-care.

## REFERENCES


Barlow, J., Wright, C., Sheasby, J., Turner, A., & Hainsworth, J. 2002. Self-management approaches for people with chronic conditions: A review. *Patient Education and Counseling* 48: 177–187.

Charles, C., Gauld, M., Chambers, L., O'Brien, B., Haynes, R., & Labelle, R. 1994. How was your hospital stay? Patients' reports about their care in Canadian Hospitals. *Canadian Medical Association Journal* 150(11): 1813–1831.

Coleman, E. 2006. *Performance measurement: Accelerating improvement.* Washington, D.C.: The National Academies Press.

Dai, Y., Chang, Y., Hsieh, C., & Tai, T. 2003. Effectiveness of a pilot project of discharge planning in Taiwan. *Research in Nursing & Health* 26: 53–63.

Doran, D., Hirdes, J., Blais, R., Baker, R., Pickard, J., & Jantzi, M. 2009. The nature of safety problems among Canadian home care clients: Evidence from the RAI-HC reporting system. *Journal of Nursing Management* 17: 165–174.

Doran, D., Sidani, S., Keatings, M. & Doidge, D. 2002. An empiricial test of the Nursing Role Effectiveness Model. *Journal of Advanced Nursing* 38(1): 29–39.

Forster, A., Clark, H., Menard, A., Dupuis, N., Chernish, R., Chandok, N., Khan, A., & Walraven, C. 2004. Adverse events among medical patients after discharge from hospital. *Canadian Medical Association Journal* 170(3): 345–349.

Hollander, M. 2008. Comparative cost analysis of home care and residential care Services. *The National Evaluation of the Cost-effectiveness of Home Care.* Hollander Analytical Services Ltd. Retrieved January, 2011 from: www.homecarestudy.com.

Hollander, M. & Tessaro, A. 2001. *Evaluation of the Maintenance and Prevention Model of Home Care.* Hollander Analytical Services Ltd. Retrieved February, 2011 from: http://www.hollanderanalytical.com/downloads/preventivehomecarereport.pdf.

Lorig, K. & Holman, H. 2003. Self-management education: History, definition, outcomes, and mechanisms. *Annal of Behavioral Medicine* 26(1): 1–7.

McCormack, D. 2003. An examination of the self-care concept uncovers a new direction for healthcare reform. *Nursing Leadership* 16(4): 48–65.

Naylor, M.D. 2003. Transitional care of older adults. *Annual Review of Nursing Research* 20: 127–147.

Naylor, M.D., Brooten, D., Jones, R., Lavizzo-Mourey, R., Mezey, M., & Pauly, M. 2004. Comprehensive discharge planning for the hospitalized elderly. A randomized clinical trial. *Annals of Internal Medicine* 120(12): 999–1006.

Ontario Assocation of Community Care Assess Centres. 2009. *Having their say and Choosing their way: Helping patients and caregivers move from hospital to home. Executive summary: Site one in south east LHIN area.* Retrieved March, 2011 from: http://www.ccac-ont.ca/Upload/oaccac/.../Executive_Summary_Site_One.pdf.

Sidani, S. 2008. Effects of patient centered care on patient outcomes: An evaluation. *Research and Theory for Nursing Practice* 22(1): 24–37.

Sidani, S. 2003. "Self-Care". In D.M. Doran, ed., *Nursing-Sensitive Outcomes: State of the Science.* Sudbury, MA: Jones & Barlett Learning.

Suter, P., Hennessey, B., Harrison, G., Fagan, M., Norman, B., & Suter, N. 2008. Home-based chronic care. *Home Healthcare Nurse* 26(4): 223–229.

The Change Foundation. 2006. *Facing the Facts. A Change Foundation Report.* Retrieved March, 2011 from: http://www.changefoundation.ca/docs/FacingFacts_HospitalToHome.pdf.

Williams, P., Lum, J., Deber, R., Montgomery, R., Kuluski, K., Peckham, A., et al. 2010. Aging at home: Integrating community-based care for older persons. *Heatlhcare Papers* 10(1): 10–23.

*Bridging Research and Good Practices towards Patient Welfare – Shih & Liang (Eds)*
 *ISBN 978-1-138-02716-9*

# Designing user interfaces of a handheld device for physical therapists

Sheau-Farn Max Liang, Li-Wei Cheng & Pei-Fang Jennifer Tsai
*National Taipei University of Technology, Taipei, Taiwan*

Chi-Lun Rau
*Taipei Medical University-Shuang Ho Hospital, Taipei, Taiwan*

ABSTRACT: The aim of this paper is to design and evaluate user interfaces on a handheld device for physical therapists in their daily practices at a rehabilitation center. Two user interfaces, patient therapy sheet and on-site therapy management, were designed according to workflow analysis and on-site observation. The overall usability of the design was examined through a modified System Usability Scale (SUS) questionnaire distributed to twenty-five physical therapists Results showed that the overall usability of these two user interfaces has been approved by the therapists. Furthermore, results from interviews revealed that the therapists considered most of the designed features to be useful to their works. Further possible improvement was also suggested by the therapists during the interviews. It is anticipated that the proposed design could enhance current practices and improve the efficiency and effectiveness of the on-site management of physical therapies.

*Keywords*: physical therapy, usability, user interface design

## 1 INTRODUCTION

Healthcare industry has become an increasingly critical sector for most developed countries, and Taiwan is no exception. After the implementation of National Health Insurance program, the tightly regulated healthcare funding approach requires hospitals and physicians to find effective ways to utilize resources. For this purpose, various healthcare information systems have been developed and applied to many hospitals and clinics. However, few of these systems have been designed for the on-site works of first-line medical staffs (Staggers and Troseth, 2011). For example, current healthcare information systems may not be appropriate for physical therapists in their continual move-around working style. While Vreeman, et al. (2009) reviewed papers about electronic health record systems in physical therapy and concluded that healthcare information systems have potential benefits for physical therapists, they emphasized the importance of integrating workflow analysis and end users' voices in the system design and development.

This study has been part of the effort to seek out opportunities for improvement at the rehabilitation center of a university-affiliated hospital. The aim of this paper is to design and evaluate proposed user interfaces on a handheld device for physical therapists in their daily practices at the rehabilitation center. A workflow analysis was applied at the design stage, and physical therapists' responses about the design was collected and examined at the evaluation stage.

## 2 USER INTERFACE DESIGN

### 2.1 *Current practices*

The area of studied rehabilitation center was about 1000 $m^2$ with more than 40 machines for various therapies, such as hydrotherapy, interferential current, muscle stimulation, infrared, transcutaneous

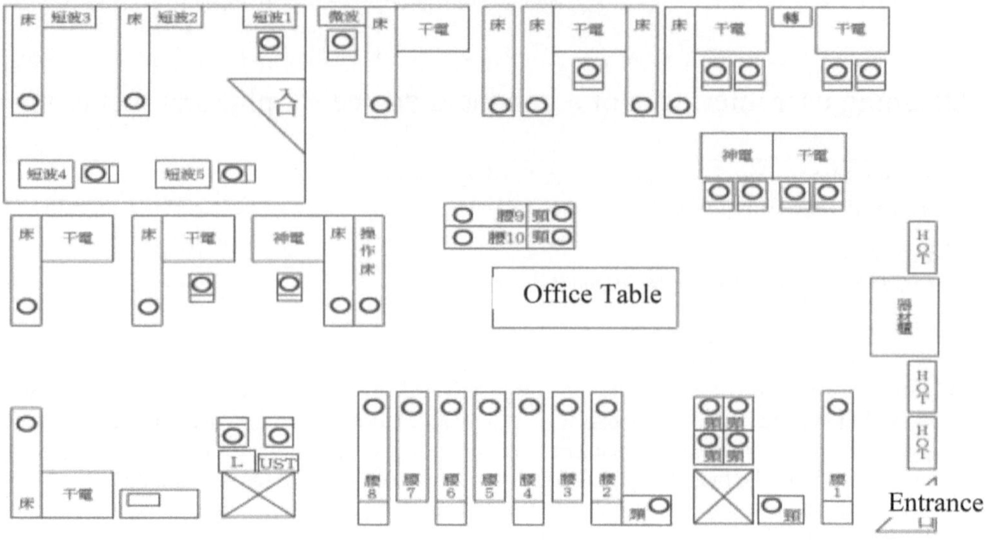


Figure 1. Layout of the rehabilitation center.

electrical nerve stimulation, circulator, shortwave diathermy, ultrasound diathermy, microwave diathermy, laser, therapeutic exercise, intermittent cervical traction, intermittent lumber traction, paraffin bath, and hot pack. The layout of the rehabilitation center is illustrated in Figure 1.

As shown in Figure 1, machines and equipment were located around an office table. There were desktop computers on the table for therapists to retrieve medical information about their patients, but patent's information was mainly provided by a paper therapy sheet. The information on the sheet included patient's personal and diagnostic information, treatment list, and date stamps of previous therapy appointments. Typically there were six appointments per prescription. For each appointment, there were usually three to five treatments. The duration for each treatment was about 15 minutes.

Patients brought their therapy sheet from the reception counter, went to the rehabilitation center, and gave the sheet to their assigned therapists to start their treatments. Since each therapist could handle up to five patients and each patient had multiple treatments, the rehabilitation center was crowded all the time. Therefore, it is important for therapists to comprehend and manage their patients' flows at the rehabilitation center.

## 2.2 *The design*

A 10-inch tablet computer was selected as the platform since this handheld device was suitable for therapists' mobile working style, and the screen size of 10-inch was large enough to clearly display necessary information.

Two user interfaces were designed based on the results of observation and workflow analysis of current practices at the rehabilitation center. One was an electronic therapy sheet of individual patients, and the other was the information of patients' treatments and locations at the center. As shown in Figure 2, the electronic therapy sheet was designed based on the same format of the paper one with several added features.

The first feature was the information of patient's weight. For the traction treatments, such as the intermittent cervical traction and the intermittent pelvic traction, a proper traction force had to be calculated and inputted to the machine by therapists. The calculation was according to patient's weight and the number of previous appointments. This information was not available on the paper sheet, and we thought it was important to include it in the new design.

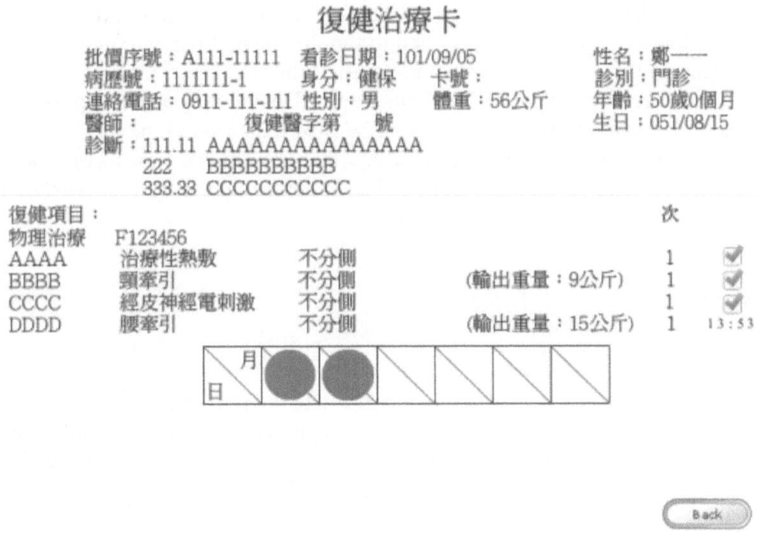


Figure 2. Electronic therapy sheet.

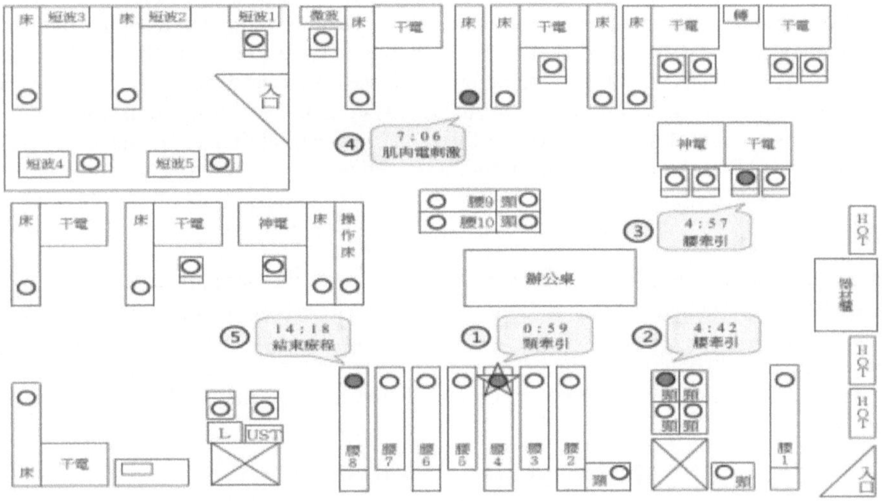


Figure 3. User interface of on-site therapy management.

To further reduce therapists' mental workloads on the calculation of correct traction forces the second feature of the user interface was the information of suggested traction forces based on the calculation formulas if traction treatments were on the list. The third added feature was the information of treatment status. At the end of each treatment item a tick symbol on a checkbox indicated the completion of that treatment, whereas a countdown time denoted the remaining time of that treatment. This feature provided dynamic information about patient's progress of the therapy. Finally, a "Back" button was displayed on the lower right corner of the screen. This was the link to the other user interface: on-site therapy management, as therapists tapped on the button.

As shown in Figure 3, the user interface of on-site therapy management was a new design for therapists to manage their patients' treatments.

The layout of the rehabilitation center was used as the background. On the user interface, red dots indicated the location of patients assigned to the therapist. An associated callout symbol for each red dot displayed the remaining time and name of the treatment. Circled numbers were also associated with red dots. They indicated patients' ranks, from 1 to 5, according to their remaining treatment times, from less to more. A flash would be shown up around the red dot if the remaining treatment time was less than one minute. The last feature was the link to the therapy sheet of individual patients as therapists tapped on any red dot. A prototype of these two user interfaces was developed for evaluation.

## 3 EVALUATION

### 3.1 *Participants*

Twenty-five physical therapists were recruited to evaluate the proposed user interfaces. Seventeen of them were female and eight were male. Their average age was 26.8 years old with a standard deviation of 3.8 years old. The range of their work experience was from two months to 12 years with the mean of 2.3 years and the standard deviation of 2.4 years.

### 3.2 *Apparatus and materials*

The prototype of user interfaces was coded by applying Adobe Flash and displayed on the 10.1-inch ASUS Transformer Pad TF300T. The System Usability Scale (SUS) questionnaire (Brooke, 1996) was used with minor modifications for the evaluation. Instead of rating the usability of the design itself, the ten questions in this modified SUS questionnaire were to evaluate the relative usability of the new design against current working practices. These questions were about the willingness of use, system complexity, ease of use, need of support, system integration, system consistency, peer learnability, obstruction of use, confidence of use, and self learnability. A five-level Likert rating scale, from strongly disagree to strongly agree, was used with corresponding scores from 1 to 5.

### 3.3 *Procedure*

For data collection, the user interfaces were demonstrated to the participants followed by a try-out session to let participants familiar with the design. Then, participants were asked to fill up the modified SUS questionnaire. Finally, a brief interview was conducted for each participant to collect their feedback about the design.

For data analysis, the overall score of the questionnaire was calculated as formula (1) listed below.

$$2.5 \times \sum[(Score_{Odd} - 1) + (5 - Score_{Even})] \quad (1)$$

In formula (1), the $Score_{Odd}$ represents the score of each odd-numbered question, and the $Score_{Even}$ represents the score of each even-numbered question. According to the formula, the lowest score was zero, and the highest was 100. Therefore, a statistical test was applied to examine whether significant difference exists between the overall score of the questionnaire and the score of 50, the middle of the score. The data collected from the interview were summarized as descriptive statistics.

## 4 RESULTS

### 4.1 *Results from questionnaire*

The mean scores and associated standard deviations of the ten modified SUS questions are presented in Table 1.

Table 1. Mean scores and standard deviations of the modified SUS questions.

| Question | 1 | 2 | 3 | 4 | 5 | 6 | 7 | 8 | 9 | 10 |
|---|---|---|---|---|---|---|---|---|---|---|
| Mean | 4.1 | 2.6 | 3.9 | 2.1 | 3.7 | 2.5 | 3.9 | 2.2 | 3.8 | 2.2 |
| S.D. | 0.6 | 0.6 | 0.6 | 0.8 | 0.8 | 0.6 | 0.6 | 0.7 | 0.6 | 0.5 |

Note: S.D.: Standard Deviation.

Table 2. Feedback of proposed features on the user interfaces.

| Feature | Percentage of the therapists who thought the feature was beneficial |
|---|---|
| Remaining treatment time indicator (O) | 88% |
| Patient location indicator (O) | 68% |
| Machine traction force information (T) | 60% |
| Treatment at the last minute indicator (O) | 52% |
| Patient's weight information (T) | 48% |
| Next treatment information (T) | 40% |
| Link between patient location and treatment list (O; T) | 36% |
| Patents' rank numbers according to their remaining treatment times (O) | 28% |
| Completed treatment indicator (T) | 12% |

Note: O: on on-site therapy management user interface; T: on therapy sheet user interface.

Table 3. Suggestions of further improvements to the user interfaces.

| Suggestion | Percentage of the therapists who made the suggestion |
|---|---|
| Operation on a smaller size of the device such as a mobile phone | 68% |
| Inclusion of vibration and sound indicators | 52% |
| Operation on current size of the device | 32% |
| Display of patient's treatment history | 32% |
| Improvement of the sensitivity of touch screen | 28% |
| Display of available treatment machines | 28% |
| Change of the last minute indicator to 30 seconds | 20% |

Note that the mean scores of the odd-numbered questions were all higher than the mean scores of the even-numbered questions The mean values and standard deviations for the overall scores collected from the 25 participants were 69.2 and 11.2. The normality of the data was checked through the Kolmogorov-Smirnov test, so a t-test against the score value of 50 was applied. Result showed that the mean overall score of the questionnaire was significantly higher than the score of 50 ($t(24) = 8.54$, $p < 0.001$), which indicates that the physical therapists were in favor of the design of user interfaces

### 4.2 *Results from interview*

The results of the interview revealed therapists' preferences and opinions about the design in detail. Table 2 shows the percentages of the therapists who thought the proposed features were beneficial to their works. As can be seen, the remaining treatment time indicator was the most favorable feature. There were also three other features that obtained more than 50% therapists' approvals: the patient location indicator, machine traction force information, and the treatment at the last minute indicator.

In addition, results from the interview suggested that further features, such as the operation on a smaller-size device and inclusion of vibration and sound as the indication signals, could be included in the design. These suggestions are listed in Table 3 with the percentage of the therapists who made the suggestions.

## 5 DISCUSSION

The overall usability of two user interfaces on a handheld device, patient therapy sheet and on-site therapy management, was approved by the physical therapists at the rehabilitation center of a university-affiliated hospitals. The designed features were useful and accepted by the therapists. While further development is necessary, it is anticipated that the proposed design could enhance current practices and improve the efficiency and effectiveness of the on-site management of physical therapies.

## ACKNOWLEDGEMENT

This research was founded by the National Taipei University of Technology-Taipei Medical University Joint Research Program (NTUT-TMU-101-09).

## REFERENCES

Brooke, J. 1996. SUS – A quick and dirty usability scale. In P.W. Jordan, B. Thomas, I.L. McClelland & B. Weerdmeester (eds.), *Usability Evaluation in Industry*: 189–194. London: Taylor & Francis.

Staggers, N., Troseth, M. R. 2011. Usability and clinical application design. In: Ball, M. J., Hannah, K. J., Newbold, S. K., Douglas, J. V. (eds.), *Nursing Informatics*: 219–241. London: Springer-Verlag.

Vreeman, D. J., Taggard, S. L., Rhine, M. D., Worrell, T. W. 2006. Evidence for electronic health record systems in physical therapy. *Physical Therapy* 86: 434–446.

*Bridging Research and Good Practices towards Patient Welfare – Shih & Liang (Eds)*
*© 2015 Taylor & Francis Group, London, ISBN 978-1-138-02716-9*

# A scoliosis survey in children with overweight at age of 6 to 13 years old

Chao Yin Wu
*Department of Industrial Engineering and Engineering Management, National Tsing Hua University, Hsinchu, Taiwan, ROC*
*Department of Rehabilitation Medicine, Mackay Memorial Hospital, Hsinchu, Taiwan, ROC*

Mao-Jiun Wang
*Department of Industrial Engineering and Engineering Management, National Tsing Hua University, Hsinchu, Taiwan, ROC*

ABSTRACT: The scoliosis screening is often used in school children health check. And childhood obesity is seen as one of the most serious public health challenges in the 21st century. There is lacking of information about whether the prevalence of scoliosis is higher among children with overweight. Thus, the prevalence of scoliosis in school children was investigated in this study. One thousand and five school children (534 boys and 471 girls) aged 6 to 13 years participated in this study. Forward bending test was used to screen whether there is a possible scoliosis occurred in each child. The prevalence of scoliosis for 6 to 13-year-old children was 12.6% with an increasing trend with age. The prevalence of scoliosis was 10.9% in boys and 14.3% in girls. The percent of overweight and obese children was 35.8% and 9%. The prevalence of scoliosis was 12.6% in children with normal weight, 16.8% in children with overweight and 6.8% in children with obesity. Children with scoliosis were found to be taller (or older) than those without scoliosis ($p < 0.05$). Further, the scoliosis effect and obesity effect were significant on some body dimensions such as shoulder height and hip girth. The prevalence of scoliosis was found to be slightly higher in children with overweight and taller children. It is important to address the issue of scoliosis by emphasizing the importance of maintaining good posture for children with overweight.

*Keywords*: scoliosis, overweight, obesity, children

## 1 INTRODUCTION

Scoliosis is a general term that describes any change of shape and position in the spine, thorax and trunk, and is defined as a "three-dimensional torsional deformity of the spine and trunk" (Negrini et al., 2012). Around eighty-five percent of people with scoliosis have the idiopathic type, which means no known cause of scoliosis. Idiopathic scoliosis may develop at any time during childhood and adolescence. It is commonly found in children between 5–8 years and 11–14 years old. Scoliosis screening in school children is recommended for early detection of spinal deformities ("School Screening Programs for the Early Detection of Scoliosis – An information statement," 2007). Adam's forward bending test is a classic screening tool, which asked the child to bend forward at waist with knee straight and any asymmetry in the contours of the back is considered as a positive sign of scoliosis. Despite the direct result of forward bending test, other physical examinations during screening were also recommended. (Janicki & Alman, 2007). Observation of standing posture to look for any asymmetry between right and left shoulder and pelvis height is often used as the first step of examination. Leg length discrepancy is an examination used once scoliosis was screened and can be a potential cause of scoliosis. Lee and Chen. (1999) reported the prevalence of scoliosis was 10.2% in children age between 10 to 14 years. They found the height-weight ratio was a significant factor for scoliosis, especially for girls. One of the most popular indexes for height-weight ratio is Body Mas Index (BMI).

BMI is widely used as a reliable indicator for body fatness and is defined as a person's weight in kilograms divided by the square of his height in meters ($kg/m^2$). Overweight has been a growing health problem in the past decade. Overweight and obesity are defined as abnormal or excessive fat accumulation that may impair health. Because the amount of body fat changes with age and the amount of body fat differs between girls and boys, the cut off point for overweight and obesity varies between girls and boys at different ages. Overweight is defined as 85th to less than the 95th percentile and obesity is defined as equal to or greater than the 95th percentile in BMI-for-age growth chart (CDC). According to WHO (WHO, 2007), overweight is 17.1 $kg/m^2$ for 6 years old girl and 21.9 $kg/m^2$ for 13 years old girl; obesity is 18.3 $kg/m^2$ for 6 years old girls and 24.3 $kg/m^2$ for 13 years old girls; whereas overweight is 16.8 $kg/m^2$ for 6 years old boys and 20.9 $kg/m^2$ for 13 years old boys; obesity is 17.8 $kg/m^2$ for 6 years old boys and 23.4 $kg/m^2$ for 13 years old boys. In their study, Nery, et al. (2010)reported the percentage of overweight is 19.8% and the percentage of scoliosis is 1.4% among 1340 students. They found no association between overweight and scoliosis. On the other hand, Chen et al. (2006) studied the relationship between BMI and scoliosis in female aged from 19 to 29. The result showed that the scoliosis detection rate in the group of BMI $< 20$ is 3 times higher than the group of BMI $> 20$.

The objectives of this study were to investigate the prevalence of scoliosis and obesity in school children, and compare the height differences of right and left shoulder and pelvis in children with scoliosis.

## 2 METHODS

### 2.1 *Subjects*

One thousand and five children (aged from 6 to13 years old)from a primary school in Taiwan were recruited in this study. The inclusion criteria were no current pain or discomfort of musculoskeletal system and the exclusion criteria were children with neurological disease orwith developmental delay. Before the study began, parents of each participant were asked to sign an informed consent form approved by the Internal Review Board of a local hospital. The means and standard deviations of demographic data for each age and gender group are listed in Table 1 and Table 2.

Instrumentation

For scoliosis screening, the Adams forward bending test was used to evaluate whether the participants had sign of scoliosis. Any asymmetry of back contour while performing the test was considered a positive sign of scoliosis. Anthropometric measurements were used to evaluate the symmetry between left and right side of body. The height differences were calculated between right and left acromion, anterior superior iliac spin (ASIS), grater trochanter (GT) of femur. To measure the height differences between right and left shoulder and pelvis, a 3D coordinate measurement system (Faro Technologies Inc.) was used to measure the 3 dimensional coordinates of the body landmarks. A digital tape was used to measure waist girth and hip girth. The accuracy of the 3D coordinate measurement system and digital tape was about 0.01 mm and 0.1 mm respectively.

### 2.2 *Procedure*

To ensure measurement accuracy, each participant was asked to wear thin clothes and barefoot during the measuring procedure. After measuring body weight and height, red circular stickers (radius: 0.5 cm) were attached on each participant's right and left ear acromion, ASIS, GT of femur. All the body landmarks were identified by an experienced physical therapist. The participants were asked to stand on both feet in front of a standing frame, the 3D coordinate measurement probe then digitize the red stickers on each participant's body to locate the relative position to the ground. A trained research assistant performed the digitization procedure and also measures the waist girth and hip girth with digital tape. After the anthropometric measurement, each participant performed the FBT and a physical therapist examined whether there was an asymmetry appeared on the back contour as a positive sign of scoliosis.

### 2.3 *Experiment design*

For statistical analysis, 3-way ANOVA was used to evaluate the effect of gender, scoliosis and obesity on the anthropometric measurements. The dependent measures included body height and weight, height differences of bilateral acromion, ASIS, and GT, waist girth and hip girth.

## 3 RESULTS

The average percentage of positive FBT was 10.9% in boys and 14.3% in girls. The percentage of positive FBT stratified with age in boys and girls are listed in table 1 and 2. The percentage of positive FBT increased with age in boys; at age of 11, the prevalence reached highest to 18.6%. In girls, the prevalence of scoliosis was found higher than boys, especially at age of 12 and 13. The prevalence was 20.7% and 30.8% at age of 12 and 13, respectively.

Table 1. The means and standard deviations of Height, Weight and BMI and percentage of scoliosis in boys.

| | Boys | | | | | | | |
|---|---|---|---|---|---|---|---|---|
| | | Height (cm) | | Weight (kg) | | BMI (kg/m$^2$) | | |
| Age (yr.) | N | Mean | SD | Mean | SD | Mean | SD | Positive FBT (%) |
| 6 | 30 | 110.24 | 7.71 | 19.19 | 3.05 | 15.82 | 2.27 | 6.7 |
| 7 | 40 | 114.53 | 5.30 | 21.32 | 3.60 | 16.16 | 1.75 | 5 |
| 8 | 56 | 120.38 | 4.85 | 24.46 | 5.17 | 16.78 | 2.76 | 7.1 |
| 9 | 88 | 126.61 | 6.27 | 27.95 | 7.09 | 17.24 | 3.11 | 12.5 |
| 10 | 117 | 134.15 | 6.21 | 33.15 | 8.05 | 18.26 | 3.44 | 14.5 |
| 11 | 118 | 141.65 | 6.33 | 38.87 | 8.85 | 19.21 | 3.33 | 18.6 |
| 12 | 57 | 148.02 | 7.56 | 44.30 | 13.07 | 19.99 | 4.41 | 14 |
| 13 | 28 | 149.71 | 7.61 | 44.03 | 11.59 | 19.40 | 3.47 | 7.1 |

Table 2. The means and standard deviations of Height, Weight and BMI and percentage of scoliosis in girls.

| | Girls | | | | | | | |
|---|---|---|---|---|---|---|---|---|
| | | Height (cm) | | Weight (kg) | | BMI (kg/m$^2$) | | |
| Age (yr.) | N | Mean | SD | Mean | SD | Mean | SD | Positive FBT (%) |
| 6 | 33 | 108.33 | 4.19 | 19.44 | 2.85 | 16.54 | 2.13 | 0 |
| 7 | 42 | 113.45 | 4.98 | 20.27 | 3.83 | 15.64 | 1.91 | 11.9 |
| 8 | 60 | 118.64 | 4.77 | 22.48 | 3.47 | 15.91 | 1.82 | 8.3 |
| 9 | 46 | 126.92 | 5.35 | 27.37 | 4.98 | 16.92 | 2.32 | 15.2 |
| 10 | 94 | 133.66 | 7.37 | 30.40 | 6.67 | 16.85 | 2.42 | 8.5 |
| 11 | 99 | 141.70 | 7.58 | 36.40 | 8.57 | 17.95 | 3.06 | 7.1 |
| 12 | 58 | 151.01 | 7.68 | 42.87 | 8.76 | 18.68 | 2.83 | 20.7 |
| 13 | 39 | 152.67 | 5.93 | 46.82 | 10.23 | 19.97 | 3.58 | 30.8 |

Table 3. Percentage of three obesity categories in different age and gender.

| Age (yr.) | Boys | | | | | | Girls | | | | | |
|---|---|---|---|---|---|---|---|---|---|---|---|---|
| | Normal weight | | Overweight | | Obese | | Normal weight | | Overweight | | Obese | |
| | N | % | N | % | N | % | N | % | N | % | N | Percentage |
| 6 | 22 | 73.3 | 3 | 10 | 5 | 16.7 | 23 | 69.7 | 4 | 12.1 | 6 | 18.2 |
| 7 | 33 | 82.5 | 3 | 7.5 | 4 | 10 | 36 | 85.7 | 3 | 7.1 | 3 | 7.1 |
| 8 | 40 | 71.4 | 8 | 14.3 | 8 | 14.3 | 53 | 88.3 | 4 | 6.7 | 3 | 5 |
| 9 | 64 | 72.7 | 8 | 9.1 | 16 | 18.2 | 35 | 76.1 | 7 | 15.2 | 4 | 8.7 |
| 10 | 80 | 68.4 | 11 | 9.4 | 26 | 22.2 | 73 | 77.7 | 16 | 17 | 5 | 5.3 |
| 11 | 67 | 56.8 | 17 | 14.4 | 34 | 28.8 | 78 | 78.8 | 6 | 6.1 | 15 | 15.2 |
| 12 | 38 | 66.7 | 7 | 12.3 | 12 | 21.1 | 46 | 79.3 | 7 | 12.1 | 5 | 8.6 |
| 13 | 18 | 64.3 | 6 | 21.4 | 4 | 14.3 | 25 | 64.1 | 8 | 20.5 | 6 | 15.4 |
| Average | | 69.5% | | 12.3% | | 18.2% | | 77.5% | | 12.1% | | 10.4% |

Table 4. Numbers and percentages of scoliosis in different obesity category and gender.

| Obesity category | Boys | | | | Girls | | | |
|---|---|---|---|---|---|---|---|---|
| | Non-scoliosis | | Scoliosis | | Non-scoliosis | | Scoliosis | |
| | N | Percentage | N | Percentage | N | Percentage | N | Percentage |
| Normal weight | 316 | 87.3% | 46 | 12.7% | 323 | 87.5% | 46 | 12.5% |
| Overweight | 51 | 81% | 12 | 19% | 47 | 85.5% | 8 | 14.5% |
| Obesity | 99 | 90.8% | 10 | 9.2% | 45 | 95.7% | 2 | 4.3% |

The average BMI was 17.86% in boys and 17.31% in girls. The BMI stratified with age in boys and girls are listed in table 1 and 2. The average BMI ranged from 15.82 to 19.99 in boys, and 15.64 to 19.97 in girls. The BMI then divided into 3 categories: normal weight, overweight and obesity. The percentages of each category in boys and girls in different age are listed in table 3. Overall, percentage of obesity was 13.3% in boys and 7.6% in girls, and percentage of overweight was 17% in boys and 14.9% in girls.

Table 4 lists the numbers and percentages of scoliosis in three obesity categories in boys and girls. The percentages of scoliosis in boys in normal weight, overweight and obesity were 12.7%, 19% and 9.2%, respectively. The percentages of scoliosis in girls in normal weight, overweight and obesity were 12.5%, 14.5% and 4.3%, respectively. Table 5 shows the ANOVA result of gender, scoliosis and obesity effect on the anthropometric measures. Height and weight were found to significantly affect by scoliosis and obesity, as well as ear lobe height differences. Ear lobe height difference was also affected by the interaction of scoliosis and obesity. In circumference measures, waist girth was affected by gender and obesity whereas hip girth was affected by obesity.

## 4 DISCUSSION

The positive rate of FBT found in school children aged from 6 to 13 was 12.6%. The percentage of positive FBT was relatively high as comparing to the previous studies. Karachalios et al. (1999) reported that the prevalence of having positive FBT for school children of 8 to 16 years old was 5.8% (156 out of 2700). Kapoor et al. (2008) reported that 2.8% children (11–13 years) were

Table 5. Anthropometric measures related to scoliosis and obesity.

| Measurements (cm) | Boys | | | | | | Girls | | | | | | Sig. |
|---|---|---|---|---|---|---|---|---|---|---|---|---|---|
| | Non-scoliosis | | | Scoliosis | | | Non-scoliosis | | | Scoliosis | | | |
| | Normal weight | Over-weight | Obese | Normal weight | Over-weight | Obese | Normal weight | Over-weight | Obese | Normal weight | Over-weight | Obese | |
| Height | 129.6 | 135.04 | 137.78 | 134.50 | 140.73 | 141.81 | 130.66 | 136.66 | 137.76 | 138.01 | 139.33 | 151.82 | S**, O** |
| Weight (kg) | 27.748 | 36.28 | 45.39 | 30.35 | 39.875 | 49.06 | 28.19 | 38.66 | 44.84 | 32.46 | 41.01 | 56.70 | S**, O** |
| Acromial height differences | 0.85 | 0.85 | 0.80 | 0.91 | 0.98 | 0.94 | 0.77 | 0.89 | 0.78 | 0.83 | 0.66 | 1.00 | |
| ASIS height differences | 0.76 | 0.85 | 0.79 | 0.87 | 0.65 | 0.81 | 0.78 | 0.73 | 0.88 | 0.73 | 0.92 | 1.19 | |
| GT height differences | 1.22 | 1.21 | 1.22 | 1.09 | 0.76 | 1.84 | 1.11 | 1.37 | 1.19 | 0.97 | 1.15 | 1.06 | |
| Waist girth | 55.89 | 63.60 | 72.32 | 56.82 | 65.54 | 74.46 | 55.18 | 63.25 | 71.01 | 57.01 | 65.84 | 72.46 | O** |
| Hip girth | 68.21 | 77.01 | 84.01 | 70.41 | 79.94 | 86.97 | 69.25 | 78.58 | 85.35 | 73.35 | 77.38 | 91.81 | S*, O** |
| Waist/Hip ratio | 0.82 | 0.83 | 0.86 | 0.81 | 0.82 | 0.86 | 0.80 | 0.81 | 0.84 | 0.78 | 0.86 | 0.79 | G*, O**, SO |

*: $p < 0.05$; **: $p < 0.01$; G: significant gender effect; S: significant scoliosis effect; O: significant obesity effect.

positive for scoliosis test. The use of FBT as a screening tool has been questioned; the positive predictive value of FBT is 0.133. (Karachalios et al., 1999), which means a high false-positive rate in screening scoliosis, and this might explain the high positive rate of FBT found in our study. The fact that we did not found any height differences in standing posture, it was assumed that the positive FBT could be a postural problem not a real spinal deformity. It is also recommended that people with positive FBT should seek further medical check-up and advices.

The girls at age of 12 and 13 had higher positive FBT rate of 20.7% and 30.8%, respectively. Scoliosis in adolescent is often reported has a high prevalence as the children are speeding up the growth. Konieczny et al. (2013) reported the epidemiology of adolescent idiopathic scoliosis and proposed that the girl to boy ratio ranges from 1.5:1 to 3:1 with increasing age. The reason that girls have higher prevalence of scoliosis is still remain unknown; genetics and hormone were proposed to be one of the reasons.

Body weight in majorityof the school children are in normal range which 69.5% in boys and 77.5% in girls. About 12% children are overweight and 14% are obese. In children aged from 2–17 years, American Center for Disease control and Prevention (CDC) (May, Freedman, Sherry & Blanck, 2013) reported about 19% male and 15% female are obese. According to WHO, in 2011, more than 40 million children under the age of five were overweight. Once considered a high-income country problem, overweight and obesity are now on the rise in low- and middle-income countries, particularly in urban settings. More than 30 million overweight children are living in developing countries and 10 million in developed countries. Nutrition, diet habit or education level might be the reason of rising weight in developing country.

In our study the percentage of positive FBT among overweight boys (19%) were twice than obese boys (9.2%); in girls the percentage were even tripled. The percentage of positive FBT were also higher in normal weight children (12%) compared to obese children (6.5%). This result was similar to those in Chen et al. (2006). Hershkovich et al. (2013) found the prevalence of spinal deformities was significantly greater among the underweight patients and proposed that increased BMI had a protective effect for developing spinal deformities. They also found subjects with spinal deformities were taller. These results were similar to our study, however, in their study, the subjects' age was 17 years old military recruits and spinal deformities included not only scoliosis but also kyphosis. Also, in our study, we did not calculate the underweight group, which might be a potential factor for scoliosis, and this should be addressed in future studies.

## 5 CONCLUSION

The percentage of positive FBT is high in our study, and it was suspected to be a postural problem. To maintain a good posture should be taught in school children. The percentage of overweight and obese ranged from 10.4% to 18.2%, which raised a caution to weight control and healthier life style should be addressed. The subjects with positive FBT were less in obese children. Although the reasons are remain to discover, it is important to clinicians to be cautious about scoliosis in less obese populations.

## REFERENCES


CDC. What is BMI percentile?, from http://www.cdc.gov

Chen, S.-C., Chen, Y.-H., & Chao, S.-Y. (2006). The relationship between Body Mass Index and Scoliosis. C J Radiologic Tech, 30(2), 115–121.

Hershkovich, O., Friedlander, A., Gordon, B., Arzi, H., Derazne, E., Tzur, D., . . . Afek, A. (2013). Association between body mass index, body height, and the prevalence of spinal deformities. Spine J. doi: 10.1016/j.spinee.2013.09.034

Janicki, J. A., &Alman, B. (2007). Scoliosis: Review of diagnosis and treatment. Paediatric Child Health, 12(9), 771–776.

Kapoor, M., Laham, S. G., & Sawyer, J. R. (2008). Children at risk identified in an urban scoliosis school screening program: a new model. J Pediatr Orthop B, 17(6), 281–287. doi: 10.1097/BPB.0b013e32831 1d4d8
Karachalios, T., Sofianos, J., Roidis, N., Sapkas, G., Korres, D., & Nikolopoulos, K. (1999). Ten-year follow-up evaluation of a school screening program for scoliosis. Is the forward-bending test an accurate diagnostic criterion for the screening of scoliosis? Spine (Phila Pa 1976), 24(22), 2318–2324.
Konieczny, M. R., Senyurt, H., &Krauspe, R. (2013). Epidemiology of adolescent idiopathic scoliosis. J Child Orthop, 7(1), 3–9. doi: 10.1007/s11832-012-0457-4
Lee, P.-H., & Chen, C.-M. (1999). School screening for scoliosis among fifth and sixth graders of two elementary schools in Taipei. Chinese Journal of Public Health, 18(5), 303–312.
May, A. L., Freedman, D., Sherry, B., &Blanck, H. M. (2013).Obesity – United States, 1999–2010. MMWR Surveill Summ, 62 Suppl 3, 120–128.
Negrini, S., Aulisa, A., Aulisa, L., Circo, A., de Mauroy, J., Durmala, J., . . . Zaina, F. (2012). 2011 SOSORT guidelines: Orthopaedic and Rehabilitation treatment of idiopathic scoliosis during growth. Scoliosis, 7(1), 3.
Nery, L. S., Halpern, R., Nery, P. C., Nehme, K. P., & Stein, A. T. (2010). Prevalence of Scoliosis Among School Students in a Town in Southern Brazil. Sao Paulo Med J., 128(2), 69–73.
School Screening Programs for the Early Detection of Scoliosis – An information statement. (2007). from http://www.srs.org/professionals/advocacy_and_public_policy/SRS-AAOS_position_statement.htm
WHO. (2007). BMI-for-age, boys or girls, 5–19 years.

*Bridging Research and Good Practices towards Patient Welfare – Shih & Liang (Eds)*
*© 2015 Taylor & Francis Group, London, ISBN 978-1-138-02716-9*

# A review of the effect on postural stability while using mobile phone

N.R. Nurwulan
*National Taiwan University of Science and Technology, Taiwan*
*Institut Teknologi Bandung, Indonesia*

H. Iridiastadi
*Institut Teknologi Bandung, Indonesia*

B.C. Jiang
*National Taiwan University of Science and Technology, Taiwan*

ABSTRACT: Using mobile phone while doing another activity is a common example of dual-task activity that we often do just about anywhere. People tend to think that multitasking is an efficient and good way to increase productivity. In fact, previous studies showed that sharing-attention between tasks takes a serious toll on productivity. The aim of this paper is to review literatures on postural stability while using mobile phone. There were 125 papers reviewed and 38 papers met the inclusion criteria. The first part of this review provides a general background of the fatalities of distracted pedestrians caused by mobile phones. The second part discusses three models of dual-task performance that proposed by Lacour et al. (2008), due to the contradictory results between dual-task studies. The last part of this paper is a review of several different methods to evaluate postural stability. Postural stability can be measured by center of pressure (COP) displacement, multiscale entropy (MSE), and multivariate multiscale entropy (MMSE). In conclusion, studies of the effect of using mobile phone on postural stability are needed to be evaluated more, in order to reduce the fatalities that could happen from dual-tasking activity.

*Keywords*: Mobile phone; Postural stability; Dual-task; COP; MSE; MMSE.

## 1 INTRODUCTION

Mobile phone has become a very crucial part in our daily lives. Using mobile phone while doing another activity is a common example of dual-task activities that we often do just about anywhere. Previous studies have shown that using mobile phones are not only jeopardizing the safety of drivers, but also the safety of pedestrians. Pedestrians' behaviors are considered to be the factor in pedestrians' injuries, because data from police department often did not mention the mistakes of the drivers (Bungum et al., 2005). Mobile phones use by pedestrians have impact on working memory (Lamberg & Muratori, 2012) and increase walking distraction that putting pedestrian at greater risk for accidents (Hatfield & Murphy, 2007; Nasar et al., 2008; Neider et al., 2010; Stavrinos et al., 2011). There are many studies about the distracted attention of pedestrians caused by mobile phones, either by doing on the road study (Bungum et al., 2005; Hatfield & Murphy, 2007; Nasar et al., 2008) or using simulator (Neider et al., 2010; Stavrinos et al., 2011; Schwebel et al., 2012). However, those studies did not examine the effect of using mobile phones on postural stability that causes fatalities.

There have been various postural stability studies related to dual-task activity using cognitive load (Marsh & Geel, 2000; Andersson et al., 2002; Huxhold et al., 2006; Simoneau et al., 2008; Fraizer & Mitra, 2008; Palluel et al., 2010; Yeh et al., 2010), physical load (Mitra, 2003; Huang et al., 2010; Hsu & Scholz, 2012), visual load (Vuillerme et al., 2000; Dault et al., 2001; Mitra,

2003; Ehrenfried et al., 2003), and auditory load (Vuillerme et al., 2000; Toppila et al., 2005; Park et al., 2011) as secondary task. For the specific study about mobile phone, Won (2012) examined the effect of mobile phone on static and dynamic postural stability. Then again, only few studies examined the effect of using mobile phone on postural stability. The purpose of this paper is to make a brief review of literatures on postural stability while using mobile phone.

## 2 METHODS

A comprehensive search strategy was developed by identifying all potentially relevant keywords, categorizing into specific phrases and subsequently combined them. There were 125 papers reviewed and 38 papers met the inclusion criteria. The inclusion criteria were: articles that were fully concerned with distracted pedestrians caused by mobile phones; articles that were fully or partially concerned with postural stability; articles that were fully concerned or partially concerned with center of pressure, multiscale entropy, and multivariate multiscale entropy; articles that were fully concerned with the effect of using mobile phone on postural stability.

Articles were limited to peer-reviewed journals' papers without restrictions regarding language. If the title and/or abstract did not provide enough information to decide whether or not the inclusion criteria were met, the full text of the article was obtained.

## 3 RESULTS AND DISCUSSIONS

### 3.1 *Previous studies*

There are contradictory findings between dual-task studies about postural stability. Some studies reported that secondary task impaired postural stability, while the other studies reported that secondary task did not affect or even enhanced postural stability.

Table 1 and Table 2 summarize several previous studies on postural stability with various secondary tasks. The studies on the effect of mobile phone were conducted by Toppila et al. (2005) and Won (2012). However, Toppila et al. (2005) only focused on the effect of hearing radiation from mobile phone on postural stability, their studies concluded that there was no difference between using mobile phone or not. Won (2012) examined the effect of mobile phone use (texting, talking on the phone, and listening to the music) on static and dynamic postural stability. Static postural stability is the ability to maintain balance on a fixed and firm support (Rieman et al.,1999). Dynamic postural stability is the ability to maintain balance while transitioning from a dynamic to a static state (Goldie et al., 1989 Cited in Wikstrom et al., 2005). The result of the study that conducted by Won (2012) was only texting impaired dynamic postural stability, while the other features of the phone did not impair the postural stability either static or dynamic.

### 3.2 *Lacour's models*

Due to the contradictory results between studies, Lacour et al. (2008) proposed three models to explain the effect of secondary task on postural stability as follows:

#### 3.2.1 *Cross-domain competition model*

In this model, attention is shared between the postural performance and the secondary task. The sharing attention in dual-task activity creates competition between posture and secondary task that will result in the decrease of the postural stability and/or decrease the performance of secondary task. Dual-task activity would be less efficient because of the attentional resource sharing.

#### 3.2.2 *U-shaped non-linear interaction model*

Performing secondary task can be either improve or diminish the postural stability, depending on the level of difficulty of the secondary task. Postural stability improvement in dual-task conditions

Table 1. Previous findings: secondary task impaired postural stability.

| Study | Method | Result |
|---|---|---|
| Shumway-Cook & Woollacott, 2000 | Used a movable force platform to perform balance and mobility function: sit, stand, reach, lean over, turn and look over each shoulder, turn in a full circle, step, and walking. Secondary task: choice reaction time auditory task. | Young adults:sway increased from 3% in the FEO (firm surface-eyes open) to 15% in the SEC (sway surface-eyes closed) condition. Elderly non-fallers: sway increased by 29% in the SEC and 24% in the SOKN (sway surface-optokinetic simulation) condition. Elderly fallers: sway increased from 32% in FEO to 35% in the SEO (sway surface-open eyes) condition. |
| Maylor et al., 2001 | Primary tasks: sitting and standing upright. Secondary tasks: Brooks spatial memory task and nonspatial memory task. | For the antero-posterior: significant effect of cognitive task, $F(2,126) = 3.06$, $P = 0.05$. The interaction between cognitive task and phase was significant, $F(2, 216) = 9.49$. For medio-lateral: significant effect of age group, $F(5, 63) = 2.75$. Interaction between cognitive task and phase was significant, $F(2,126) = 7.07$. |
| Teasdale & Simoneau, 2001 | Primary tasks: sitting, standing upright. For sitting task, subjects did trials in normal vision and normal proprioception; for standing task, subjects did five combinations of vision – no vision and perturbed proprioception – reintegrated proprioception. Secondary task: probe-reaction time. | Proprioception reintegration under no-vision: significant effect of age group $F_{1,14} = 7.2$, $P < 0.05$; condition $F_{33,42} = 70.8$, $P < 0.001$; and significant interaction of group X condition $F_{3,42} = 4.9$, $P < 0.005$. Proprioception reintegration under vision: significant effect of age group $F_{1,14} = 8.16$, $P < 0.01$; condition $F_{3,42} = 30.4$, $P < 0.001$; and significant interaction of group X condition $F_{3,42} = 4.4$, $P < 0.01$. |
| Dault et al., 2003 | Primary tasks: sitting, standing on stable surface, and standing on unstable surface. Secondary tasks: silent, combination (repeating letter immediately after hearing it), articulation, and motor task. | Articulation and combination task increased sway path: $F_{4,72} = 5.41$, $P < 0.003$; root mean square (RMS) AP: $F_{4,72} = 6.68$, $P < 0.0001$; mean power frequency (MPF) AP: $F_{4,72} = 5.07$, $P < 0.01$; root mean square ML: $F_{4,72} = 4.28$, $P < 0.01$; mean power frequency ML: $F_{4,72} = 2.88$, $P < 0.02$. |
| Mitra, 2003 | Primary tasks: standing still with open stance (heels 10 cm apart) and closed stance (feet pressed together). Secondary tasks: visual search tasks with different combinations of reference frames (world and head), search loads (low and high), and target conditions (present and absent). | ML sway was influenced by stance: $F_{1,16} = 224.24$, $p < 0.05$; search load: $F_{1,16} = 4.289$, $P=0.055$. AP sway was influenced by stance: $F_{1,16} = 36$, $P < 0.05$; search load: $F_{1,16} = 11.101$, $P < 0.05$. |
| Pellechia, 2003 | Primary task: standing on foam pad. Secondary tasks: digit reversal, 2-bit classification, and counting backwards by 3s. | Length of COP was greater for the counting backwards than other conditions ($P < 0.01$). Cognitive task influenced sway range for AP: $F_{3,57} = 9.84$, $P < 0.001$ and ML: $F_{3,57} = 3.03$, $P < 0.005$; sway variability for AP: $F_{3,57} = 5.70$, $P < 0.01$. |

Table 2. Previous findings: secondary task did not impair postural stability.

| Study | Method | Result |
|---|---|---|
| Vuillerme et al., 2000 | Primary task: standing upright. Secondary tasks: verbal reaction time (VRT) task to indicate the LED color with two conditions (easy and difficult). | COP range was smaller during than before the secondary task (7.01 vs. 9.17 mm, $P < 0.05$). COP speed was slower during than before the secondary task (8.22 vs. 9.74 mm/s, $P < 0.05$). COP range was smaller after than before the secondary task (7.54 vs. 9.17 mm, $P < 0.05$). COP speed was slower after than before the secondary task (8.23 vs. 9.74 mm/s, $P < 0.05$). |
| Marsh & Geel, 2000 | Primary tasks: standing still with combinations of eyes open and eyes closed on hard or foam surface. Secondary tasks: verbal reaction time (VRT). | Main effect of the task was not significant: $F_{3,81} = 0.703$, $P = 0.553$. No significant interaction between group X task interaction: $F_{3,81} = 1.82$, $P = 0.15$. |
| Andersson et al., 2002 | Primary tasks: standing upright with and without vibratory simulation on half of the trials. Secondary tasks: silent backward counting and silent rating scale of the balance. | Subjects swayed less while doing the mental task compared with standing only $t(29) = 2.6$, $P = 0.014$. Subjects swayed less while doing the silent counting + stimulation compared with stimulation only $t(19) = 2.06$, $P = 0.0554$. |
| Riley et al., 2003 | Primary task: standing upright. Secondary tasks: digit memory tasks with three levels of difficulty (easy, medium, and hard). | Sway variability was significantly less in the medium and difficult conditions than in the easy condition ($P = 0.0004$ and 0.008, respectively). No significant effect of the cognitive task on COP path length. |
| Huxhold et al., 2006 | Primary tasks: sitting, standing still. Secondary tasks: choice-reaction time, delayed-digit recall (2-back), and spatial working memory (2-back) tasks. | The effect of cognitive task on postural stability followed the U-shaped model: $F_{1,37} = 20.99$, $P < 0.05$, $\eta_p^2 = 0.36$ (raw COP area), $F_{1,37} = 20.79$, $P < 0.05$, $\eta_p^2 = 0.36$ (log-transformed COP area). For raw COP area, there was interaction of quadratic age X cognitive demand: $F_{1,37} = 5.15$, $P < 0.05$, $\eta_p^2 = 0.1$. For log-transformed COP area, there was interaction of linear age X cognitive demand: $F_{1,37} = 8.35$, $P < 0.05$, $\eta_p^2 = 0.18$. |
| Simoneau et al., 2008 | Primary tasks: standing still with open stance (heels 40 cm apart) and tandem stance. Secondary task: digit memory task. | In older adults, COP velocity was slower during the difficult secondary task ($P < 0.001$). In younger adults, the performance of secondary task improved in the complex posture ($P = 0.002$). |

can be observed in the low-demanding secondary task, the easy cognitive task can shift the attention from postural stability that results in the improvement of postural stability automatically. However, increasing the difficulty level of secondary task induces the increase in the postural sway.

3.2.3 *Task prioritization model*
The last model is the task prioritization model, which hypothesize that older adults tend to do the "posture first", because they prioritize postural control over cognitive performance. This model is referring into the SOC (Selection of the crucial goal; Optimization of the performance of the selected goal; and Compensation for the alternative strategies) model of Baltes (1990). According to this model, older adults predominantly use the hip strategy(using the hip joints to maintain balance) to keeps the center of gravity to avoid falls. This strategy is more demanding than the ankle strategy (using the ankle muscles to maintain balance) that often used by the younger adults. However, the hip strategy is safer for the older adults.

Regardless of the differences in the parameters to evaluate postural performance in the previous studies, three models proposed by Lacour et al. above are more likely cover the contradictory findings of the postural stability studies.

## 3.3 *Postural stability measurements*

Postural stability has been studied in various ways with a range of different measurements. In the past, the most typical analysis was usually using linear methods. However, the output signals from human body are dynamic, non-linear and non-stationary (Costa et al., 2005).

3.3.1 *Center of Pressure*
Center of Pressure (COP) is "the position of the applied force vector that is influenced by the shear forces produced by body segment accelerations" (Hasan et al., 1996). The COP data are often used to analyze postural stability because it can be obtained from a force platform directly (Hasan et al., 1996). The signal of the COP has bivariate distribution and defined by ML (medio-lateral, x-axis) and AP (antero-posterior, y-axis). The COP is defined as the arithmetic means of $y_0(n)$ and $x_0(n)$.

$$y(n) = y_0(n) - \bar{y}, \quad x(n) = x_0(n) - \bar{x}$$
$$\text{where,} \quad \bar{y} = \frac{1}{N}\sum y_0(n), \quad \bar{x} = \frac{1}{N}\sum x_0(n), \quad n = 1, \ldots, N \tag{1}$$

The COP stabilogram that often used are: mean resultant distance (MDIST), total excursion (TOTEX), mean displacement velocity (MVELO), sway area (AREA-SW), the 95% confidence circle area (AREA-CC), the 95% confidence ellipse area (AREA-CE), square root of the sum of the displacement variances in x (Rsw) and y (Rfa), and planar deviation (PD).

3.3.2 *Multiscale Entropy*
As mentioned above, the signal data from force platform are dynamic, non-linear, and non-stationary. Thus, it is needed to find another method that more appropriate to analyze postural stability. Costa et al. (2002) proposed multiscale entropy (MSE) to measure complexity by calculating SampEn as a function of scale, converting the original time series into coarse-grain data and then compute the entropy for every scale and map its dependence on the scale.

$$y_j^{(\tau)} = {}^{1}/_{\tau} \sum_{i=(j-1)\tau+1}^{j\tau} x_i \tag{2}$$

To filter the noise of the signal data, Huang et al. (1998) developed empirical mode decomposition (EMD), a technique of signal processing with the assumption that the signal consists of groups of intrinsic mode functions (IMFs) that is defined as a function that satisfies: (1) In the whole time series, the number of extrema and the number of zero-crossings must either be equal or differ at most by one; (2) At any signal point, the mean value of the envelope defined by the local maxima and the envelope defined by the local minima is zero.

$$X(t) = \sum_{i=1}^{n} c_i + r_n \tag{3}$$

Jiang et al. (2013) evaluated the potential of using EMD-enhanced MSE on postural stability study. The results showed that MSE can be applied to analyze COP data. The advantage of using EMD-enhanced MSE compared to the traditional methods is that EMD can decompose COP time series data into different frequency signals to characterize higher and lower frequency signals. However, the physical meaning of the IMF is unclear.

### 3.3.3 *Multivariate Multiscale Entropy*

MSE analysis can only consider data channel separately which is only appropriate if the multivariate signals are independent and there is no correlation statistically (Ahmed & Mandic, 2012). Multiscale multivariate entropy (MMSE) can evaluate the structural complexity of multivariate systems by calculating the relative complexity of the multichannel signals through the plot of the multivariate sample entropy (Ahmed & Mandic, 2011).

Multivariate extension of EMD was proposed to cater for a general class of multivariate signals (Rehman & Mandic, 2009). Multivariate empirical mode decomposition (MEMD) is able to extract common rotational modes across the signal components.

Previous studies by Wei et al. (2012) and Huang et al. (2013) showed that MEMD-enhanced MMSE is a very promising method to analyze postural stability. MEMD-enhanced MMSE is more sensitive to the changes and able to distinguish different sways more clearly. However, there is difficulty in determining which IMF includes the valuable physiological features in bio-signals, bias problem when increasing the scales of MSE, and the calculation is time consuming and memory-intensive (Wei et al., 2012).

## 4 CONCLUSIONS

The suggestions for future studies are: (1) Studies of the effect of using mobile phone on postural stability are needed to be evaluated more to reduce the risk of falls; (2) Studies should be directed towards testing the Lacour's models, in order to evaluate which model is the most suitable for postural stability studies; (3) Using the same parameters of assessments or even the similar protocols to get reliable comparison between studies; (4) Using various task conditions and measurements to get better understanding of the effect of secondary task on postural stability; (5) The difficulties between tasks should be made in the sufficiently different levels to make the attention-sharing cost of the main task measureable.

## REFERENCES


Ahmed, M.U. & Mandic, D.P. 2012.Multivariate multiscale entropy analysis. *IEEE Signal Processing Letters, 19(2)*, 91–94.

Ahmed, M.U. & Mandic, D.P. 2011. Multivariate multiscale entropy: A tool for complexity analysis of multichannel data. *Physical Review, 84*, 061918.

Andersson, G., Hagman, J., Talianzadeh, R., Svedberg, A., Larsen, H.C. 2002. Effect of cognitive load on postural control. *Brain Research Bulletin, 58(1)*, 135–139.

Costa, M., Goldberger, A.L., Peng, C.K. 2002.Multiscale entropy analysis of complex physiologic time series. *Physical Review Letters, 89(6)*, 068102-1-4.

Costa, M., Goldberger, A.L., Peng, C.K. 2005.Multiscale entropy analysis of biological signals. *Physical Review E, 71*, 021906-1-18.

Bungum, T.J., Day, C., Henry, J. 2005. The association of distraction and caution displayed by pedestrians at a lighted crosswalk. *Journal of Community Health, 30(4)*, 269–278.

Dault, M.C., Frank, J.S., Allard, F. 2001. Influence of a visuo-spatial, verbal and central executive working memory task on postural control. *Gait and Posture, 14*, 110–116.

Dault, M.C., Yardley, L., Frank, J.S. 2001.Does articulation contribute to modifications of postural control during dual-task paradigms? *Cognitive Brain Research, 16*, 434–440.

Ehrenfried, T., Guerraz, M., Thilo, K.V., Yardley, L., Gresty, M.A. 2003.Posture and mental task performance when viewing a moving visual field. *Cognitive Brain Research, 17*, 140–153.

Fraizer, E.V. & Mitra, S. 2008. Methodological and interpretive issues in posture-cognition dual-tasking in upright stance. *Gait and Posture, 27*, 271–279.

Hasan, S.S., Robin, D.W., Szurkus, D.W., Ashmead, D.H., Peterson,S.W., Shiavi, R.G. 1996. Simultaneous measurement of body center of pressure and center of gravity during upright stance. Part I: methods. *Gait and Posture, 4*, 1–10.

Hatfield, J. & Murphy, S. 2007. The effects of mobile phone use on pedestrian crossing behavior at signalised and unsignalised intersections. *Accident Analysis and Prevention, 39*, 197–205.

Hsu, W.L. & Scholz, J.P. 2012. Motor abundance supports multitasking while standing. *Human Movement Science, 31*, 844–862.

Huang, C.Y., Cherng, R.J., Hwang, I.S. 2010. Reciprocal influences on performances of a postural-suprapostural task by manipulating the level of task-load. *Journal of Electromyography and Kinesiology, 20*, 413–319.

Huang, C.W., Sue, P.D., Abbod, M.F., Jiang, B.C., Shieh, J.S. 2013. Measuring center of pressure signals to quantify human balance using multivariate multiscale entropy by designing a force platform. *Sensors, 13*, 10151–10166.

Huxhold, O., Li, S.C., Schmiedek, F., Lindenberger, U. 2006. *Brain Research Bulletin, 69*, 294–305.

Jiang, B.C., Yang, W.H., Shieh, J.S., Fan, J.S.Z., Peng, C.K. 2013. Entropy-based method for COP data analysis. *Theoretical Issues in Ergonomics Science,14(3)*, 227–246.

Lacour, M., Bernard-Demanze, L., Dumitrescu, M. 2008. Posture control, aging, and attention resources: Models and posture-analysis methods. *Clinical Neurophysiology, 119*, 411–421.

Lamberg, E.M. & Muratori, L.M. 2012. Cell phones change the way we walk. *Gait and Posture, 36(3)*, 655.

Maylor, E.A., Allison, S., Wing, A.M. 2001. Effects of spatial and nonspatial cognitive activity on postural stability. *British Journal of Psychology, 92*, 319–338.

Marsh, A.P. & Geel, S.E. 2000.The effect of age on the attentional demands of postural control. *Gait and Posture, 12*, 105–113.

Mitra, S. 2003. Postural costs of suprapostural task load. *Human Movement Science, 22*, 253–270.

Nasar, J., Hecht, P., Wener, R. 2008. Mobile telephones, distracted attention, and pedestrian safety. *Accident Analysis and Prevention, 40*, 69–75.

Neider, M.B., McCarley, J.S., Crowell, J.A., Kaczmarski, H., Kramer, A.F. 2010.Pedestrians, vehicles, and cell phones. *Accident Analysis and Prevention, 42*, 589–594.

Palluel, E., Nougier, V., Olivier, I. 2010. Postural control and attentional demand during adolescence. *Brain Research, 1358*, 151–159.

Park, S.H., Lee, K., Lockhart, T., Kim, S. 2011.Effect of sound on postural stability during quiet standing. *Journal of Neuroengineering and Rehabilitation, 8*:67.

Pellecchia, G.L. 2003. Postural sway increases with attentional demands of concurrent cognitive task. *Gait and Posture, 18*, 29–34.

Prado, J.N., Stoffregen, T.A., Duarte, M. 2007. Postural sway during dual-tasks in young and elderly adults. *Gerontology, 53*, 274–281.

Rehman, N. & Mandic, D.P. 2009. Multivariate empirical mode decomposition. *Proceedings of the Royal Society*, 10.1098/rspa.2009.0502.

Rehman, N. & Mandic, D.P. 2010. Filter bank property of multivariate empirical mode decomposition. *IEEE Trans. Signal Proceeding, 59*, 2421–2426.

Rieman, B.L., Caggiano, N.A., Lephart, S.M. 1999. Examination of a clinical method of assessing postural control during a functional performance task. *Journal of Sport Rehabilitation, 8*, 171–183.

Riley, M.A., Baker, A.A., Schmit, J.M. 2003.Inverse relation between postural variability and difficulty of a concurrent short-term memory task. *Brain Research Bulletin, 62*, 191–195.

Schwebel, D.C., Stavrinos, D., Byington, K.W., Davis, T., O'Neal, E.E., de Jong, D. 2012. Distraction and pedestrian safety: How talking on the phone, texting, and listening to music impact crossing the street. *Accident Analysis and Prevention, 45*, 266–271.

Shumway-Cook, A. & Woollacott, M. 2000. Attentional demands and postural control: The effect of sensory context. *Journal of Gerontology, 55A(1)*, M10–M16.

Simoneau, E.M., Billot, M., Martin, A., Perennou, D., Van Hoecke, J.2008. Difficult memory task during postural tasks of various difficulties in young and older people: A pilot study. *Clinical Neurophysiology, 119*, 1158–1165.

Stavrinos, D., Byington, K.W., Schwebel, D.C. 2011. Distracted walking: Cell phones increase injury risk for college pedestrians. *Journal of Safety Research, 42*, 101–107.

Teasdale, N., Simoneau, M. 2001. Attentional demands for postural control: the effects of aging and sensory reintegration. *Gait and Posture, 14*, 203–206.

Toppila, E., Forsman, P., Pyykko, I., Starck, J. 2005. Use of mobile phones and its acute effects on postural stability and hearing. *Gait and Posture, 21*, S16.

Vuillerme, N., Nougier, V., Teasdale, N. 2000. Effects of a reaction time task on postural control in humans. *Neuroscience Letters, 291*, 77–80.

Wei, Q., Liu, D.H., Wang, K.H., Liu, Q., Abbod, M.F., Jiang, B.C., Chen, K.P., Wu, C., Shieh, J.S. 2012. Multivariate multiscale entropy applied to center of pressure signals analysis: An effect of vibration stimulation of shoes. *Entropy, 14*, 2157–2172.

Wikstrom, E.A., Tillman, M.D., Smith, A.N., Borsa, P.A. 2005. A new force-plate technology measure of dynamic postural stability: The dynamic postural stability index. *Journal of Athletic Training, 4*, 305–309.

Woollacott, M. & Shumway-Cook, A. 2002. Attention and the control of posture and gait: a review of an emerging area of research. *Gait and Posture, 16*, 1–4.

Won, J.I. 2012.Effects of using a mobile phone on postural control. *Physical Therapy Korea, 19(3)*, 61–70.

*Bridging Research and Good Practices towards Patient Welfare – Shih & Liang (Eds)*
 *ISBN 978-1-138-02716-9*

# A heuristic evaluation on the usability of health information websites

Y.N. Chen & S.L. Hwang
*Department of Industrial Engineering and Engineering Management, National Tsing Hua University, Hsinchu, Taiwan*

ABSTRACT: A heuristic evaluation on the usability of three existing health information websites in Taiwan was conducted in this study. Interface problems of health information websites were identified through evaluation against five evaluation items we generalized from two recognized rules. Design suggestions for health information websites are proposed so that common weaknesses can be avoided in the future. By improving the overall usability, users' satisfaction with the websites will increase and knowledge of health issues can be promoted.

*Keywords*: Usability, health Information, heuristic evaluation

## 1 INTRODUCTION

### 1.1 *Background*

According to Research, Development and Evaluation Commission (RDEC)'s 2013 investigation in Taiwan, more than 85 percent of households have Internet access and the percentage of individuals with access to the Internet exceeds 75 percent. With the Internet's rapid development, websites have been widely used to convey information. The field of human-computer interaction (HCI), i.e. the interaction between users and computers, has become a major issue for the web designers.

As population structure has changed, the government has put the health and well-being as one of the most important policies. Taiwan's official completed the restructuring of the Department of Health as the Ministry of Health and Welfare in 2013. To keep the public informed of its health knowledge and vision, Taiwan's ministry of Health and Welfare and other institutions have established many relative websites about medical and healthcare knowledge. However, most of these websites carry a high information load and sometimes bore users. Hence, to increase the public's interest, website designers should consider design guidelines and user opinions to achieve a user friendly interface. Poor usability and lack of attractiveness may be barriers to user satisfaction (Flavián et al., 2006). Enhancing usability and increasing user interest are critical to web pages that carry a high information load, such as healthcare and medical knowledge websites.

### 1.2 *Objectives*

The health-related websites in Taiwan have some interface usability problems. Thus, the objectives of this research are as follows:

(1) Enhance the usability of web pages through heuristic method.
(2) Propose design suggestions for such high information load web pages.

In the information age, digital eye strain is concerned due to a variety of consumer electronics and over-time digital device use. To reduce the cognitive load, it is important to find a better way to arrange the information well for the websites which have high information load, especially the educational and healthcare promotional websites.

### 1.3 *Heuristic evaluation and website design principle*

Many techniques can be employed to evaluate the usability of websites. One of them is heuristic evaluation. The advantages of heuristic evaluation include the intuitiveness and rapidity. It's a cheap and easy way to learn and to conduct. A few general guidelines can correct for the majority of usability problems and it can be done by a few usability experts (Holzinger, 2005).

Heuristic evaluation involves having a small set of expert evaluators judge whether the interface elements follow established usability principles. Ten usability heuristics for user interface design proposed by Nielsen (1994) and eight golden rules of interface design proposed by Shneiderman (2010) are commonly used for evaluating website usability.

a. Nielsen's Ten Usability Heuristics for User Interface Design
   (1) Visibility of system status.
   (2) Match between system and the real world.
   (3) User control and freedom.
   (4) Consistency and standards.
   (5) Error prevention.
   (6) Recognition rather than recall.
   (7) Flexibility and efficiency of use.
   (8) Aesthetic and minimalist design.
   (9) Help users recognize, diagnose, and recover from errors.
   (10) Help and documentation.
b. Shneiderman's Eight Golden Rules of Interface Design
   (1) Strive for consistency.
   (2) Cater to universal usability.
   (3) Offer informative feedback.
   (4) Design dialogs to yield closure.
   (5) Prevent errors.
   (6) Permit easy reversal of actions.
   (7) Support internal locus of control.
   (8) Reduce short-term memory load.

## 2 METHODS

Five evaluation items are developed in this research which extracted from 10 Usability Heuristics for User Interface Design and 8 Golden Rules of Interface Design. In this paper, we use the heuristic evaluation to evaluate the existing health information websites to point out the pros and cons, and then propose design suggestions to improve the usability of health information websites.

### 2.1 *Five evaluation items for website design*

There are some similarities between these two recognized usability principles described in previous section. We generalized 5 evaluation items from 10 heuristics and 8 golden rules. Table 1 shows the five evaluation items and corresponding heuristics and rules.

The five evaluation items for website design are important issues which designers should consider. The evaluation items fit most conditions and can be applied to evaluation of multiple types of websites. The items and related explanations are as follows:

(1) Consistency
    Use consistent modes and actions in similar situations, including words, fonts, colors, menus, and layout.

Table 1. Evaluation items for website design.

| Evaluation items | Nielsen's ten heuristics | Shneiderman's eight golden rules |
|---|---|---|
| Consistency | a4. Consistency and standards. | b1. Strive for consistency. |
| Usability | a7. Flexibility and efficiency of use.<br>a10. Help and documentation. | b2. Cater to universal usability. |
| Controllability | a1. Visibility of system status.<br>a3. User control and freedom. | b3. Offer informative feedback.<br>b6. Support internal locus of control. |
| Error prevention | a5. Error prevention.<br>a9. Help users recognize, diagnose, and recover from errors. | b5. Prevent errors.<br>b7. Permit easy reversal of actions. |
| Cognitive load | a2. Match between system and the real world.<br>a6. Recognition rather than recall.<br>a8. Aesthetic and minimalist design. | b8. Reduce short-term memory load. |

(2) Usability
Consider the needs of diverse users. Use the methods that are most appropriate for all users. For example, provide shortcuts for experts and guidance for novices.

(3) Controllability
Let users control the system. Keep users informed about the system through appropriate feedback so that they understand what is happening on the website.

(4) Error prevention
Ensure that the interface prevents users from making serious errors and provides reversible actions for recovery.

(5) Cognitive load
Minimize users' memory load by following real-world conventions and making instructions, objects, actions, options and web-site locations visible so that users do not have to remember too many things.

### 2.2 *Heuristic evaluation*

A heuristic evaluation of the usability of health information websites was conducted in this study by three evaluators with experience in designing web pages. The three usability evaluators received a heuristic evaluation checklist containing five evaluation items for website design.

Three health information websites as our case study are:

a. Formosans' e Medical School: http://fms.mohw.gov.tw/
b. Health Resources Center: http://health.ntuh.gov.tw/health/hrc_v3/
c. 24Drs.com: http://www.24drs.com/

## 3 RESULTS

The results from the heuristic evaluations can be compared as a reference for improving the website usability. All of them have some good designs, such as: (i) the attribute of hyperlink object will change when cursor moving on, (ii) provide site map and (iii) show the date that the website last updated. Nevertheless, there are some usability problems among the three interfaces.

a. Formosans' e Medical School
Formosans' e Medical School (Figure 1) provides online courses for e-learning and consultation service. The interface problems identified by heuristic evaluation are listed in Table 2.

Figure 1. Formosans' e Medical School.

Table 2. Interface problems of Formosans' e Medical School.

| Evaluation items | Problems |
|---|---|
| Consistency | – Users cannot return to the home page through clicking the logo at the top left of the page.<br>– Alignment types used for menu items are not consistent.<br>– The style of home page is not consistent with other pages.<br>– The integration of two resources, "Formosans' e Medical School" and "Taiwan e Doctor", is not effectively shown. |
| Usability | – Staff's message board which is not open to the public should not be put in one of the most important position of menu. |
| Controllability | – No indicating where the hyperlink is linked to, a file or an external website.<br>– Most of links in the menu open the page in new window, but not in the original one. |
| Error prevention | – The content of menu item is split into two lines.<br>– No checking the accuracy of required data when member registering.<br>– Voting repeatedly is allowed in voting system and overdue vote is not closed. |
| Cognitive load | – Web page locations are not visible.<br>– Five different bulletin boards make user confused. |

b. Health Resources Center
   Health Resources Center (Figure 2) provides health knowledge to the public and patients. The interface problems identified by heuristic evaluation are listed in Table 3.
c. 24Drs.com
   The website 24Drs.com (Figure 3) provides diverse medical news and health knowledge to the public. The interface problems identified by heuristic evaluation are listed in Table 4.

We took from the list of five evaluation items on heuristic evaluation. The number of problems identified against the heuristics is shown in Table 5.

Figure 2. Health Resources Center.

Table 3. Interface problems of Health Resources Center.

| Evaluation items | Problems |
|---|---|
| Consistency | – Users cannot return to the home page through clicking the logo at the top left of the page. |
| | – Alignment types used for menu items are not consistent. |
| | – The style of home page is not consistent with other pages. |
| | – The layout, font-size, and font-type of each page are not consistent. |
| | – The dates the articles posted are not in a consistent way. |
| | – Single-byte punctuation marks are used in Chinese sentence. |
| Usability | – No shortcuts provided. |
| Controllability | – No indicating where the hyperlink is linked to, a file or a page. |
| | – No indicating the page number of the health knowledge. |
| Error prevention | – Overdue lectures are not marked. |
| | – The font-color of content text and hyperlink text are too similar to distinguish. |
| Cognitive load | – No redundant pictures are provided. |
| | – There is no logical grouping of items. |

The problem which is discovered in more than one interface can be regarded as common weakness. For example, designers often neglect the consistency of alignment type, date format, font type and the website style. The common weaknesses are as in Table 6.

Based on the results from heuristic evaluation, we propose some design suggestions for web designers to follow. Designers can practice the suggestions on the elements, including layout, menu, date, hyperlink, navigation, use for graphics, and professional knowledge. Design suggestions for health information websites are shown in Table 7.

Figure 3. 24Drs.com.

Table 4. Interface problems of 24Drs.com.

| Evaluation items | Problems |
|---|---|
| Consistency | – The dates the articles posted are not in a consistent way. |
| Usability | – No consider the users using different browser. |
| Controllability | – N/A |
| Error prevention | – The font-color of icons and their background are too similar to distinguish. |
| | – Partial functions and pages are not available. |
| | – No checking the accuracy of required data when member registering. |
| | – The explanation of account register is not precise. |
| Cognitive load | – Web page locations are not visible. |
| | – Too much items shown in the left menu. |
| | – Screens are too cluttered with fields. |

Table 5. The number of problems found in each of the three interfaces.

| Name of interface | e Medical School | Resources Center | 24Drs |
|---|---|---|---|
| Problems found | 12 | 13 | 9 |
| Evaluation items: | | | |
| Consistency | 4 | 6 | 1 |
| Usability | 1 | 1 | 1 |
| Controllability | 2 | 2 | N/A |
| Error prevention | 3 | 2 | 4 |
| Cognitive load | 2 | 2 | 3 |

Table 6. Common weaknesses found to be lacking on the websites.

| Evaluation items | Common weaknesses |
|---|---|
| Consistency | – Users cannot return to the home page through clicking the logo at the top left of the page.<br>– Alignment types used for menu items are not consistent.<br>– The style of home page is not consistent with other pages.<br>– The dates the articles posted are not in a consistent way. |
| Usability | – N/A |
| Controllability | – The destination or file type of hyperlink is not well described.<br>– No appropriate feedback for every action. |
| Error prevention | – No checking the accuracy of required data when member registering. |
| Cognitive load | – Web page locations are not visible. |

Table 7. The design suggestions for health information websites.

| Design elements | Suggestions |
|---|---|
| Layout | – The style of each web page should be consistent, including the font type and the arrangement of content in similar situation.<br>– The logo with the name of website is usually at the top left of every page, and the logo should be linked to home page. |
| Menu | – Make logical grouping of items and avoid showing too much items in a menu.<br>– Consistent alignment type should be used.<br>– Avoid splitting the content of menu item into two lines.<br>– Avoid using external links in menu. |
| Date | – State the dates the articles posted in a consistent way.<br>– Show the date that the website last updated. |
| Hyperlink | – Avoid using similar color for text content, hyperlink.<br>– Provide appropriate feedback for every user actions.<br>– Indicating where the hyperlink is linked to, a file or an external website.<br>– Not to open internal links in new window, unless necessary. |
| Navigation | – Provide shortcuts, search functionality and site map.<br>– Web page locations should be visible, including page title and page number. Let users know where they are. |
| Use for graphics | – Provide redundant pictures or icons which are meaningful. Avoid using irrelevant images. |
| Professional knowledge | – FAQ or consultation is necessary.<br>– The reference should be provided. |

## 4 DISCUSSION

Based on the five evaluation items which integrating the most commonly used rules, interface problems of three health information websites have been identified through heuristic evaluation. Some usability problems are discovered in more than one websites. These common weaknesses can be avoided in the future according to the design suggestions. The results in this article provide design suggestions for health information websites, and these can be a reference for other knowledge websites as well. These suggestions can be refined or extended for adapting to different type of websites. By improving the overall website usability, user satisfaction will increase and the knowledge of health can be promoted.

## ACKNOWLEDGEMENT

Partial of this study was supported by Advanced Manufacturing and Service Management Research Center at National Tsing Hua University through Toward World-Class Universities Project (ICT Product Interface Design and System Innovation, Project Number: 101N7071E1).

## REFERENCES

24Drs.com 2014. Available at: http://www.24drs.com/ (Accessed: 9 Jan 2014).

Flavián C., Guinalíu M. & Gurrea R. 2006. The role played by perceived usability, satisfaction and consumer trust on website loyalty, Information & Management 43(1):1–14.

Formosans' e Medical School 2014. Available at: http://fms.mohw.gov.tw/ (Accessed: 9 Jan 2014).

Health Resources Center 2014. Available at: http://health.ntuh.gov.tw/health/hrc_v3/ (Accessed: 9 Jan 2014).

Holzinger, A. 2005. Usability engineering methods for software developers. Comm. of ACM. 48(1):71–74.

Nielsen, J. & Mack, R. L. (eds.) 1994. Usability Inspection Methods. New York: John Wiley & Sons, Inc.

Shneiderman, B. & Plaisant, C. 2010. Designing the User Interface: Strategies for Effective Human-Computer Interaction: Fifth Edition, Boston: Addison-Wesley Publ. Co.

*Bridging Research and Good Practices towards Patient Welfare – Shih & Liang (Eds)*
*© 2015 Taylor & Francis Group, London, ISBN 978-1-138-02716-9*

# Investigation on lifestyle factors that influence obesity of adult workers

N. Takayama
*University of KinDAI Himeji, Himeji, Hyogo, Japan*

T. Amemiya
*Faculty of Sociology, Kansai University, Suita, Osaka, Japan*

K. Nishikawa
*Graduate School of Psychology, Kansai University, Suita, Osaka, Japan*

Y. Suzaki
*Japanese Red Cross Kyushu International College of Nursing, Munakata, Fukuoka, Japan*

H. Ariyoshi
*Saga University, Saga, Saga, Japan*

ABSTRACT: We conducted a survey to clarify the stress factors through eating behaviors and the lifestyle influencing adult workers' obesity. We used the questionnaire by Tanbo and Inagaki (2009). The number of the subjects was 483 males and 119 females. The mean age was 32.9. The factors were exercise habit, stress control, health control, weight concern and leftover eating. The reliability values were above 0.63 respectively. These of seven scales with eating behavior and sleep were above 0.68 respectively (Takayama et al. 2012). DEBQ scales were used (Van Strien et al. 1986). BMI showed positive correlation with weight concern, eating of leftover, quick eating and high calorie food, showing negative correlation with eating between meals, dietary likes and dislikes and insomnia. Eating of leftover, quick eating and high calorie food were significantly high in male and eating between meals was significantly high in female. Restrained eating showed positive correlation with health control.

## 1 BACKGROUND AND PURPOSE

In Japan the life of adult workers is usually job-centered and continue to work in that way until middle ages, notwithstanding with changes of her/his personal and public environment around her and him. According to Tanbo and Inagaki (2009), the Japanese workers continue to work until 35 years old without being given the opportunity to take the company's annual medical checks that should include blood tests, or an ECG test or ventral girth measurement after taking medical checks at the time of employment. The research reveals how the health of young workers are neglected by employers, although these are the ages when people often suffer from the lifestyle-related diseases.

Sleep, dietary, exercises, smoking and alcohol drinking in daily life have a great influence on their health. They have very much to do not only with the three fatal diseases as cancer, heart diseases and cerebral diseases but also with the serious health problems as arterial sclerosis, hyperglycemia, diabetes and metabolic syndrome. Proper exercises are effective for both prevention and cure and also for better quality of life. The concern about lifestyle and health leads the workers to appropriate lifestyle.

The purpose of this research is to clarify the lifestyle factors that influence on obesity of adult workers.

Table 1. Number in age groups and BMI, n = 602.

| | BMI Category | | | |
|---|---|---|---|---|
| | Underweight | Normal weight | Over weight | Total (%) |
| 10~19 | 0 | 4 | 0 | 4 (0.6%) |
| 20~29 | 21 | 159 | 39 | 219 (36.4%) |
| 30~39 | 21 | 216 | 50 | 287 (47.7%) |
| 40~49 | 5 | 48 | 18 | 71 (11.8%) |
| 50~59 | 1 | 14 | 3 | 18 (3.0%) |
| 60 | 0 | 3 | 0 | 3 (0.5%) |
| Males | 23 | 356 | 104 | 483 (80.2%) |
| Females | 25 | 88 | 6 | 119 (19.8%) |

## 2 METHODS

A survey was conducted on the intranet of a manufacturing company, from November 7 to 25, 2011. Participants were 602 adult workers. 483 of them were males (80.2%) and 119 were females (19.8%). The mean age was 32.9 years old. The researchers explained the purpose of the research and promised to observe the privacy protection policy. Every subject consented to participate in this survey and to use the information gained from it as promised.

Based on BMI (Body Mass Index), participants were classified as underweight (BMI ≦ 18.5), normal weight (18.5 > BMI ≦ 25) and overweight (BMI > 25) according to the criterion of obesity by Japanese Society for the Study of Obesity. 48 of them were classified as underweight and 110 were classified as overweight. We used 48 items of the lifestyle questionnaire by Tanbo and Inagaki (2009) developed to investigate the factors leading to the increase of body weight of young workers. We revised a part of it so that it could be applied to both sexes. Each question had six choices such as 'never', 'rarely', 'yes & no', 'sometimes', 'often' and 'always'.

IBM SPSS Statistics Version 20 was used for data analysis. We used the eating and sleep questionnaire with 37 question items developed by Takayama et al. (2012). We also used the Japanese version of Dutch Eating Behavior Questionnaire (DEBQ) which consists of three scales such as 'Emotional Eating', 'Restrained Eating' and 'External Eating' (Van Strien et al. 1986), with 33 question items.

## 3 RESULTS

### 3.1 *The subjects*

The age groups and the number of BMI groups are shown in Table 1. The overweight was 50 (80.3%) and the age group was the thirties.

### 3.2 *Result of factor analysis*

Factor stress 0.35 and more was adopted as item standard and five factor analyses were adopted. The accumulative contribution ratio was 35.0%. The results of factor analysis are shown as follows.

The results of the factor analysis of lifestyle survey are shown in Table 2. The 1st factor was exercise, the 2nd was stress control, the 3rd was health control, the 4th was weight concern and the 5th was eating of leftover. As for the reliability value, exercise was $\alpha = 0.820$ (7 items), stress control $\alpha = 0.728$ (8 items), health control $\alpha = 0.713$ (6 items), weight concern $\alpha = 0.635$ (4 items), eating of leftover $\alpha = 0.627$ (3 items), which showed enough reliability.

Table 2. The results of factor analysis and the lifestyle.

| Factor | Items | Factors | $\alpha$ value |
|---|---|---|---|
| 1st factor | 7 items | Exercise | $\alpha = .820$ |
| 2nd factor | 8 items | Stress Control | $\alpha = .728$ |
| 3rd factor | 6 items | Health Control | $\alpha = .713$ |
| 4th factor | 4 items | Weight Concern | $\alpha = .635$ |
| 5th factor | 3 items | Eating of Leftover | $\alpha = .627$ |

Table 3. Correlation of BMI, lifestyle and gender, n = 602.

| BMI | Exercise | Stress control | Health control | Weight concern | Leftover eating |
|---|---|---|---|---|---|
| BMI (n = 602) | −0.018 | 0.038 | 0.045 | **0.319**** | **0.290**** |
| Male (n = 483) | −0.001 | 0.030 | 0.085 | **0.356**** | **0.216**** |
| Female (n = 119) | 0.123 | **−0.189*** | 0.136 | **0.279**** | **0.252**** |

*$p < 0.05$, **$p < 0.01$.

Table 4. Gender comparison of life style scales.

| | Mean value (*SD*) | | |
|---|---|---|---|
| | Male (n = 483) | Female (n = 619) | *p* value |
| Exercise | 3.6 (1.24) | 3.9 (1.21) | 0.001 |
| Stress Control | 3.5 (0.92) | 3.2 (0.95) | 0.001 |
| Health Control | 3.5 (0.96) | 3.8 (0.88) | 0.001 |
| Weight Concern | 3.2 (1.11) | 3.2 (1.13) | ns |
| Eating of Leftover | 3.9 (1.17) | 3.1 (1.14) | 0.001 |

### 3.3 *Correlation of BMI, lifestyle and gender*

The correlation of BMI, lifestyle and gender is shown in Table 3.

BMI had a slight positive correlation with weight concern and leftover eating especially with the males.

## 4 GENDER COMPARISON OF LIFESTYLE SCALES

Gender comparison of lifestyle scales is shown in Table 4. In stress control and leftover eating, males showed meaningful high values ($p < 0.001$). In exercise habit and health control, females showed high meaningful values ($p < 0.001$ respectively).

## 5 DISTRIBUTION ANALYSIS OF LIFESTYLE BY BMI CLASSIFICATION

The factor distribution analysis with the BMI classes as independent variable and each lower measurement as dependent variable showed slight difference in stress control and meaningful difference in weight concern and eating of leftover. $F(2.599) = 31.482, p < 0.01, F(2.599) = 28.010, p < 0.01$. No meaningful difference was noticed in exercise and health control.

The distribution analysis of lifestyle by BMI classification is shown in Table 5.

Table 5. Distribution analysis of lifestyle by BMI classification, n = 602.

| | 1 Underweight mean (*SD*) | 2 Normal weight mean (*SD*) | 3 Overweight mean (*SD*) | *F* value | Multiple analysis |
|---|---|---|---|---|---|
| Exercise | 3.9 (1.14) | 3.6 (1.26) | 3.7 (1.21) | 0.935 | |
| Stress Control | 3.4 (0.88) | 3.4 (0.93) | 3.7 (0.99) | **2.922**$^{\dagger}$ | **2 < 3**$^{\dagger}$ |
| Health Control | 3.5 (1.14) | 3.6 (0.94) | 3.6 (0.95) | 0.608 | |
| Weight Concern | 2.5 (1.07) | 2.1 (1.06) | 3.8 (1.06) | **31.482*** | **1 > 2***, 1 < 3*** 2 < 3*** |
| Left-over Eating | 2.8 (1.26) | 3.6 (1.16) | 4.3 (1.12) | **28.010*** | **1 < 2***, 1 < 3*** 2 < 3*** |

$^{\dagger}p < 0.1$, $^{***}p < 0.001$.

Table 6. Correlation of lifestyle with EBSQ and DEBQ, n = 602.

| | Exercise | Stress control | Health control | Weight concern | Leftover eating |
|---|---|---|---|---|---|
| Balanced Diet | 0.061 | **−0.187** | **0.318** | 0.061 | 0.052 |
| Between Meals | 0.068 | 0.087* | −0.015 | 0.074 | −0.017 |
| High Calorie Food | **0.156** | **0.334** | **−0.290** | **0.185** | **0.208** |
| Convenience Food | **0.106** | **0.342** | **−0.219** | 0.067 | 0.080* |
| Quick Eating | **0.112** | **0.174** | −0.080* | **0.193** | **0.214** |
| Dietary Likes and Dislikes | **0.124** | 0.081* | **−0.196** | 0.005 | **−0.298** |
| Insomnia | **0.297** | 0.071 | **−0.157** | 0.036 | 0.019 |
| Emotional Eating | 0.085* | 0.067 | 0.080* | **0.200** | 0.085* |
| Restrained Eating | **−0.244** | **−0.184** | **0.568** | 0.052 | −0.096* |
| External Eating | **0.166** | **0.287** | −0.038 | **0.119** | **0.118** |

$^{*}p < .05$, $^{**}p < .01$.

## 6 CORRELATION OF LIFESTYLE SCALE WITH EBSQ AND DEBQ

The correlation of lifestyle with 'exercise', 'stress control', 'weight concern', and 'eating of leftover' is shown in Table 6.

## 7 DISCUSSION

With the factor analysis, five factors such as 'exercise', 'stress control', 'health control', 'weight concern', 'eating of leftover' came up. These factors came up as what control lifestyle, which is the same result of Tanbo and Inagaki (2009). BMI correlates with 'weight concern' and 'eating of leftover'.

Leftover Eating correlates with quick eating and high calorie food, which suggests the relationship with obesity. BMI correlates with 'weight concern'. 'Weight concern' and eating of leftover' correlate more in female. This is probably because females cook at homes. When we realize that in very many Japanese homes, wives are still supposed to cook for family members at home. There is a supposition that gender equality not only at working places but also at home among family members may help establish the healthy lifestyle of working people. 'Health control' correlates with 'balanced dietary' and showed negative correlation with 'high calorie food' and 'convenience food'. Health concern' is important for preventing lifestyle diseases by controlling lifestyle.

'Stress control' was high in the group of overweight, which suggests the relationship of stress with obesity. It also leads to 'high calorie food and 'convenience food'. This clarifies the eating behavior in stressful condition.

Mental influence on obesity was clarified by the relationship of 'restrained dietary' of DEBQ with 'health concern'.

This study with adult workers clarified that 'eating of leftover' has great influence on obesity. The study about the relationship of eating behavior with the length of working years, working conditions and employment conditions is our future task.

## 8 CONCLUSIONS

(1) The eating behaviors that lead to obesity were found to be 'eating of leftover, 'quick eating' and 'high calorie food'.
(2) The subjects who showed 'health control' had a tendency of 'balanced diet' and tend to avoid 'high calorie food' and 'convenience food'.
(3) The overweight group showed high in 'stress control' and showed a tendency of the eating behaviors of 'high calorie food' and 'convenience food'.
(4) Those who showed 'restrained diet' of DEBQ are found to have a tendency of 'health control'.
(5) In this research of eating behavior, the recognition of the gender equality is found not to be ignored, though it was not the first expectation.

## REFERENCES


Kato, K. & Roth, R. (2009). Behavioral Difference between Japanese and Australian University Students. *Journal of Yasuda University* 37, 209–220.

Takayama, N., Amemiya, T., Nishikawa, K., Yoshizu, J., Ariyoshi, H., Suzaki, Y. & Nakamura, T. (2012). Investigation on Eating Behaviors of Young Students in Relation to Obesity *Japan Health Medicine Association*, 21(1), 28–35.

Takayama N., Amemiya T., Nishikawa K., Yoshizu J., Ariyoshi H., Suzaki Y. & Nakamura T. (2012). Investigation on Eating Behavior of Adult Workers and Young Students Using, Japanese Version of Dutch Eating Behavior Questionnaire, *Japan Health Medicine Association*, 21(2), 87–94.

Tanbo, K. & Inagaki, M. (2009). A Study on Factors and the Background Contributing to Weight Gain and Obesity in Young Male Employees after Employment – Classification of Young Obese Male Employees by Characteristic Patterns of Thinking and Behaviors –, *Japanese journal of nursing research*, 32, 39–48.

Van Strien, T., Frijters. J.E.R., Gerard, P. A., Bergers, G.P.A. & Defares, P.B. (1986). The Dutch Eating Behavior Questionnaire (DEBQ) for Assessment of Restrained Emotional, and External Eating Behavior. *International Journal of Eating Disorders* 5, 295–315.

*Bridging Research and Good Practices towards Patient Welfare – Shih & Liang (Eds)*
*© 2015 Taylor & Francis Group, London, ISBN 978-1-138-02716-9*

# Evaluation of the accuracy of mobile healthcare applications

Chuan-Jun Su & Shi-Feng Huang
*Department of Industrial Engineering & Management, Yuan Ze University, Taiwan*

ABSTRACT: As the mobile applications for health monitoring exploding in cheaper and smarter ways, they enable physiological data to be acquired pervasively, and present potential in classifying users' health levels based on the variation trend of the acquired data for important reference of Health Self-Management (HSM). However, as mentioned, in order to achieve a reliable assessment on heath levels, the issue of accuracy of physiological data must be addressed. Without accurate physiological data as input, the quality assessment of health level cannot be achieved regardless the analysis algorithm used. We are mainly concerned with the issue of trustworthiness of popular mobile based health monitoring apps available in the market. We address this issue by analyzing physiological data generated by five popular apps. The results of this research can be useful for users to understand the boundaries and limitations of mobile based health monitoring apps and for developers to make improvements on future developments.

## 1 INTRODUCTION

### 1.1 *Research background and motivation*

Mobile devices, such as smart phones and tablet PCs, have evolved over recent years. They come with camera(s) and a variety of embedded sensors such as GPS, gyroscope, accelerometer, etc., which enable the development of novel physiological signs monitoring applications. Various vital signs can be measured conveniently and ubiquitously including heart rate, blood pressure, body temperature, electrocardiogram (ECG), etc. Many innovative measuring technology have been developed, for examples, SpiroSmart (Larson, et al., 2012) which uses a Microphone to Measure Lung Function on a Mobile Phone, Cardiograph (MacroPinch Ltd., 2013) measures Heart Rate Pulse using iPhone & iPad Camera, etc. A recent announcement made by Fujitsu Laboratories that real-time pulse monitoring can be achieved by measuring variations in the brightness of the person's face and using just a camera on one's smartphone or tablet (Adario, 2013). Details of this technology had been presented at the 2013 General Conference of the Institute of Electronics, Information, and Communication Engineers in Gifu, Japan (Chen, Yalu, & Cunwei, 2013).

With the massive availability and convenience of smart mobile devices based vital sign monitoring applications, people are generally willing to continuously collect their physiological data which can then be used to classify their health levels based on the variation trend. However, in order to have a reliable assessment on heath levels, the accuracy of physiological data must be assured. Without accurate physiological data as input, the quality assessment of health level cannot be achieved regardless the analysis algorithm used. In this paper, we are mainly concerned with the issues of the accuracy and reliability of physiological data acquired from popular smart phones based apps available in the market before making substantial efforts on the analysis of health level assessment. We look into these issues through an empirical study by analyzing physiological data generated by top-ranked five heart rate monitoring apps on the App Store and Google Play at the time of this study.

## 2 LITERATURE REVIEW

### 2.1 *Acquiring physiological data using smartphones*

Various HSM strategies help people understand relationships between their health status, behavioral factors, and environmental factors, such as diet, exercise habit, physical activity, air quality, and so on. Therefore, personal information software help people track and visualize as mentioned above information that can help improve their health outcomes and quality of life. Such tools, which include ubiquitous smartphones, network-aware medical devices, digital pedometers, online meal diaries, and a wide diversity of other hardware and software that have attracted much attention from academic circles and industry in recent years. Research has shown that using personal health information applications can improve health outcomes (Frost & Smith, 2003).

Due to smartphone's unique features of versatility and universality, researchers have been using smartphones as tools for encouraging physical activity and healthy diets, for symptom monitoring in asthma and heart disease, for sending patient reminders about upcoming appointments, for supporting smoking cessation, and for a range of other health problems. Smartphone apps allow physicians to monitor patients with chronic heart failure (Scherr, et al., 2006) and to detect early signs of arrhythmia or ischemia that can indicate an imminent heart attack (Rubel, et al., 2005). In addition, patients themselves can use smartphone apps and sensing and measurement devices to keep track of their physical activities (Consolvo, et al., 2008) and to monitor physiological markers relevant to their health status (Mohan, Marin, Sultan, & Deen, 2008).

### 2.2 *Evaluating the usability of mobile healthcare applications*

With the proliferation of smartphones, there are new possibilities for using mobile phones as tools for health promotion. There have been previous studies that set out to examine the content of the 47 iPhone apps for smoking cessation that were distributed through the online iTunes store. They come to the conclusion that iPhone apps for smoking cessation were found to have low levels of adherence to key guidelines from the US Public Health Service's 2008 Clinical Practice Guidelines (Abroms, Padmanabhan, Thaweethai, & Phillips, 2011). A research was conducted for reviewing diabetes apps running on Android smartphones (Demidowich, Lu, Tamler, & Bloomgarden, 2012). A list of 80 free and paid apps was compiled by searching the Android Market in April 2011 and 42 of them were filtered for this study. Despite the variations in functionality, usability, price, and number of downloads, only 4 of the 42 apps studies had up to standard usability score. The research revealed that few apps available in the market provide effective means for diabetes management.

With the popularity and ubiquity of mobile devices, there are massive mobile apps developed in relation to health management. According to a USA Today's report, there are as many as 40,000 medical applications currently on the market for smartphones and tablets and the market is still growing (Gold, 2012).

This research aims to evaluate the quality of physiological data generated by top ranked five heart rate monitor apps at the point of study, which utilize only smartphone embedded sensors without any other external devices. This work intends to provide some insight into the appropriateness of using smartphone as a device for HSM.

## 3 RESEARCH METHODOLOGY

### 3.1 *Experimental subject*

There are 30 volunteer adults participate in this study serving as experimental subjects. They would be asked to fill out a health survey to ensure their healthy level prior to taking the experiment. Moreover, subject was required to read subject conditions to understand the purpose of the experiment, experimental procedure and potential risks involved. Following informed written consent, subjects had their heights and weights measured and provided a brief health history of chronic diseases and medication usage.

### 3.2 *Experimental design*

The lowest acceptable heart rate acquisition method is Borg's RPE scale. Because of the Borg's RPE scale provides a fairly good estimate of the actual heart rate, the Borg's RPE scale results would be the first group heart rate data in this experiment. The second group heart rate data is measured by five HRM apps that is the objective of the evaluation in this research. The third group heart rate data is came from heart rate medical device which would be the benchmark data. These three data groups will be assessed by the following methods: ANOVA, Pearson correlation coefficients, and Bland-Altman plots.

In order to reduce the effects of exercise intensity on heart rate readings, we use U.S. athlete Sally Edwards's heart rate partitioning methods. In her research, the heart rate can be divided into five zones, and the difference between the five zones is according to the proportion of Max HR response. The subjects will be in accordance with our instructions that execute a corresponding movement with each zone. When the medical device confirms the heart rate up to zone standard, we will use heart rate monitor apps to measure the subject's heart rate and let them to complete the Borg's RPE scale. Finally, we can get the experimental data in different zones, and analyze these data whether it will affect heart rate monitor apps to measure heartbeats.

In the course of pre-test, we found the most of subjects were difficult to achieve the required heart rate standard in zone 3, even if a few subjects finally reached standard, but just only kept a few minutes. At first we slowly increased exercise intensity, but when subjects achieved the standard that they had to pay a lot of effort, and also just kept a few time that we could only finish one or two apps measurement. This problem is not only to be the physiological burden on subjects, but also brings the pressure of time and safety to our research, therefore we decide to adjust the experiment for first three zone: Normal, take a rest for 5 minutes.Zone 1, fifty meters walking. Zone 2, jogging one hundred meters.

As an initial phase, the subjects were asked to complete the exercise prescribed in one of the three activities. A data acquisition process that comprises three types of heart rate measurements was subsequently performed to attain the benchmark, Borg, and app heart rate data as illustrated in Figure 1.

The HRM apps selected include Heart Rate, Heart Fitness, Heart Beats, Heart Rate Calculator, and Instant Heart Rate. The selection was made on 10th of May, 2013, by searching applications featured in apple app stores and Google Play.

All data were analyzed with the SPSS Advanced Statistics software package and MedCalc complete statistical program. In this study, one-way ANOVA was utilized to examine heart rate differences between the measuring instruments. The differences between the mean values for all the heart rates at the three measuring instruments were tested for significance by means of ANOVA for repeated measurements with post hoc Scheffe's test which is the more appropriate test to use when predicted differences are small, and the consequences of a Type II error outweigh the consequences of a Type I error. The Scheffe's test assumes equal sized experimental and control groups in the ANOVA. An alpha level of 0.05 was set prior to analysis.

The Pearson's correlation coefficient has been the most common measure of correlation in statistics for assessing reliability. The idea is that if a high (>0.8) and statistically significant correlation coefficient is obtained, the equipment is deemed to be sufficiently reliable (Coolican, 2009). The Pearson correlation coefficients (r) comparison were done in pair between two groups among benchmark data, Borg data, and app data.

In the limits of agreement analysis, the first step is to present and explore the test-retest data with a Bland-Altman plot, which is the individual subject differences between the tests plotted against the respective individual means (Bland & Altman, 1995).As tools are expected to be highly correlated measuring the same signal agreement of the benchmark data and app data were assessed for each activity using Bland-Altman plots for repeated measures with 95% limits of agreement.

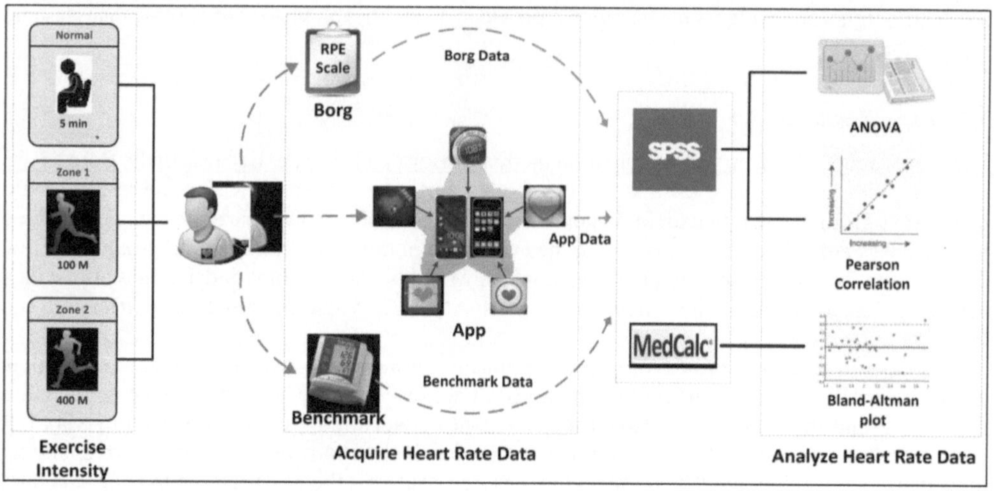


Figure 1. The overall process of acquiring and analyzing heart rate data.

## 4 STATISTICAL ANALYSIS AND RESULTS INTERPRETATION

A total of 30 healthy adults, aged from 20 to 33, volunteered to participate in this study. The participants consist of 15 males and 24 of them have been using smartphone. Among these 24 people, there are 9 people who have experience in using health related Apps in the past. About 10% of the subjects regularly measure their vital sign and half of the subjects perform regular exercise. Following informed written consent, subjects had their weights measured and revealed their information regarding age and sex. The Sally Edward's formula is then used for estimating their MHR (Max Heart Rate). The mean of MHR is 192.67 ± 2.006 (bpm).

### 4.1 *Result and analysis*

ANOVA Post hoc comparisons using Dunnett's T3 or Scheffe procedures were used to determine which pairs of the three group means differed. These results are given in Table 1 and indicate that heart rate data which measured by RPE (M = 110.00) significantly higher on the standardized test than heart rate data which measured by App: What's My Heart Rate in iPhone (M = 103.73) and in Android (M = 103.73) in zone 1 situation. The heart rate data which measured by BC20 (M = 125.93) significantly higher on the standardized test than heart rate data which measured by App: What's My Heart Rate in iPhone (M = 121.47) and in Android (M = 119.90) in zone 2 situation. For App: Cardiograph, the heart rate data which measured by App: Cardiograph in Android (M = 116.10) significantly lower on the standardized test than heart rate data which measured by BC20 (M = 124.10) and RPE (M = 123.33) in zone 2 situation. For App: Instant Heart Rate, the heart rate data which measured by BC20 (M = 110.20) significantly higher on the standardized test than heart rate data which measured by App: Instant Heart Rate in Android (M = 106.80) in zone 1 situation. The heart rate data which measured by App: Instant Heart Rate in Android (M = 116.27) significantly lower on the standardized test than heart rate data which measured by BC20 (M = 123.97) and RPE (M = 127.00) in zone 2 situation.

The Pearson's correlation coefficient provides information about the magnitude of correlation as well as the direction of the relationship. The heart rates obtained from these five HRM apps were highly correlated with those from BC20 in normal situation. However, the correlation will decrease as exercise intensity increase. The app is modestly correlated with RPE scale's data during all conditions.

Table 1. Summary of result and analysis of five HRM apps.

| | | ANOVA Post hoc | Pearson Correlation | The range of the 95% LoA | The width of the 95% LoA |
|---|---|---|---|---|---|
| App: What's My Heart Rate | | | | | |
| Normal | BC20 versus iPhone | 0.133 | −0.821** | −0.1±8.75 | 17.5 |
| | BC20 versus Android | 0.367 | −0.731** | −0.4±8.75 | 17.5 |
| | RPE versus iPhone | 5.000 | −0.094 | – | – |
| | RPE versus Android | 4.767 | −0.096 | – | – |
| Zone 1 | BC20 versus iPhone | 0.500 | −0.676** | −0.5±7.30 | 14.6 |
| | BC20 versus Android | 0.500 | −0.593** | −0.5±8.40 | 16.8 |
| | RPE versus iPhone | 6.267* | −0.062 | – | – |
| | RPE versus Android | 6.267* | −0.304 | – | – |
| Zone 2 | BC20 versus iPhone | −4.467* | −0.457** | 4.5±11.15 | 22.3 |
| | BC20 versus Android | −6.033** | −0.564** | 6.0±8.90 | 17.8 |
| | RPE versus iPhone | −0.133 | −0.006 | – | – |
| | RPE versus Android | 1.433 | −0.204 | – | – |
| App: Cardiograph | | | | | |
| Normal | BC20 versus iPhone | −0.967 | −0.896** | 1.0 ± 6.25 | 12.5 |
| | BC20 versus Android | −1.333 | −0.869** | 1.3 ± 7.05 | 14.1 |
| | RPE versus iPhone | 4.767 | −0.114 | – | – |
| | RPE versus Android | 5.133* | −0.106 | – | – |
| Zone 1 | BC20 versus iPhone | 0.333 | −0.741** | −0.3 ± 8.20 | 16.4 |
| | BC20 versus Android | −2.900 | −0.709** | 2.9 ± 8.80 | 17.6 |
| | RPE versus iPhone | 5.133 | −0.378* | – | – |
| | RPE versus Android | 8.367* | −0.060 | – | – |
| Zone 2 | BC20 versus iPhone | N | −0.664** | 0.3 ± 9.65 | 19.3 |
| | BC20 versus Android | N | −0.445* | 8.0 ± 9.60 | 19.2 |
| | RPE versus iPhone | N | −0.205 | – | – |
| | RPE versus Android | N | −0.012 | – | – |
| App: Heart Beat Rate | | | | | |
| Normal | BC20 versus iPhone | N | −0.773** | 1.9 ± 8.15 | 16.3 |
| | BC20 versus Android | N | −0.760** | 1.7 ± 9.10 | 18.2 |
| | RPE versus iPhone | N | −0.072 | – | – |
| | RPE versus Android | N | −0.200 | – | – |
| Zone 1 | BC20 versus iPhone | N | −0.789** | −1.1 ± 6.30 | 12.6 |
| | BC20 versus Android | N | −0.780** | 0.5 ± 6.80 | 13.6 |
| | RPE versus iPhone | N | −0.416* | – | – |
| | RPE versus Android | N | −0.472* | – | – |
| Zone 2 | BC20 versus iPhone | N | −0.681** | −0.1 ± 10.0 | 20.0 |
| | BC20 versus Android | N | −0.476* | 0.9 ± 12.6 | 25.2 |
| | RPE versus iPhone | N | −0.146 | – | – |
| | RPE versus Android | N | −0.183 | – | – |
| App: Runtastic | | | | | |
| Normal | BC20 versus iPhone | N | −0.892** | 2.0 ± 7.00 | 14.0 |
| | BC20 versus Android | N | −0.803** | 1.0 ± 9.25 | 18.5 |
| | RPE versus iPhone | N | −0.278 | – | – |
| | RPE versus Android | N | −0.073 | – | – |
| Zone 1 | BC20 versus iPhone | N | −0.715** | −1.7 ± 8.30 | 16.6 |
| | BC20 versus Android | N | −0.872** | −1.8 ± 5.40 | 10.8 |
| | RPE versus iPhone | N | −0.307 | – | – |
| | RPE versus Android | N | −0.323 | – | – |
| Zone 2 | BC20 versus iPhone | N | −0.517** | −2.3 ± 10.20 | 20.4 |
| | BC20 versus Android | N | −0.638* | 1.6 ± 8.80 | 17.6 |
| | RPE versus iPhone | N | −0.508** | – | – |
| | RPE versus Android | N | −0.411* | – | – |

*(Continued)*

Table 1. Continued.

| | | ANOVA Post hoc | Pearson Correlation | The range of the 95% LoA | The width of the 95% LoA |
|---|---|---|---|---|---|
| App: Instant Heart Rate | | | | | |
| Normal | BC20 versus iPhone | N | −0.861** | 0.7 ± 8.55 | 17.1 |
| | BC20 versus Android | N | −0.803** | 1.0 ± 10.0 | 20.0 |
| | RPE versus iPhone | N | −0.432* | – | – |
| | RPE versus Android | N | −0.675** | – | – |
| Zone 1 | BC20 versus iPhone | N | −0.416** | −0.6 ± 7.05 | 14.1 |
| | BC20 versus Android | −3.400* | −0.344 | 3.4 ± 7.50 | 15.0 |
| | RPE versus iPhone | N | −0.098 | – | – |
| | RPE versus Android | 4.533 | −0.218 | – | – |
| Zone 2 | BC20 versus iPhone | N | −0.649** | −3.1 ± 9.30 | 18.6 |
| | BC20 versus Android | −7.700** | −0.099 | 7.7 ± 16.2 | 32.4 |
| | RPE versus iPhone | N | −0.071 | – | – |
| | RPE versus Android | 10.733* | −0.180 | – | – |

**: $p < 0.01$; *: $p < 0.05$; LoA: Limits of Agreement; N: No significant differences.

## 5 CONCLUSIONS

According to the results of a single factor analysis of variance, most of the significant differences appeared in the higher exercise intensity, but also had two Apps didn't have any significant differences, which were App: Heart Beat Rate and App: Runtastic. In Pearson correlation, the lower exercise intensity had much higher correlation coefficient, but the correlation decreased as exercise intensity increased. For agreement analysis, the App: Runtastic in Android versus BC20 showed a good agreement that mean difference closed to 0 and the width of the 95% limits of agreement was the smallest.

Although each App's width of the 95% limits of agreement is not smaller than clinically acceptable error, the convenience and ubiquity are indispensable factors for measurement vital signs. When people use this kind of health care applications for the convenience, they also lose the accuracy of the data. The value of this thesis, is that it provides a reference to the users who use the heart rate monitoring apps that they must take into consideration of error values.

## REFERENCES


Abroms, L. C., Padmanabhan, N., Thaweethai, L., & Phillips, T. (2011). iPhone apps for smoking cessation: A content analysis. American Journal of Preventive Medicine, 40(3), pp. 279–285.

Adario, S. (2013, March 18). Fujitsu smartphone reads your face to check your pulse. Retrieved from http://www.dvice.com/2013-3-18/fujitsu-smartphone-reads-your-face-check-your-pulse

Bland, J. M., & Altman, D. G. (1995). Comparing methods of measurement: Why plotting difference against standard method is misleading. Lancet, 346(8982), pp. 1085–1087.

Chen, X., Yalu, Y., & Cunwei, L. (2013). Feature Measurement ofThe Face Based on Infrared Pattern Projection Technique. 2013 IECIE General Conference. Gifu, Japan.

Consolvo, S., Klasnja, P., McDonald, D., Avrahami, D., Froehlich, J., Legrand, L., & Landay, J. (2008). Flowers or a robot army?: Encouraging awareness & activity with personal, mobile displays. UbiComp 2008 – Proceedings of the 10th International Conference on Ubiquitous Computing, (pp. 54–63).

Coolican, H. (2009). Research Methods and Statistics in Psychology. New York: Routledge.

Demidowich, A. P., Lu, K., Tamler, R., & Bloomgarden, Z. (2012). An evaluation of diabetes self-management applications for android smartphones. Journal of Telemedicine and Telecare, 18(4), pp. 235–238.

Frost, J., & Smith, B. K. (2003). Visualizing health: Imagery in diabetes education. Proceedings of the 2003 Conference on Designing for User Experiences, DUX '03.

Gold, J. (2012, June 27). FDA regulators face daunting task as health apps multiply.

Larson, E. C., Goel, M., Boriello, G., Heltshe, S., Rosenfeld, M., & Patel, S. N. (2012). SpiroSmart: Using a microphone to measure lung function on a mobile phone. UbiComp'12 – Proceedings of the 2012 ACM Conference on Ubiquitous Computing, (pp. 280–289).

MacroPinch Ltd. (2013, March 18). Cardiograph: Heart Rate Pulse Measurement using your iPhone & iPad Camera – Track the Cardio Fitness of your Friends and Family. Retrieved from https://itunes.apple.com/cn/app/cardiograph-heart-rate- pulse/id441079429?l=en&mt=8

Mohan, P., Marin, D., Sultan, S., & Deen, A. (2008). MediNet: Personalizing the self-care process for patients with diabetes and cardiovascular disease using mobile telephony. 30th Annual International Conference of the IEEE Engineering in Medicine and Biology Society, (pp. 755–758).

Rubel, P., Fayn, J., Nollo, G., Assanelli, D., Li, B., Restier, L., & Chevalier, P. (2005). Toward personal eHealth in cardiology. Results from the EPI-MEDICS telemedicine project. Journal of Electrocardiology, 38(4), pp. 100–106.

Scherr, D., Zweiker, R., Kollmann, A., Kastner, P., Schreier, G., & Fruhwald, F. M. (2006). Mobile phone-based surveillance of cardiac patients at home. Journal of Telemedicine and Telecare, 12(5), pp. 255–261.

# *Part C: Safety, hazards and MSDs*

*Bridging Research and Good Practices towards Patient Welfare – Shih & Liang (Eds)*
*© 2015 Taylor & Francis Group, London, ISBN 978-1-138-02716-9*

# Importance of occupational hazards in nurses MSD symptoms

F. Serranheira & A. Sousa-Uva
*Escola Nacional de Saúde Pública, Universidade Nova de Lisboa, Lisboa, Portugal*
*CMDT – Centro de Investigação em Malária e Doenças Tropicais, Lisboa, Portugal*

M. Sousa-Uva
*Escola Nacional de Saúde Pública, Universidade Nova de Lisboa, Lisboa, Portugal*
*INSA – Instituto Nacional Dr. Ricardo Jorge, Lisboa, Portugal*

ABSTRACT: One of the most common occupational health issue troubling nurses' is work-related musculoskeletal disorders (WRMSDs). This study (Portuguese) analyzes the prevalence of musculoskeletal symptoms, and its association to nurse's tasks. All registered nurses in Portugal were invited to answer a web-questionnaire, in 2011, with four major dimensions: socio-demographic; musculoskeletal symptoms; nurse's tasks relation with symptoms; and health status. The most frequent and worst intensity-wise self-reported symptoms were low back symptoms. Tasks had a major effect on the increased likelihood of WRMSDs symptoms in upper back, low back, wrist/hands and in ankles/feet. Invasive procedures were the only working task producing simultaneously effects on lumbar, dorsal, wrist/hand and ankles/feet musculoskeletal symptoms. Occupational characteristics such as nurses professional category and specific working tasks are the most important risk factors for musculoskeletal symptoms and therefore prevention should focus also on interventions aimed at workplaces and work tasks.

*Keywords*: Work-Related Musculoskeletal Disorders (WRMSDs); Nurses; Occupational Health; Ergonomics; Occupational Health and Safety.

## 1 INTRODUCTION

Work-related musculoskeletal disorders (WRMSDs) are a major occupational concern in the relationships between work and disease (Ando et al., 2000; Trinkoff, Brady, & Nielsen, 2003; Smith, Mihashi, Adachi, Koga, & Ishitake, 2006). Over the past few years there has been an increasing interest on studying the associations between work and MSDs (musculoskeletal disorders).

WRMSDs are a large group of medical conditions that affect the musculoskeletal system which include, among others, diseases of the muscles, tendons, nerves, and also joint injuries (Fonseca & Serranheira, 2006).

WRMSDs could be divided into two major groups related to: (i) load mobilization and transport, mainly with low back needs and (ii) the performance of static or repetitive tasks, with or without application of force, frequently using the upper limbs.

Nurses are considered the group of health care professionals more often affected by WRMSDs, mainly low-back pain (LBP) (Trinkoff, Brady & Nielsen, 2003). Actually nursing involves heavy physical work tasks such as lifting and transferring patients, which require sudden movements, most of the time in non-neutral postures (mostly awkward postures), working in organizational strain, and operating hazardous equipment (Kee & Seo, 2007).

The etiology of WRMSDs is multifactorial (Ariëns et al., 2001; Larsson, Sogaard, & Rosendal, 2007) and many studies conducted in nurses have found associations between work-related musculoskeletal symptoms and: i) poor posture (including awkward and static postures) (Engels & Ackermann, 1996); ii) lifting patients and other manual handling; iii) transferring patients out of bed (Pheasant & Stubbs, 1992; Hignett, 1996); iv) work-related stress (Estryn-Behar et al., 1990); v) organizational factors, like reduction of nursing staff (Camerino, Estryn-Behar, Conway, van Der Heijden, & Hasselhorn, 2008); and also vi) some intrinsic or individual factors (Lagerström, Wenemark, Hagberg, & Hjelm, 1996).

The most frequent reported complains associated with nursing demands are on the back, neck and shoulders (Feyer et al., 2000; Gonge, Jensen, & Bonde, 2001; Alexopoulos, Burdorf, & Kalokerinou, 2003).

In occupational medicine it is generally assumed that perceived muscular discomfort and/or pain may be an early sign of musculoskeletal disorders (Wahlstrom, 2005; Reenen, van der Beek, Blatter, van der Grinten, Mechelen & Bongers, 2008) relevant for WRMSDs risk assessment (Genaidy, Delgado, & Bustos, 1995).

Studies focused on the relations between nursing work demands and WRMSDs symptoms expose differences in each ward, hospital, and among different countries (Ando, et al., 2000; Trinkoff, Lipscomb, Geiger-Brown, Storr, & Brady, 2003; Alexopoulos, Burdorf, & Kalokerinou, 2006; Smith, Mihashi, Adachi, Koga & Ishitake, 2006; Fonseca & Serranheira, 2006; Kee & Seo, 2007; Serranheira, Cotrim, Rodrigues, Nunes, & Sousa-Uva, 2012).

The present study aimed to identify and analyze the prevalence of musculoskeletal symptoms in Portuguese hospital nurses and its association with different nursing tasks.

## 2 METHODS

This was a cross-sectional national wide study that targeted all 62.566 Registered Nurses (RN) at the Portuguese Nursing Council (PNC) with 26.920 (43%) of them working in hospitals.

Portuguese nurses were invited to participate in this self-report WRMSDs symptoms questionnaire used as the survey instrument. The request was made at the Portuguese Nursing Council (PNC) website and it was available online on the Survey Monkey Platform between June 2010 and February 2011.

Questionnaire has been adapted from the Nordic Questionnaire on Musculoskeletal disorders (NMQ - Nordic Musculoskeletal Questionnaire) previously validated and accepted as a survey instrument (Kuorinka et al., 1987; Warming, Precht, Suadicani, & Ebbehoj, 2009) and already used in Portugal (Fonseca & Serranheira, 2006; Serranheira, Pereira, Santos, & Cabrita, 2003; Serranheira, Uva, & Lopes, 2008). It is mainly divided in four sections intended to characterize: (i) demographic profile such as age, gender weight and height; (ii) the presence or absence of musculoskeletal symptoms in nine anatomical sites in the past 12 months and in the past 7 days; (iii) details on the respondent's job and its relationship with symptoms, such as information about work status, work setting, years of practice and nursing activities and (iv) general health status. Body segments evaluated includes trunk (cervical, dorsal and low back), upper limbs (shoulders, elbows and wrist/hand) and lower limbs (thighs, knees and ankles/feet).

Statistical analysis was performed using the software Statistical Package for Social Sciences (SPSS) version PASW Statistics 21.

To determine the increased likelihood of having WRMSDS symptoms in the past 12 months was estimated the Odds Ratio and the respective 95% confidence interval through the method of Logistic Regression: Forward LR. A level of significance of 5% was set for all statistical tests.

The Chi-square Test of Independence was used to evaluate if the prevalence of low-back symptoms depends on gender, age, Body Mass Index (BMI) and nurses professional category.

## 3 RESULTS

A total of 2.140 registered nurses answered the questionnaire (3.42% of all Portuguese registered nurses). The questionnaire was answered by 1.396 hospital nurses which, accordingly to

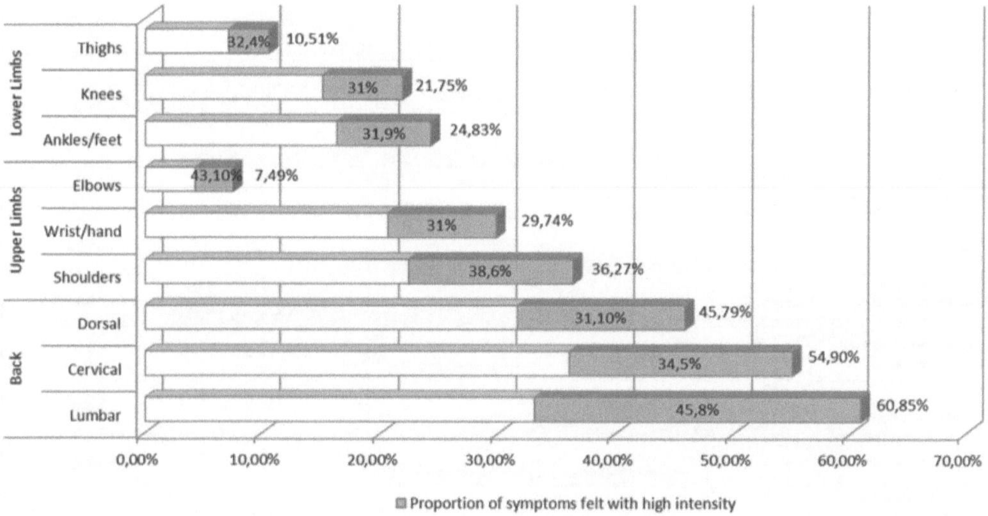


Figure 1. Prevalence of WRMDs symptoms in the last 12 months in hospital nurses and proportion for each of these felt with high intensity.

the Portuguese Nursing Council (2010), represent up to 5.19% of all nationwide hospital nurses (n = 26.920).

The sample was quite largely female (75.8%), having averages figures for age (37.2 years); weight (67.02 kg), and height (1.65 m).

Findings clearly showed a high frequency of musculoskeletal symptoms in the last 12 months in hospital nurses (Figure 1).

The most prevalent complain was in the Lumbar region (60.85%). Among the 1.396 hospital nurses, 69.6% referred to work night-shifts, and 88% reported to have more than 1 musculoskeletal symptom.

Other relevant aspects were reported, like symptoms intensity in which about 19% of the total sample says that feel one or more WRMSDs with high intensity. Low back symptoms were also the worst intensity-wise self-reported symptoms (45.8%). In what concerns about WRMDs frequency, 33% of all nurses' affirm to feel pain 6 times per year.

Working tasks studied had no effect on the increased likelihood of having shoulders, elbows, cervical, thighs and knees musculoskeletal symptoms.

Nevertheless, there were tasks increasing the probability of having WRMSDs symptoms in upper back, low back, wrist/hands and in ankles/feet areas. These effects were more evident when those tasks were performed more than 10 times a day, highlighting the importance of the frequency of such symptoms (Figure 2).

Tasks performed more than 10 times a day, such as invasive procedures (Odds Ratio = 2.142); care of hygiene and patient comfort in bed (Odds Ratio = 2.484); patient mobilization in bed (Odds Ratio = 2.022); and patient feeding (Odds Ratio = 2.186) had an effect on Dorsal and Low Back Pain ($p < 0.05$).

Invasive procedures were the only working task producing simultaneously effects on lumbar, dorsal, wrist/hand and ankles/feet musculoskeletal symptoms.

Low back pain (LBP) was the most common symptom and was associated with a greater number of working tasks. It was also analyzed the relationship with some individual (age, body mass index (BMI), gender) and job factors (professional category). The Chi-square Test of Independence revealed that low back pain was independent of gender ($\chi^2 = 2.37$, $p = 0.123$), age ($\chi^2 = 3.86$, $p = 0.27$) and BMI ($\chi^2 = 1.663$, $p = 0.197$), but was dependent of job category ($\chi^2 = 18.86$, $p = 0.001$). Individuals with a professional category of "nurses" and "graduate nurses" had a

| | Classes | Back: Dorsal Odds ratio (IC95%) | Back: Dorsal p value | Back: Lumbar Odds ratio (IC95%) | Back: Lumbar p value | Upper Limbs: Wrist/hand Odds ratio (IC95%) | Upper Limbs: Wrist/hand p value | Lower Limbs: Ankles/feet Odds ratio (IC95%) | Lower Limbs: Ankles/feet p value |
|---|---|---|---|---|---|---|---|---|---|
| Invasive procedures | 2-5 times per day | | | — | 0.08 | 1.43 (1.017 - 2.012) | 0.04 | 1.563 (1.08 - 20264) | 0.018 |
| | 6-10 times per day | >0.05 | | — | 0.198 | 1.632 (1.078 - 2.47) | 0.021 | 1.762 (1.13 - 2.748) | 0.012 |
| | More than 10 times per day | | | 2.149 (1.432 - 3.225) | <0.001 | 1.724 (1.134 - 2.619) | 0.011 | 1.943 (1.244 - 3.034) | 0.003 |
| Drugs administration | 2-5 times per day | — | 0.36 | — | 0.188 | — | 0.286 | | |
| | 6-10 times per day | 1.909 (1.242-2.932) | 0.003 | 1.682 (1.095 - 2.584) | 0.017 | — | 0.07 | >0.05 | |
| | More than 10 times per day | — | 0.083 | 1.857 (1.264 - 2.728) | 0.02 | 1.677 (1.079 - 2.608) | 0.022 | | |
| Blood pressure monitoring and glucose finger prick testing | 2-5 times per day | | | — | 0.614 | | | | |
| | 6-10 times per day | >0.05 | | — | 0.146 | >0.05 | | >0.05 | |
| | More than 10 times per day | | | 1.595 (1.094 - 2.326) | 0.015 | | | | |
| Care of Hygiene and comfort of patients in bed | 2-5 times per day | — | 0.508 | 1.363 (1.020 - 1.820) | 0.037 | | | | |
| | 6-10 times per day | — | 0.107 | 1.629 (1.121 - 2.368) | 0.011 | >0.05 | | >0.05 | |
| | More than 10 times per day | 1.787 (1.067-2.991) | 0.027 | 2.484 (1.395 - 4.424) | 0.002 | | | | |
| Positioning and mobilization of patients in bed | 2-5 times per day | — | 0.238 | — | 0.52 | — | 0.294 | | |
| | 6-10 times per day | — | 0.073 | — | 0.29 | — | 0.721 | >0.05 | |
| | More than 10 times per day | 1.691 (1.152-2.481) | 0.007 | 2.022 (1.354 - 3.019) | ≤0.001 | 1.555 (1.024 - 2.361) | 0.038 | | |
| Tranfer and transport of patients | 2-5 times per day | — | | — | 0.535 | | | | |
| | 6-10 times per day | — | 0.058 | — | 0.119 | >0.05 | | >0.05 | |
| | More than 10 times per day | — | | 1.751 (1.102 - 2.782) | 0.018 | | | | |
| Patient lifting without mechanical assistance | 2-5 times per day | — | 0.126 | — | 0.124 | | | | |
| | 6-10 times per day | 1.603 (1.111-2.313) | 0.012 | — | 0.064 | >0.05 | | >0.05 | |
| | More than 10 times per day | 2.129 (1.396-3.343) | 0.001 | 1.709 (1.066 - 2.739) | 0.026 | | | | |
| Feeding patients | 2-5 times per day | — | 0.246 | 1.355 (1.014 - 1811 | 0.04 | | | | |
| | 6-10 times per day | — | 0.224 | — | 0.318 | >0.05 | | >0.05 | |
| | More than 10 times per day | 1.653 (1.022-2.674) | 0.041 | 2.186 (1.283 - 3.725) | 0.004 | | | | |
| Care of Hygiene and comfort of patients in WC | 2-5 times per day | — | 0.335 | — | 0.137 | | | | |
| | 6-10 times per day | 1.669 (1.011-2.737) | 0.045 | 1.779 (1.039 - 3.048) | 0.036 | >0.05 | | >0.05 | |
| | More than 10 times per day | — | 0.072 | — | 0.4 | | | | |
| Wound treatment | 2-5 times per day | | | | | | | — | 0.371 |
| | 6-10 times per day | >0.05 | | >0.05 | | >0.05 | | 2.11 (1.391 - 3.202) | <0.001 |
| | More than 10 times per day | | | | | | | — | 0.152 |

*Reference category: 0-1 times per day*

Figure 2. Results from Logistic Regression to quantify nursing tasks effects on the presence of musculoskeletal symptoms in hospital nurses. The squares with ">0.05" means that was not found any significant association.

higher prevalence of low back pain (23.44% and 22.88% respectively), and professionals with a category of "nurse-specialists", "chief nurse" and "nurse supervisors" had a lower prevalence (9.42%, 4.86% and 0.37% respectively).

## 4 DISCUSSION

Findings clearly showed, similarly to other previous studies (Jensen, Ryholt, Burr, Villadsen & Christensen, 2002; Alexopoulos, Burdorf & Kalokerinou, 2003, 2006; Fonseca & Serranheira, 2006; Kee & Seo, 2007; Serranheira, Cotrim, Rodrigues, Nunes, & Sousa-Uva, 2012), a higher prevalence of musculoskeletal symptoms in the last 12 months (cervical = 48.5%; dorsal = 45.9%; lumbar = 60.85%; shoulder = 36.27%; elbows = 7.49%; wrist/hands = 29.74%; hips/thighs = 19.51%; knees = 21.75%; ankles/feet = 24.83%).

Low back symptoms were the most frequent and worst intensity-wise self-reported symptoms. These findings were also similar to other previous studies, where low back pain was described as a very frequent occupational problem in hospital nurses with prevalence ranging between 40% and 60% (Alexopoulos, Burdorf & Kalokerinou, 2006; Smith, et al., 2006; Kee & Seo, 2007; Lorusso, Bruno, & L'Abbate, 2007; Tinubu, Mbada, Oyeyemi, & Fabunmi, 2010; Holtermann, Clausen, Jørgensen, Burdorf, & Andersen, 2012).

Results suggest that musculoskeletal symptoms in upper back, low back, wrist/hand and in ankles/feet body regions were associated to specific nursing working tasks, which probably means they could be prevented with the implementation of adequate risk management measures. These associations were more evident when those tasks were performed more than 10 times a day, highlighting the importance of the frequency in which tasks are performed for the presence of such symptoms. The most prevalent musculoskeletal symptoms were low back and upper back symptoms, and the less affected symptoms were those from the wrist/hand and ankles/feet areas.

Working tasks had no effect on the increased likelihood of having shoulders, elbows, cervical, thighs and knees musculoskeletal symptoms. These results suggest that those working tasks seemed not to play an important role for the presence of those symptoms, perhaps being much more influenced by other work occupational hazards, like work-related stressors, organizational factors and some individual factors, like age, gender or BMI (Estryn-Behar, et al., 1990; Lagerström, Wenemark, Hagberg & Hjelm, 1996; Camerino, et al., 2008).

Neck symptoms have been described in literature as frequent WRMSDs' symptoms in nurses (Parot-Schinkel, et al., 2012), and this was confirmed in the present study. However, it was not identified any significant effect of working tasks on these symptoms. Further research must be developed in order to identify other possible risk factors, or confounding factors, that possibly didn't allowed identifying here relations between neck symptoms and working tasks.

Invasive procedures were just the only working task producing simultaneously effects on all lumbar, dorsal, wrist/hand and ankles/feet musculoskeletal symptoms. Perhaps this could be explained by specific task characteristics which could involve demanding procedures for all body segments and, on the other hand, because it may involve some occupational stressors' exposure.

## 5 CONCLUSIONS

Nurses' professional category and nurses' ward specific tasks are the most important organizational risk factors for the typically prevalent nurses' musculoskeletal symptoms (low-back symptoms), unlike individual factors such as the age, gender, BMI which in this study were not associated with low-back pain. Therefore from now on that should focus mainly on interventions aimed at preventing these WRMSDs and safeguard nurses' occupational health and safety acting on those specific working tasks.

Nurses' musculoskeletal disorders are most of times preventable and intervention in work process, in work equipment's and in organizational factors must be encouraged. This study highlights the high prevalence of WRMSDs symptoms in Portuguese registered nurses and the need for the development of occupational prevention programs, particularly in hospitals. Occupational Health and Safety are important for healthcare professionals but also for patient safety.

In addition to identify musculoskeletal symptoms and their related occupational hazards in this study, namely occupational low back pain, it was illustrated the need of a better knowledge about these interactions. That knowledge will contribute for better design of occupational risk management programs (Sousa-Uva, 2010) concerning WRMSD's prevention.

## REFERENCES


Alexopoulos, E.C., Burdorf, A., & Kalokerinou, A. 2003. Risk factors for musculoskeletal disorders among nursing personnel in Greek hospitals. *Int Arch Occup Environ Health, 76*(4), 289–294.

Alexopoulos, E., Burdorf, A. & Kalokerinou, A. 2006. A comparative analysis on musculoskeletal disorders between Greek and Dutch nursing personnel. *International Archives of Occupational and Environmental Health, 79*(1), 82–88.

Ando, S., Ono, Y., Shimaoka, M., Hiruta, S., Hattori, Y., Hori, F., & Takeuchi, Y. (2000). Associations of self estimated workloads with musculoskeletal symptoms among hospital nurses. *Occup Environ Med, 57*(3), 211–216.

Ariëns, G., Bongers, P., Douwes, M., Miedema, M., Hoogendoorn, W., Van der Wal, G. & Van Mechelen, W. (2001). Are neck flexion, neck rotation, and sitting at work risk factors for neck pain? Results of a prospective cohort study. *Occupational and Environmental Medicine, 58*(3), 200–207.

Camerino, D., Estryn-Behar, M., Conway, P., van Der Heijden, B., & Hasselhorn, H. 2008. Work-related factors and violence among nursing staff in the European NEXT study: a longitudinal cohort study. *International Journal of Nursing Studies, 45*(1), 35–50.

Engels, M., & Ackermann, M. 1996. Pathogenesis of ruminant herpesvirus infections. *Vet Microbiol, 53*(1–2), 3–15.

Estryn-Behar, M., Kaminski, M., Peigne, E., Bonnet, N., Vaichere, E., Gozlan, C. & Giorgi, M. (1990). Stress at work and mental health status among female hospital workers. *British Journal of Industrial Medicine, 47*(1), 20–28.

Feyer, A., Herbison, P., Williamson, A., de Silva, I., Mandryk, J., Hendrie, L. & Hely, M. 2000. The role of physical and psychological factors in occupational low back pain: a prospective cohort study. *Occupational and Environmental Medicine, 57*(2), 116–120.

Fonseca, R. & Serranheira, F. 2006. Sintomatologia músculo-esquelética auto-referida por enfermeiros em meio hospitalar. *Revista Portuguesa de Saúde Pública, Volume Temático*, 37–44.

Genaidy, A., Delgado, E. & Bustos, T. 1995. Active microbreak effects on musculoskeletal comfort ratings in meatpacking plants. *Ergonomics, 38*(2), 326–336.

Gonge, H., Jensen, L. & Bonde, J. 2001. Do psychosocial strain and physical exertion predict onset of low-back pain among nursing aides? *Scandinavian journal of work, environment & health*, 388–394.

Hignett, S. 1996. Work-related back pain in nurses. *Journal of Advanced Nursing, 23(6), 1238–1246.*

Holtermann, A, Clausen, T, Jørgensen, M, Burdorf, A. & Andersen, L. 2013. Patient handling and risk for developing persistent low-back pain among female healthcare workers. *Scandinavian Journal of Work, Environment & Health, 39*(2):164–169.

Jensen, C., Ryholt, C., Burr, H., Villadsen, E. & Christensen, H. 2002. Work-related psychosocial, physical and individual factors associated with musculoskeletal symptoms in computer users. *Work & Stress, 16*(2), 107–120.

Kee, D. & Seo, S. 2007. Musculoskeletal disorders among nursing personnel in Korea. *International Journal of Industrial Ergonomics, 37*(3), 207–212.

Kuorinka, I., Jonsson, B., Kilbom, A., Vinterberg, H., Biering-Sørensen, F., Andersson, G. & Jørgensen, K. 1987. Standardised Nordic questionnaires for the analysis of musculoskeletal symptoms. *Appl Ergon, 18*(3), 233–237.

Lagerström, M., Wenemark, M., Hagberg, M. & Wigaeus Hjelm, E. 1996. Occupational and individual factors related to musculoskeletal symptoms in five body regions among Swedish nursing personnel. *International Archives of Occupational and Environmental Health, 68*(1), 27–35.

Larsson, B., Sogaard, K. & Rosendal, L. 2007. Work related neck-shoulder pain: a review on magnitude, risk factors, biochemical characteristics, clinical picture and preventive interventions. *Best Practice & Research Clinical Rheumatology, 21*(3), 447–463.

Lorusso, A., Bruno, S. & L'Abbate, N. 2007. A review of low back pain and musculoskeletal disorders among Italian nursing personnel. *Industrial health, 45*(5), 637–644.

Parot-Schinkel, E., Descatha, A., Catherine, P., Audrey, L., Annette, & Roquelaure, Y. 2012. Prevalence of multisite musculoskeletal symptoms: a French cross-sectional working population-based study. *BMC Musculoskeletal Disorders, 13*(1), 122.

Pheasant, S. & Stubbs, D. 1992. Back pain in nurses: epidemiology and risk assessment. *Appl Ergon, 23*(4), 226–232. doi: 000368709290150T [pii]

Reenen, H., van der Beek, A., Blatter, B., van der Grinten, M., Mechelen, W. & Bongers, P. 2008. Does musculoskeletal discomfort at Work predict future musculoskeletal pain? *Ergonomics*. 51 (5), 637–648.

Serranheira, F., Cotrim, T., Rodrigues, V., Nunes, C. & Sousa-Uva, A. 2012. Nurses' working tasks and MSDs back symptoms: results from a national survey. *Work: A Journal of Prevention, Assessment and Rehabilitation, 41*, 2449–2451.

Serranheira, F., Pereira, M., Santos, C. & Cabrita, M. 2003. Auto-referência de sintomas de lesões músculo-esqueléticas ligadas ao trabalho (LMELT) numa grande empresa em Portugal. *Revista Portuguesa de Saúde Pública, 2*, 37–48.

Serranheira, F., Uva, A. & Lopes, F. 2008. *Lesões músculo-esqueléticas e trabalho: alguns métodos de avaliação do risco.* (Vol. 5). Lisboa: Sociedade Portuguesa de Medicina do Trabalho.

Serranheira, F., Cotrim, T., Rodrigues, V., Nunes, C. & Sousa-Uva, A. 2012. Lesões musculoesqueléticas ligadas ao trabalho em enfermeiros portugueses: <<ossos do ofício>> ou doenças relacionadas com o trabalho? *Revista Portuguesa de Saúde Pública*, 30(2), 193–203.

Smith, D., Mihashi, M., Adachi, Y., Koga, H. & Ishitake, T. 2006. A detailed analysis of musculoskeletal disorder risk factors among Japanese nurses. *Journal of safety research, 37*(2), 195–200.

Sousa-Uva, A. 2010. Diagnóstico e Gestão do Risco em Saúde Ocupacional. Lisboa: ACT.

Tinubu, B., Mbada, C., Oyeyemi, A. & Fabunmi, A. 2010. Work-related musculoskeletal disorders among nurses in Ibadan, South-west Nigeria: a cross-sectional survey. *BMC Musculoskelet Disord, 11*, 12. doi: 1471-2474-11-12 [pii]10.1186/1471-2474-11-12

Trinkoff, A., Brady, B. & Nielsen, K. 2003. Workplace prevention and musculoskeletal injuries in nurses. *J Nurs Adm, 33*(3), 153–158.

Trinkoff, A., Lipscomb, J., Geiger-Brown, J., Storr, C. & Brady, B. 2003. Perceived physical demands and reported musculoskeletal problems in registered nurses. *American Journal of Preventive Medicine, 24*(3), 270–275. doi: 10.1016/S0749-3797(02)00639-6

Wahlstrom, J. 2005. Ergonomics, musculoskeletal disorders and computer work. *Occup Med (Lond), 55*(3), 168–176. doi: 55/3/168 [pii]10.1093/occmed/kqi083

Warming, S., Precht, D., Suadicani, P. & Ebbehoj, N. 2009. Musculoskeletal complaints among nurses related to patient handling tasks and psychosocial factors–based on logbook registrations. *Appl Ergon, 40*(4), 569–576. doi: S0003-6870(08)00125-7 [pii]10.1016/j.apergo.2008.04.021

*Bridging Research and Good Practices towards Patient Welfare – Shih & Liang (Eds)*
*© 2015 Taylor & Francis Group, London, ISBN 978-1-138-02716-9*

# The survey of sharps injuries by unsealing the neck of ampoule and safety tool design

Ching-Wen Lien
*Nursing Department, Taipei Veterans General Hospital, Taipei, Taiwan*

Bor-Shong Liu
*Department of Industrial Engineering and Management, St. John's University, New Taipei City, Taiwan*

Jun-Ting Lee
*Nursing Department of Jen-Chi Senior Home, Taipei Jen-Chi Relief Institution, New Taipei City, Taiwan*

ABSTRACT: Present study was survey the operating behavior for ampoule sharps injuries in an emergency room in a New Taipei Hospital. Participants of 21 nurses reported that the most commonly used are 1 ml (57%) and 20 ml (20%) of ampoules. In addition, there are 19 sharps injuries by unsealing the neck of ampoule during one month study period. Present study provided the new method to snap-off the neck of sealed ampoule. The purpose of present study was evaluated the methods of snap-off and sizes of ampoule on forearm muscle activities, wrist postures and operating performance. Each subject performed 45 trials of snap-off the ampoules on three methods (i.e., grinding slice, grinding slice with gloves and novel tool) and three sizes of ampoule (1 ml, 5 ml, and 20 ml) for five repetition. Results of analysis showed that the pronator teres muscle were heavier load as increased the sizes of ampoule. For performance analysis, time of snap-off were longer in 20 ml of ampoule. For wrist posture, the ranges of motion for wrist were significant lower while using the novel tool. Present study provided the novel tool effectively to snap-off the ampoule to decrease forearm workload and sharp injuries.

*Keywords*: sharps injuries, ampoule, occupational safety and health, product design, EMG

## 1 INTRODUCTION

Unsafe injection practices place at risk not only patients, but also healthcare workers, who practice universal precautions inconsistently and are commonly exposed to blood in the course of their work. The 2011 statistic data from EPINet (The Exposure Prevention Information Network) in Taiwan showed that the rate of needlestick and sharps injuries is 3.5 (100 people × year). Health care workers (medical, nursing, technical, and supporting personnel) were surveyed to identify factors associated with non-reporting behavior of sharps injuries. The 87.3% of health care workers reported to have experienced the sharps injuries. However, literatures review revealed that a total of 81.8% of sharps injuries were not reported, with job category significantly affecting reporting behavior. Nurses are at risk acquiring infections in the workplace. While many infections can be transmitted through a sharps injury, hepatitis B virus (HBV), hepatitis C virus (HCV), and human immunodeficiency virus (HIV) are the most consequential (Lanphear, 1994; Trim & Eliott, 2003). However, the number of sharps injuries sustained by healthcare workers (HCW) is still unclear, primarily due to under-reporting (OSHA, 1999). The reliability of sharps injury data is ambiguous due to under-reporting. First documented in 1983 (Hamory, 1983), up to 40% of all sharps injuries were not reported. Subsequent studies have estimated the rate of under-reporting to be between 26 and 85% (Burke and Madan, 1997; Dobie et al., 2002).

From reports of AHEPA University Hospital in Greece, representing an overall rate of 2.4% reported injuries per 100 HCWs/year (Pournaras et al., 1999). Trim and Elliott (2003) reported that mean rate of 4.0% (range 1.0–6.2%) sharps injuries per 10000 HCW was calculated from eight studies involving more than 7000 HCW in UK. Nurses were most commonly exposed (91%) compared with phlebotomistsn (3%) and doctors (6%), which not only reflects the proportion of nurses in healthcare but the number of sharp devices used by them in clinical practice (PHLS, 1999). The epidemiology of needlestick and sharps injuries (NSIs) was investigated among a complete cross-section of 1162 nurses from a large hospital in southern Japan (Smith et al., 2006). Forty-six percent had experienced an NSI in the previous year. Most were caused by ampoules or vials, which injured 32.3% of all nurses and accounted for 42.9% of all NSI events. From Hong Kong reported that there were a total of 51 reported cases of NSIs (43 needlestick injuries and 8 sharps injuries) (Cheung et al., 2010). The annual prevalence of NSIs in four academic years from 2002–2003 to 2005–2006 ranged from 0.6 to 1.6 cases while the incidence rate was one new case per 100 nursing students per academic year. Injection needles accounted for the highest rate of needlestick injuries at 86%. Broken glass from opening ampoules (62.50%) was responsible for most sharps injuries. Shiao et al. (1999) surveyed 10,469 full-time medical, nursing, technical, and supporting personnel employed at 16 randomly selected hospitals from 132 available accredited teaching hospitals in Taiwan. Questionnaires were completed by 82.6% (8645) of their samples, of whom 87.3% reported to have experienced a recent NSI. Medical staff had the highest non-reporting rate. Shiao et al. (2001) were also undertaken to measure the incidence of NSIs in Taiwanese support personnel. Most of the injuries were sustained by cleaners handling sharps inappropriately disposed of by clinical staff. Of the 61.0% of SIs sustained by support personnel, 65.7% were cleaners, 24.5% central supply workers, 5.8% porters and 4.1% laundry workers. No nationally accepted standardized method for collecting and calculating sharps injuries has been developed. However, the Exposure Prevention Information Network (EPINet) is widely used in many countries. EPINet does provide healthcare establishments with a standardized method for recording percutaneous and mucocutaneous injuries using a pre-programmed form. Customized reports, statistical analysis and tracking of injuries are provided, and information can be linked to national and international databases as required. Present study was survey the operating behavior for ampoule sharps injuries in an emergency room.

## 2 METHOD

Present study consisted of two stages. In the first stage, a questionnaire survey was used to examine the operating behavior for ampoule sharps injuries in an emergency room. In the second stage, the evaluated experiment was conducted to examine the effects of snap-off methods and sizes of ampoule on forearm muscle activities, wrist postures and operating performance.

### 2.1 *Survey of operating behavior for ampoule sharps injuries*

Present study was survey operating behavior for ampoule sharps injuries in an emergency room in a New Taipei City hospital. The survey involved frequency of snap-off ampoule, sharps injuries of snap-off ampoule, conditions of inappropriate use, and usage of Exposure Prevention Information Network.

### 2.2 *Evaluated experiment*

#### 2.2.1 *Participants*

Ten nurses and emergency medical technician were recruited in present study. All subjects were healthy and reported no musculoskeletal problems that might influence performance detrimentally.

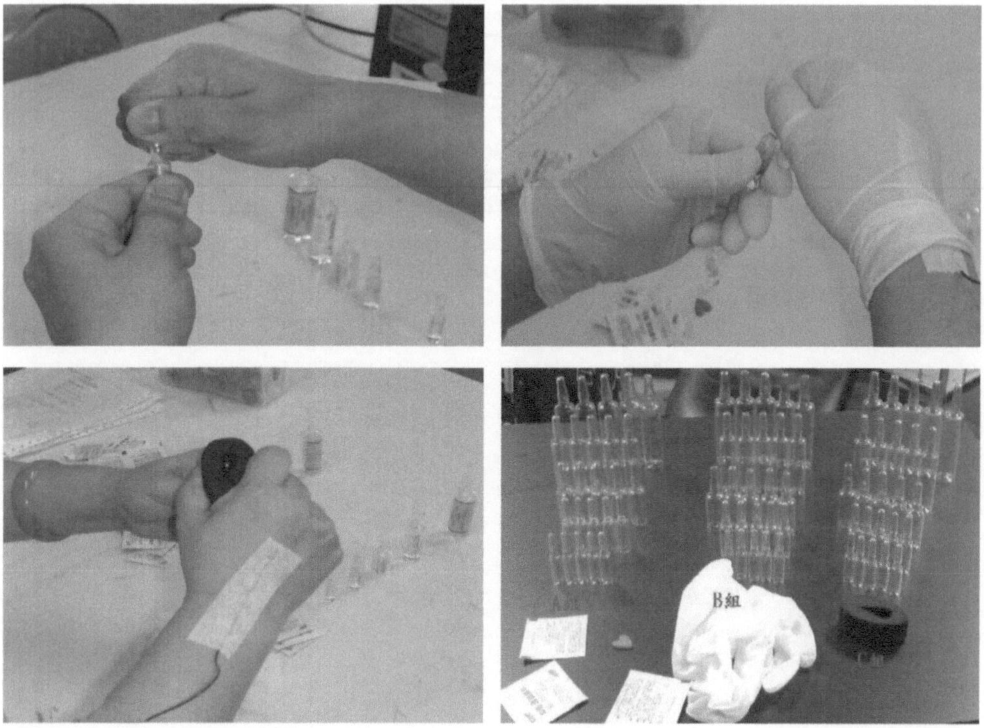

Figure 1. Evaluated experiment trials

### 2.2.2 *Apparatus and materials*

Each subject performed 45 trials of snap-off the ampoules on three methods (i.e., grinding slice, grinding slice with gloves and novel tool) and three sizes of ampoule (1 ml, 5 ml, and 20 ml) for five repetition (Figure 1). The experimental empty ampoules made by GL Glass Company (Hsinchu, Taiwan) and sizes dimensions of ampoule presented in Table 1. The idea of novel opener tool is prepare the grinding slice inside rubber ring, so users could put the ampoule inside the rubber ring and then rotated the rubber ring to snap-off the neck of ampoule.

Electromyography (EMG) activities of the target muscle groups were recorded by means of bipolar direct measurements (MP150, Biopac Systems Inc, California, USA). Pairs of Ag/AgCl surface electrodes (Blue Sensor M-00-S, Medicotest, Denmark) separated by 20 mm were placed on Pronator Teres of dominantly hand. The signals were amplified, the band-pass was filtered at 20-500 Hz, and the A/D was converted and sampled at 1000 Hz. Reference voluntary contraction for normalization was obtained during maximum isometric voluntary contraction (MVC) of the specific muscle group for two trials of approximately 5 seconds each.

In addition, the miniature joint angle sensor (S720, Measurand Inc., Canada) had been mounted on the neutral axis of wrist. Further, wrist angles have been measured during experiment.

### 2.2.3 *Data analysis*

A factorial design with block factor (participants) and two factors (sizes of ampoule and methods of snap-off) was applied in present study. The dependent variables were angles of wrist, EMG (%) for Pronator Teres and operating time (sec). Further, analysis of variance (ANOVA) was utilized to identify significant differences between conditions for dependent variables. Statistical significance was set at a probability level of 0.05. Where statistically significant differences were identified, the Duncan post hoc test was used to examine the differences.

Table 1. Dimensions of ampoules (mm).

| Ampoules | 1 ml | 5 ml | 20 ml |
|---|---|---|---|
| Length | 48 | 65 | 105 |
| Diameter | 9 | 17 | 22 |
| Neck length | 19 | 24 | 25 |
| Neck diameter | 5 | 7 | 10 |

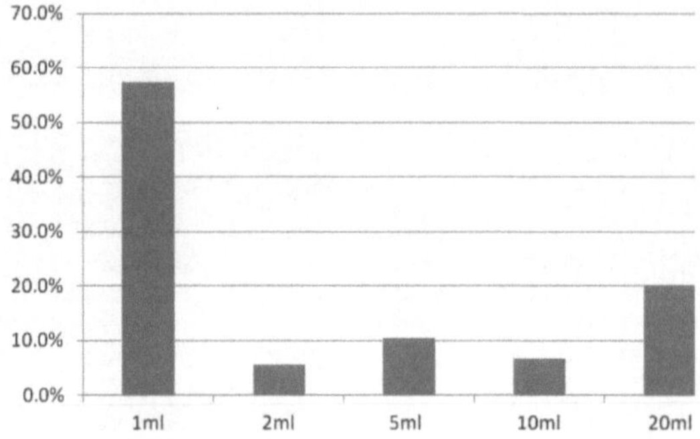


Figure 2. Frequency of snap-off ampoule between sizes.

## 3 RESULTS

### 3.1 *Survey of ampoule sharps injuries*

There are twelve one nurses in an emergency room in a New Taipei City hospital. The survey involved frequency of snap-off ampoule, sharps injuries of nap-off ampoule, conditions of inappropriate use, and usage of Exposure Prevention Information Network during one shift.

There are 386 times of snap-off ampoule during survey period. Figure 2 showed the frequency of snap-off ampoule between sizes. The most usage sizes of ampoule are 1 ml (n = 164, 57.3%) and 20 ml (n = 57, 19.9%). The 19 times of sharps injuries occurred by snap-off ampoule. Half sharps injuries events had occurred by 1 ml (n = 10, 52.6%). In addition, there are 19 times crushed ampoules (e.g. Figure 3). However, all incidents have not been notified. The possible reason might that interviewees considered no infection problem and the notification process is too complicated. For survey of applied Exposure Prevention Information Network (EPINet), only 20% of nurses understood this network system.

### 3.2 *Evaluated experiment*

#### 3.2.1 *Effects of snap-off methods*

Results of ANOVA showed that these were not significantly differences between snap-off methods in EMG of Pronator Teres ($F(2, 72) = 2.17$, $p = 0.12$). However, the ranges of motion for wrist were significant lower while using the novel tool. ($F(2, 72) = 3.05$, $p < 0.05$). In addition, range of motion for wrist angles were 34.4 degrees, 38.3 degrees, and 42.3 degrees in using novel tool, grinding slice and grinding slice with glove respectively (Figure 4). For operating performance analysis (Figure 5), mean time of snap-off was better significantly while using novel tool (20.9 sec) than in using grinding slice with glove (32.2 sec) and grinding slice only (35.8 sec).

#### 3.2.2 *Effects of ampoule sizes*

Results of ANOVA showed that these were significantly differences between ampoule sizes in EMG of Pronator Teres ($F(2, 72) = 23.9$, $p < 0.001$) and operating time ($F(2, 72) = 6.3$, $p < 0.05$).

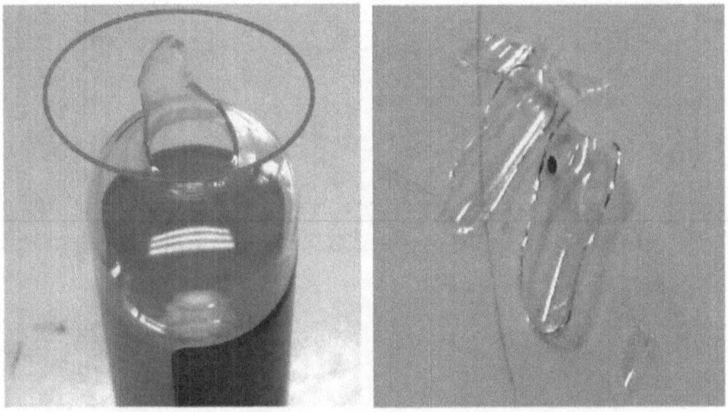

Figure 3. The crushed ampoules.

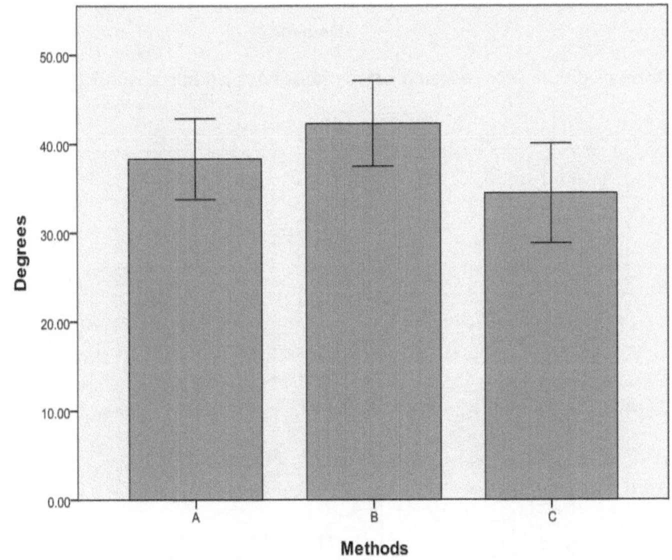


Figure 4. The ranges of motion for wrist between snap-off methods (A: grinding slice; B: grinding slice with glove; C: novel tool).

The pronator teres muscle were heavier load as increased the sizes of ampoule. These are 67.1%, 54.5%, and 46% of maximum isometric voluntary contraction (MVC) for 20 ml, 5 ml, and 1 ml of ampoule respectively (Figure 6). The mean wrist flexion angles were not significant difference between sizes of ampoule.

For performance analysis, Duncan's multiple range test indicated that time of snap-off were longer in 20 ml (35.7 sec) than 5 ml (32.2 sec) and 1 ml (20.9 sec) ampoule (Figure 7).

## 4 DISCUSSION AND CONCLUSION

Percutaneous exposure incidents (PEI) are a broad descriptive term that includes needlestick and sharps injuries, as well as cutaneous and mucous exposures to blood and serum. Smith et al. (2005) investigated all reported percutaneous exposure incidents among staff from a large Australian hospital in the 3-year period (2001–2003). There were a total of 373 PEI, of which 38.9% were

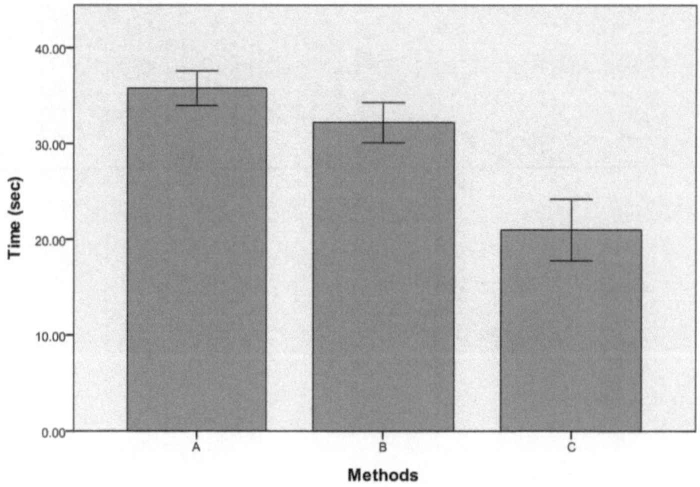


Figure 5. The performance time between snap-off methods (A: grinding slice; B: grinding slice with glove; C: novel tool).

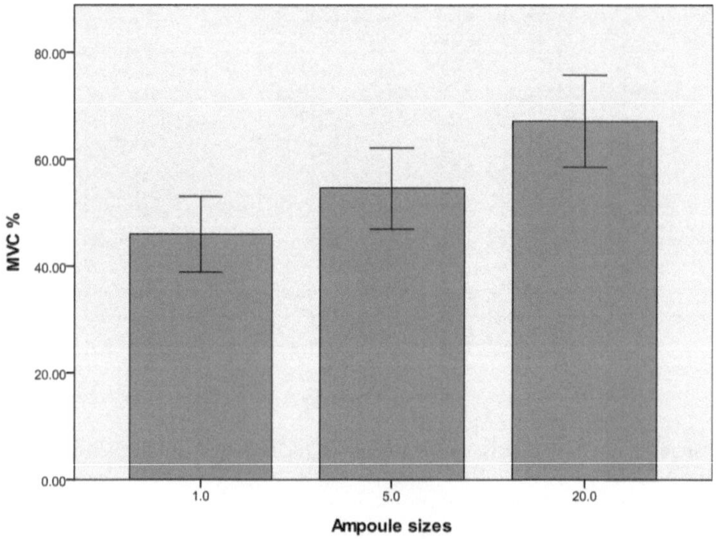


Figure 6. The pronator teres muscle activities between ampoule sizes.

needlestick injuries, 32.7% were cutaneous exposures and 28.4% sharps-related injuries. Needlestick and Sharps Injuries (NSI) represent an important workplace issue in contemporary health care. Australian study of Peggat (1987), NSI as a cause of injury among nurses was exceeded only by musculoskeletal injuries. In a study of Chinese nurses, almost all had experienced at least one NSI in their professional lifetime (Smith et al., 2004). Of the devices involved, 67% were contaminated (previously used on a patient), 15% were not, and the status of the remaining 18% was unknown. Nurses who indicated their departments were understaffed were 7.6 times more likely to have incurred a NSIs than nurses who said their departments were adequately staffed.

In addition, Japanese study conducted in 2004 also found that ampoules or vials, hollow bore needles and insulin needles were common causes of needlestick and sharps Injuries (Smith et al. 2010). Almost half the Japanese nurses had sustained an NSI in the previous year, which is slightly

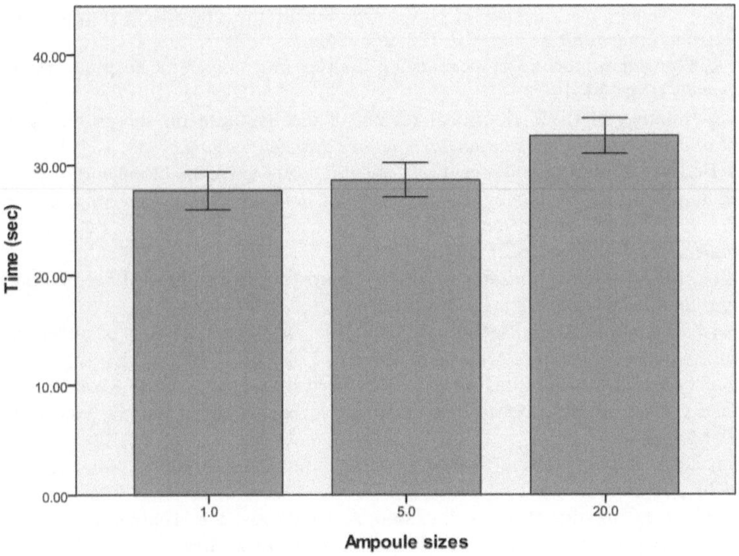


Figure 7. The performance time between ampoule sizes.

higher than that reported from Australia (29%), and UK (38%). However, some other study conducted among Taiwan hospital support personnel (61%), Chinese nurses (82%) (Elmiyeh et al., 2004; Knight & Bodsworth, 1998; Shiao et al., 2001; Phipps et al., 2002). Similar to results, research from Singapore (Ng et al., 2002), Taiwan (Guo et al., 1999), and Korea (Smith et al., 2006) has indicated that disposable syringes and medication ampoules continue to present NSI risks for Health Care Workers. Results of analysis for risk factors of sharp injury showed that 36.9% of nurses had suffered sharp injury. The risk factors to sharp injuries were opening the glass ampoule with gauze and lack of the standard ampoule precautions. The main risk factors leading to the sharp injuries were mistakes or deficiencies in operation.

Prevention of sharps injuries should be aimed at nurses and doctors, and include safe handling of hollow-bore needles. Previous strategies to prevent sharps injuries have included general training and education, whereas more recently the introduction of needle protective devices has become the focus for prevention (Asai et al., 2002; Hanrahan & Reutter, 1997). Thus, present study provided the novel tool effectively to snap-off the ampoule. In addition, results of EMG and wrist postures suggest that novel tool could reduce the risk of needlestick injuries, such as ease of use and saving of time. Evaluating efficacy with regard to reducing sharps injuries and usability in the clinical setting should be examined in further study. Present study provided the novel tool effectively to snap-off the ampoule to decrease forearm workload and sharp injuries.

## REFERENCES

Asai, T. Hidaka, I. Kawashima, A. Miki, T. Inada, K. & Kawachi, S. 2002. Efficacy of catheter needles with safeguard mechanisms. *Anaesthesia* 57(6): 572–577.

Burke, S. & Madan, I. 1997. Contamination incidents among doctors and midwives: reasons for non-reporting and knowledge of risks. *Occupational medicine* 47(6): 357–360.

Cheung, K. Ho, S.C. Ching, S.S.Y. & Chang, K.K.P. 2010. Analysis of needlestick injuries among nursing students in Hong Kong. *Accident Analysis & Prevention* 42(6): 1744–1750.

Dobie, D.K. Worthington, T. Faroqui, M. & Elliott TSJ. 2002. Avoiding the point. Lancet, 259: 1254.

Elmiyeh, B. Whitaker, I.S. James, M.J. Chahal, C.A.A. Galea, A. & Alshafi, K. 2004. Needle-stick injuries in the National Health Service: a culture of silence. JRSM 97(7): 326–327.

Guo, Y.L. Shiao, J. Chuang, Y.C. & Huang, K.Y. 1999. Needlestick and sharps injuries among health-care workers in Taiwan. *Epidemiology infection* 122: 256–265.
Hamory, B.H. 1983. Underreporting of needlestick injuries in a university hospital. *American journal of infection control* 11(5): 174–177.
Hanrahan, A. & Reutter, L. 1997. A critical review of the literature on sharps injuries: epidemiology, management of exposures and prevention. *Journal of advanced nursing* 25(1): 144–154.
Knight, V.M. & Bodsworth, N.J. 1998. Perceptions and practice of universal blood and body fluid precautions by registered nurses at a major Sydney teaching hospital. *Journal of Advanced Nursing* 27(4): 746–751.
Lanphear, B.P. 1994. Trends and patterns in the transmission of bloodborne pathogens to health care workers. *Epidemiologic reviews* 16(2): 437.
Leggat, P.A. 1987. Microbiological hazards posed by 'sharps': complications of needle puncture injuries to nurses. *Australian Microbiologist* 8 (1987): 311–314.
Ng, L.N. Lim, H.L. Chen, Y.H. & Bin Bachok, D. 2002. Analysis of sharps injury occurrences at a hospital in Singapore. *International Journal of Nursing Practice* 8: 274–281.
Occupational Safety and Health Administration. 1999. Record Summary of the Request for Information on Occupational Exposure to Bloodborne Pathogens due to Percutaneous Injury: Executive Summary. US Department of Labor.
Phipps, W. et al. 2002. Risk of medical sharps injuries among Chinese nurses. *American Journal of Infection Control* 30: 277–282
Pournaras, S. Tsakris, A. Mandraveli, K. Faitatzidou, A. Douboyas, J. & Tourkantonist, A. 1999. Reported needlestick and sharp injuries among health care workers in a Greek general hospital. *Occupational Medicine* 49(7): 423–426.
Public Health Laboratory Service AIDS & STD Centre. 1999. Occupational Transmission of HIV: Summary of Published Reports, PHLS.
Shiao, J. C. McLaws, M.L. Huang, K.Y. & Guo, Y.L. 2001. Sharps injuries among hospital support personnel. *Journal of Hospital Infection* 49(4): 262–267.
Shiao, J.S.C. McLaws, M.L. Huang, K.Y. Ko, W.C. & Guo, Y.L. 1999. Prevalence of nonreporting behavior of sharps injuries in Taiwanese health care workers. *American journal of in control*, 27(3): 254–257.
Smith, D.R. Mihashi, M. Adachi, Y. Nakashima, Y. & Ishitake, T. 2006. Epidemiology of needlestick and sharps injuries among nurses in a Japanese teaching hospital. *Journal of Hospital Infection* 64(1): 44–49.
Smith, D.R. Choe, M.A. Jeong, J.S. Jeon, M.Y. Chae, Y.R. & An. G.J. 2006. Epidemiology of needlestick and sharps injuries among professional Korean nurses. *Journal of Professional Nurse* 22: 359–366.
Smith, D.R. Leggat, P.A. & Takahashi, K. 2006. Percutaneous exposure incidents among Australian hospital staff. *International Journal of Occupational Safety and Ergonomics* 11: 323–330.
Smith, D.R. Muto, T. Sairenchi, T. Ishikawa, Y. Sayama, S. Yoshida, A. & Townley-Jones, M. 2010. Hospital safety climate, psychosocial risk factors and needlestick injuries in Japan. *Industrial Health* 48: 85–95.
Smith, D.R. Wei, N. & Wang, R.S. 2004. Needlesticks and sharps injuries among Chinese hospital nurses. *Advances in Exposure Prevention* 7: 11–12.
Trim, J.C. & Elliott, T.S.J. 2003. A review of sharps injuries and preventative strategies. *Journal of Hospital Infection* 53: 237–242.

*Bridging Research and Good Practices towards Patient Welfare – Shih & Liang (Eds)*
*© 2015 Taylor & Francis Group, London, ISBN 978-1-138-02716-9*

# Effect of different wrist positions on median nerve cross-sectional area at proximal carpal tunnel

Ping Yeap Loh & Hiroki Nakashima
*Graduate School of Design, Kyushu University, Fukuoka, Japan*

Satoshi Muraki
*Faculty of Design, Kyushu University, Fukuoka, Japan*

ABSTRACT: Musculoskeletal disorders such as carpal tunnel syndrome can be caused by repeated, awkward wrist postures during daily work. The objective of this study was to identify the changes of median nerve cross-sectional area at different wrist angles. Twelve right-handed males participated in this study. Ultrasound examination was performed on the right wrist. A total of seven angles were examined, namely, wrist neutral (0°), wrist extension 15°, 30°, 45° and wrist flexion 15°, 30°, 45°. The results showed that the reduction of median nerve cross-sectional area was significant when wrist neutral changed to wrist extension and wrist flexion ($p < 0.001$). Our findings showed that the reduction of median nerve cross-sectional area was greater at wrist extension 45° and wrist flexion 45° than at wrist neutral and other wrist extension and flexion angles.

## 1 INTRODUCTION

Computers have become essential tools in daily life. Computer systems have been introduced in many industries to improve system output. There is a paradigm shift from paper-based documentation to electronic-based documentation in daily work, for example, the use of electronic medical records in healthcare. With the high acceptance of mobile and fixed computer station usage in a clinical setting, nurses and other medical staff are required to perform basic computer operations such as keyboard typing and mouse use in their daily work routine (Ammenwerth et al., 2003; Andersen et al., 2009).

Musculoskeletal disorders (MSD) are known to be increasing among workers in the healthcare sector, such as nurses and sonographers. Disabilities arising from work-related MSD may affect worker performance and quality of life. Various risk factors contribute to MSD, such as physical work demands and awkward postures during work. In the daily occupational activities of nurses, repeated wrist and finger motion occurs in up to 68% (Freimann et al., 2013). A high rate of MSD of the upper limbs, involving wrist and finger joints, has also been reported. Good ergonomic practices in workplaces are thus important to reduce the risk factors that contribute to repetitive strain disorder of the upper limbs.

Carpal tunnel syndrome (CTS) is one of the most commonly reported MSD. Bonfiglioli et al. (2007) suggested that biomechanical stress and intensive manual work associated with inadequate recovery time might generate impairment of the median nerve at the wrist level, and the risk increases proportionally with increasing duration of hand use. Thus, awkward wrist joint posture and repetitive wrist joint movement are considered as factors associated with a high risk of CTS. Mechanical stress caused by different wrist angles on the median nerve may lead to peripheral nerve compression syndrome.

The rapid advancement in ultrasound (US) imaging technology has allowed portable US machines to be used in different places and the production of high-quality images. The advantages of US are that it is non-invasive and provides real-time imaging at low cost. US has been

Table 1. Demographic data of the participants (n = 12).

| | Minimum | Maximum | Mean ± SD |
|---|---|---|---|
| Age (years) | 21.6 | 29.3 | 24.3 ± 2.3 |
| Height (cm) | 165.0 | 186.0 | 171.4 ± 5.6 |
| Weight (kg) | 51.6 | 112.6 | 69.5 ± 17.6 |
| BMI (kg/m$^2$) | 17.5 | 38.4 | 23.6 ± 5.6 |
| Wrist circumference (cm) | | | |
| Right | 15.0 | 18.3 | 16.3 ± 0.9 |
| Left | 15.0 | 17.8 | 16.0 ± 0.9 |

widely applied in musculoskeletal diseases such as peripheral neuropathy (Wong et al., 2009). Structures under the skin can be observed by using high-resolution US imaging. For example, compression of the median nerve and the degree of displacement during tendon motion at the carpal tunnel can be observed through real-time US.

The objective of this study is to investigate the effect of wrist extension and flexion motion on the median nerve cross-sectional area (MNCSA) at the proximal carpal tunnel level by using US imaging.

## 2 MATERIALS AND METHODS

### 2.1 *Participants*

Twelve right-handed healthy males were recruited for this study (Table 1). All the participants were free from signs and symptoms of CTS, as indicated by screening tests such as Phalen's test, CTS Tinel test and Boston Carpal Tunnel Questionnaire (BCTQ) (Bakhsh et al., 2012; LaJoie et al., 2005; Sambandam et al., 2008). This study was approved by the Ethics Committee, Faculty of Design, Kyushu University.

### 2.2 *Ultrasound examination*

Ultrasound (US) examinations were performed using *GE Healthcare Ultrasound System* (Model LOGIQ e) with a 6–13 mHz transducer (Model 12L-RS). A sonarpad (thickness 7.0 mm; Nippon BXI Inc., Japan) was used as the coupling medium during these examinations.

The MNCSA was determined by a direct measurement method, a tracing method (TM) (Duncan et al., 1999), using *ImageJ* software (Schneider et al., 2012). Specifically, the MNCSA was measured by tracing the outer hypoechogenic rim of the median nerve (Fig. 1). The average of measurements of three ultrasound images was used to represent the MNCSA at each wrist position. The MNCSA at wrist neutral was used as a reference point.

The participants sat upright with their forearm resting on a forearm support on a table and were instructed to relax their wrist and fingers during US examination. Seven wrist positions were examined, comprising wrist neutral (0°), wrist extension 15°, 30°, 45° and wrist flexion 15°, 30°, 45° (Fig. 2). The median nerve of the right wrist was examined at the proximal carpal tunnel level in the transverse plane. Pisiform bone mark was used as a landmark of the proximal carpal tunnel during US examination.

### 2.3 *Statistical analysis*

All results are expressed as mean ± standard deviation (SD). One-way repeated ANOVA was performed to identify the effect of wrist neutral changing to other wrist positions on MNCSA. Post hoc pairwise Bonferroni-corrected comparisons were carried out to determine the mean differences

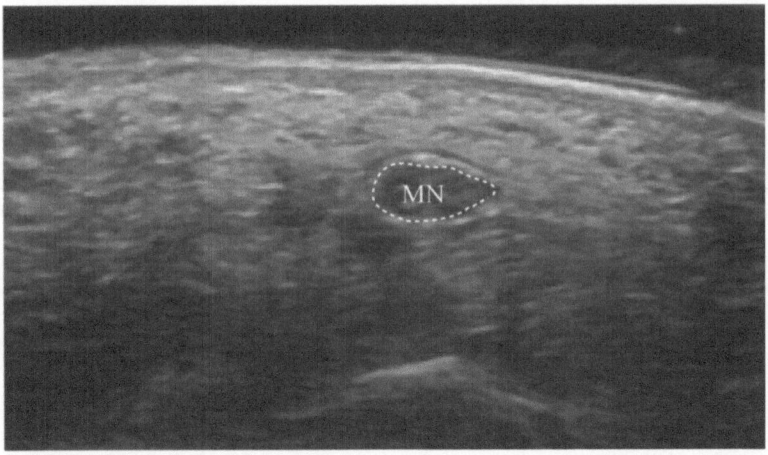


Figure 1. Tracing method on median nerve (MN).

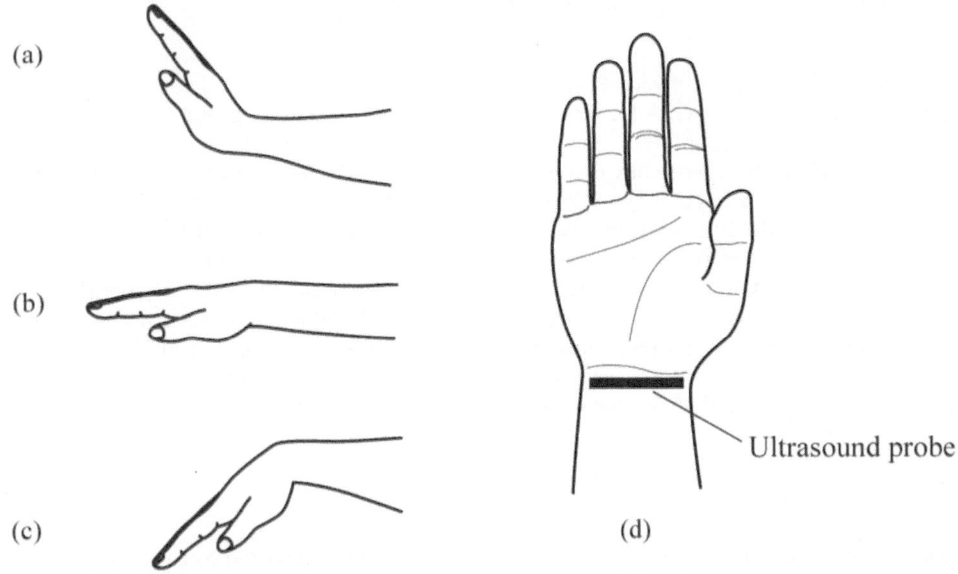


Figure 2. Wrist positions for US examination and ultrasound probe location: (a) wrist extension; (b) wrist neutral; (c) wrist flexion; (d) ultrasound probe.

among the different wrist positions. All statistical analyses were performed using SPSS version 21.0 software (SPSS Inc., Chicago, IL).

## 3 RESULT

### 3.1 *MNCSA changes with different wrist positions*

The mean value of MNCSA at right wrist neutral (0°) was $8.44 \pm 1.40$ mm$^2$. There was a significant effect when wrist neutral was changed to wrist extension ($p < 0.001$) (Figure 3). The mean values of MNCSA of right wrist extension at 15°, 30° and 45° were $7.78 \pm 1.44$ mm$^2$, $7.07 \pm 1.12$ mm$^2$

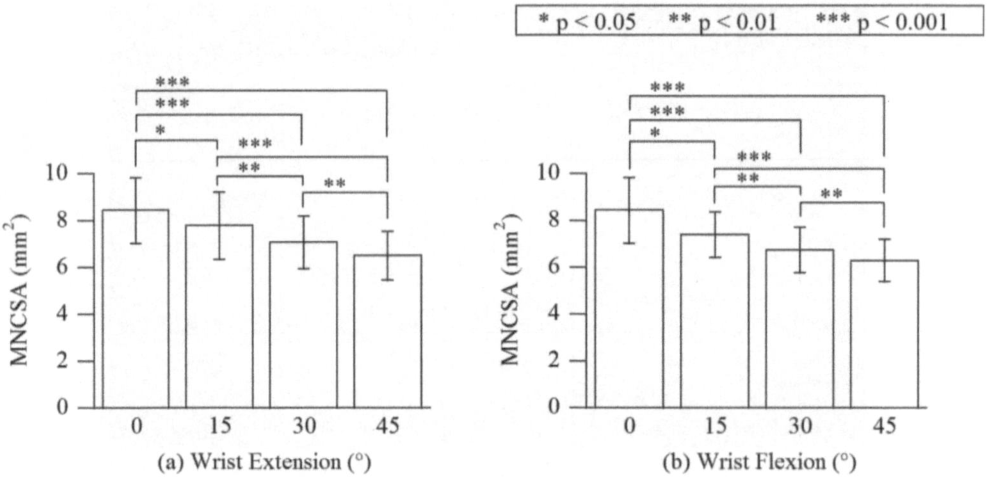


Figure 3. Median nerve cross-sectional area (MNCSA) at different wrist positions: (a) wrist neutral change to wrist extension; (b) wrist neutral change to wrist flexion.

and $6.50 \pm 1.04\,mm^2$, respectively. Post hoc pairwise Bonferroni-corrected comparisons showed significant reduction in MNCSA for wrist extension 15° ($p < 0.05$), along with wrist extension 30° and wrist extension 45° ($p < 0.001$), compared with that at wrist neutral.

Similarly, the effect of wrist flexion on MNCSA of the right wrist was significant ($p < 0.001$) (Figure 3). The mean values of the MNCSA at wrist flexion 15°, 30° and 45° were $7.39 \pm 0.98\,mm^2$, $6.74 \pm 0.97\,mm^2$ and $6.29 \pm 0.90\,mm^2$, respectively. The reductions in MNCSA compared with that at wrist neutral were significant for wrist flexion 15° ($p < 0.05$), 30° and 45° ($p < 0.001$).

A comparison of MNCSA percentage changes was made among wrist extension and flexion positions with wrist neutral as a reference point (Table 2).When the wrist extension increased to 15°, 30° and 45°, the MNCSA reductions were approximately 7.8%, 16.2% and 23.0%. A similar trend in MNCSA reduction was observed when wrist neutral changed to wrist flexion positions, namely, 12.4%, 20.1% and 25.5% for wrist flexion 15°, 30° and 45°.

## 4 DISCUSSION

Individual and multiple finger extension and flexion motion was shown to cause median nerve compression (Lopes et al., 2011). However, the effect of passive wrist motion at different angles on MNCSA was unknown. Three different angles were examined for both wrist extension and flexion, namely, 15°, 30° and 45° . As shown in Figure 3, reduction of MNCSA was observed when the wrist changed from wrist neutral to wrist extension and flexion angles. There were significant differences when the wrist angle changed from 15° to 30°, 15° to 45° and 30° to 45° for both wrist extension and flexion motion. The reduction in MNCSA could be caused by compression stress within the carpal tunnel during wrist extension and flexion motion. Therefore, computer users should avoid working with larger wrist extension and flexion angles during keyboard typing because they can lead to higher compression stress on the median nerve.

Mogk & Keir (2009) described that predicted reductions of the carpal tunnel volume when wrist neutral was changed to wrist extension 30° and wrist flexion 30° were −7.0% and −6.4%, respectively. Although the reductions in carpal tunnel volume were similar for extension 30° and flexion 30°, from Table 2, the percentage reductions of MNCSA for wrist extension 30° and flexion 30° motion were −16.2% and −20.1%, respectively. Wrist flexion may thus cause higher compression stress on the median nerve.

Table 2. Median nerve cross-sectional area ($mm^2$) at different wrist angles and percentage changes compared with wrist neutral (n = 12).

| | Wrist angle | | | |
|---|---|---|---|---|
| Wrist motion | Neutral (0°) | 15° | 30° | 45° |
| Extension (MNCSA*) | 8.44 ± 1.40 | 7.78 ± 1.44 | 7.07 ± 1.12 | 6.50 ± 1.04 |
| Percentage changes (%) | N/A# | −7.8 | −16.2 | −23.0 |
| Flexion (MNCSA*) | 8.44 ± 1.40 | 7.39 ± 0.98 | 6.74 ± 0.97 | 6.29 ± 0.90 |
| Percentage changes (%) | N/A# | −12.4 | −20.1 | −25.5 |

* MNCSA = median nerve cross-sectional area
#N/A = Not applicable

Most occupations require workers to perform repetitive tasks in their daily work routines. A cross-sectional survey showed that 17% of workers, such as nurses and office workers, reported discomfort at the wrist and hand region after the use of a keyboard and repeated wrist and hand motion (Coggon et al., 2013).

In this pilot experiment, the effect of wrist angle changes on MNCSA was determined for the right dominant hand. However, the differences of MNCSA between dominant and non-dominant hands at different wrist motions are unknown. Therefore, further research is needed to understand the effect of dynamic wrist angles on MNCSA changes during daily computer work.

## 5 CONCLUSION

In this study, a trend of a reduction in MNCSA was observed when wrist neutral was changed to wrist extension and flexion. The wrist neutral position showed the largest MNCSA among all wrist positions. In comparison, both wrist extension 45° and flexion 45° showed greater MNCSA reduction than other wrist positions. Thus, higher wrist extension and flexion may cause higher compression stress on the median nerve.

## REFERENCES


Ammenwerth, E., Mansmann, U., Iller, C., & Eichstadter, R. (2003). Factors affecting and affected by user acceptance of computer-based nursing documentation: Results of a two-year study. *Journal of the American Medical Informatics Association: JAMIA, 10*(1), 69–84.

Andersen, P., Lindgaard, A. M., Prgomet, M., Creswick, N., & Westbrook, J. I. (2009). Mobile and fixed computer use by doctors and nurses on hospital wards: Multi-method study on the relationships between clinician role, clinical task, and device choice. *Journal of Medical Internet Research, 11*(3), e32.

Bakhsh, H., Ibrahim, I., Khan, W., Smitham, P., & Goddard, N. (2012). Assessment of validity, reliability, responsiveness and bias of three commonly used patient-reported outcome measures in carpal tunnel syndrome. *Ortopedia, Traumatologia, Rehabilitacja*, 14(4), 335–340.

Bonfiglioli, R., Mattioli, S., Fiorentini, C., Graziosi, F., Curti, S., & Violante, F. S. (2007). Relationship between repetitive work and the prevalence of carpal tunnel syndrome in part-time and full-time female supermarket cashiers: A quasi-experimental study. *International Archives of Occupational and Environmental Health, 80*(3), 248–253.

Coggon, D., Ntani, G., Palmer, K. T., Felli, V. E., Harari, R., Barrero, L. H., et al. (2013). Disabling musculoskeletal pain in working populations: Is it the job, the person, or the culture? *Pain, 154*(6), 856–863.

Duncan, I., Sullivan, P., & Lomas, F. (1999). Sonography in the diagnosis of carpal tunnel syndrome. *AJR. American Journal of Roentgenology, 173*(3), 681–684.

Freimann, T., Coggon, D., Merisalu, E., Animagi, L., & Paasuke, M. (2013). Risk factors for musculoskeletal pain amongst nurses in estonia: A cross-sectional study. *BMC Musculoskeletal Disorders, 14*, 334-2474-14-334.

LaJoie, A. S., McCabe, S. J., Thomas, B., & Edgell, S. E. (2005). Determining the sensitivity and specificity of common diagnostic tests for carpal tunnel syndrome using latent class analysis. *Plastic and Reconstructive Surgery*, 116(2), 502–507.

Lopes, M. M., Lawson, W., Scott, T., & Keir, P. J. (2011). Tendon and nerve excursion in the carpal tunnel in healthy and CTD wrists. *Clinical Biomechanics (Bristol, Avon), 26*(9), 930–936.

Mogk, J. P. M., & Keir, P. J. (2009). The effect of landmarks and bone motion on posture-related changes in carpal tunnel volume. *Clinical Biomechanics, 24*(9), 708–715.

Sambandam, S. N., Priyanka, P., Gul, A., & Ilango, B. (2008). Critical analysis of outcome measures used in the assessment of carpal tunnel syndrome. *International Orthopaedics*, 32(4), 497–504.

Schneider, C. A., Rasband, W. S., & Eliceiri, K. W. (2012). NIH image to ImageJ: 25 years of image analysis. *Nature Methods, 9*(7), 671–675.

Wong, D. C., Wansaicheong, G. K., & Tsou, I. Y. (2009). Ultrasonography of the hand and wrist. *Singapore Medical Journal, 50*(2), 219–25; quiz 226.

*Bridging Research and Good Practices towards Patient Welfare – Shih & Liang (Eds)*
*© 2015 Taylor & Francis Group, London, ISBN 978-1-138-02716-9*

# Relationships of musculoskeletal disorder symptoms and perceived workload among hospital workers

Taebeum Ryu
*Hanbat National University, Daejeon, South Korea*

Sanghyun Kwon, Soo-Chan Jee, Myoung-Don Oh & Myung Hwan Yun
*Seoul National University, Seoul, South Korea*

ABSTRACT: This study attempts to investigate the prevalence of musculoskeletal disorders (MSD) among Korean hospital workers and to analyze the relationship of MSD symptoms and workload perceived by workers. This study designed a survey based on Nordic questionnaire to obtain MSD symptoms and the degree of four perceived workloads: work repetition, urgency, physical exertion and satisfaction. In this survey, more than 1,800 workers in a hospital participated. The prevalence of MSD was analyzed for each body part, and MSD cases, which were predetermined in this study, were identified. The relationship between the MSD cases and each perceived workload was analyzed using chi-square test. The pain in the shoulder was the most prevalent among the workers as 52%, and the low back (37%) and leg discomfort (36%) followed. The MSD cases of this study were also the most prevalent in the shoulder (13%). Female workers had higher rate of MSD cases than the males. Among the four workload variables, the physical exertion was the statistically related to MSD cases for all the body parts. In addition, the others also had significant relation to MSD cases except one or two body parts. This study found that Korean hospital workers had MSD symptoms mainly in the shoulder, low back and legs in order, which was different from those in North America or Europe. This study provides another evidence that subjective physical exertion perceived by workers is an important factor to explain MSD cases as same as the objective one.

*Keywords*: Musculoskeletal disorder, Hospital workers, Psychosocial factors, Perceived workload

## 1 INTRODUCTION

Tasks performed by hospital workers include most of risk factors related to musculoskeletal disorders (MSD) such as high repetition, awkward posture, prolonged body motion, and forceful manual exertion. Hospital workers do their simple and repetitive works for long time. Most of their works are atypical and require awkward postures and high force for handling heavy patients and objects. Due to these reasons, hospital workers have been known to be highly exposed to MSD incidents (Park, et al., 2005).

It have been reported that hospital workers have high prevalence and incidence of MSD by many previous studies. The rate of MSD incidence among health care workers was the highest as 3.2% in the United States comparing other industries (Bernard, et al., 1997). Caillard and Iwatsubo (Cha, et al., 2007) summarized that the rate of low back pain among the heath care workers was about 50% compared to 20–50% among workers in the other industries. Other related studies also reported that the prevalence rate of low back pain was 30–85% among health care workers (Murtomaa, 1982). In addition, the upper limb and neck pain rate among Nebraskan dentists was 29% (Fish & Morris-Allen, 1998), the shoulder and hand pain rates 47% and 27% respectively (Nakladalova, et al., 1995), and the carpal tunnel syndrome symptoms rate among cardiac sonographer 63% (Vanderpool, et al., 1993).

Despite of high exposure to the MSD risk factors and high MSD symptom prevalence among health care workers, there were few studies of MSD prevalence among the hospital workers in South Korea. Although many studies investigated the MSD prevalence among the Korean workers in automobile manufacturing, shipbuilding, steel and electronics, the studies for hospital workers are limited (Kim, et al., 2010). Additionally, there are many MSD surveys for Korean nurses (JK P, et al., 2010), (Seo and Kee, 2005), only the study of (Park, et al., 2008) looked into the MSD prevalence among the hospital workers.

Though there were several evidences about the relationship between the MSD prevalence and some psychosocial factors, more studies are needed to find the more various psychosocial variables related to MSD. In the MSD prevalence study among hospital laundry workers, Village (Village, 2000) analyzed the effects of psychosocial variables such as work interesting, work variation, relationship with coworkers, ability to influence work condition on MSD. The study found some significant relationships such the variety in work related to less pain and mentally exhausted after work did to more pain. For nurses, the psychosocial variables related to MSD symptoms for some body parts (Lagerström, et al., 1996), and physical exertion as well as job stress had a significant effect on MSD symptoms (Josephson, 2011). Bergqvist et al. (Bergqvist, et al., 1995) reported that in video display terminal (VDT) the influence of organizational variables such as flexible rest break, extreme peer contact, and task flexibility had significant effects on MSD symptoms. Lee et al. (Lee, et al., 2007) also insisted that the MSD symptoms and perceived work-related stress were related each other among automobile manufacturing workers. However, the perceived workload by workers such as perceived repetition, urgency and physical exertion was not studied yet as psychosocial variables related to MSD.

For the two reasons above, this study investigated the prevalence of MSD among Korean hospital workers and attempted to analyze the relationship of MSD symptoms and detailed perceived workload and work satisfaction. To do this, this study prepared a questionnaire to survey MSD symptoms, perceived workload, and work satisfaction for hospital workers. One thousand and eight hundred workers of a hospital participated in this MSD survey.

## 2 MATERIALS AND METHODS

### 2.1 *Questionnaire*

A questionnaire was prepared to obtain 1) basic demographic information, 2) perceived workload and satisfaction 3) symptoms and its degree of each body part. First part of the questionnaire included demographic questions such as gender, age, stature, weight, and type of worker. In the second section, the degree of work repetition, urgency, physical exertion and work satisfaction were measured with 5-point Likert scale from 'strongly disagree' to 'strongly agree'. Last part of the questionnaire contained questions about the occurrence of pain for six body parts during the last 12 months according to Nordic questionnaire. And the intensity of pain was included to rate the severity of symptoms with 4-point scales: "mild", "moderate", "severe", and "unbearable" based on Pennsylvania State University MSD Survey form (You et. al., 1999). Cases of this study were defined as symptoms which occurred in last 12 months with more than severe intensity.

### 2.2 *Subjects*

Hospital workers in a large-size hospital participated in this survey and the questionnaire was self-administered. Of a total of 3,008 hospital workers, 1,852 workers (62% of total employee) responded to the survey (Table 1). Of the responded workers, most of them were female; 78.5% (1454) was female and 21.5% (398 workers) was male. Moreover, of them, 43% were in their 20s, 36% in their 30s, 16% in their 40s and 5% in their 50s. The mean age of them was 32: the males averaged 36.5 years, and the females 30.9. The mean of working period was 7 years and 9 months: the males' mean was 8 years and 7 months, and the females' was 7 years and 6 months.

Table 1. Composition of responded hospital workers.

| | Male | | Female | | Total | |
|---|---|---|---|---|---|---|
| Age | Frequency | % | Frequency | % | Frequency | % |
| ≤29 | 105 | 26.4 | 692 | 47.6 | 797 | 43.0 |
| 30–39 | 138 | 34.7 | 531 | 36.5 | 669 | 36.1 |
| 40–49 | 108 | 27.1 | 184 | 12.7 | 292 | 15.8 |
| ≥50 | 43 | 10.8 | 34 | 2.3 | 77 | 4.2 |
| unclear | 4 | 1.0 | 13 | 0.9 | 17 | 0.9 |
| Total | 398 | 100 | 1454 | 100 | 1852 | 100 |

Table 2. Distribution of perceived workload and satisfaction.

| | Repetition | | Urgency | | Physical exertion | | Satisfaction | |
|---|---|---|---|---|---|---|---|---|
| | Frequency | Ratio (%) | Frequency | Ratio | Frequency | Ratio | Frequency | Ratio |
| Strongly Disagree | 19 | 1 | 13 | 1 | 41 | 2 | 23 | 1 |
| Disagree | 135 | 7 | 184 | 10 | 356 | 19 | 153 | 8 |
| Neutral | 423 | 23 | 589 | 32 | 601 | 33 | 975 | 53 |
| Agree | 885 | 48 | 811 | 44 | 630 | 34 | 624 | 34 |
| Strongly agree | 384 | 21 | 249 | 13 | 218 | 12 | 71 | 4 |

### 2.3 *Statistical analysis*

This study analyzed MSD prevalence, MSD cases and the relationship of MSD cases and the perceived workload. The occurrence rate of MSD symptoms was analyzed for each body part, and MSD cases of this study, pains occurred in last 12 months with more than severe intensity, were identified. The cases in each body part were compared between male and female and among worker types. The relationship between the MSD cases and the each detailed perceived workload was analyzed using chi-square test.

## 3 RESULTS

### 3.1 *Distribution of perceived workload and work satisfaction*

The frequency and proportion according to the degree of perceived workload and work satisfaction were summarized in Table 2. About 70% of the responded hospital workers perceived their works as repetitive or more. On the question of if their works are repetitive, 21% of the workers were in 'strongly agree', and 48% in 'agree'. On the perceived work urgency, about 60% of the hospital workers reported their works were urgent or more. On the question of if their works are urgent, 13% of the workers were in 'strongly agree', and 44% in 'agree'.

For the perceived physical exertion, about 46% of the hospital workers perceived that they used manual force during their work. For the question of if they use physical exertion in their work, 12% of the workers were in 'strongly agree', and 34% in 'agree'. For the work satisfaction, while only 9% of the workers were satisfied with their works, 38% were not. For the question of if they are satisfied with their work, 1% of the workers were in 'strongly agree', 8% in 'agree', 53% in 'neutral', 34% in 'disagree' and 4% in 'strongly disagree'. The satisfaction of the hospital workers was correlated low with the other three perceived workload. The correlation coefficients between satisfaction and repetition, urgency and physical exertion were 0.23, 0.31 and 0.23, respectively.

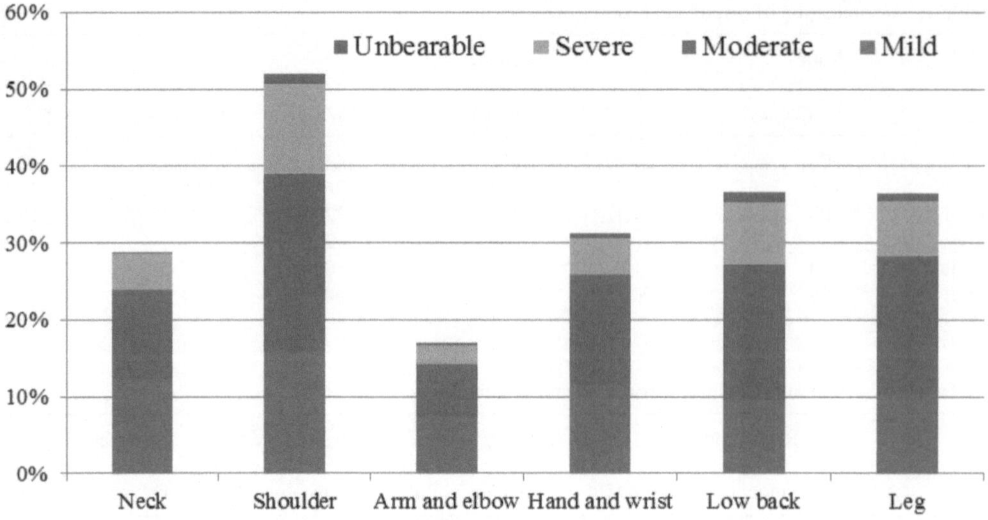


Figure 1. Prevalence of MSD symptom and its severity among hospital workers.

Table 3. Frequency of MSD cases of six body parts between male and female workers.

| | Neck | | Shoulder | | Arm/Elbow | | Hand/Wrist | | Low back | | Leg | |
|---|---|---|---|---|---|---|---|---|---|---|---|---|
| | Ob. | Ex. | Ob. | Ex. | Ob. | Ex. | Ob. | Ex. | Ob. | Ex. | Ob. | Ex. |
| Male | 11 | 19.5 | 18 | 51.5 | 1 | 11.1 | 5 | 21.2 | 17 | 37.0 | 6 | 31.9 |
| Female | 79 | 71.1 | 220 | 188.3 | 51 | 40.6 | 94 | 77.3 | 155 | 135.2 | 143 | 116.4 |
| $\chi^2$ | 4.6 | | 27.2 | | 11.9 | | 15.9 | | 13.7 | | 27.0 | |
| $p$ | <0.05 | | <0.05 | | <0.05 | | <0.05 | | <0.05 | | <0.05 | |

Abbreviations: Ob, Observed frequency of cases; Ex,. Expected frequency of cases

### 3.2 *MSD symptoms of hospital workers*

Figure 1 shows the proportion of the hospital workers reporting pain in each body part within the last 12 months. The pain in the shoulder was the most prevalent among the hospital workers as 52%, and the low back (37%) and leg discomfort (36%) followed. The prevalence of neck and arm/elbow pain was relative small as 29% and 17%, respectively.

The MSD cases of this study, pain which occurred in last 12 months with more than severe intensity, were also the most prevalent in the shoulder (13%), and the low back (9%) and legs (8%) in order. The prevalence of cases in hand/wrist, neck, and arm/elbow was less than 5%.

As results of Chi-square test, the occurrence of MSD cases among the hospital workers was significantly different by gender for the six body part (Table 3). The observed frequency of cases among the female workers was more than the expected for all the body parts, while vice versa among the male workers. Especially, the difference between the observed and expected frequency was large in the shoulder and legs compared to the other body parts.

The major painful body parts were also different by worker types. The proportion of MSD cases was shown along with body parts in Figure 2. Among the nurses, the cases appeared mainly in the shoulder (12%), low back (12%) and legs (11%), and among the administrative assistants, the cases were distributed almost evenly in the neck (9%), shoulder (9%), hand/wrist (9%), low back (9%) and legs (11%). Among the pharmacists, while the cases were not in the legs, there were

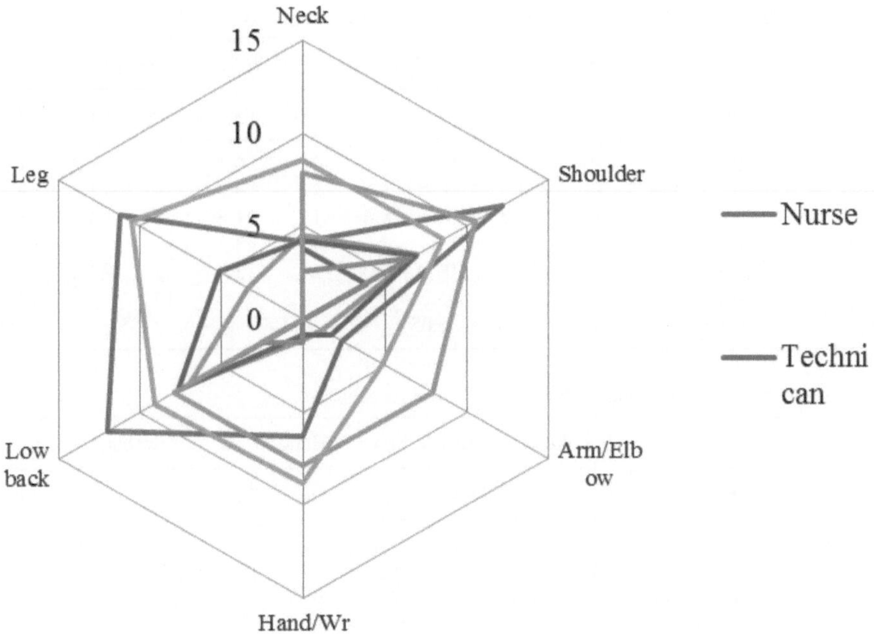


Figure 2. Proportion of MSD cases by worker types.

Table 4. Frequency of cases along with perceived work repetition.

| | Neck | | Shoulder | | Arm/Elbow | | Hand/Wrist | | Low back | | Leg | |
|---|---|---|---|---|---|---|---|---|---|---|---|---|
| | Ob. | Ex. | Ob. | Ex. | Ob. | Ex. | Ob. | Ex. | Ob. | Ex. | Ob. | Ex. |
| Strong disagree | 0 | 0.9 | 1 | 2.5 | 0 | 0.5 | 0 | 1.0 | 0 | 1.8 | 0 | 1.5 |
| Disagree | 1 | 6.6 | 4 | 17.6 | 3 | 3.8 | 1 | 7.2 | 7 | 12.6 | 6 | 10.9 |
| Neutral | 13 | 20.6 | 38 | 55.2 | 7 | 11.9 | 12 | 22.7 | 25 | 39.4 | 19 | 34.1 |
| Agree | 40 | 43.2 | 113 | 115.7 | 21 | 25.0 | 45 | 47.5 | 83 | 82.5 | 74 | 71.5 |
| Strong agree | 36 | 18.7 | 85 | 50.1 | 21 | 10.8 | 41 | 20.6 | 57 | 35.7 | 50 | 30.9 |
| $\chi^2$ | 8.7 | | 16.8 | | 3.4 | | 11.6 | | 9.5 | | 10.5 | |
| $p$ | <0.05 | | <0.05 | | >0.05 | | <0.05 | | <0.05 | | <0.05 | |

considerable cases in the other body parts. Among the other worker types, the frequency of cases was low.

### 3.3 *Relationship of MSD cases and perceived workload*

Chi-square test was conducted to test if distribution of the cases is different with perceived work repetition for the six body parts (Table 4). In all the body parts except the arm/elbow, the case distribution was statistically different along with the degree of perceived work repetition, and there were more cases among the workers who perceived their works repetitive than not. The observed frequency of cases among the workers who at least agreed with the work repetition was more than the expected, while vice versa among ones who was neutral at most.

Likewise, chi-square test was conducted to test if the case distribution is different with perceived work urgency (Table 5). In all the body parts except the neck and legs, the distribution was significantly different along with the degree of perceived work urgency. The observed case frequency

Table 5. Frequency of cases along with perceived work urgency.

| | Neck | | Shoulder | | Arm/Elbow | | Hand/Wrist | | Low back | | Leg | |
|---|---|---|---|---|---|---|---|---|---|---|---|---|
| | Ob. | Ex. | Ob. | Ex. | Ob. | Ex. | Ob. | Ex. | Ob. | Ex. | Ob. | Ex. |
| Strong disagree | 0 | 0.6 | 0 | 1.7 | 0 | 0.4 | 0 | 0.7 | 2 | 1.2 | 1 | 1.0 |
| Disagree | 4 | 9.0 | 14 | 24.0 | 6 | 5.2 | 8 | 9.9 | 9 | 17.1 | 9 | 14.8 |
| Neutral | 23 | 28.7 | 49 | 76.9 | 3 | 16.6 | 17 | 31.6 | 35 | 54.9 | 35 | 47.2 |
| Agree | 40 | 39.5 | 119 | 105.9 | 24 | 22.8 | 45 | 43.5 | 83 | 75.6 | 67 | 65.0 |
| Strong agree | 23 | 12.1 | 59 | 32.5 | 19 | 7.0 | 29 | 13.4 | 43 | 23.2 | 36 | 20.0 |
| $\chi^2$ | 4.5 | | 17.6 | | 11.7 | | 7.8 | | 12.3 | | 5.5 | |
| $p$ | >0.05 | | <0.05 | | <0.05 | | <0.05 | | <0.05 | | >0.05 | |

Table 6. Frequency of cases along with perceived physical exertion.

| | Neck | | Shoulder | | Arm/Elbow | | Hand/Wrist | | Low back | | Leg | |
|---|---|---|---|---|---|---|---|---|---|---|---|---|
| | Ob. | Ex. | Ob. | Ex. | Ob. | Ex. | Ob. | Ex. | Ob. | Ex. | Ob. | Ex. |
| Strong disagree | 5 | 2.0 | 6 | 5.3 | 2 | 1.1 | 1 | 2.2 | 2 | 3.8 | 2 | 3.3 |
| Disagree | 11 | 17.2 | 23 | 46.3 | 3 | 9.8 | 9 | 18.7 | 13 | 33.0 | 8 | 28.7 |
| Neutral | 17 | 29.0 | 53 | 78.1 | 7 | 16.6 | 14 | 31.6 | 31 | 55.7 | 28 | 48.5 |
| Agree | 39 | 30.4 | 98 | 81.9 | 28 | 17.4 | 45 | 33.1 | 78 | 58.4 | 71 | 50.9 |
| Strong agree | 17 | 10.5 | 60 | 28.3 | 11 | 6.0 | 28 | 11.5 | 47 | 20.2 | 40 | 17.6 |
| $\chi^2$ | 14.2 | | 23.0 | | 17.4 | | 19.7 | | 30.5 | | 32.1 | |
| $p$ | <0.05 | | <0.05 | | <0.05 | | <0.05 | | <0.05 | | <0.05 | |

among the workers who perceived their work urgent or more was more than the expected, while vice versa among ones who was neutral at most. Though this trend was found in the neck and legs, it was not statistically significant.

If the case distribution is different with perceived physical exertion was also tested using chi-square test (Table 6). In all the body parts, the distribution was statistically different along with the degree of perceived physical exertion. The observed frequency of cases among the workers who at least agreed with the physical exertion was more than the expected, while vice versa among ones who was neutral at most.

Lastly, if the case distribution is different with work satisfaction was tested (Table 7). In all the body parts except the arm/elbow, the distribution was significantly different along with the degree of work satisfaction. The observed frequency of cases among the workers who were dissatisfied with their work was more than the expected, while vice versa among ones who was neutral or satisfied.

## 4 DISCUSSION

Agreed to the previous studies, the present study found that the shoulder pain was most prevalent (52%) among the hospital workers and it may be related to the fact that the nurses are the majority of the hospital workers. In the study, the most prevalent pain among nurses was found in the shoulder, low back and legs in order. That was exactly same as the prevalence order among the whole hospital workers. Smith et al. (Smith, et al., 2003) conducted a MSD survey among Japanese nurses and found that shoulder pain was the most prevalent. In the study of (Bejia, et al., 2005), nurses' the main body parts where pain occurred had similar order: the shoulder, low back, neck and legs.

It was also found that the case prevalence among female workers was higher than among the male in this study, which means that the hospital workloads are still challenging to the female.

Table 7. Frequency of cases along with work satisfaction.

| | Neck | | Shoulder | | Arm/Elbow | | Hand/Wrist | | Low back | | Leg | |
|---|---|---|---|---|---|---|---|---|---|---|---|---|
| | Ob. | Ex. | Ob. | Ex. | Ob. | Ex. | Ob. | Ex. | Ob. | Ex. | Ob. | Ex. |
| Strong agree | 3 | 3.5 | 7 | 9.2 | 1 | 2.0 | 1 | 3.8 | 2 | 6.6 | 2 | 5.7 |
| agree | 22 | 30.4 | 55 | 81.1 | 10 | 17.6 | 22 | 33.1 | 52 | 57.8 | 33 | 50.0 |
| Neutral | 48 | 47.5 | 143 | 126.8 | 34 | 27.5 | 58 | 51.8 | 89 | 90.3 | 90 | 78.2 |
| disagree | 16 | 7.5 | 29 | 19.9 | 6 | 4.3 | 16 | 8.1 | 23 | 14.2 | 19 | 12.3 |
| Strong disagree | 1 | 1.1 | 6 | 3.0 | 1 | 0.6 | 1 | 1.2 | 5 | 2.1 | 4 | 1.8 |
| $\chi^2$ | 12.2 | | 15.2 | | 6.0 | | 14.2 | | 9.3 | | 13.7 | |
| $p$ | $<0.05$ | | $<0.05$ | | $>0.05$ | | $<0.05$ | | $<0.05$ | | $<0.05$ | |

The observed frequency of cases among the female was more than the expected, while vice versa among the male in most body parts. This implies that female workers were highly exposed to MSD symptoms compared to males. The gender has been appointed as the significant factor affecting MSD symptoms in the previous studies (Bergqvist, Wolgast, Nilsson & Voss, 1995) (Cha, Ryu, Choi, Lee, Kim, Chung & Jeong, 2007).

Among the work types of hospital workers, nurses, administrative assistants and pharmacists had higher rate of case occurrence than the other worker types, but the main body parts including cases were different each other. Among nurses, the main body parts were the shoulder, low back and legs, among administrative assistants the neck and hand/wrist, and among pharmacists the arm/elbow. It may indicate the body parts used mostly in the worker types. Due to their long walking distance in hospital, the cases were prevalent in the legs among nurses and administrative assistants, while there were no leg cases among pharmacists. Additionally, the cases in the shoulder and low back among nurses can be accounted by their work characteristics such as prolonged standing posture, handling heavy objects and patients (Jeong & Koo, 2006) .

Among the perceived workload variables of this study, the physical exertion was the highly related to MSD cases. The observed frequency of cases in each body part was significantly more than the expected one among the workers who thought they used much force in their work. Josephson et al. (Josephson, et al., 1997) also concluded in their study that perceived exertion combined with job strain was the factor to increase MSD cases. Moreover, Menzel et al. (Menzel, et al., 2004) analyzed that the frequency of high-risk patient-handling tasks and the number of heavy patient handling, which require much physical exertion, had a significant effect on MSD cases. This study also supported that subjective physical exertion perceived by workers was an important factor to explain MSD cases as same as the objective one.

The perceived work repetition, satisfaction and urgency also were statistically related to MSD cases. The cases of all the body parts except the arm/elbow were significantly different with the perceived repetition and satisfaction. The perceived work urgency also had significant relation to the cases of all the body parts except the neck and legs. Therefore these can be also used as subjective factors to explaining MSD cases with further in-depth investigation.

This study analyzed the prevalence of MSD symptoms among hospital workers through a survey and tested the relationship of MSD cases and perceived workload. The prevalent MSD symptoms were mainly found in the shoulder, low back and legs in order. It was also found that female workers were more highly exposed to MSD than males. Last the perceived workload variables in this study had significant relationship with MSD cases.

## 5 CONFLICT OF INTEREST

No potential conflict of interest relevant to this article was reported.

## REFERENCES


Bejia, I., Younes, M., Jamila, H. B., Khalfallah, T., Ben Salem, K., Touzi, M., Akrout, M. & Bergaoui, N. 2005. Prevalence and factors associated to low back pain among hospital staff. *Joint Bone Spine* 72 (3): 254–259.

Bergqvist, U., Wolgast, E., Nilsson, B. & Voss, M. 1995. Musculoskeletal disorders among visual display terminal workers: Individual, ergonomic, and work organizational factors. *Ergonomics* 38 (4): 763–776.

Bernard, B., Putz-Anderson, V. & Burt, S. 1997. A critical review of epidemiologic evidence for work-related musculoskeletal disorders of the neck, upper extremity, and low back. *Cincinnati: Centers for Disease Control and Prevention National Institute for Occupational Safety and Health publication*: 97–141.

Cha, J., Ryu, T., Choi, H., Lee, J., Kim, M., Chung, M. & Jeong, C. 2007. Survey of musculoskeletal disorders in korean dentists 26(2): 137–147.

Fish, D. & Morris-Allen, D. 1998. Musculoskeletal disorders in dentists. *The New York state dental journal* 64 (4): 44.

Jeong, E.-H. & Koo, J.-W. 2006. Analysis of musculoskeletal burdened work among nurses at a university hospital. *Journal of the Ergonomics Society of Korea* 25 (3): 97–103.

JK, P., SH, J., DS, K., KH, H., HY, L., EY, C., JH, C. & HS, W. 2010. Musculoskeletal disorders and job stress risk factors in general hospital nurses: Nursing tasks and musculoskeletal discomfort symptoms. *Proceedings of Fall Conference of Ergonomic Society of Korea*: 86–89.

Josephson, M., Lagerström, M., Hagberg, M. & Hjelm, E. W. 1997. Musculoskeletal symptoms and job strain among nursing personnel: A study over a three year period. *Occupational and environmental medicine* 54 (9): 681–685.

Kim, K.-S., Park, J.-K. & Kim, D.-S. 2010. Status and characteristics of occurrence of work-related musculoskeletal disorders. *Journal of the Ergonomics Society of Korea* 29 (4): 405–422.

Lagerström, M., Wenemark, M., Hagberg, M. & Hjelm, E. W. 1996. Occupational and individual factors related to musculoskeletal symptoms in five body regions among swedish nursing personnel. *International archives of occupational and environmental health* 68 (1): 27–35.

Lee, K., Chun, Y., Kim, J. & Jung, M. 2007. A study on relationships of subjective intensity of work, job stress and self-reported musculoskeletal symptoms. *Fall Conference of Ergonomic Society of Korea 2007*: 18–23.

Menzel, N. N., Brooks, S. M., Bernard, T. E. & Nelson, A. 2004. The physical workload of nursing personnel: Association with musculoskeletal discomfort. *International journal of nursing studies* 41 (8): 859–867.

Murtomaa, H. 1982. Work-related complaints of dentists and dental assistants. *International Archives of Occupational and Environmental Health* 50 (3): 231–236.

Nakladalova, M., Fialova, J., Korycanova, H. & Nakladal, Z. 1995. State of health in dental technicians with regard to vibration exposure and overload of upper extremities. *Central European journal of public health* 3: 129.

Park, J., Boyer, J., Tessler, J., Perez, G. & Punnett, L. 2005. Exposure assessment of musculoskeletal disorder risk factors in hospital work: Inter-rater reliability of path observations. *Proceedings of the Human Factors and Ergonomics Society Annual Meeting*.

Park, J., Kim, D. & Seo, K. 2008. Musculoskeletal disorder symptom features and control strategies in hospital workers. *Journal of the Ergonomics Society of Korea* 27 (3): 81–92.

Seo, S. & Kee, D. 2005. Survey of musculoskeletal disorders among nurses in a general hospital. *Journal of the Ergonomics Society of Korea* 24 (2): 17–24.

Smith, D. R., Ohmura, K., Yamagata, Z. & Minai, J. 2003. Musculoskeletal disorders among female nurses in a rural japanese hospital. *Nursing & Health Sciences* 5 (3): 185–188.

Vanderpool, H. E., Friis, E. A., Smith, B. S. & Harms, K. L. 1993. Prevalence of carpal tunnel syndrome and other work-related musculoskeletal problems in cardiac sonographers. *Journal of Occupational and Environmental Medicine* 35 (6): 604–610.

Village, J. 2000. A comparison of "case definitions" to estimate prevalence of work-related musculoskeletal disorders in hospital laundry workers and the influence of psychosocial variables. *Proceedings of the Human Factors and Ergonomics Society Annual Meeting*.

You, H., Simmons, Z., Freivalds, A., Kothari, M. J. & Naidu, S. H. 1999. The development of a risk assessment model for carpal tunnel syndrome. *Advances in Occupational Ergonomics and Safety*: 179–186.

*Bridging Research and Good Practices towards Patient Welfare – Shih & Liang (Eds)*
*© 2015 Taylor & Francis Group, London, ISBN 978-1-138-02716-9*

# A survey study on musculoskeletal pain: Prevalence, regional distribution and risk factors among Indian physiotherapists

D. Sharan, M. Mohandoss, R. Ranganathan & J.S. Rajkumar
*RECOUP Neuromusculoskelatal Rehabilitation Centre, Bangalore, Karnataka, India*

ABSTRACT: Work Related Musculoskeletal Disorders (WRMSD) are common in physiotherapist and are continually growing. The objective of the study was to evaluate the prevalence, risk factors and regional distribution of MSD among Indian physiotherapists. It is a prospective survey study in which 225 physiotherapists in India were asked to fill in a structured questionnaire. The mean age of the subjects was 28.6 ± 2.96 years. On an average the subjects worked for 9.29 ± 1.08 hours per day for 6 days a week. 81.25% of the subjects complained of pain or discomfort within the past year. The commonest site of pain was neck, upper back, lower back, wrist and hand, shoulder and ankle. Ergonomic risk factors noted were prolonged standing and adopting static postures, frequent application of force with hands or arms, frequent bending or twisting of upper body, working in awkward postures, and lifting or transferring heavy patients.

## 1 INTRODUCTION

WRMSD can be defined as a musculoskeletal injury resulting from a work-related event. Physiotherapist are recognized to be knowledgeable in prevention and treatment of WRMSD, yet they are themselves susceptible to sustaining WRMSD because they are often involved in physically demanding and intense, repetitive tasks in their practices performing intensive work tasks related to the delivery of patient care (Bork et al., 1996; Cromie et al., 2000; Glover et al., 2005). WRMSD are common in physiotherapists and are continually growing. Numerous studies indicated that the WRMSD prevalence is higher (approximately 80%) in physiotherapists below the age of 35 and they typically develop symptoms in the initial five years of their occupation (Adegoke et al., 2008; Cromie et al., 2000; West et al., 2001; Girbig et al., 2013; McPhail et al., 2014). Ironically, these therapists those who treat patients with musculoskeletal disorders are at a high risk of developing WRMSD themselves. Cromie et al found that even though 91% of Physical therapists in their study were affected by WRMSD at some time in their career, only 7.1% claimed for worker's compensation (Cromie et al., 2000). Although they did not claim for compensation, most of the therapists in the study reported high level of discomfort. High prevalence of pain and disorder in the physical therapist has forced them to reconsider their career and leave their current job in many countries (Wolpert et al., 1992; Noh et al., 1993; Higgs et al., 2001; McPhail et al., 2014). Physiotherapists are a common target because of the nature of their work, involving more physical contact and exertion. The major risk factors contributing to MSD among Physiotherapists are treatments such as manual therapy, exerting force, repetitive movements, continuous bending, patient transfer, sudden or un-anticipated movements, restricted work area and lack of physiotherapists. While a number of studies have reported the prevalence of musculoskeletal disorders among physiotherapists, a very limited number of studies have been done in the Indian context. The objective of the study was to evaluate the prevalence, risk factors and regional distribution of MSD among Indian physiotherapists.

Table 1. Regional distribution of pain.

| Body part | Percentage involved |
|---|---|
| Neck | 56.25% |
| Upper Back | 56.25% |
| Lower Back | 50.07% |
| Wrist & Hand | 25% |
| Shoulder | 22.65% |
| Ankle | 17.18% |

## 2 METHODOLOGY

### 2.1 *Study design and samples*

It is a prospective survey study in which physiotherapist from various part of India were contacted through email and telephone to fill a questionnaire. A total of 225 physiotherapist participated in the survey. All the participants were explained about the need for the study and asked to fill in the questionnaire after getting an informed consent from all. All the participants who fit the inclusion criteria and are willing to participate in the study were considered for the study. Inclusion Criteria included minimum one year of work experience, working as a full time physiotherapists, not involved in any other occupation (full time or part time). Exclusion criteria are any other systemic health problems, suffering from any other musculoskeletal disorders.

### 2.2 *Materials and intervention*

A structured questionnaire was filled by the physiotherapists, which included the demographic data such as age, sex, height, weight, total work experience, number of working hours in a day, type or department of work, questions related to regular exercise, physical risk factors associated with working condition, present health status, Nordic musculoskeletal questionnaire, questions related to various workplace risk factors and the 32 items short form of workstyle questionnaire addressing psychosocial factors (Feuerstein et al., 2005). Descriptive statistical analysis was used to calculate the frequencies and percentages. The descriptive analysis of data presented as tables. Results of continuous measurements are presented as Mean (±SD) and results of categorical measurements are presented in Number (%).

## 3 RESULTS

### 3.1 *Demography*

Among the participants 132 were male and 93 were female. The mean age of the subjects was $28.6 \pm 2.96$ years. The mean height and weight of the subjects was $164 \pm 9.6$ cm and $65.95 \pm 13.85$ kg respectively. The average years of experience of the subjects was $3.21 \pm 2.90$ years. On an average the subjects worked for $9.29 \pm 1.08$ hours per day for 6 days a week. Only 23.4% of the subjects exercised regularly.

### 3.2 *Pain and regional distribution*

81.25% of the subjects complained of pain or discomfort within the past year. The commonest site of pain was neck, upper back and lower back. Percentage of involvement of different body regions is shown in Table 1.

### 3.3 *Workstyle risk*

Analysis of short form of workstyle questionnaire revealed that 67.18% of the subjects reported an adverse workstyle risk (total score >28). Correlation analysis of workstyle risk and presence of MSD revealed that there was a strong positive correlation ($P < 0.05$) between adverse workstyle and predisposition of MSD.

### 3.4 *Risk factors associated with working condition*

The other documented ergonomic risk factors were prolonged standing and adopting static postures, frequent application of force with hands or arms, frequent bending or twisting of upper body, working in awkward postures, and lifting or transferring heavy patients.

## 4 DISCUSSION

The prevalence of WRMSD was 81% which was less than the prevalence reported Australia about 91% (Molumphy et al., 1985), and Nigeria about 91.3% (Adegoke et al., 2008) but higher than prevalence reported in in the United States about 61% (Borg et al., 1996), and Kuwait about 47.6% (Alrowayeh et al., 2010).

The study shows that the participants of younger age and less years of experience had a increased incidence of WRMSDs, as the younger age and less experienced physiotherapists might be a victim of MSDs due their inappropriate handling and treatment techniques as reported in earlier studies. (Passier et al., 2011; Girbig et al., 2013; Bork et al., 1996). The other reason might be development of positive coping strategies by older therapists due to experience, which make them less susceptible to injuries when compared to younger therapists.

As the working hours are more, the incidence of exposure to the WRMSDs is also more and the working hours reported in our study are more than reported in previous studies (Salik et al., 2004 and Molumphy et al., 1985).

The study also revealed that lack of exercises is also a possible factor for WRMSDs among physiotherapists. Physiotherapist though do a lot of physical exertion lack of a structured routine exercise is considered to be one of the factors predisposing WRMSD which was already noted in a previous study (McPhail et al., 2014).

Manual therapy methods which involves frequent application of force with hands and arms and adapting awkward postures along with lifting and transferring of heavy patients was found to be the risk for predisposing WRMSD among physiotherapist which coincides with most of the studies done previously (West et al., 2001; Darragh et al., 2012; Passier et al., 2011; Nordin et al., 2011).

Increased adverse work style can be considered to be the reason behind occurrence of WRMSDs among physiotherapists. Considering psychosocial risk factors is essential when planning a prevention strategy. There was a previous study done by Campo et al stated that therapist with high levels of demands and low levels of control were at increased risk for both turnover and work-related pain (Campo et al., 2009).

The results suggested the common sites of WRMSDs for physiotherapists were neck, back and hand regions which is consistent with the previous studies (Alrowayeh et al., 2010; Darragh et al., 2012; Passier et al., 2011; Glover et al., 2005; Nordin et al., 2011).

## 5 CONCLUSION

The study concluded that physiotherapist are prone to develop musculoskeletal disorders as there are various risk factors present in the workplace and the working conditions in Indian scenario. More research and studies might help to clearly understand the risk factors of WRMSD and develop preventive strategies for Physiotherapists. Eliminating all the risk factors associated with the development of WRMSD in Physical therapist may not be possible due to the nature of the job, but based on the number of prevention techniques being suggested, it is possible to reduce the risk factors leading to WRMSD.

## REFERENCES


Adegoke B, Akodu A, Oyeyemi A: Work-related musculoskeletal disorders among Nigerian Physiotherapists. BMC Musculoskelet Disord 2008, 9:112.

Alrowayeh HN, Alshatti TA, Aljadi SH, Fares M, Alshamire MM, Alwazan SS: Prevalence, characteristics, and impacts of work-related musculoskeletal disorders: a survey among physical therapists in the State of Kuwait. BMC Musculoskeletal Disorders 2010,11:116.

Bork BE, Cook TM, Rosecrance JC, Engelhardt KA, Thomason ME, Wauford IJ, Worley RK: Work-related musculoskeletal disorders among physical therapists. Phys Ther 1996, 76:827–835.

Campo M,Weiser S, Koenig K: Job Strain in Physical Therapists.Phys Ther. Sep 2009; 89(9): 946–956

Cromie JE, Robertson VJ, Best MO: Work-related musculoskeletal disorders in physical therapists: prevalence, severity, risks, and responses. Phys Ther 2000, 80:336–351.

Darragh A, Campo M, King P: Work-Related Activities Associated with Injury in Occupational and Physical Therapists.Work. 2012; 42(3). doi:10.3233/WOR-2012-1430.

Feuerstein M, Nicholas RA, Huang GD, Haufler AJ, Pransky G,Robertson M: Work style: development of a measure of response to work in those with upper extremity pain. J Occup Rehabil 2005, 15:87–104.

Girbig M, Deckert S, Kopkow C, Latza U, Dulon M, Nienhaus A, Groneberg D, Seidler A: Work-related complaints and diseases of physical therapists – protocol for the establishment of a "Physical Therapist Cohort" (PTC) in Germany. Journal of Occupational Medicine and Toxicology 2013, 8:34.

Glover W, McGregor A, Sullivan C, Hague J: Work- related musculoskeletal disorders affecting members of the Chartered Society of Physiotherapy. Physiother 2005, 91:138–147.

Higgs J, Refshauge K, Ellis E: Portrait of the physiotherapy profession. J Interprof Care 2001, 15:79–89.

McPhail and Waite: Physical activity and healthrelated quality of life among physiotherapists: a cross sectional survey in an Australian hospital and health service. Journal of Occupational Medicine and Toxicology 2014, 9:1.

Molumphy M, Unger B, Jensen GM, Lopopolo RB: Incidence of work-related low back pain in physical therapists. Phys Ther 1985, 65: 482–486.

Noh S, Beggs C: Job turnover and regional attrition among physiotherapists in northern Ontario. Physiother Can 1993, 45:239–244.

Nordin NAM, Leonard JH, Thye Ng C: Work-related injuries among physiotherapists in public hospitals-a Southeast Asian picture. Clinical Science 2011, 66:373–378.

Passier L, McPhail S: Work related musculoskeletal disorders amongst therapists in physically demanding roles: qualitative analysis of risk factors and strategies for prevention. BMC Musculoskelet Disord 2011, 12:24.

Salik Y, Özcan A: Work-related musculoskeletal disorders: a survey of physical therapists in Izmir-Turkey. BMC Musculoskeletal Disor 2004, 5:27.

Wolpert R, Yoshida K: Attrition survey of physiotherapists in Ontario. Physiother Can 1992, 44:17–24.

West DJ, Gardner D: Occupational injuries of physiotherapists in North and Central Queensland. Aust J Physiother 2001, 47:179–186.

*Bridging Research and Good Practices towards Patient Welfare – Shih & Liang (Eds)*
*© 2015 Taylor & Francis Group, London, ISBN 978-1-138-02716-9*

# Risk factors for musculoskeletal disorders in healthcare professionals: A systematic review

D. Sharan, M. Mohandoss, R. Ranganathan & J.S. Rajkumar
*RECOUP Neuromusculoskelatal Rehabilitation Centre, Bangalore, Karnataka, India*

ABSTRACT: The objective of the study was to review the current literature in order to learn the methods of study, measurement tools used, different disciplines evaluated, risk factors and other findings related to musculoskeletal disorders (MSD) among healthcare professionals (HCP). A systematic review of the published literature was conducted to find out the ergonomic risk factors among HCPs and suggested recommendations. A 4 step review was done. 284 articles were found by a keyword search in MEDLINE which was reduced to 29 after a 4 step review. Almost all the HCPs had a high prevalence of MSD irrespective to their field and specialty. The majority of studies focused on a self administered questionnaire. Very few experimental studies and no randomised controlled trials were found during the review. Most of the studies attempted to identify the risk factors of MSD in HCPs.

## 1 INTRODUCTION

The demand for healthcare is on the rise in the present era due to population growth, immigration and increasing expectations of people. On the other side, healthcare costs continue to rise faster than inflation and consequently, increasing attention is focused on the operation of healthcare organisations. The management is looking for ways to economise through cost savings and cost cutting leading to concerns over the occupational health of healthcare professionals (HCP). There has been a steady increase in number of HCPs, irrespective of the field, affected with musculoskeletal disorders (MSD) due to ergonomic risk factors. A study revealed that prevalence rate was 76% for low back pain and 60% for neck, shoulder pain among non specialised nurses, ICU nurses, operation room nurses and X-ray technicians (Bos et al., 2007) which made it clear that prevalence of MSD was high among HCPs regardless of the specialty of work. Health care work is known to be high-risk for MSDs (Campo et al., 2008; Lorusso et al., 2007; Ylipaa, 2002). It is estimated that almost one-third of all cases of sick leave among HCPs was related to MSDs (Meijsen & Knibbe, 2007). Even in developed countries, it appears that MSDs are under reported among HCPs (Menzel, 2007). Back, neck, shoulder and knee problems are the most common complaints among medical, dental, and nursing students (Smith et al., 2005; Thornton et al., 2008; Smith et al., 2004). Studies of MSDs among HCPs have mainly focused on dentists, physical therapists, and nurses (Alexopoulos et al., 2004; Mehrdad et al., 2012). The common risk factors which contribute to the development of MSDs in HCPs can be categorised as those related to personal background variables like anthropometric characteristics, age, hereditary factors, etc. and those related to work like repetitive motion, static posture, force, awkward position, vibration, temperature, biological factors, chemical irritating or toxic factors, radiation, etc. (Ligget et al., 2007; Pandis et al., 2007; Kierklo et al., 2011). Though many studies are available there is a lacuna in knowing about risk factors for MSD among HCPs, its measurement and preventive methods. The objective of the study was to review the current literature in order to learn the methods of study, various measurement tools used, different disciplines evaluated, important risk factors and other findings related to MSD among HCPs.

## 2 METHODOLOGY

A systematic review of the published literature was conducted to learn the various risk factors, assessment methods and preventive methods for HCPs. The systematic review process involved 4 steps and step 1 was a keyword search in the MEDLINE (Pubmed) with set lower and higher limits. Keywords used were "health professionals," "ergonomics" and "musculoskeletal disorders" and a total of 284 articles were found. Trials were made on finding out the appropriate keywords like healthcare workers, healthcare personnel, health care practitioners, hospital workers and ergonomic risk factors. But the results obtained for them were not adequate and appropriate. Limits were set for the studies published between the years 1995 to 2013. Three reviewers conducted the systematic review based on the EPOC criteria and any disagreements were dissolved through discussion among them.

Step 2 was a title review and studies which had "health professional" or any other titles like physicians, surgeons, dentist, physiotherapist etc in the title of the study were only considered for further review. 96 studies were selected by step 2 review for further review.

In Step 3 review, language of the article and presence of contents were reviewed. After step 3 review, 16 articles which did not have any text or articles in languages other than English were excluded and 80 studies were further reviewed.

Step 4 review was the content review and only research papers which had full articles and abstracts with clear details of study design, study population, outcome measures, results and conclusion were considered for review. Studies involving students were excluded and studies involving practicing professionals only were included for the review. Case studies, case series and review articles were excluded. Scientific and systematic reviews were included. Based on the eligibility criteria the articles were finalised and the rest of the articles were excluded from the systematic review. The healthcare professionals included in the review ranged from hospital cleaners or to home healthcare aides to paramedical workers and medical practitioners, so that most HCPs were taken into account. A total of 29 studies were included in the study after step 4 review.

All the 29 studies were analysed to see the common research design used, study population, measurement tools used, outcomes, conclusions and recommendations in healthcare ergonomics. Prevalence in different HCPs, common risk factors, region affected and preventive strategies were reviewed.

## 3 RESULTS

A total of 284 articles were found after keyword search in MEDLINE, which were reviewed twice by three reviewers which led to a final tally of 29 research articles. The method of review of the research articles is shown in figure 1.

After step 1 review, a total of 96 studies involving health professionals were identified. The maximum number of studies was found involving nurses and dentist. Sonologists, professionals involved in patient lifting and handling, and hospital workers were studied in few studies. A limited number of studies were found involving physicians, surgeons, physiotherapist, occupational therapist, radiologist, lab technicians, pharmacist and other HCPs. Table 1 gives the details about the number of studies found involving various HCPs during the review.

After 4 steps of reviewing a total 29 research articles were finally analysed. Details about the year of publication, professionals studied, study design and samples used, measurement tools and the final outcome that could be noted from the study are listed in table 2 and table 3. Most of the studies were surveys and they used a structured questionnaire to measure the problem and outcome. No randomised controlled trials (RCTs) were found.

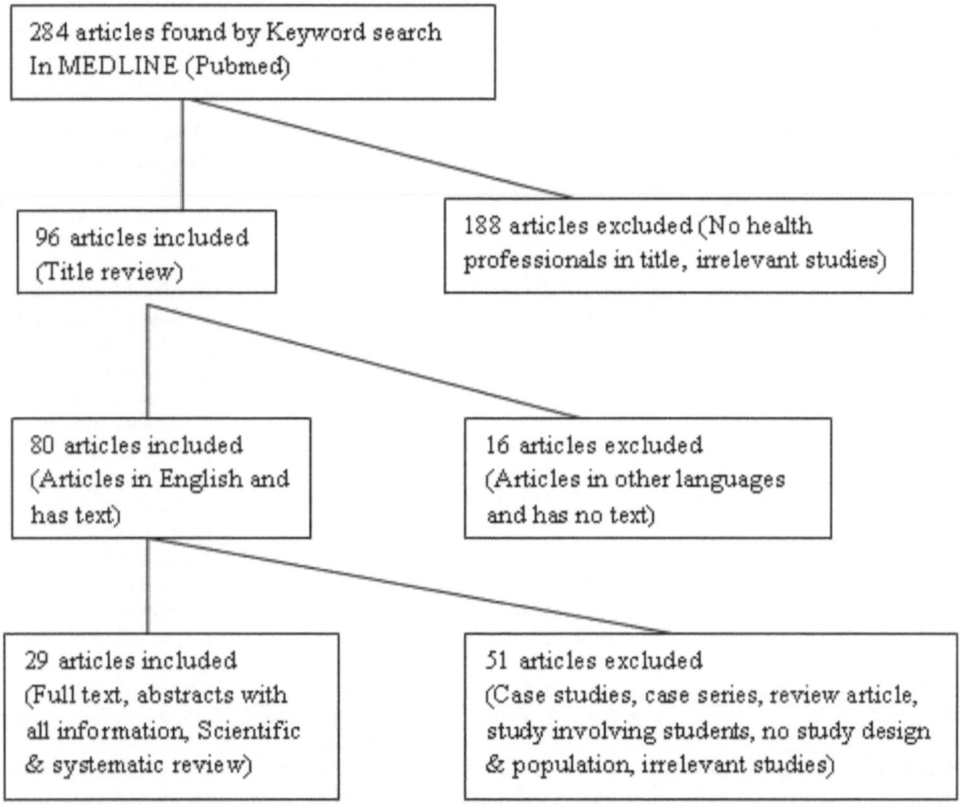


Figure 1. Schema of reviewing of the articles.

Table 1. Number of studies found involving various health professionals (n = 96).

| Health professionals studied | Number of studies found |
|---|---|
| Dentist | 28 |
| Nurses & aides | 34 |
| Physiotherapist | 04 |
| Occupational therapist | 01 |
| Physicians | 03 |
| Ophthalmologist | 01 |
| Lapro surgeons | 03 |
| Thoracic surgeons | 01 |
| Sonographers | 05 |
| Electrophysiologist | 01 |
| Vascular technologist | 01 |
| OT technician | 01 |
| X ray technician | 03 |
| Patient handling | 07 |
| Health care worker | 05 |
| Lab technicians | 04 |

Table 2. Review of all full articles (n = 13).

| Author & year | Professional | Study design & studied | Measurement sample | Outcome size tool |
|---|---|---|---|---|
| Mehrdad et al. (2012) | Physicians | Cross sectional survey, 405 | Self administered questionnaire Physicians from 4 hospitals | Distribution of Pain & Risk factors |
| Darragh et al. (2012) | PT, OT | Cross sectional survey, 1436 OT & 1861 PT | Self administered questionnaire | Association between work activities & distribution of pain |
| Sharan et al. (2012) | PT | Scientific review, | – | Prevalence, risk factors, preventive measures |
| Kierklo et al. (2011) | Dentist | Descriptive survey, 220 Dentist | Self administered questionnaire | Distribution of Pain, Association with experience, gender, posture & breaks |
| Lindfors et al. (2006) | Female Dental health workers | Cross sectional survey, 945 female | Self administered questionnaire Dental workers | Predictors of upper extremity disorder, association with Pschosocial work Charecteristics |
| Rolander et al. (2005) | Dentist | Cross sectional analysis, 27 dentist | Questionnaire & portable ergonomic Observation (PEO) | Co-relation between subjective, objective evaluation & MSD |
| Droeze et al. (2005) | Dentist | Experimental survey, 57 dentist | Self administered questionnaire after intervention | Effectiveness of ergonomic intervention |
| Violante et al. (2004) | Female nurse | Cross sectional survey, 901 female nurses | Self administered questionnaire | Physical, individual & psychosocial risk factors for LBP |
| Kjellberg et al. (2005) | Nurse | Experimental survey, 224 nurses | Questionnaire, Video observation & work technique assessment | Association between personal factors & work technique & LBP |
| Szymańska (2002) | Dentist | Descriptive survey, 268 dentist | Self administered questionnaire | Use of prophylactic measures & its effect |
| Alexandre et al. (2001) | Nurse aides | Case control study, 56 nursing aides | Self administered questionnaire & VAS after Intervention | Effectiveness of education programme in preventing MSD |
| Ando et al. (2000) | Nurse | Cross sectional survey, 314 nurses | Self administered questionnaire | Distribution of pain & association of work, Demographic conditions |
| Josephson et al. (1997) | Female nurse | 3 year follow up survey, 565 nurses | Karasek & Theorell's model for job strain, NMQ & VAS | Job strain is a risk factor for MSD and risk increases with physical exertion |

## 4 DISCUSSION

The review identified several HCPs who were at risk of MSD due to their work. Among the HCPs, the most commonly affected were nursing staffs followed by dentists, physiotherapists and radiographers, which was perhaps due to the higher physical contact with the patient and equipments and

Table 3. Review of all Abstracts (n = 16).

| Author & Year | Professional | Study design & studied | Measurement sample size | Outcome tool |
|---|---|---|---|---|
| Gopinadh et al. (2012) Sharma et al. (2011) Booyens et al. (2009) | Dentist | Cross sectional | Self administered survey, 170, 102 & 362 dentist | Distribution of pain, questionnaire risk factors and its association with MSD, awareness. |
| Morandeira et al. (2012) | Lapro Surgeon | Cross sectional survey, 78 surgeons | Self administered questionnaire | Association between work activities & MSD |
| Roll et al. (2012) | Sonologist, Vascular | Cross sectional survey, 2163 technician | Self administered questionnaire DMS & VT | Co-morbidities, Association of work activities in MSD |
| Harcombe et al. (2010) Tinubu et al. (2010) | Nurse | Cross sectional survey, 280 & 128 nurses | Self administered questionnaire | Distribution of Pain, Association with activities |
| Kułagowska (2008) | Nurse anesthetist | Cross sectional analysis, 57 tasks | Questionnaire & OWAS | Distribution of pain, postural risk for various tasks |
| Simon et al. (2008) | Nurse | Cross sectional analysis, 21516 Nurses from 7 countries | Self administered questionnaire | Psychosocial & Physical risk factors for back & neck pain |
| Bos et al. (2006) | Nurse & X ray technician | Cross sectional survey, 3169 samples | Self administered questionnaire | Pain distribution, risk factors |
| Battevi et al. (2006) | Patient handling & transfer | Validation study, 2603 sample | MAPO for risk assessment | Valid tool for risk assessment in hospitals |
| Nelson et al. (2005) | Nurse | pre- post experiment study, nurses of 23 high risk units | Survey, weekly progress, injury & cost logs | Effectiveness of ergonomic intervention |
| Corona et al. (2005) | Nurse & PT | cross sectional survey, 100 nurse + 100 PT | Self administered questionnaire | Pain prevalence, organisational risk factors |
| Berguer et al. (2001) | Lapro Surgeon | Cross sectional survey, 726 surgeons | Self administered questionnaire | Hand size & difficulty using instruments. |
| Anton et al. (2002) | Dental hygienist | Cross sectional analysis, 95 dentist | Questionnaire & NCS study | Role of NCS in diagnosis of Carpel tunnel Syndrome |
| Mirk et al. (1999) | Sonographer | Descriptive survey, | Self administered 340 sonographers | Physical risk factors questionnaire |

higher number of working professionals in the nursing field. Some other notable professionals with higher percentage of risks were surgeons, radiologists and ophthalmologists, which was possibly due to the reduced operating space for a prolonged period of time. The Workplace Health Indicator Tracking and Evaluation (WHITE) tracked about 84,318 HCPs between the years of 2004 to 2008 and reported about 35,927 individuals recorded ergonomic hazards. Female HCPs were affected more commonly compared to their male counterparts. The most affected body regions due to their work were lower back, followed by neck and shoulders. The perceived source of ergonomic hazards for the HCPs was reported as increased physical load, patient handling, physical strain and economic pressure. In addition to the physical factors, psychosocial risk factors contributed to the development of MSD among the HCPs. HCPs working in rehabilitation centers were associated with increased lower back symptoms while those working in out-patient clinics had increased prevalence of thumb or wrist symptoms. The tools used for the data collection of the articles were mostly self reported questionnaires and those studies were qualitative in nature.

## 5 CONCLUSION

A large proportion of studies involved dentist and nurses. The prevalence rates of MSD was high and similar in all HCPs irrespective to the field and specialisation of work. Very few experimental studies and scientific reviews were found and no RCT's were found. Few measurement tools were found to identify risk facts. The review recommends further research on ergonomics in HCPs with use of more valid and reliable measurement tools in order to have a better understanding of the risk of MSD and to recommend effective preventive strategies for the HCPs.

## REFERENCES


Alexandre NM, de Moraes MA, Corrêa Filho HR, Jorge SA. Evaluation of a program to reduce back pain in nursing personnel. Rev Saude Publica. 2001 Aug; 35(4): 356–61.

Alexopoulos EC, Stathi IC, Charizani F. Prevalence of musculoskeletal disorders in dentists. BMC Musculoskelet Disord. 2004; 5: 16.

Ando S, Ono Y, Shimaoka M, Hiruta S, Hattori Y, Hori F, Takeuchi Y. Associations of self estimated workloads with musculoskeletal symptoms among hospital nurses. Occup Environ Med. 2000 Mar; 57(3): 211–6.

Anton D, Rosecrance J, Merlino L, Cook T. Prevalence of musculoskeletal symptoms and carpal tunnel syndrome among dental hygienists. Am J Ind Med. 2002 Sep; 42(3): 248–57.

Battevi N, Menoni O, Ricci MG, Cairoli S. MAPO index for risk assessment of patient manual handling in hospital wards: a validation study. Ergonomics. 2006 Jun 10; 49(7): 671–87.

Berguer R, Hreljac A. The relationship between hand size and difficulty using surgical instruments: a survey of 726 laparoscopic surgeons. Surg Endosc. 2004 Mar; 18(3): 508–12.

Bos E, Krol B, van der Star L, Groothoff J. Risk factors and musculoskeletal complaints in non-specialized nurses, IC nurses, operation room nurses, and X-ray technologists. Int Arch Occup Environ Health. 2007 Jan; 80(3): 198–206.

Booyens SJ, van Wyk PJ, Postma TC. Musculoskeletal disorders amongst Practising South African oral hygienists. SADJ. 2009 Oct; 64(9): 400–3.

Campo M, Weiser S, Koenig KL, Nordin M. Work-related musculoskeletal disorders in physical therapists: a prospective cohort study with 1-year follow-up. Phys Ther. 2008; 88: 608–619.

Corona G, Amedei F, Miselli F, Padalino MP, Tibaldi S, Franco G. Association between relational and organizational factors and occurrence of musculoskeletal disease in health personnel. G Ital Med Lav Ergon. 2005 Apr-Jun; 27(2): 208–12.

Darragh AR, Campo M, King P. Work-related activities associated with injury in occupational and physical therapists. Work. 2012; 42(3): 373–84.

Droeze EH, Jonsson H. Evaluation of ergonomic interventions to reduce musculoskeletal disorders of dentists in the Netherlands. Work. 2005; 25(3): 211–20.

Gopinadh A, Devi KN, Chiramana S, Manne P, Sampath A, Babu MS. Ergonomics and musculoskeletal disorder: as an occupational hazard in dentistry. J Contemp Dent Pract. 2013 Mar 1; 14(2): 299–303.

Harcombe H, McBride D, Derrett S, Gray A. Physical and psychosocial risk factors for musculoskeletal disorders in New Zealand nurses, postal workers and office workers. Inj Prev. 2010 Apr; 16(2): 96–100.

Josephson M, Lagerström M, Hagberg M, Wigaeus Hjelm E. Musculoskeletal symptoms and job strain among nursing personnel: a study over a three year period. Occup Environ Med. 1997 Sep; 54(9): 681–5.

Kierklo A, Kobus A, Jaworska M, Botuliński B. Work-related musculoskeletal disorders among dentists – a questionnaire survey. Ann Agric Environ Med. 2011 Jun; 18(1): 79–84.

Kjellberg K, Lagerström M, Hagberg M. Work technique of nurses in patient transfer tasks and associations with personal factors. Scand J Work Environ Health. 2003 Dec; 29(6): 468–77.

Kułagowska E. The musculoskeletal system load during work performed by nurse anesthetists. Med Pr. 2008; 59(4): 287–92.

Leggat PA, Kedjarune U, Smith DR: Occupational health problems in modern dentistry: a review. Ind Health 2007, 45, 611–621.
Lindfors P, von Thiele U, Lundberg U. Work characteristics and upper extremity disorders in female dental health workers. J Occup Health. 2006 May; 48(3): 192–7.
Lorusso A, Bruno S, L'Abbate N. Musculoskeletal complaints among Italian X-ray technologists. Ind Health. 2007; 45: 705–708.
Mehrdad R, Dennerlein JT, Morshedizadeh M. Musculoskeletal disorders and ergonomic hazards among Iranian physicians. Arch Iran Med. 2012 Jun; 15(6): 370–4.
Meijsen P, Knibbe HJ. Work-related musculoskeletal disorders of perioperative personnel in the Netherlands. AORN J. 2007; 86: 193–208.
Menzel NN. Underreporting of musculoskeletal disorders among health care workers: research needs. AAOHN J. 2008; 56: 487–494.
Mirk P, Magnavita N, Masini L, Bazzocchi M, Fileni A. Frequency of musculoskeletal symptoms in diagnostic medical sonographers. Results of a pilot survey. Radiol Med. 1999 Oct; 98(4): 236–41.
Morandeira-Rivas A, Millán-Casas L, Moreno-Sanz C, Herrero-Bogajo ML, Tenías-Burillo JM, Giménez-Salillas L. Ergonomics in laparoendoscopic single-site surgery: survey results. J Gastrointest Surg. 2012 Nov; 16(11): 2151–9.
Nelson A, Matz M, Chen F, Siddharthan K, Lloyd J, Fragala G. Development and evaluation of a multifaceted ergonomics program to prevent injuries associated with patient handling tasks. Int J Nurs Stud. 2006 Aug; 43(6): 717–33.
Pandis N, Pandis BD, Pandis V, Eliades T: Occupational hazards in orthodontics: A review of risks and associated pathology. Am J Orthod Dentofacial Orthop 2007, 132, 280–292.
Rolander B, Karsznia A, Jonker D, Oberg T, Bellner AL. Perceived contra observed physical work load in Swedish dentists. Work. 2005; 25(3): 253–62.
Roll SC, Evans KD, Hutmire CD, Baker JP. An analysis of occupational factors related to shoulder discomfort in diagnostic medical sonographers and vascular technologists. Work. 2012; 42(3): 355–65.
Sharan D, Ajeesh PS. Injury prevention in physiotherapists – a scientific review. Work. 2012; 41 Suppl 1: 1855–9.
Sharma P, Golchha V. Awareness among Indian dentist regarding the role of physical activity in prevention of work related musculoskeletal disorders. Indian J Dent Res. 2011 May–Jun; 22(3): 381–4.
Simon M, Tackenberg P, Nienhaus A, Estryn-Behar M, Conway PM, Hasselhorn HM. Back or neck-pain-related disability of nursing staff in hospitals, nursing homes and home care in seven countries–results from the European NEXT-Study. Int J Nurs Stud. 2008 Jan; 45(1): 24–34.
Smith DR, Leggat PA. Musculoskeletal disorders among rural Australian nursing students. Aust J Rural Health. 2004; 12: 241–245.
Smith DR, Wei N, Ishitake T, Wang RS. Musculoskeletal disorders among Chinese medical students. Kurume Med J. 2005; 52: 139–146.
Szymańska J. Disorders of the musculoskeletal system among dentists from the aspect of ergonomics and prophylaxis. Ann Agric Environ Med. 2002; 9(2): 169–73.
Tinubu BM, Mbada CE, Oyeyemi AL, Fabunmi AA. Work-related musculoskeletal disorders among nurses in Ibadan, South-west Nigeria: a cross-sectional survey. BMC Musculoskelet Disord. 2010 Jan 20; 11: 12.
Violante FS, Fiori M, Fiorentini C, Risi A, Garagnani G, Bonfiglioli R, Mattioli S. Associations of psychosocial and individual factors with three different categories of back disorder among nursing staff. J Occup Health. 2004 Mar; 46(2): 100–8.
Ylipaa V, Szuster F, Spencer J, Preber H, Benko SS, Arnetz BB. Health, mental well-being, and musculoskeletal disorders: a comparison between Swedish and Australian dental hygienist. J Dent Hyg. 2002; 76: 47–58.

*Bridging Research and Good Practices towards Patient Welfare – Shih & Liang (Eds)*
*© 2015 Taylor & Francis Group, London, ISBN 978-1-138-02716-9*

# Factors affecting the perceived slipperiness rating

Wen-Ruey Chang, Mary F. Lesch & Simon Matz
*Liberty Mutual Research Institute for Safety, Hopkinton, MA, USA*

Chien-Chi Chang
*Department of Industrial Engineering and Engineering Management,*
*National Tsing Hua University, Hsinchu, Taiwan, ROC*

ABSTRACT: Perceived slipperiness rating (PSR) has been widely used to assess walkway safety. In this experiment, 29 participants were exposed to 15 different floor conditions. The relationship between their perceived slipperiness rating and all other objective measurements, such as utilized coefficient of friction (UCOF), kinematics and available coefficient of friction (ACOF), was explored with a backward step-wise regression analysis. The results show that the UCOF and ACOF were the major contributors to perceived slipperiness, which suggests that the participants might have relied on foot slip to form their PSR.

## 1 INTRODUCTION

Data from the Liberty Mutual Workplace Safety Index (Liberty Mutual Research Institute for Safety, 2013) show that the costs for disabling workplace injuries in 2011 due to falls on the same level in the USA were estimated to be approximately 8.6 billion US dollars or 15.4% of the total cost burden. Slippery floors, typically caused by contaminants, are a critical factor for falls on the same level (Chang et al. 2001). Bell et al. (2008) indicated that liquid contamination was the most common cause (24%) of slip, trip and fall incidents in healthcare workers. Various approaches have been used to assess slipperiness. Chang et al. (2003) outlined biomechanics, human-centered approaches, available coefficient of friction (ACOF) and surface roughness as major elements in the measurement of slipperiness.

Perceptions can be considered measurements of slipperiness. Perceptions, based on both visual cues and proprioceptive feedback, can be used prospectively or retrospectively to assess slipperiness and can supplement objective measurements of slipperiness. In a prospective study conducted in limited service restaurants, a more slippery rating was associated with a higher rate of subsequent slipping (Courtney et al. 2013). In validating a particular measure of slipperiness, different measurements are usually compared. Quite often, a correlation between the perception and objective measures of slipperiness has been reported. However, human responses to surfaces with different degrees of slipperiness are very complicated, so two parameters are not likely to be fully representative of complex human sensing mechanisms and their relationship might not reveal additional components involved. Additional elements could be involved in forming perceptions of slipperiness.

In this experiment, the participants were exposed to walkways of different slipperiness and were asked to avoid a slip while walking on these surfaces. The relationship between perceived slipperiness ratings and different objective measures, including ACOF, kinetics and kinematics that might affect their outcome, was explored using a backward step-wise regression analysis. This relationship could identify critical factors that help form the final outcome of perceived slipperiness rating.

## 2 METHODS

The five floor types used in this experiment, referred to as floor types A to E, were: (A) a standard quarry tile with raised-profiled tread lines perpendicular to the walking direction, (B) standard flat quarry tile, (C) vinyl composition sheet, (D) marble tile and (E) glazed porcelain tile.

A multiple floor moveable walkway system was constructed, as shown in Figure 1, which consisted of two sections of five different moveable walkways, a stationary force plate area with covers made to match the walkways, and two stationary straight extensions at both ends. The extensions were connected and aligned with the moveable walkways and were covered with floor type B, the standard flat quarry tiles. In both moveable sections, each of the five walkways was covered with one of the floor types selected. Two force plates (Model 9281C, Kistler Instrument Corporation, Amherst, New York, USA) were installed in the middle of the walkway to measure the ground reaction forces. These two force plates were installed along the length of the walkway, one right after the other, in the right hand side of the walkway in the direction of walking when biomechanical data were collected. Although there were no force plates on the left side of the walkway in the force plate area, additional force plate covers were made to cover this area. When a particular floor type was selected for data collection, both sections of the desired walkway were moved to align with the extensions at both ends and four force plate covers of the same floor type were used to cover the force plate area. These four force plate covers were secured to the walkway. Each section of the moveable walkway was approximately 2.44 m long. Each of the four force plate covers was approximately 60 cm by 40 cm. Once the floor changes were completed, a straight walkway of

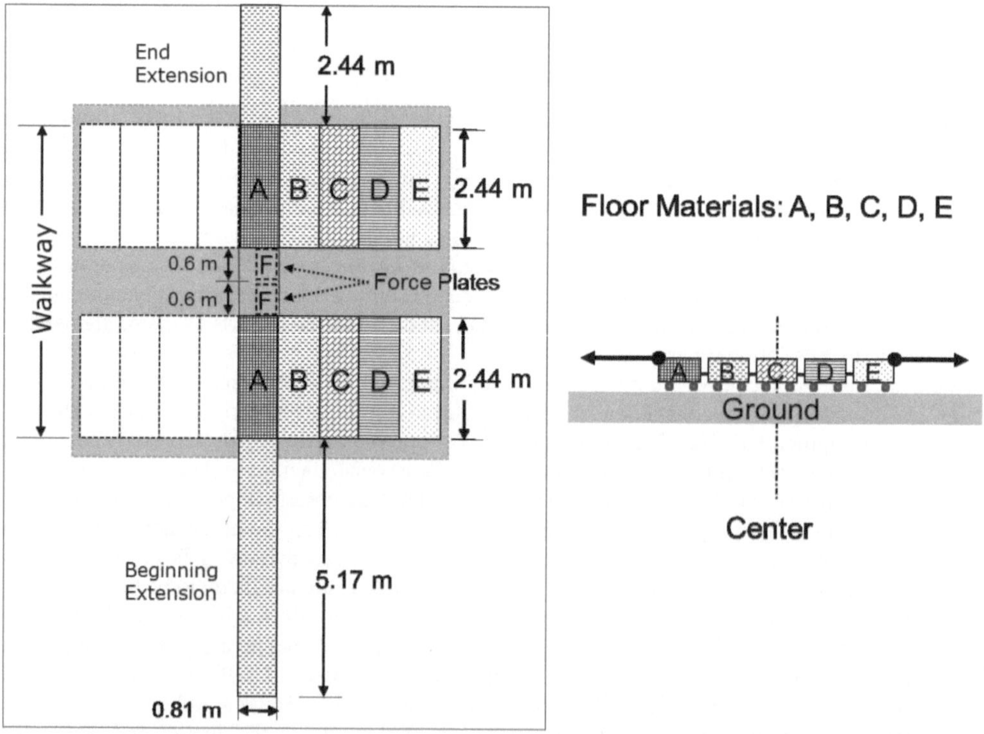


Figure 1. Experimental apparatus of a moveable walkway system with multiple floor types: (A) a standard quarry tile with raised-profiled tread lines perpendicular to the walking direction, (B) standard flat quarry tile, (C) vinyl composition sheet, (D) marble tile and (E) glazed porcelain tile.

approximately 6.08 m long and 0.81 m wide for a particular floor type was ready for data collection. The extension at one end of the walkway where the participants exited after completion of the walking trials was approximately 2.44 m long, shown as the top section of Figure 1. The other extension, where the participants initiated the walking trials, was approximately 5.17 m long, shown as the bottom section in Figure 1.

Surface conditions were dry, water and glycerol. A fine water mist was applied with a garden sprayer. A mixture of 45% glycerol with water by weight was used as another contaminant and was applied with a paint roller.

Thirteen males and 16 females took part in this experiment. The average and standard deviation of weight, height and age were $62.0 \pm 9.18$ kg, $163.0 \pm 7.79$ cm and $30.9 \pm 10.79$ years, respectively, for females, and were $80.3 \pm 14.02$ kg, $179.3 \pm 8.07$ cm and $24.3 \pm 9.41$ years, respectively, for males. The protocol used was approved by an institutional review board and written, informed consent was obtained from each participant. Leather loafers were supplied as the standard footwear in this experiment. The heels of the loafers were replaced with smooth Neolite test liners, a standard test material in the USA, in order to standardize the shoe sole materials. A safety harness was used. Prior to the start of the first trial, the contaminant, if required by the condition, was applied to the entire walkway. The surface condition started from dry and then wet, and was completed with glycerol. Under each surface condition, the five different floor types were randomized. The participants had a clear view of the entire walkway prior to walking on it. They started from one end of the walkway and went to the other end, then turned around and walked back to the beginning of the walkway. This was considered one trial. Data were collected only on the first leg of this trial. They repeated walking back and forth on the walkway sections four more times. After each fifth trial, participants rated perceived slipperiness.

The sampling rate for the force plates was 1000 Hz. All the force plate data were processed with the fourth order zero-lag Butterworth low pass filter with a cut-off frequency of 24 Hz. Utilized coefficient of friction (UCOF) obtained from the vector sum of the transverse and longitudinal components of the ground reaction force divided by the normal force at the same instant was identified and extracted from each successful strike.

Kinematics was collected with a motion tracking system (Motion Analysis Corp., Santa Rosa, CA, USA) with a sampling rate of 200 Hz. Seventeen passive reflective markers were attached to each participant on the sacral vertebrae, and right and left of the oscalcani, ankle, anterior superior iliac spine, heel, knee, shank, thigh and toe. All the kinematic data were processed with the fourth order zero-lag Butterworth low pass filter with a cut-off frequency of 12 Hz.

The participants rated their perceptions of slipperiness immediately after the fifth trial for each walkway and surface condition (for a total of 15 times) by drawing a horizontal line across a vertical line representing the range of slipperiness from 'not at all slippery' at one end to 'extremely slippery' at the other end. A ratio was obtained by dividing the distance from the line drawn by the participants to 'not at all slippery' by the total distance from 'extremely slippery' to 'not at all slippery' on the survey form. This ratio was then multiplied by 100 to obtain the final rating used in the subsequent analyses. Based on these calculations, the perception rating of 100 represented 'extremely slippery', while a rating of zero represented 'not at all slippery'.

Fifty tiles or locations along the walking path on each walkway were selected for friction measurements. Under each surface condition, two friction measurements were conducted on each selected tile or location, one in each walking direction, for a total of 100 measurements. The portable inclinable articulated strut slip tester (PIAST), also known as the Brungraber Mark II, was used in this experiment. A standard test method for this slipmeter is published by the American Society for Testing and Materials (ASTM) (ASTM F-1677-05, 2005). The measurement protocol described by Chang et al. (2012) was used in this experiment. During a measurement with this slipmeter, the operator systematically impacts the floor with multiple strikes at different angles to determine the angle at which a non-slip becomes a slip. Only a fast movement was considered as a slip as described in Chang (2002).

The ACOF was measured in this experiment with smooth Neolite test liners, the same material as the shoe heels. The material was cut into squares approximately 7.6 by 7.6 cm to be attached to the test foot holder of the slipmeter as a footwear test pad. Before the first measurement at each tile or location under each floor type and surface condition, the surface of the Neolite pad was sanded using hand pressure with 400 grit silicon carbide abrasive paper. The sanding protocol described by Chang et al. (2014) was used in this experiment. A brush was used to remove particles on the Neolite surface due to sanding. The sanding procedure was repeated until all visible scratches or dirt had been removed from the Neolite surface. Before friction measurements were conducted, floor surfaces were cleaned with a solution of 50% ethanol diluted with de-ionized water and applied with clean paper towels. During a friction measurement on a contaminated surface, water or glycerol was applied to the measurement areas using the methods described in the surface condition section and replenished throughout repeated impacts on the floor surface. The three surface conditions were randomized in the friction measurements. Under each surface condition, the sequence of friction measurements was randomized among the five floor types. Under each floor type, the sequence of friction measurements was randomized among the selected locations.

The biomechanical measurements, including UCOF and seven kinematic variables, of the fifth trial and the perceived ratings for each condition were extracted at the individual participant level. Seven kinematic parameters were step length, walking speed and, at heel contact, heel angle with the floor, heel angular velocity, horizontal heel velocity, vertical heel velocity and ankle angle. The ACOF values measured on each walkway under each surface condition were averaged across all the measurements to reflect an overall condition of the walkway under that particular surface condition. This ACOF value was assigned to each individual participant under the same conditions. A backward step-wise regression analysis with the perceived slipperiness rating as the dependent variable, and UCOF, ACOF and all kinematic variables as the independent variables was used to explore their relationship.

## 3 RESULTS AND DISCUSSION

The means and standard deviations of the ACOF, perception rating, UCOF and all kinematic variables for each floor type and surface condition are shown in Table 1.

Table 1. Averaged perceived rating, available coefficient of friction (ACOF), utilized coefficient of friction (UCOF) and gait kinematic parameters for each floor type and surface condition.

| FT | SC | PR | SL (mm) | WS (m/s) | HA (deg) | HAV (deg/s) | HV (m/s) | VV (m/s) | AA (deg) | UCOF | ACOF |
|---|---|---|---|---|---|---|---|---|---|---|---|
| A | D | 6.6 | 803 | 1.99 | 28.3 | −1.43 | 0.398 | −0.0691 | 88.2 | 0.227 | 0.822 |
| | W | 14.2 | 799 | 1.95 | 26.9 | −1.51 | 0.402 | −0.0257 | 89.2 | 0.228 | 0.677 |
| | G | 13.9 | 793 | 1.88 | 26.5 | −1.36 | 0.396 | −0.0605 | 89.4 | 0.213 | 0.557 |
| B | D | 7.7 | 805 | 1.97 | 27.5 | −1.54 | 0.392 | −0.0282 | 89.2 | 0.240 | 0.838 |
| | W | 22.8 | 790 | 1.90 | 26.2 | −1.50 | 0.357 | 0.0004 | 89.8 | 0.204 | 0.35 |
| | G | 17.3 | 775 | 1.82 | 24.7 | −1.39 | 0.361 | −0.0347 | 90.6 | 0.214 | 0.37 |
| C | D | 42.4 | 781 | 1.92 | 25.1 | −1.59 | 0.353 | 0.0411 | 90.9 | 0.213 | 0.907 |
| | W | 37.7 | 781 | 1.87 | 26.2 | −1.50 | 0.383 | 0.0131 | 90.1 | 0.215 | 0.169 |
| | G | 30.6 | 744 | 1.75 | 21.4 | −1.49 | 0.301 | 0.0729 | 92.6 | 0.191 | 0.081 |
| D | D | 18.3 | 786 | 1.92 | 26.3 | −1.47 | 0.404 | −0.0086 | 89.9 | 0.220 | 0.208 |
| | W | 83.3 | 694 | 1.68 | 18.0 | −1.41 | 0.296 | 0.0851 | 93.0 | 0.139 | 0.002 |
| | G | 93.5 | 570 | 1.34 | 7.5 | −0.85 | 0.202 | 0.0340 | 97.6 | 0.112 | 0.001 |
| E | D | 66.4 | 755 | 1.88 | 23.7 | −1.45 | 0.337 | 0.0072 | 90.6 | 0.185 | 0.573 |
| | W | 91.6 | 682 | 1.66 | 18.1 | −1.29 | 0.342 | 0.0872 | 92.7 | 0.127 | 0.032 |
| | G | 94.0 | 567 | 1.30 | 8.1 | −0.85 | 0.181 | 0.0365 | 96.6 | 0.112 | 0.012 |

Abbreviations: FT: floor type, SC: surface condition, D: dry, W: wet, G: glycerol, PR: perceived rating, SL: step length, WS: walking speed, HA: heel angle, HAV: heel angular velocity, HV: horizontal velocity, VV: vertical velocity, AA: ankle angle

The adjusted $R^2$ value at the beginning of the backward step-wise regression analysis in which UCOF, ACOF and all seven kinematic variables were included was 0.519. Among the independent variables, UCOF and ACOF were the only variables with statistical significance in the regression equation with p values of $<0.001$ for UCOF and 0.0018 for ACOF. In the final analysis, the adjusted $R^2$ value was 0.524 with the remaining variables UCOF, ACOF and ankle angle and horizontal heel velocity at heel contact. The adjusted $R^2$ did not change much throughout the analysis and both UCOF and ACOF were statistically significant in every step of the analysis.

The UCOF and ACOF were the major contributors to the outcome of perceived slipperiness rating. The UCOF represents the friction needed for locomotion on a floor surface, while the ACOF represents the friction that could be supported at the shoe and floor interface without a slip. These two variables also determined whether a slip would happen at the shoe and floor interface. A slip is more likely to occur when the UCOF exceeds the ACOF. It has been reported in the literature that increases in stance and stride times, and step width, as well as decreases in stride length, walking speed, heel horizontal velocity, heel horizontal and vertical accelerations, heel and floor angle and UCOF are used to avoid a slip on slippery surfaces (Swensen et al. 1992, Bunterngchit et al. 2000, Fong et al. 2005, Lockhart et al. 2007, Menant et al. 2009, Cappellini et al. 2010). The results appear to suggest that the participants in this experiment may have relied on foot slip to form their perceived slipperiness rating.

## 4 CONCLUSION

The results of the backward step-wise regression analysis indicated that the UCOF and ACOF were the major contributors to the outcome of perceived slipperiness rating. These two variables also determined whether a slip would happen at the shoe and floor interface. The results show that PSR might be a useful tool for determining the degree of slipperiness of a surface area when objective field measurements aren't available.

## REFERENCES


American Society for Testing and Materials F-1677-05, 2005. Standard method of test for using a portable inclinable articulated strut slip tester (PIAST). In *Annual Book of ASTM Standards, 15.07*. West Conshohochen, PA: American Society for Testing and Materials.

Bell, J.L., Collins, J.W., Wolf, L., Grönqvist, R., Chiou, S.S., Chang, W.R., Sorock, G.S., Courtney, T.K., Lombardi, D.A. & Evanoff, B. 2008. Evaluation of a comprehensive slip, trip, and fall prevention program for hospital employees. *Ergonomics* 51(12): 1906–1925.

Bunterngchit, Y., Lockhart, T., Woldstad, J.C. & Smith, J.L. 2000. Age related effects of transitional floor surfaces and obstruction of view on gait characteristics related to slips and falls. *International Journal of Industrial Ergonomics* 25: 223–232.

Cappellini, G., Ivanenko, Y.P., Dominici, N., Poppele, R.E. & Lacquaniti, F. 2010. Motor patterns during walking on a slippery walkway. *Journal of Neurophysiology* 103(2): 746–760.

Chang, W.R. 2002. The effects of slip criteria and time on friction measurements. *Safety Science* 40: 593–611.

Chang, W.R., Grönqvist, G., Leclercq, S., Myung, R., Makkonen, L., Strandberg, L., Brungraber, R.B., Mattke, U. & Thorpe, S.C. 2001. The role of friction in the measurement of slipperiness, Part 1: friction mechanisms and definition of test condition. *Ergonomics* 44: 1217–1232.

Chang, W.R., Courtney, T.K., Grönqvist, R., &Redfern, M.S. 2003. *Measuring slipperiness: human locomotion and surface factors*. Taylor & Francis, London, ISBN 0-415-29828-8.

Chang, W.R., Matz, S. and Chang, C.C. 2014. The stochastic distribution of available coefficient of friction for human locomotion of five different floor surfaces. *Applied Ergonomics* 45: 811–815.

Courtney, T.K., Verma, S.K., Chang, W.R., Huang, Y.H., Lombardi, D.A., Brennan, M.J. & Perry, M.J. 2013.Perception of slipperiness and prospective risk of slipping at work. *Occupational and Environmental Medicine* 70: 35–40.

Fong, D.T.P., Hong, Y. & Li, J.X. 2005. Lower-extremity gait kinematics on slippery surfaces in construction worksites. *Medicine & Science in Sports & Exercise* 37(3), 447–454.

Liberty Mutual Research Institute for Safety. 2013. 2013 Workplace Safety Index. In *From Research to Reality* 16(3): 6–7. Available at http://www.libertymutual.com/researchinstitute. Hopkinton, MA: Liberty Mutual Research Institute for Safety.

Lockhart, T.E., Spaulding, J.M. & Park, S.H. 2007. Age-related slip avoidance strategy while walking over a known slippery floor surface. *Gait & Posture* 26: 142–149.

Menant, J.C., Steele, J.R., Menz, H.B., Munro, B.J. & Lord, S.R. 2009. Effects of walking surfaces and footwear on temporo-spatial gait parameters in young and older people. *Gait & Posture* 29: 392–397.

Swensen, E.E., Purswell, J.L., Schlegel R.E. & Stanevich, R.L. 1992. Coefficient of friction and subjective assessment of slippery work surface. *Human Factors* 34: 67–77.

*Bridging Research and Good Practices towards Patient Welfare – Shih & Liang (Eds)*
*© 2015 Taylor & Francis Group, London, ISBN 978-1-138-02716-9*

# An integrated programme to promote safe surgery best practices in Italian hospitals

R. Onofrio, P. Trucco & C. De Capitani
*Politecnico di Milano, Milano, Italy*

F. Lorenzi
*Astir s.r.l., Milano, Italy*

M. Montinari & E. Pernazza
*Associazione Chirurghi Ospedalieri Italiani (ACOI)*

ABSTRACT: The paper discusses preliminary results of an integrated programme aiming to support continuous improvement processes in Italian hospitals, leveraging on the correct selection, adoption and monitoring of validated best practices covering the entire surgical process. A multidisciplinary panel of experts adopted a multi centric process analysis approach to define the reference surgical pathway and a set of best practices that are characterized by their ability to ensure the effectiveness of one or more key safety tasks (KST). The study adopted a process-based approach to systematically identify: i) the role of each single KST in sustaining safety capabilities and objectives of the surgical process; ii) the contribution of best practices to the effective execution of KST. The pilot application in 41 surgical units demonstrated the ability of the programme to effectively support the coherent introduction of safe surgery best practices.

## 1 INTRODUCTION

Patient safety represents a priority area of intervention in healthcare policies worldwide. The publication of "To Err is Human: Building a safer health care system" reported by the Institute of Medicine (Kohn 2000) represented a cultural shift in the scientific literature in the field of patient safety and it was the trigger for further epidemiological studies regarding analysis of adverse events in different countries such as in Australia (Kable 2002), New Zealand (Davis 2002, 2003), United Kingdom (Vincent 2001), Denmark (Schioler et al. 2001), Canada (Baker 2004), Spain (Aranaz-Andrés et al. 2007, 2008, 2011) and Sweden (Soop et al. 2009). Although it is difficult to compare largely heterogeneous data from different national health systems and contexts, data regarding preventability of adverse events seem to be in agreement among all the studies. About half of the adverse events could be prevented and therefore considered as "organizational errors". These epidemiological studies have been applied to different clinical settings and the data demonstrated that the operating room is one of the most affected by preventable adverse events.

Accordingly, the healthcare sector has been recognised as one of the most complex socio-technical systems because of its intrinsic characteristics. Indeed, in healthcare organisations even the simplest procedure needs the integration of articulated processes, requires different skills and expertise in a context under rapid evolution also from a technological point of view. In terms of complexity, the operating room has some key peculiarities: an environment characterised by increasingly complex procedures, multi-professional staff with a very tight interactions, sophisticated technological equipment and devices, time constraints, and occurrence of unexpected critical conditions since the patient is likely to be under unstable and critical conditions, even induced. As a consequence, the operating room requires well-planned and shared actions and behaviours, aiming at both the prevention of perioperative accidents and the successful completion of the procedure.

Pioneer studies (Kohn 2000) introduced a new approach to the analysis of adverse events in healthcare, grounded on human error theory, and characterised by an integrated consideration of cognitive and organizational factors. James Reason's studies (Reason 1990) in aviation and other High Reliability Organisations paved the way and promoted a rapid diffusion of this new approach along with the evolution of human factors studies in healthcare. Lately issues concerning the need to define a new framework for patient safety have been raised in the Healthcare sector. The new framework is no longer based on a prescriptive view but on a cultural shift which is, in turn, determined by:

- the assumption that human error is unavoidable and should be treated as a starting point for organisational learning; the need of a system view to build safer healthcare processes, by adopting a process rather than a functional approach;
- considering the adverse event as something which has to do with the interactions of technical, organizational and process-related features not only caused by a pure out-of-the-context human error;
- monitoring adverse and sentinel events to design new policies of prevention and risk management in a logic of seamless improvement;
- experimenting organizational models and innovative technological solutions able to better support the execution of complex processes and procedures;
- the key role assumed by EBM (Evidence-based Medicine) in the process of updating clinical practices and healthcare delivery processes.

In the last 10 years a lot of industrial risk management techniques have been adapted and applied to clinical contexts, thanks to an increased consensus around the "need for systemic perspectives in healthcare risk management", the view of "errors like source of learning" and the belief that "healthcare can learn a lot from high-risk industry" (Vincent 2001, Cagliano et al. 2011, Verbano & Turra, 2010).

This new cultural view emphasises the need for systemic approaches and best practices able to drastically reduce preventable errors in surgery. Great effort has been directed to specific international research projects (Haynes 2009, Chang 2010), as well as the issuing of national and international guidelines and recommendations to establish best practices for a safer surgery (Age.na.s 2008, WHO 2009, Italian Ministry of Health 2009, 2010, 2011-Ministero della Salute).

Promoting the culture of quality and safety in surgery is also the goal of a joint initiative started in 2010 by ACOI (Italian Association of Hospital Surgeons) and the School of Management of Politecnico di Milano (Italy). Its main objective is the development and implementation of an integrated programme for supporting the continuous improvement in surgery,based on the correct selection, adoption and monitoring of validated best practices covering the entire surgical process.

## 2 METHODOLOGY

A multidisciplinary panel of 15 experts adopted a multi centric process analysis approach to define the reference surgical pathway. The panel involved senior surgeons, anaesthetists, nurses, risk managers, quality managers, patient representatives, risk analysts. As for a process-driven analysis, it allows to study the organization not as sum of activities, functions, hierarchically interlinked, but as a set of homogeneous assets related to each other beyond functional boundaries and governed by the coordination mechanisms, beyond the hierarchical structure. Assuming a cross-functional view of the organization puts the efforts for improving operators' behaviours and clinical practices both into a more consistent system perspective, opening to widest improving potentials. Furthermore, the patient who is involved in a surgical process, gives great importance not only to the final result of surgery, but also to the entire healthcare delivery process. Based on this view, the logical sequence of activities has been recovered, taking into account the primary goals of the process, and coherently with six process safety capabilities, namely:

– patient identification;
– patient clinical management;
– team communication;
– clinical documentation and traceability;
– management of equipment and medications;
– patient empowerment.

Finally, the surgical process has been modelled into 15 phases, from first visit until discharge after surgical operation.

The analysis also led to the identification of the key safety tasks (KSTs) that ensure high patient safety standard throughout the entire surgical process. These KSTs have been then matched with the 6 general process safety capabilities; to better fit with the standards and recommendations set up for the Italian healthcare system KSTs have also been matched with the16 specific objectives defined by the Italian Ministry of Health in the "Manual for patient safety in the operating room – recommendations and check-list" (Italian Ministry of Health 2009).

The panel of experts also performed a systematic literature review to identify those safety practices for which large consensus or empirical evidences are reported on their effectiveness in surgery. The identification of a list of "best practices" was thus based on the Evidence Based Medicine approach, that is "the conscientious, explicit and judicious use of current best evidence in making decisions about the care of the individual patient. It means integrating individual clinical expertise with the best available external clinical evidence from systematic research" (Sackett et al. 1996). Following this approach, in order to draft the best practices, all the issues related to the surgical process have been defined, the best available scientific evidence has been selected, and the information gathered has been critically reviewed.

There is not only one definition of the term best practice in medicine. Within the project, the term best practice means: " a coherent and consolidated body of knowledge (expressed in the form of guidelines, recommendations, procedure or set of technological capabilities) that, within a given context (clinical, organizational, relational) proves to have a strong positive contribution to the achievement of quality and safety objectives – measured in terms of efficiency and effectiveness – and can therefore be taken as a model (standard), or it may be generalized or adapted to other contexts". The relevance of a best practice should thus refer to the likelihood and extent of patient health state improvement. The relevance would be higher if supported by strong scientific evidence. In the context of our project, best practices are characterized by their ability to effectively support one or more KSTs of the surgical process.

Twenty-five best practices have been identified and divided into three main categories: i) clinical protocols: ii) personal or team procedures and safety rules; iii) technological and eHealth solutions for safety. By reviewing the selected literature, the multidisciplinary panel of experts also defined a minimum set of requirements for the correct implementation of each best practice, along with a set of assessment indicators.

The project's knowledge base, featuring the integration of the above mentioned pieces of information, relied on the creation of an online database which allows to get an immediate picture of the relationships between the single phases of the surgical pathway, the corresponding KSTs, the associated best practices and objectives. In short, the project's knowledge base is composed by:

– description of the 15 phases of the reference surgical pathway: Specialist visit, Pre-surgical hospitalization, surgical hospitalization check in, pre-operative hospital stay, check-out from the ward, operating room and surgical team preparation, patient check-in at the surgical unit, patient preparation, anaesthesia induction, operation execution and patient monitoring, closure of the intervention, patient check-out from surgical unit, check out from the operating room, postoperative hospital stay, discharge (Figure 1);
– description of the 6 main process safety capabilities;

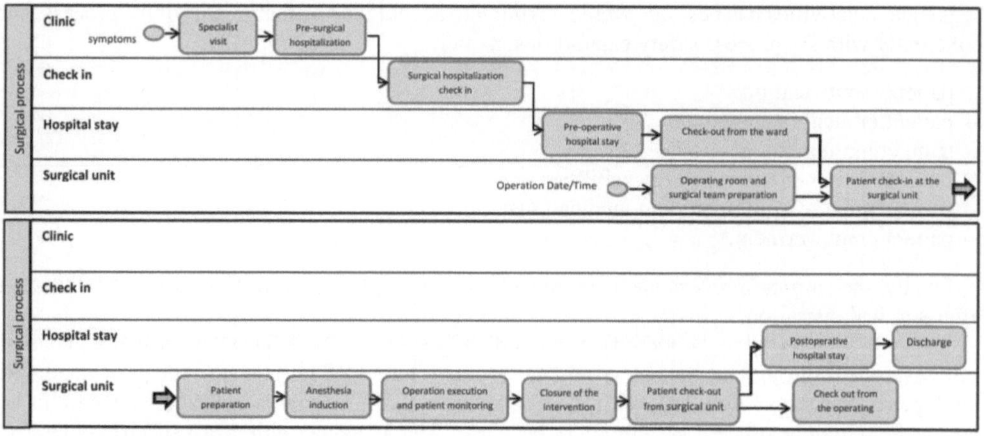


Figure 1. Reference surgical pathway.

Table 1. Selected 25 safe surgery best practices.

| Best practice category | Best practice typology | Best practice denomination |
|---|---|---|
| Clinical protocols | Diagnostic/therapeutic recommendations for safety | Surgical and Anaesthetic Risk Assessment |
| | | Postoperative Deep Venous Thrombosys (DVT) prophilaxis |
| | | Antibiotic prophilaxis |
| | | Safe and rational use of blood in surgery |
| | | Operative bleeding prevention |
| | | Patient preparation |
| Personal or team procedures and safety rules | Executive Procedure | Surgical hand hygiene |
| | | Social and antiseptic handwashing |
| | | Surgical patient identification |
| | | Patient positioning |
| | | Pathological samples management |
| | | Post operative pain management (Acute Pain Service) |
| | Checking procedure | Surgical Safety Check-list |
| | | Patient Check-in and check-out |
| | | Prevention of Retained Surgical Items (RSI) |
| | | Proper documentation of surgical procedures |
| | Communication Procedure | Incidents and Sentinel Events Resporting System |
| | | Integrated prescribing system |
| | | Time out |
| | | Informed consent process |
| Technological and eHealth solutions for safety | Biomedical Technology | Auto-controlled surgical instruments |
| | ICT solutions | Electronic Patient Identification |
| | | Electronic RSI prevention |
| | | Peri-operative workflow management |
| | | Medical Decision Support Systems |

- description of the 16 safety objectives in surgery (Italian Ministry of Health 2009).
- list of 85 KSTs to assure high safety capabilities in the surgical process, associated to each phase and objective;
- descriptions of 25 safe surgery best practices (Table 1) with associated scientific documentation, minimum requirements for the proper implementation and assessment indicators.

Finally, the first version of the knowledge base was validated by a different panel of 8 experts, selected among ACOI senior members. The final reviewed version of the knowledge base was made available on-line to surgeons, anaesthetists, nurses and risk managers of 41 Surgical Units (SUs), distributed over the whole Italian territory, involved in a two-years safety improvement programme (from 2011 to 2012) articulated into two phases:

- Phase 1: mapping of the actual surgical process, compared to the reference process and the 85 KSTs; review of best practices adopted in the SU. The implementation of Phase 1 was supported by a standar checklist, built on the reference knowledge;
- Phase 2: assessment of minimum requirements for the implementation of the best practices declared by each SU (supported by a standard checklist) and pilot implementation of a set of indicators to monitor the impact on process safety performance (supported by a MSExcel™ template for data collection and reporting).

In each phase, all the participants received a two days training – organised into lessons, case discussions and role playing exercises-along with the toolkit (online knowledge base, checklists and data collection worksheet). In the present paper,we only discuss the aggregated results of the first phase of the programme, based on the following information, covering all the 41 SUs involved:

- dimensional and organizational characteristics of the hospital and Surgical Unit;
- description of the surgical process implemented by the SU, i.e. surgical phases and corresponding key safety tasks (KSTs);
- type and mode of implementation of best practices adopted by the Surgical Unit, with particular attention to the 25 best practices identified within the project.

## 3 DISCUSSION OF RESULTS

The analysis of aggregated results has been structured as follows:

- Comparison and gap analysis of the actual surgical pathways implemented in the41 SUs, with respect to the reference process and the 85 KSTs;
- Assessment of the 6 safety process capabilities and degree of coverage of the 16 specific objectives defined in (Italian Ministry of Health 2009);
- Documentation of the level of adoption of safe surgery best practices

Beside, each SU has been provided with a personalised benchmark report where its profile has been compared to the overall sample of SUs participating in the project.

The analysis of surgical process was focused on the qualitative assessment of the degree of alignment of actual surgical pathways to the referenced process; note that a different configuration of the surgical pathway cannot be regarded as poor quality or safety condition *per se*, thus this first analysis was only used to appreciate the degree of alignment/misalignment of current practices among different SUs.

On the other hand, the analysis of KSTs, showed a significant degree of conformity among the SUs: the 29% (12 out of 41) of the sample claimed not to have discrepancy with the modelled surgical pathway, the 41% (17 out of 41) of the SUs presented no more than 5 discrepancies (i.e. KSTs executed in a different phase with respect to the reference process)out of the 85 KSTs (6%); the 9% (4 out of 41) of the SUs reported 15 or more discrepancies. Figure 2 shows the distribution of discrepancies along the 15 phases of surgical process. The phases with more discrepancies are those before the entry to the SU, as "Pre-surgical hospitalization" and "surgical hospitalization check in". The reason mainly lies in the absence of the "Pre-surgical hospitalization" in more efficient SUs and consequently a different allocation of the corresponding KSTs. Furthermore the phases of "patient preparation for surgery" and "patient check-out from surgical unit" present discrepancies of 16% & 20.12% respectively, due to cases of missed surgical team briefing and debriefing. In brief, collected data demonstrated that more than the 80% of the SUs regularly execute more than

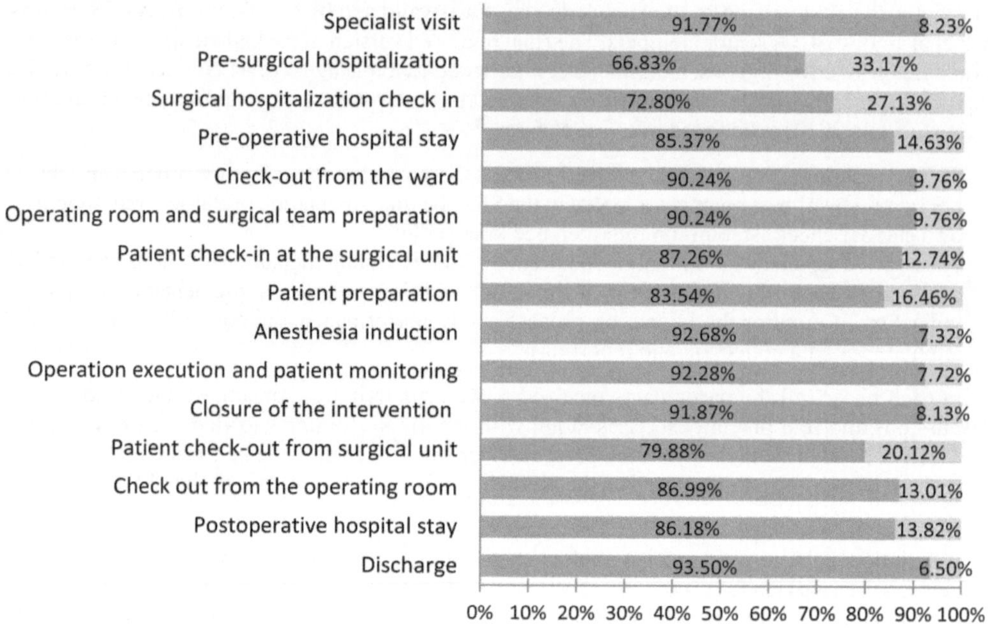


Figure 2. Average degree of conformity and discrepancy in the surgical process (executed, missed, postponed Key Safety Tasks) for the entire sample (41 SUs).

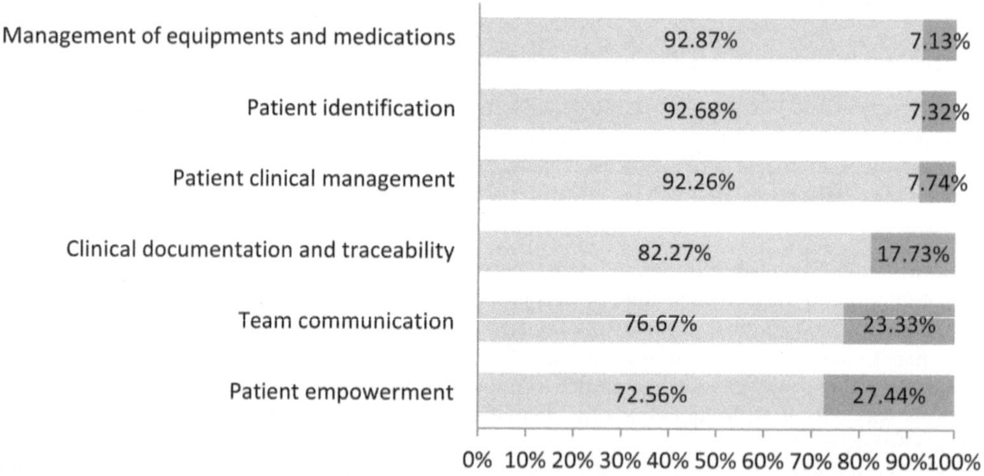


Figure 3. Average level of deployment of the six process safety capability for the entire sample (41 SUs).

the 85% of the KSTs with a high degree of alignment even between SUs with different process arrangements.

The second type of analysis concerned the 6 main process safety capabilities. On average the sample returned a level of deployment that is higher than 90% for the following three capabilities: patient identification, patient clinical management, and management of equipment and medications. Indeed, as shown in Figure 3, the requirements for a complete and correct team communication (communication capability) and the correct documentation and traceability of the care treatment (documentation and traceability capability) resulted relatively less developed. One of the

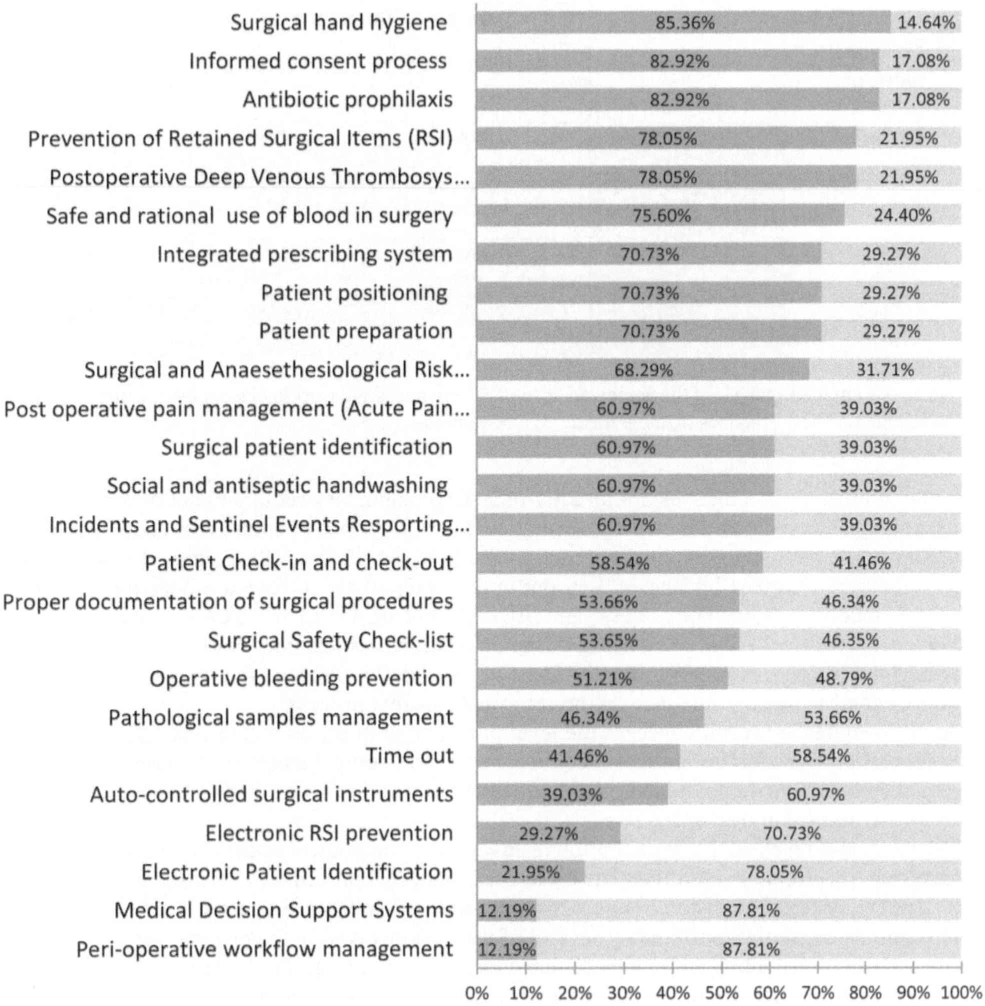


Figure 4. Degree of adoption of safe surgery best practices in the sample (41 SUs).

main causes of a lack in communication among all the actors involved in the surgical process is the low prevalence of briefing and debriefing practices. Also patient empowerment, i.e. the direct involvement of patients in safety procedures, is a process safety capability that still needs to be fully developed. Thanks to the knowledge base provided by the project, these results can be matched with the phases "patient preparation for surgery" and "patient check-out from surgical unit", in which the level of conformity on KSTs directly connected to those capabilities is lower than 85% (see Figure 2).

The third level of analysis concerned the degree of adoption (diffusion) of the selected 25 safe surgery best practices (Figure 4). Clinical related best practices are those with the highest level of diffusion within the 41 SUs, coherently with the relatively higher level of documentation and consensus they receives in literature. Best practices regularly adopted by more than the 75% of the sample are: Surgical hand hygiene, Prevention of Retained Surgical Items (RSI), Postoperative Deep Venous Thrombosys (DVT) prophilaxis, Safe and rational use of blood in surgery, Antibiotic prophilaxis.

| Perceived Workload | Perceived Effectiveness | | | |
|---|---|---|---|---|
| | High | Medium-High | Medium-Low | Low |
| High | 13,32% | 1,90% | 0,00% | 0,00% |
| Medium-High | 22,99% | 17,28% | 0,15% | 0,00% |
| Medium-Low | 16,25% | 9,66% | 1,90% | 0,00% |
| Low | 13,47% | 2,49% | 0,59% | 0,00% |

Figure 5. Combined analysis of the perceived level of effectiveness and workload associated with the adoption of the selected 25 best practices (sample mean: 575 responses).

Although these best practices are directly related to mandatory requirements or recommendations, internal procedures adopted by several SUs are still not coherent with documented best practices or international guidelines. In fact, the 15% (6 out of 41) of SUs claimed to have a procedure for hands hygiene but the 21% (9 out of 41) claimed not to have a procedure for Venous Thromboembolism prophylaxis. Less than 70% (25 out of 41) of the SUs claimed to have an Incidents and Sentinel Events Reporting System (for which a mandatory National Protocol exists; Italian Ministry of Health 2009). Finally, 7 out of 41 SUs (17%) stated that they do not adopt (or partially adopt) national guidelines on the Informed Consent process.

Another interesting result concerns the level of diffusion of the so-called procedural best practices; in particular, as for the surgical checklist and time out, only the 60% of the sample declared a systematic implementation. Notably, also these best practices are prescribed by the Italian Ministry of Health to assure surgical safety.

The study also revealed that the adoption of advanced technological solutions for patient safety in surgery, e.g. Electronic Patient Identification, Electronic RSI prevention, Peri-operative workflow management, Medical Decision Support Systems, are extremely rare, mainly due to economic or organisational constraints suffered by several of the SUs involved in the programme.

In partial contrast with the scenario depicted by previous figures, all the professionals involved in the surgical process demonstrated a quite positive attitude towards the systematic adoption of safe surgery best practices: even though the perceived additional workload is not negligible, the judgement on the level of perceived effectiveness of the adopted best practices is clearly high (Figure 5).

## 4 CONCLUSIONS

Thanks to this study and its achievements it has been possible to provide a systematic view and documentation of the surgical pathway, which is no longer understood as a sequence of activities or a mix of coordinated functions, but as a complex socio-technical process that requires high safety capabilities.

Furthermore, the identification of a set of well documented and widely recognised best practices provides SUs with state-of-the-art knowledge on quality and safety in surgery. A set of tools has also been developed and tested: along with the knowledge base it can support the diffusion of continuous improvement programmes in SUs.

Key findings of the first pilot implementation of the programme revealed a persisting gap between state of the art knowledge on safe surgery and current practices in Italian SUs. This is mainly due to organisational constraints and lack of effective change management process.

However, results returned by the first phase of the study and discussed in the present paper suffer some limitations. Indeed, the first phase of the study neglected the analysis on "how" the best practices are actually implemented by the SUs, thus the coherence of real practice with the reference guidelines and recommendations has not been checked by the experts but only declared by practtioners; this is the focus of the second phase of the programme, covering the evaluation of the degree of conformity of actual practices to a set of minimum requirements, derived by international guidelines, protocols and scientific literature, for each one of the 25 safe surgery best practices.

The pillars which this project is based on, i.e. problem-driven training activity, on-the-job assessment of practice and methodological support, can be considered as the platform for a generic continuous improvement programme, also applicable to safety capability and performance of the surgical process. On this regard, both the knowledge base and the toolkit are under further development and have been integrated into a web-based platform named "Safe Surgery Online". Thanks to this online tool, single SUs or entire hospitals may carry out independent analyses of their own surgery processes, benchmark safety performances anonymously and share experiences on a regular basis and with real time feedbacks. On the other side, the analysis and reporting tools, also implemented in the platform, allow users to work on the actual gaps of the SU through different views: process, safety objectives, safety capabilities, Key Safety Tasks (KST), Best Practices, professional profiles.

In conclusion, the systematic approach implemented in the pilot programme, and tested by 41 SUs distributed over the Italian territory, demonstrated to positively contribute to increase safety awareness among professionals, to effectively support continuous improvement and the change management process in surgery, by promoting the coherent selection and implementation of best practices towards safer surgery.

## REFERENCES


Aranaz-Andrés JM, Ruiz-López P, Aibar-Remón C et al. 2007. Adverse events in general and digestive surgery departments in Spanish hospitals. Cir Esp 2007; 82.

Baker GR, Norton PG, Flintolf V, et al. 2004. The Canadian Adverse events Study: the incidence of adverse events among hospital patients in Canada. Canadian Medical Association Journal, 2004, 179(11): 1678–1686.

Berwick 2005. Five years after To err is Human: what have we learned? JAMA 2005, 293: 2384–23.

Cagliano, A. C., Grimaldi, S., & Rafele, C. 2011. A systemic methodology for risk management in healthcare sector. *Safety Science*, *49*(5), 695–708. doi:10.1016/j.ssci.2011.01.006.

Davis P., Lay-Yee R, Briant R et al. 2002. Adverse events in New Zealand public hospitals I: occurrence and impact. New Zealand Medical Journal, 2002, 115 (1167):U271.

Davis P., Lay-Yee R, Briant R et al. 2003. Adverse events in New Zealand public hospitals II: occurrence and impact. New Zealand Medical Journal, 2003, 116 (1183):U624.

Gawande AA. 1999. When doctors make mistakes. The New Yorker, 1st February 1999.

Gawande AA., Zinner MJ, Studdert DM et al. 2003. Analysis of errors reported by surgeons at three teaching hospitals. Surgery 2003.

Haynes Alex B. et al. 2009. A Surgical Safety Checklist to Reduce Morbidity and Mortality in a Global Population. The New England Journal of Medicine January 29, 2009 -360:491-9.

Italian Ministry of Health. 2009. Manuale per la Sicurezza in sala operatoria: raccomandazioni e checklist. Dipartimento della qualità, Direzione Generale della Programmazione sanitaria, dei livelli di assistenza e dei principi etici di sistema (in Italian).

Italian Ministry of Health. 2009. Protocollo per il Monitoraggio degli Eventi Sentinella. Luglio 2009. Dipartimento della qualità, Direzione Generale della Programmazione sanitaria, dei livelli di assistenza e dei principi etici di sistema (in Italian).

Italian Ministry of Health. 2011. Protocollo per il monitoraggio degli eventi sentinella – Terzo rapporto. Settembre 2005- Dicembre 2010. Luglio 2011. (in Italian).

Italian Ministry of Health. 2010. Sicurezza dei Pazienti e Gestione del Rischio Clinico: Manuale per la formazione degli operatori sanitari. Dipartimento della qualità, Direzione Generale della programmazione sanitaria, dei livelli di assistenza e dei principi etici di sistema (in Italian).

Kohn LT, Corrigan JM, Donaldson MS, editors. 2000. To err is human – building a safer health system. Institute of Medicine – National Academy Press 2000.
Leape LL., Brennan TA, Laird N et al. 1991. The nature of adverse events in hospitalized patients. Results of the Harvard Medical Practice Study II. New England Journal of Medicine, 1991, 324 (6):377–84.
Reason, James. 1990. Human Error. New York, NY, Cambridge University Press.
Sackett DL. et al. 1996. Evidence based medicine: what it is and what isnt't. BMJ 1996.
Schioler T, Lipezak H, Pedersen BL et al. 2001. Danish Adverse Event Study. Incidence of adverse events in hospital. A retrospective study of medical records, Ugeskr laeger, 2001, 163 (39):5370–8.
Senders W, Moray NP (eds). 1991 Human error (cause, predic- tion, and reduction): analysis and synthesis. Hillsdale (NJ), Law-rence Erlbaum.
Thomas EJ, Studdert DM, Runchiman, WB et al. 2000. A comparison of iatronic injury studies in Australia and the USA I: context, method, casemix, population, patient and hospital characteristics. International Journal of Quality in Health Care, 2000,12 (5):371–378.
Vincent C., Neale G, Woloshynowych M. 2001. Adverse events in British hospitals: preliminary retrospective record review. British Medical Journal, 2001, 322:517–9.
Vincent C., 2003. Understanding and responding to adverse events.N Eng J Med, 348, 11 pp. 1051–1056.
Wilson RM, Runciman WB, Gibberd RW et al. 1995. The Quality in Australian Health Care Study. Medical Journal of Australia, 1995, 163:458–71.
Whorld Health Organization. 2011. Guidelines for Safe Surgery. 2009. Safe Surgery Saves Lives. Website: "http://whqlibdoc.who.int/publications/2009/9789241598552_eng.pdf".

*Bridging Research and Good Practices towards Patient Welfare – Shih & Liang (Eds)*
*© 2015 Taylor & Francis Group, London, ISBN 978-1-138-02716-9*

# Consideration of occupational stressors that influence senses of satisfaction at work in Japan

Ayako Kajimura, Nanae Shintani & Shigeki Tatsukawa
*Faculty of Health Sciences, Department of Nursing Kansai University of International Studies, Japan*

ABSTRACT: This study aims to find the factor of occupational stressors that influence "Job Satisfaction" and to consider the needs of improvement of the environment for employees to carry "Job Satisfaction". The Survey Method was to collect questionnaire papers from 63 respondents who worked in a food manufacturer. And items of hypostatic factor of "Occupational Stressor" in NIOSH under factor analysis as explanatory variable and "Job Satisfaction" as objective variable were analyzed by multiple regression analysis. The results of Multiple Regression Analysis of Job Satisfaction and Occupational Stressors, there were 4 factors that showed significant differences that influenced for Job Satisfaction. The factor of "Friendly Environment ($\beta = 0.348$ $p < 0.01$)" was the one that influenced the most. In the second place, the factor of "Well-developed Relationship ($\beta = 0.287$ $p < 0.05$)", "Job Discretion ($\beta = 0.272$ $p < 0.05$)", and "Workload ($\beta = 0.197$ $p < 0.05$)" also influenced Job satisfaction. On the other hand, the factors of "Prehension on work ($\beta = 0.029$ $p > 0.05$)" and "Skill Utilization ($\beta = 0.048$ $p > 0.05$)" did not influence significantly. Factors of "Friendly Environment" and "Well-developed Relationship" are essential to Job Satisfaction. It is expected to build improvement of the environment for employees' health and safety by themselves and occupational health nurses as collective work. And occupational health nurses should be responsible to improve the environment as one strategy of mental health supports.

## 1 INTRODUCTION

### 1.1 *Employment pattern*

According to "The National Survey of Lifestyle Preferences in Japan", which Japanese Cabinet Office releases every 3 years, the number of people with satisfaction of their work has been decreasing since 1978 on a long-term basis. Especially there is lack of a sense of satisfaction for meaningful items of "Employment Stability", "Increase in Their Incomes", and "Worthwhile Job". Stagnant economic growth since 1990's in Japan made pattern of employment as "the custom of long-term employment" re-thought. So that, number of none-permanent workers has increased. The item of "Worthwhile Job" from "The National Survey of Lifestyle Preferences in Japan" decreased to 16.6% in 2005, which used to be about 30% in 1978 (Ministry of Health, Labour and Welfare in Japan, 2005). Moreover, the number of employees on sick leave due to depression has been increasing currently, which assumed that there is a certain amount of "stressor" surrounding employees in Japan.

### 1.2 *Occupational stress and countermeasure of mental health at work*

During prolonged economic stagnation, Occupational Stress and Countermeasure of mental health at work have drawn increasing attention these days.

The results of the National Survey of Lifestyle Preferences in Japan in 2005, 50.9% of all the people in the survey felt "Worthwhile Job" which was related to the Interest of work.

On the contrary, 51.8% of all the people in the survey also felt "Lots of Stress at work" while they were working. This is opposing and exceeding result from "Worthwhile Job (50.9%)"(Ministry of Health, Labour and Welfare in Japan, 2005).

Workers having occupational stress more than Worthwhile Job, get less motivated for their work. Over-load stress could be able to make workers have worse condition of their mental health as well. So that, it is very important to develop the countermeasure of mental health at work to decrease workers with Over-load stress.

According to former study of WATNABE's (2002), there are 2 general types of Countermeasure of mental health at work, "Each Worker to acquire thrive on stressors" and "creating the environment to decrease stress and worker not to response the stress" (WATANABE, 2002).

Ministry of Health, Labour and Welfare in Japan suggested some strategies of Countermeasure of mental health at work with "Mental health plan guideline for worker" in 2000, which includes "mental care support for employees, supervisors and managers" from Occupational health professions. It revealed that occupational health nurses have an important role to improve the environment for employees as Countermeasure of mental health at work. For Countermeasure of mental health at work, it is very important to find the occupational stressors that influence Job Satisfaction at work currently. And decreasing those occupational stressors should make the Worthwhile job to increase for workers (Ministry of Health, Labour and Welfare in Japan, 2000).

## 2 PURPOSE

This study aims to find the factor of occupational stressors that influence "Job Satisfaction" and to consider the needs of improvement of the environment for employees to carry "Job Satisfaction".

## 3 METHOD

The Survey Method was to collect questionnaire papers with "Occupational Stressors" in NIOSH (National Institute of Occupational Safety and Health) and "Job Satisfaction" from workers who work in a food manufacturer under analysis procedure below.

1. Factor analysis of NIOSH (National Institute of Occupational Safety and Health) in the questionnaire to find hypostatic factors.
2. The hypostatic factors from NIOSH (National Institute of Occupational Safety and Health) in the questionnaire as explanatory variable and "Job Satisfaction" as objective variable were analyzed by multiple regression analysis.
3. SPSS Statistics Version21 was used as the analyzing tool.

## 4 RESULTS

There were 63 respondents age from 20 to 63 year old who worked in a food manufacturer with 100% of the collection rate. 45 of them were male and the other 18 were female, average age; $43.5 \pm 10.75$.

### 4.1 *Hypostatic factors of NIOSH*

Under analysis of the date from collected questionnaires by Principal Factor Method (Promax Rotation), 8 Hypostatic factors of NIOSH were found and named as factors of 1: "Well-developed Relationship", 2: "Workload", 3: "Non-opposing work", 4: "Prehension on work", 5: "Job discretion", 6: "Skill Utilization", 7: "Competence at work", 8: "Friendly environment". The confidence coefficients of those factors were shown as below (Table 1).

Table 1. Factor Analysis of Occupational Stressors from workers work in a food manufacturer.

| Itmes | Factors | | | | | | | |
|---|---|---|---|---|---|---|---|---|
| | 1 | 2 | 3 | 4 | 5 | 6 | 7 | 8 |
| Stressor32. How much does your immediate boss go out of his way to do things to make your work life easier for you? | **0.906** | 0.039 | 0.09 | -0.102 | 0.133 | 0.028 | -0.15 | 0.052 |
| Stressor34. How much can your immediate boss be relied on when things get tough at work? | **0.891** | 0.146 | -0.011 | 0.019 | -0.01 | 0.046 | 0.032 | -0.098 |
| Stressor33. How much do other people at work go out of his way to do things to make your work life easier for you? | **0.786** | 0.042 | 0.082 | 0.047 | 0.027 | 0.024 | -0.174 | 0.164 |
| Stressor36. How much does your immediate boss listen to your personal problem? | **0.777** | -0.029 | 0.088 | 0.072 | -0.126 | 0.033 | 0.234 | -0.085 |
| Stressor35. How much can other people at work be relied on when things get tough at work? | **0.767** | -0.012 | -0.028 | 0.131 | -0.095 | -0.017 | -0.037 | 0.08 |
| Stressor37. How much do other people at work listen to your personal problem? | **0.601** | -0.104 | 0.013 | 0.131 | -0.015 | -0.038 | 0.108 | 0.086 |
| Stressor18. How often is there a great deal to be done? | 0.052 | **0.893** | 0.061 | -0.078 | 0.038 | -0.114 | 0.09 | 0.063 |
| Stressor17. How often does your job leave you with little time to get things done? | 0.09 | **0.863** | -0.013 | -0.128 | -0.021 | -0.046 | 0.065 | -0.057 |
| Stressor19. How often is there a marked increase in the work load? | 0.03 | **0.67** | 0.119 | 0.001 | -0.134 | 0.061 | 0.121 | 0.061 |
| Stressor30. There are clashes between subgroups within my group | 0.264 | -0.084 | **0.877** | -0.123 | 0.127 | -0.084 | 0.148 | -0.106 |
| Stressor27. There is difference of opinion among the members of my group | -0.028 | -0.028 | **0.682** | 0.139 | -0.026 | 0.093 | -0.116 | -0.262 |
| Stressor 3. I receive incompatible requests from two or more people. | 0.02 | 0.006 | **0.671** | 0.178 | -0.207 | 0.098 | -0.084 | -0.125 |
| Stressor 2. I work with two or more groups who operate quite differently. | 0.05 | -0.036 | **0.546** | -0.112 | -0.142 | 0.046 | -0.108 | 0.073 |
| Stressor31. There is lack of mutual assistance between my group and other groups | -0.031 | 0.197 | **0.516** | -0.049 | 0.101 | -0.216 | 0.009 | -0.011 |
| Stressor 4. I work on ennecessary things. | -0.304 | -0.001 | **0.499** | -0.001 | 0.204 | 0.014 | -0.093 | 0.04 |
| Stressor 1. I receive an assignment without the help I need to complete it. | -0.144 | 0.363 | **0.493** | 0.081 | 0.043 | -0.175 | 0.078 | 0.009 |
| Stressor25. You sometimes receive awful complaines from clients and the public or comflict with them at work | 0.039 | -0.266 | **0.456** | -0.22 | 0 | 0.189 | 0.28 | 0.174 |
| Stressor 6. I know that I have devided my time proterly. | -0.039 | -0.285 | 0.111 | **0.701** | 0.138 | -0.164 | -0.156 | 0.17 |
| Stressor 7. I know what my responsibilities are. | -0.045 | 0.118 | -0.099 | **0.698** | 0.225 | 0.034 | 0.028 | -0.089 |
| Stressor 8. I know exactly what is expected of me. | 0.056 | -0.133 | 0.044 | **0.697** | 0.036 | 0.089 | 0.19 | -0.094 |
| Stressor 5. There are clear, planned goals and objectives of my job. | 0.225 | 0.219 | -0.009 | **0.668** | -0.033 | -0.131 | -0.074 | 0.053 |
| Stressor 9. Explanation is clear about what has to be done on my job. | 0.275 | -0.041 | -0.116 | **0.562** | 0.02 | 0.041 | 0.067 | 0.014 |
| Stressor12. I can freely decide your work pace | 0.028 | -0.141 | 0.065 | 0.136 | **0.792** | 0.056 | 0.035 | -0.002 |
| Stressor11. I can freely decide your work load | -0.114 | -0.076 | 0.058 | 0.012 | **0.755** | 0.023 | 0.271 | 0.15 |
| Stressor10. I can freely decide your sequence of work | -0.004 | 0.026 | -0.058 | 0.306 | **0.634** | 0.126 | 0.046 | -0.087 |
| Stressor24. How often can you use the skills from your previous experience and training? | -0.066 | -0.032 | -0.051 | 0.044 | -0.003 | **0.812** | 0.076 | -0.093 |
| Stressor23. How often you give a change to do the things you do the best? | 0.139 | 0.035 | -0.028 | -0.229 | 0.165 | **0.752** | -0.033 | 0.001 |
| Stressor22. How often does your job let you use the skills and knowledge you learned in school? | 0.046 | 0.016 | 0.038 | 0.068 | -0.003 | **0.563** | -0.007 | 0.126 |
| Stressor13. I can make decision about which task I should assign to | -0.007 | 0.134 | 0.035 | 0.03 | 0.157 | 0.013 | **0.787** | 0 |
| Stressor15. I have authority to educate or train other employees | -0.085 | 0.28 | -0.045 | -0.029 | 0.084 | 0.183 | **0.642** | -0.014 |
| Stressor14. I can decide when to finish your work | 0.044 | 0.09 | -0.039 | 0.024 | 0.407 | -0.096 | **0.641** | 0.015 |
| Stressor28. The members of my group are supportive of each other's ideas | 0.046 | 0.026 | -0.128 | 0.052 | 0.026 | -0.052 | 0.028 | **0.791** |
| Stressor29. There is friendliness among the members of my group | 0.177 | -0.022 | -0.175 | -0.082 | 0.071 | 0.053 | 0.007 | **0.699** |

**Principal Factor Method (Promax Rotation)**

Table 2. The factor name of Occupational Stressors by factor analysis.

| | Name of the factor | $\alpha$ coefficient in Cronbach |
|---|---|---|
| 1st factor | Well-developed Relationship | .919 |
| 2nd factor | Workload | .896 |
| 3rd factor | Non-opposing work | .819 |
| 4th factor | Prehension on work | .842 |
| 5th factor | Job discretion | .878 |
| 6th factor | Skill Utilization | .770 |
| 7th factor | Competence at work | .856 |
| 8th factor | Friendly environment | .805 |

Table 3. Fundamental statistics of Job Satisfaction.

| | Frequency | Minimum | Maximum | Average | SD |
|---|---|---|---|---|---|
| Job Satisfaction | 63 | 1.0 | 4.6 | 2.771 | .7325 |

### 4.2 *Occupational stressors that influence job satisfaction*

In the questionnaire of this study, "Environment and comfort of facilities in workplace", "Skill Utilization", "Relationship at work", "Amount of work", "Satisfaction of position and role at work" were used as Job Satisfactions. And I assessed them based on Likert scale of one to five. The basic statistics (Table 3) and the results of Multiple Regression Analysis of Job Satisfaction and Occupational Stressors (Table 4) were as below.

The results of Multiple Regression Analysis of Job Satisfaction and Occupational Stressors, there were 4 factors that showed significant differences that influenced for Job Satisfaction. The factor of "Friendly Environment ($\beta = 0.348$ $p < 0.01$)" was the one that influenced the most. In the

Table 4. Multiple Regression Analysis of Job Satisfaction and Occupational Stressors.

| | $\beta$ for standardized coefficients | t value | p value |
|---|---|---|---|
| Well-developed Relationship | .287 | 2.558 | .013** |
| Workload | .197 | 2.107 | .040* |
| Prehension on work | .029 | .263 | .794 |
| Job discretion | .272 | 2.580 | .013** |
| Skill Utilization | .048 | .497 | .621 |
| Friendly environment | .348 | 3.219 | .002*** |
| $R^2$ | .524 | | |

*p < 0.5, **p < .03, ***p < .01.

second place, the factor of "Well-developed Relationship ($\beta = 0.287$ $p < 0.05$)", "Job Discretion ($\beta = 0.272$ $p < 0.05$)", and "Workload ($\beta = 0.197$ $p < 0.05$)" also influenced Job satisfaction.

On the other hand, the factors of "Prehension on work ($\beta = 0.029$ $p > 0.05$)" and "Skill Utilization ($\beta = 0.048$ $p > 0.05$)" did not influence significantly.

## 5 DISCUSSIONS

### 5.1 *Hypostatic factors of NIOSH*

According to NIOSH Model of Job Stress, non-work factors, individual factors and buffer factors adjust and buffer job stressors to promote acute stress reactions.

Factor analysis by NIOSH Model of Job Stress was done, and I found that Occupational stressors of the workers who work in a food manufacturer were built with 8 factors as "Well-developed Relationship", "Workload", "Non-opposing work", "Prehension on work", "Job discretion", "Skill Utilization", "Competence at work", "Friendly environment".

### 5.2 *Relation between occupational stressor and job satisfaction*

Although the result of Job Satisfaction was from 1.0 to 4.6, Job Satisfaction in average was likely low. The respondents of this study were workers who worked in a food manufacturer. And most of those take care of production lines there. So that, they repeat the same simple work in a part of the production lines everyday. In addition, it's said that the time period of familiarization program workers learn to work in production lines is normally short time such as one week (Tomuro, 2007). The job in production lines is a simple task that can familiarize in a short period of time, is difficult to make workers feel the effort of what they have done. That is probably why workers cannot get Worthwhile Job for Job Satisfaction. And the result of Job Satisfaction also showed Likely low as $2.77 \pm 0.732$.

According to Health, Labour and Welfare Ministry in Japan(2004), Worthwhile Job is most essential factor to be connected with Job Satisfaction. And Their Incomes, Skill Development, and Working Hours are also essentials for it. To increase Job Satisfaction, Countermeasure for workers to carry Worthwhile Job should be necessary (Ministry of Health, Labour and Welfare in Japan, 2004).

Multiple regression analysis was used in this study to find the factors of occupational stressors, which influences the respondents' Job Satisfaction. And it was revealed that the factor of "Friendly Environment ($\beta = 0.348$ $p < 0.01$)" was the one that influenced the most. And the factor of "Well-developed Relationship ($\beta = 0.287$ $p < 0.05$)" also influenced Job satisfaction as the second most. These results showed that the Factors of "Friendly Environment" and "Well-developed Relationship" are essential to "Job Satisfaction" more than the factor of "Job Contents".

The Survey of Working person in 2012 (Figure 1) also showed that people thought the Friendly Environment was the most important factor. Although Worthwhile Job is a great influence for Job

Figure 1. Important Essences for Work (combined 1 to 3).

Satisfaction, employees demand "Friendly Environment" more in the view of relation between Occupational Stress and Job Satisfaction (Recruit Works Institute 2012).

In the second place, the factor of "Job Discretion ($\beta = 0.272$ $p < 0.05$)", and "Workload ($\beta = 0.197$ $p < 0.05$)" also influenced Job satisfaction significantly. These factors are related to "Job Contents". The respondents of this study are mostly workers who take care of production lines. So that, they repeat the same simple task work in a part of the production lines everyday. The Job Contents should make the factor of "Job Discretion and "Workload" lower.

This result can possibly be a distinction of workers with production lines in a food manufacture.

On the other hand, the results showed that the factors of "Prehension on work ($\beta = 0.029$ $p > 0.05$)" and "Skill Utilization ($\beta = 0.048$ $p > 0.05$)" did not influence Job Satisfaction significantly.

The respondents of this study are mostly workers who work on ensuring strict job assigned to each and everyone, and be requested to produce the exact amount of food product by company.

Repeating the same simple task work everyday make less opportunities of "Prehension on work" and "Skill Utilization" for workers. This should be why those factors did not show much influence for Job Satisfaction significantly.

### 5.3 *Improvement of the environment for employees to carry less occupational stress*

The four factors of Occupational Stressors "Friendly environment", "Well-developed Relationship", "Job discretion", and "Workload" indicated the influence Job Satisfaction. Despite the fact that it is almost impossible to get rid of all the stressors surrounding employees, it is expected for occupational health nurses to build improvement of the environment for employees' job Satisfaction and decreasing these stressors.

It's said that the Strength of Companies in Japan is "Genba-ryoku, empowerment faculty and sensibility in practice (Ikeda, 2006)". To increase the "Genba-ryoku", implementation of improvement activities in small groups at work is predictably-effective as much as the supports for workers' activities and training, to solve the issue in workplace. One of the proposes to increase the "Genba-ryoku" is that self-initiative work process with workers using originality and ingenuity make them achieve the self-fulfillment, be respected the human qualities. Moreover, it makes the workplace fulfilled with worthwhile.

To order the Job Contents as in "top -down decision-making" way does not have to be the only one. It is necessary to use the procedure of implementation of improvement activities in small groups at work for workers carrying "Job discretion", even if the workers are under production lines. In addition, "Friendly environment" and "Well-developed Relationship" are also essential for implementation of improvement activities in small groups at work.

It is very important to develop organization with improvement including exchanges of opinions in a constructive way. Therefore, occupational health nurses get necessarily into the small groups to consider the implementation of improvement activities.

It is expected to build improvement of the environment for employees' health and safety by themselves and occupational health nurses as collective work. And occupational health nurses should have a view as a third person to evaluate the environment at workplace.

## 6 CONCLUSION

This study aims to find the factor of occupational stressors that influence "Job Satisfaction" and to consider the needs of improvement of the environment for employees to carry "Job Satisfaction". And the factors of occupational stressors are categorized as "Well-developed Relationship", "Workload", "Prehension on work", "Job discretion", "Skill Utilization", and "Friendly environment".

It is revealed that the factors of "Friendly environment", "Well-developed Relationship", "Job discretion", and "Workload" influenced Job Satisfaction. To decrease those stressors, occupational health nurses should be responsible to improve the environment as one strategy of mental health supports. However, these stressors are not the only ones that influence All Job Satisfaction.

According to "NIOSH Model of Job Stress", there are following other factors of non-work factors, individual factors and buffer factors, health behaviors, and relations among family members. Workers are surrounded in Job and Non-job related stressors.

It is expected to build improvement of the environment for employees' health and safety by themselves and occupational health nurses as collective work. And occupational health nurses should have a view as a third person to evaluate the environment at work.

The improvement of the environment to decrease the occupational stressors is not only for decreasing employees stressors and increasing Job Satisfaction. But it is also a countermeasure development of mental health care. As an important task of occupational health nurses with managing health care for all employees, it is extremely meaningful for occupational health nurses to assess and improve employees' health and environment at work in multilateral vision.

Moreover, the factors of "Friendly environment" and "Well-developed Relationship" which is revealed in this study are essential for employees continuously working. Those influenced Job Satisfaction no matter what type of job they have. But the factors of "Job discretion" and "Workload" can possibly be characteristic results that employees with production lines. And it is recommended to research for relation between occupational stressors and Job Satisfaction in many types of job in the future.

## REFERENCES


Japanese Cabinet Office 2005: The National Survey of Lifestyle Preferences in Japan, Available at: http://www5.cao.go.jp/seikatsu/senkoudo/senkoudo.html Accessed December, 2013

Ministry of Health, Labour and Welfare in Japan, 2000: Guideline of Mental Health at work, Available at: http://www2.mhlw.go.jp/kisya/kijun/20000809_02_k/20000809_02_k_shishin.html Accessed December, 2013

Ministry of Health, Labour and Welfare in Japan, 2004: An economic white paper in 2004; Analysis of labor economy, Available at: http://www.mhlw.go.jp/wp/hakusyo/roudou/04/ Accessed December, 2013

Mitsuho Ikeda. 2006. Genba-ryoku, empowerment faculty and sensibility in practice. Available at: http://www.cscd.osaka-u.ac.jp/user/rosaldo/060518genba.html#author. Accessed December, 2013

NISHO; National Institute of Occupational Safety and Health. Available at: http://www.cdc.gov/niosh/ Accessed December, 2013

Recruit Works Institute 2012: The Survey of Working Person Available at: http://www.works-i.com/?active_action=repository_view_main_item_detail&page_id=17&block_id=302&item_id=1090&item_no=1 Accessed December, 2013

Tomuro Kensaku 2007: Contract laborers and the workplace division of labor of the Japanese automobile industry, *Journal of The Ohara Institute for Social Research*, Hosei University 585: 57–70.

Watanabe Naoto 2002: Occupational Stress, *Psychology of career progression – Career Organization Lifelong Development-*, Kawashima Shoten, ISBN-10: 4761007575: 201–228.

# *Part D: Simulation, modeling and decision making*

*Bridging Research and Good Practices towards Patient Welfare – Shih & Liang (Eds)*
*© 2015 Taylor & Francis Group, London, ISBN 978-1-138-02716-9*

# Integration of biophysiological information with a point-of care decision support system to promote patient safety at home

Winnie Sun & Diane Doran
*University of Toronto, Toronto, Ontario, Canada*

ABSTRACT: There is increasing concern to support older adults who want to live independently in their own homes. The purpose of this study is to investigate the integration of two non-intrusive approaches to monitoring home care clients activity level, along with access to best practice guidelines for clinicians at the point of care. A prototype Remote Activity Monitoring and Guidelines System has been developed that uses a GPS-equipped BlackBerry to monitor a person's mobility outside the home. The System includes a pressure-sensitive mat that is placed under a regular bed mattress and can monitor sleep disturbances, and how long it takes to enter and exit the bed. This research study provided valuable information about the feasibility and effectiveness of integrating bio-mobility information obtained through remote monitoring with decision support technologies to support safer client care for seniors.

## 1 INTRODUCTION

The purpose of this paper is to describe the development of remote patient monitoring on promoting safer client care for older adults in the community. The focus of the research study was specifically to investigate the user interface design and data fusion strategies for integrating bio-mobility information, about home care clients as they go about their daily lives, with a point-of-care (POC) decision support system for health care providers.

The research study focused on the integration of 1) non-intrusive monitoring approaches, using binary and analogue sensors, developed at Carleton University, and 2) a cell phone-based GPS-supported monitoring system to track patients' daily life events and health conditions developed by at Wilfrid Laurier University (WLU), with 3) a prototype web-based software application developed by Dr. Doran and her team at the University of Toronto that equips clinicians with handheld units to assist in the collection and utilization of bedside patient data and provide real-time feedback, such as suggested best-practice guidelines, tailored to an individual patient's needs. That software was adapted specifically for the homecare clinicians' environment. This study focused the knowledge gained through those three initiatives to evaluate a wireless patient monitoring and data collection system for use by health professionals for point-surveillance (e.g., falls risk), to record patients' vital signs data, to provide immediate feedback to clinicians, and to facilitate virtual healthcare management. The goal is to improve the efficiency and safety of homecare services through technological innovations that will reduce the demand on scarce and expensive human resources.

## 2 BACKGROUND

There are three compelling trends that must be addressed in healthcare planning: 1) Impact of demographics on community care; 2) Emerging health care technologies; and 3) Safety concerns among home care clients. Firstly, there is increasing concern in the health care community about the need to provide ways to support older adults who want to live independently in their homes as they age. A large percentage of clients receiving home care support are elderly and live alone (Lang & Edwards, 2006). Home is an unpredictable environment for care of older people because the healthcare provider visits each home for relatively short periods (an average of 45 minutes) and only a few times per week (Martigan & Fortisky, 1999). Furthermore, home health providers are relatively isolated in their work. This context of relative isolation for providers and clients poses unique challenges for health service delivery, particularly with respect to issues of coordination and communication (Lang & Edward, 2006). Compounding these challenges, the rate of increase of home care expenditures in Canada is more rapid than all other areas of the publicly funded health care system, with the exception of medication (Coyte & McKeever, 2001). The demographics in Canada suggest that the costs will escalate significantly over the next two decades. In Ontario, for example, the government has projected that the 65+ population will swell from 1,680,440 in 2007 to 2,282,630 by 2017, and it will almost double to 3,189,390 by 2027. Under the current delivery model, this increase would require a substantial number of additional home care nurses, which, given the current and anticipated shortage of nurses, is a goal that cannot be attained. Clearly, cost-effective solutions are needed to manage the spiraling costs and to address the social challenges of providing care to older adults at home.

Secondly, technology is providing a variety of alternatives and opportunities specific to healthcare issues and costs. It is imperative to move ahead with addressing the options pertinent to home care through reliable research that will support proven technologies, such as remote patient monitoring. Lastly, there is an increasing concern about the safety of clients in home care settings. A recent study in the United States investigated patient safety concerns among 3,013,287 home care clients. The results indicated that 13% of all home care clients had experienced an adverse event (Martigan & Fortisky, 1999). Clients who experienced such events were generally older. The types of adverse events experienced were unexpected death, urinary tract infection, fall or accident at home, wound deterioration, unexpected nursing home admission, increase in the number of pressure ulcers, improper medication administration or side effects, and hypo/hyper-glycemia. The results of a recent consensus conference suggest that Canada experiences a similar range of patient safety concerns among the home care population (Masotti et al, 2007). Remote patient monitoring interventions have the potential to address a number of these concerns through technological tools that will improve monitoring of client care in the home settings, and improve communication among providers, particularly at point of transition within the health system. This research study addressed each of the above challenges by investigating solutions for integrating bio-mobility information on client status with a point-of-care decision-support system that will provide home care clinicians with better scheduling capability, improved assessment of client mobility/activity in and out-of-home, and more timely access to decision support.

## 3 METHODS

This research study was divided into three phases that were conducted over 32 months: (1) Prototype Development; (2) Usability Testing; and (3) Pilot Field Testing.

### 3.1 *Phase one (prototype development)*

During the phase one of this research study, the two objectives for the prototype development phase were: (1) to determine the most appropriate user interface features for data extractions by mapping out information flow from the various information sources in real-life point-of-care (POC) scenarios, as well as engaging homecare nurses in researching most informative way of presenting the information; and (2) to develop a prototype Web-based platform for presenting information to healthcare providers.

In order to determine which client groups would benefit from bio-mobility monitoring, focus groups were conducted with homecare clinicians from Saint Elizabeth Health Care and Central

East Community Care Access Centre, including nurses, occupational therapists, physiotherapists and case managers. Twenty health care professionals participated in a total of 3 focus groups. The focus groups were led jointly by a member of the research team and students from both Engineering and Nursing. This arrangement ensured a truly multidisciplinary perspective and students from the two different programs have benefited from the collaborative training opportunity.

The focus group discussions were audio taped, transcribed, and the transcripts were jointly analyzed by a team of researchers and students from Nursing and Engineering. Over this period of time, monthly research meetings were conducted with the research team and students to ensure effective communication and to coordinate the activities of the three research centres. Collaboration among the different institutions occurred both in person and by teleconference to discuss the priorities, objectives and planning logistics of conducting the focus groups. In addition quarterly meetings were held with the research team and industry partners, who have been engaged in contributing ideas for data collection and analysis. For example, we collaborated with a human factors expert from Nortel in planning the focus group interview questions. We have also collaborated with HInext (IT application developer) in planning the prototype for the user interface. TELUS, a leading service provider has provided the wireless service to support the prototype development and testing. RIM has provided wireless handhelds that were both used by clinicians for demonstration purposes and by the research team for development work.

### 3.2 *Phase two (usability testing)*

After the completion of prototype development in phase one, the phase two of this study was guided by the following objectives: To evaluate the feasibility, safety, and acceptability of the bio-physiological monitoring and decision-support system (the prototype system) to end users in a laboratory setting by assessing whether the prototype system performs to technical expectations, as well as assessing the nurse users' satisfaction with the system.

After the research team conducted the focus groups with home care providers about a proposed client monitoring system in Phase I, the information that was learned had to be compiled in order to begin designing a prototype client monitoring and decision support system. The research team agreed that a human-centered design approach was appropriate because it emphasizes user involvement and applies an iterative approach to the design and testing of the prototype. Therefore, the goal of this milestone was to be able to prepare the team for conducting a usability test on low-fidelity potential interfaces of the system. The results from the testing would then help the research team resolve conceptual and detailed design issues regarding the prototype displays.

The research team then conducted the low-fidelity usability testing with 9 homecare clinicians (1 physiotherapist, 1 occupational therapist, 3 nurses and 4 case managers). Each session was one hour in length and was audio and video recorded. The audio recording was transcribed and used together with the video-recording in the usability analysis. The sessions were conducted with clinicians who had also been participants in phase one study. The purpose of the interviews was to solicit feedback from potential end-users about the specific design and application of a client monitoring and decision support system.

### 3.3 *Phase three (pilot testing)*

Followed by the completion of usability testing in phase two, the phase three of this study is the pilot field test of the prototype System (Remote Activity Monitoring and Guidelines System) in the community. The Pilot Field Testing of the Remote Activity Monitoring and Guideline System was guided by three objectives:

1. Determine the value of the mobility-related data to home care clinicians when they plan care
2. Assess the usefulness and placement of the Best Practice Guidelines (BPGs) in the electronic user interface.
3. Assess the clinician and client experience with the prototype System.

The Remote Activity Monitoring and Guideline System could collect the following bio-mobility data:

1. BlackBerry: Length of time in the home, time away from home, time walking outside the home, time using vehicular transportation, distance travelled. Although the actual locations visited were potentially identifiable by GPS, only the time and distance were retained as study data.
2. Pressure sensitive mattress: length of time in bed; time spent entering and exiting bed. The mat could also detect additional data such as pulse and respirations. These additional data elements were not the focus of the current study, but the research team will plan to evaluate the quality of this additional data and to test algorithms for processing the data into usable information.

Proxy client who was over the age of 55 and was requested to carry a BlackBerry (e.g. Storm II) for a period of 4 days. The BlackBerry passively logs GPS co-ordinates accurately within 5 meters, and provides valuable data about the physical activity of the study participant. The participant was asked to wear the device clipped at their hip (using a standard BlackBerry "holster" style clip) as much as possible during the day, and was taught how to recharge it overnight. Proxy client has been informed that they can turn off the BlackBerry at any time if they wish. Client has also been asked to allow the research team to place a pressure sensitive mat under their existing mattress, and to place a research laptop computer nearby. A member of the research team was available to provide ongoing technical support during the pilot field testing.

Beginning after 4 days of data collection, 4 different home care clinicians made home visits to the proxy client. A member of the research team performed a joint visit to observe clinicians interacting with the prototype System in the proxy client's home. Each clinician assessed the client's needs and created a care plan based on the mobility related data available from the System, in consultation with the client. The research team provided the proxy client with an enhanced health history in order to make the home care visit more realistic and to evaluate various features of the System. The client was able to see the graphs of his/her own activity during the nursing visit. The clinicians were asked to complete a questionnaire about how the mobility information influenced their clinical decision making, the usefulness and location of the links to the best practice guidelines, their opinions about how the System enabled their clinical workflow, and their "wish list" of other bio-physiological data that they would like to be able to monitor remotely in the future. The client has been asked to comment on his/her experience being monitored by the BlackBerry device and the bed mat.

## 4 RESULTS

### 4.1 *Phase one (prototype development)*

During Phases one and two of the study, the research team consulted with home care clinicians to determine their requirements for information flow and their preferences for types of client information that they would like to be able to monitor remotely. A prototype Remote Activity Monitoring and Guidelines System has been developed that uses a GPS-equipped BlackBerry to monitor a person's mobility outside the home (e.g. how far they walk each day in the community, etc.). The System includes a pressure-sensitive mat that is placed under a regular bed mattress and can monitor sleep disturbances, and how long it takes to enter and exit the bed. Depending on specific bed characteristics, it can also monitor pulse and respirations, although this was not the focus of the current study.

During the phase one of prototype development, four themes were explored in the focus groups: 1) the client groups for whom it would be beneficial to have remote monitoring technologies; 2) the type of client data or resources that would be helpful to clinicians; 3) the format and frequency of data to be received by the clinicians; and 4) the potential users of remote monitoring technologies.

The findings from the focus group interviews indicated that the client groups that would benefit the most from remote monitoring technologies are those with mobility issues, with cognitive impairments or who require continuous monitoring due to high risk conditions. A wide range of client data were considered helpful and the need for specific data is context dependent. For example, the type of data providers use to support decision making varies with the clinical expertise and role of the caregiver. Graphical representation of monitoring data was preferred to numeric presentation, with the importance of displaying trends over time and the value of including methodology of alerting clinicians of deviation in trends. In addition to the different types of client groups were identified by the interview participants, different target end-user groups were also identified. These included other formal healthcare providers, but also informal providers such as family members.

Home care clinicians reported that detailed information about client mobility would be helpful when they are planning care for clients who are trying to increase their activity levels (e.g. following hip or knee surgery, to help manage chronic health conditions such as diabetes and COPD, or during cardiac rehab, etc). It would also provide reliable data about how often a client's sleep is interrupted which can be helpful when managing pain. In addition, the prototype System includes embedded links for clinicians to access Best Practice Guidelines such as pain management, diabetes management, COPD, pressure ulcers, prevention of falls, etc. Clinicians were consulted in an iterative process during the development of the System to determine the characteristics of the graphical user interface, and the display was developed and modified based on user feedback.

### 4.2 *Phase two (usability testing)*

From the findings of the focus groups, homecare client scenarios were developed and refined to match the identified client groups. Scenarios were generated for each homecare clinician type and described such aspects as what the clinician might intend to do and why, in what context and how, and what the outcomes and consequences might be during typical client visits. These scenarios helped to guide the design of the prototype displays in understanding what features might be important to display and where they should be placed. These same scenarios were then slightly modified for use during the usability testing. The Phase two of the study included usability testing of the System with home care nurses and therapists, and integrating their recommendations about the User Interface into the prototype.

The prototype was tested in a controlled lab setting in order to assess whether it performed to technical specifications. The Remote Monitoring and Guideline System was ready for implementation in a field study after the safety and feasibility have been demonstrated and confirmed in a lab setting.

It is important to understand the value of the mobility data to the clinician's decision making process. Understanding the end user's needs and interaction with the prototype in actual home care settings is important to making the prototype successful. It is critical that the technology provides clinical support to assist clinicians when planning care. The Phase three of this study tested the Remote Monitoring and Guideline System with a proxy client in his/her home. This pilot test highlighted design issues, implementation challenges, and clinical benefits that will be addressed when planning a future clinical trial.

### 4.3 *Phase three (pilot testing)*

During the pilot field testing, data were collected on usability and acceptability for both the client and nurses. Semi-structured interviews and a questionnaire was used to identify questions that arose about the client or care-giving situation as a result of having access to the bio-mobility information, to describe the types of clinical issues for which the monitoring information was helpful, and to describe its perceived impact on clinical decision making, client outcomes, and the quality of nurses' work life. Interviews with the client and the nurses during the field test revealed that the Remote Monitoring and Guideline System was helpful for all of the stakeholders and the system was not perceived as being intrusive by the client.

During the interview, the elderly described: "The remote capture of mobility data provided detailed information that allowed me to learn more about my own health status, and made it possible to track my ongoing progress towards health goals, such as being more physically active". Another added benefit from remote capture of health information was that the client "felt empowered by the increased understanding of own health condition that helped in the management and care of chronic diseases". The client suggested that the use of remote monitoring has the potential to enhance client safety, especially for older adults in the community who are living alone.

Following one week of monitoring, four nurses made independent visits to review the mobility data in consultation with the proxy client, and created a plan of care using the remote monitoring system. During the interviews, nurses described the usefulness of mobility and bedmat data in supporting the care of their home care client. The nurses indicated that the remote capture of data was helpful to the management of client care including assessment, client teaching, evaluation and care planning. For nursing assessment, the remote monitoring data was helpful to assess client status, such as "determining client's risk for falls; level of pain based on client's mobility level and amount of time spent in bed, as well as using bedmat data to assess quality of sleep pattern, such as insomnia and urinary frequency at night. Nurses also described: "The facts provided by the System can be used as evidences to evaluate client's condition, and the trending graphs and numbers can be used an educational tool to help explain client progress and set targets for future health goals".

For care planning, nurses found the point-of-care access to best practice guidelines provided by the System was useful in the development of individualized care plan tailored to client's needs. In particular, the nurses indicated that "the use of practice guidelines helped the development of client care plan during the initial visit, and then the evaluation of nursing interventions for subsequent client visits". When real-time information about mobility was integrated with access to practice guidelines to complement standard assessments, nurses felt that they were "better able to make informed decisions about client care". For example, the nurses explained: "the remote capture of data could help them decide if a telephone call is appropriate, or if a home visit is warranted". Overall, nurses revealed that remote monitoring system was an excellent communication tool that could help facilitate inter-professional collaboration when being used by different health care professionals. Proper clinician education, training and support regarding the use of System are important to its successful implementation in home care settings.

## 5 CONCLUSION

According to Statistics Canada, the fastest growing segment of the Canadian population is the "over 65" age category. There is increasing concern to provide ways to support older adults who want to live independently in their homes as they age. Many home care clients are elderly and live alone with only a few visits a week from healthcare providers. As a result, both clients and health providers feel isolated with unique challenges for health service delivery. Creative health care solutions are needed in order to reduce the demands on the health care system from this older population and to support older adults who want to live independently in their homes as they age.

In this study, Remote patient monitoring intervention was found to have the potential to address a number of these concerns through technological tools that would improve monitoring of client care in the home settings, and improve communication among providers. The Remote Monitoring and Guideline System that integrates bio-mobility information on client status with a point-of-care decision-support system would provide home care clinicians with better scheduling capability, improved assessment of client mobility/activity in and out-of-home, and more timely access to decision support. Although this study was conducted with a limited pilot field-testing in the community, the results provided valuable knowledge about the integration of bio-physiological monitoring technologies into the routine clinical practice of homecare practitioners. In the future, more and larger field-testing is needed for such applications and for research purposes. For instance, costs/benefits analysis needs to be performed to examine the cost-effectiveness of remote client monitoring. Pilot field-testing could also investigate the potential system application that extends beyond home care populations, such as long-term care and rehabilitation settings. Finally, it is imperative that future research could generate evidence-based knowledge about the efficacy and efficiency of remote monitoring technologies, and their influence on quality of care and client outcomes.

## REFERENCES

Coyte, P.C. & McKeever, P. 2001. Home care in Canada: passing the buck. Canadian Journal of Nursing Research 33: 11–25.

Lang, A., & Edwards, N. 2006. Safety in Home Care: Broadening the Patient Safety Agenda to Include Home Care Services. Canadian Patient Safety Institute: Edmonton, Alberta: pp. 1–42.

Madigan, E., & Fortinsky, A. 1999. Alternative measures of resource consumption in home care episodes. Public Health Nursing 16: 198–204.

Masotti, P., et al. 2007. Adverse events in communication care: developing a research agenda. Healthcare Quarterly 2007. 10(3): 63–69.

*Bridging Research and Good Practices towards Patient Welfare – Shih & Liang (Eds)*
*© 2015 Taylor & Francis Group, London, ISBN 978-1-138-02716-9*

# Simulation study on the impact of working time systems on the work-life balance of employees in a hospital

G. Zülch, P. Stock, D. Schmidt & M. Leupold
*ifab-Institute of Human and Industrial Engineering, Karlsruhe Institute of Technology, Karlsruhe, Germany*

ABSTRACT: The organization of working time in the hospital sector is a particular planning problem due to 24/7 service. The consequences of a change in working times can only be assessed using the simulation especially in view of the work-life balance of employees. For this purpose, the simulation procedure *OSim-GAM* is presented here in brief. To account for the influence of the working time on work-life balance, employee types are defined, whereby they differ in their stress situation and working time preferences. To model the types of employees, a list of time slices is deposited where this type of employee must pursue or want to pursue private obligations and therefore gives rise to conflicts between work and private life. In addition, for each conflict its severity is assessed. The application of his procedure is shown here based on the case study of an intensive care unit.

## 1 NEED FOR COMPUTER-AIDED ORGANIZATION OF WORKING TIME

When planning a working time system, i.e. the totality of the used working time models in an operational unit, there are a variety of ways that can hardly be overlooked by a planner without proper planning and evaluation methods. In addition, a number of legal, contractual and ergonomic requirements must be observed. In this way, the organization of working time sets high demands on planners, as operational requirements must also be adequately taken into account (Knauth 2002, pp. 51).

In addition, due to atypical or long working hours, interactions often occur between the professional and non-professional live. In 2010, out of 13,767 employees in Germany 34.6% stated that overtime, long working hours and commutes constitute a frequent workload, and 20.8% named shift work as a burden (Kroll, Müters & Dragano 2011, Table 2). Against this background, the evaluation of a working time system not only plays a role for the operational goals, but the effects of workload and work-life balance of employees must also be taken into account (for further details see e.g. Stock & Zülch 2013, pp. 205). A typical example of this is the hospital area, with its requirement for the permanent presence of staff.

A systematic approach to planning and implementation of working time systems is useful to evaluate the variety of options adequately. For this purpose, in practice checklists, expert assessments, utility, sensitivity, and cost considerations as well as cost-benefit analyses are frequently used for the evaluation of working time systems (cf. Bogus 2002, pp. 40). The use of software enables a more detailed evaluation of work-time systems, although a static target-performance comparison is usually performed based on a set of criteria (e.g. length of the working day, maintaining rest periods; cf. e.g. Böker 2011).

However, no mechanism is currently available with which the planner prospectively, i.e. before realization, can review the impact of a working time system, taking into account the operational dynamics and effects on the work-life balance of employees. An agent-based method has been developed purely for short-term swapping of individual time slices to better avoid time conflicts between work and private life (Zülch & Gamber 2011).

## 2 THE *OSim-GAM* SIMULATION PROCEDURE

Against this background, the ifab-Institute of Human and Industrial Organization of the Karlsruhe Institute of Technology in Germany developed the simulation procedure *OSim-GAM* (German acronym for "Object-simulator for the design of working time models"; Bogus 2002, pp. 160) in order to support a prospective organization of working time. To account for the influence of the working time system on work-life balance, employee types are defined, which differ in workload and working time preferences (for the methodology refer to e.g. Schmidt, Stock & Zülch 2012). Each employee of the regarded organisational unit, in addition to his qualifications, is associated with a type of employee. For each type of employee, a list of time periods is stored in which conflicts between work and private life may occur because the employee must pursue or want to pursue private obligations. In addition, for each conflict a conflict severity can be specified. For further details the reader is referred to Leupold, Stock & Zülch (2013).

In the following, a simulation study is presented in an intensive care unit. Another pilot study from the retail sector is documented in Zülch, Stock, Schmidt & Leupold (2012).

## 3 SIMULATION-SUPPORTED ORGANIZATION OF WORKING TIME IN A HOSPITAL

### 3.1 *Initial situation of the Intensive Care Unit*

The intensive care unit investigated in this simulation study has 20 beds and treats approximately 5,500 patients annually. In nursing service, there are 52 mainly female employees, of whom 30 are employed full-time and 22 work part-time. The employees are able to master all activities that occur in the intensive care unit.

Seven types of patients can be distinguished, who differentiate themselves from each other in terms of their condition and their necessary treatment. This results in different retention times for the types of patients in the intensive care unit, which can last three to four hours for newly operated but stable patients, and up to several days for unstable patients. In addition to direct activities on the patients, there are several indirect activities, such as ordering material and medication. Since the patients are mainly taken into the intensive care unit as a result of surgery, the duties of the staff in the nursing service are relatively easy to plan, however unplanned emergencies can also occur.

In the initial situation, a three-shift model with early, late and night shifts as well as various intermediate services is practiced. Here, however, there are some problems, such as too short rest periods or too many night shifts in succession. In the past, as a result of short-term adjustments to planned working times, rapid changes of shifts occurred, which led to increased stress among employees and poor planning of their working times.

### 3.2 *Types of employees in terms of their working time preferences*

In a survey, the working time preferences of employees from different areas of hospitals were determined. Herebey, the interviewee had to name the hourly time slots in which time conflicts could possibly occur if the employee would be employed. Furthermore, the severity of the conflict was assessed. With the help of the cluster analysis types of employees have been derived that are similar in terms of their working time preferences. For this, the k-means algorithm was used (Bacher, Pöge & Wenzig 2010, pp. 299), one of the partitioned methods of cluster analysis. Altogether, the following five groups could be identified, of which only three are described here in more detail in order to preserve anonymity (Figure 1):

Cluster 1 (n = 25): average age 33.0 years; youngest group of people; most of the time spent with partner and children compared to other clusters; most of the time spent on hobbies and animal care; highest prioritization of maintenance of partnership; least amount of time for housework.

Cluster 2 (n = 23): average age 46.1 years; oldest group of people; highest amount of time spent on religious interests (though on low level); compared to other clusters least time spent on

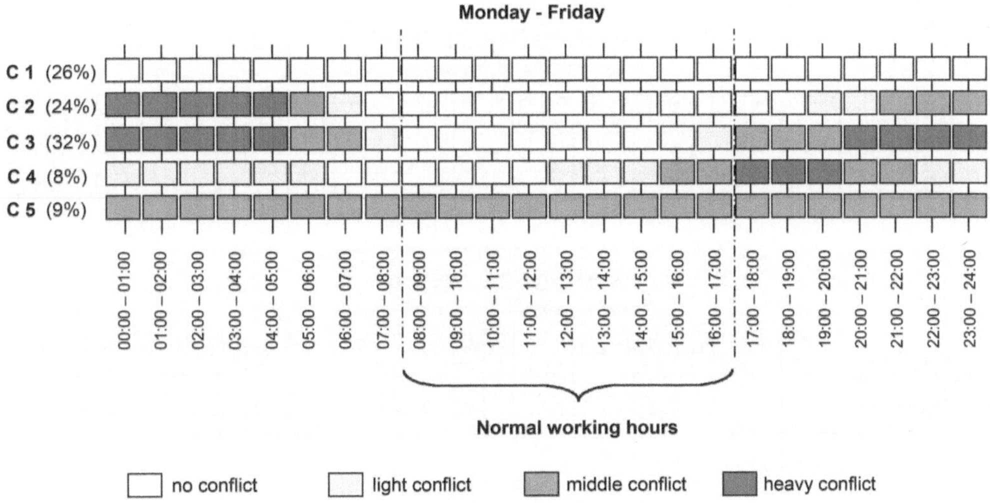


Figure 1. Time conflicts of hospital employees from Monday to Friday.

housework, craft activities, contact with friends, education, sports or fitness and hobbies; lowest prioritization of the maintenance of partnerships and artisanal activities; highest prioritization of secondary activities.

Cluster 3 (n = 30): average age 41.3 years; lowest time spent on volunteer activities, contact with friends, hobbies and animal care; lowest prioritizing for the care of relatives, secondary and voluntary activities; highest prioritization of wellness in comparison to the other clusters.

Cluster 4 (n = 8) and cluster 5 (n = 9): no detailed description to ensure anonymity.

In addition to the working time preferences from Monday to Friday (Figure 1) these people also were questioned about their weekends. As expected, this produced the most time conflicts.

### 3.3 *Experimental design of the simulation study*

The objective of the simulation study presented here is to investigate the influence of different working time systems on the conflict situation of employees. Against this background, in addition to the three-shift system of the initial situation two more working time systems were examined, namely, each a three-shift system with longer (working time system 1) and shorter shift cycles (working time system 2; for details see Figure 2). Both shift systems are considered in accordance with Grzech-Šukalo, Hänecke, Jäger & Nachreiner (2003, pp. 23) to be advantageous in such continuous work (for the following see also Stock, Schmidt, Leupold & Zülch 2013).

The identified types of employees were distributed stochastically according to their frequency on the 52 modelled people. A total of 20 different distributions were randomly generated for the types of employees, so as to examine the influence of the distribution of the types of employees to the conflicts that may arise. Thus for the three working time systems a total of 60 combinations arose; for each 20 simulation repetitions were performed in order to safeguard the results in terms of mean values, standard deviations, etc.

### 3.4 *Results of the simulation study by OSim-GAM*

The results of the simulation study were evaluated by organizational, monetary and employee-oriented objectives. For this purpose the concept of the degree of goal achievement was used (Wedemeyer 1989). According to this concept an optimal situation obtains a value of 100%, and a pessimal situation a value of 0%.

**Working time system 1**

| | Mo | Tu | We | Th | Fr | Sa | Su | Mo | Tu | We | Th | Fr | Sa | Su | Mo | Tu | We | Th | Fr | Sa | Su | Mo | Tu | We | Th | Fr | Sa | Su |
|---|---|---|---|---|---|---|---|---|---|---|---|---|---|---|---|---|---|---|---|---|---|---|---|---|---|---|---|---|
| Shift group 1 | L | L | L | | N | N | N | N | | | L | L | L | L | | N | N | N | | E | E | E | E | E | E | E | | |
| Shift group 2 | N | | | L | L | L | L | | N | N | N | | E | E | E | E | E | E | E | | | L | L | L | | N | N | N |
| Shift group 3 | | N | N | N | | E | E | E | E | E | E | E | | | L | L | L | | N | N | N | N | | | L | L | L | L |
| Shift group 4 | E | E | E | E | E | | | L | L | L | | N | N | N | N | | | L | L | L | L | | N | N | N | | E | E |

**Working time system 2**

| | Mo | Tu | We | Th | Fr | Sa | Su | Mo | Tu | We | Th | Fr | Sa | Su | Mo | Tu | We | Th | Fr | Sa | Su | Mo | Tu | We | Th | Fr | Sa | Su |
|---|---|---|---|---|---|---|---|---|---|---|---|---|---|---|---|---|---|---|---|---|---|---|---|---|---|---|---|---|
| Shift group 1 | E | E | L | L | N | N | N | | | E | E | L | L | L | N | N | | | E | E | E | L | L | N | N | | | |
| Shift group 2 | | | E | E | L | L | L | N | N | | | E | E | E | L | L | N | N | | | | E | E | L | L | N | N | N |
| Shift group 3 | N | N | | | E | E | E | L | L | N | N | | | | E | E | L | L | N | N | N | | | E | E | L | L | L |
| Shift group 4 | L | L | N | N | | | | E | E | L | L | N | N | N | | | E | E | L | L | L | N | N | | | E | E | E |

**Legend**

E early shift L late shift N night shift

Figure 2. Examined working time systems (according to Grzech-Šukalo, Hänecke, Jäger & Nachreiner (2003, pp. 23).

The service degree does not change with the three working time systems and each reaches the optimum value of 100%. This is due to the existing situation where sufficient nursing staff is available in the intensive care unit. The degree of throughput time as the ratio of actual and planned residence time of the patients improves in both working time variants by about 15% points. In contrast the utilization of staff and the process costs, i.e. the cost of residence time of patients in the intensive care unit, worsens by approximately 20% points.

With the employee-oriented goal achievement degrees, the physical stress on employees does not change when changing the working time system (Figure 3). However, the change means a time stress decreased by about 30% points, measured here by the number of queued activities of the nursing staff. Furthermore, in both working time variants the mean number of time conflicts rises numerically by about 50% for all conflict severity levels compared to the initial situation. This is explained by the fact that employees in the initial situation have already swapped shifts according to their working time preferences.

The simulation results show that the choice of the working time system has a significant impact on the organizational, monetary and employee-oriented objectives. For the various distributions of employee types, significant differences in the number of time conflicts occurring could also be identified. For this reason, the working time system reacts most strongly to changes in the distribution of the types of employees.

## 4 CONCLUSIONS AND FURTHER ACTION

When designing a working time system thus not only the operating conditions, but also the personnel structure must be considered. Since this can change over the period of validity of a working time system, there remains the challenge of developing robust working time systems, which hardly change even with a variable personnel structure in terms of their impact on work-life balance of employees. In order to estimate the effects, the mean of the simulation is suitable.

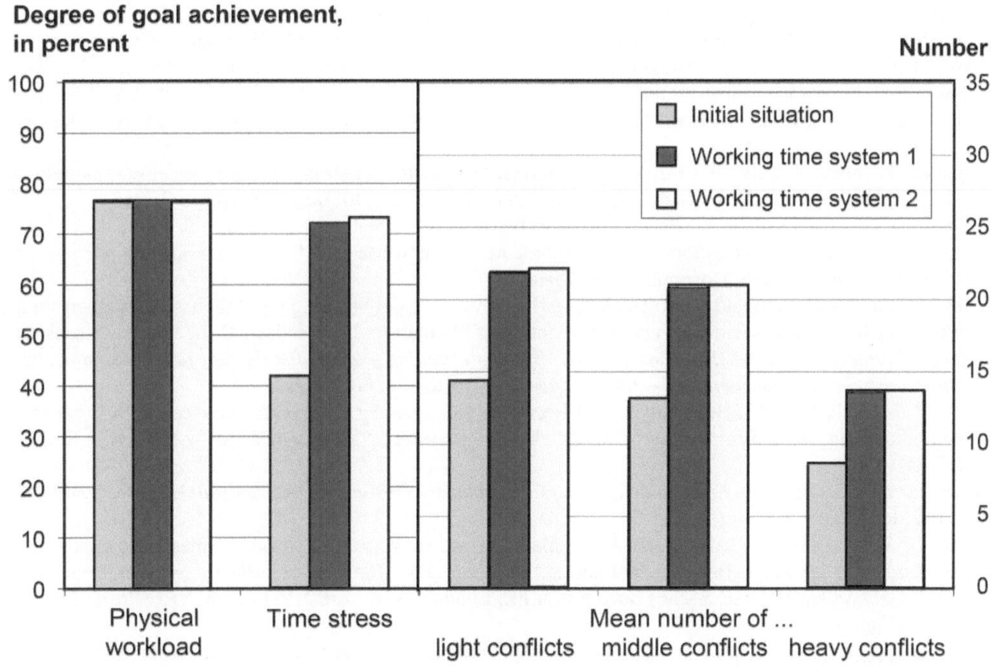


Figure 3. Employee-oriented goal achievements.

Currently, a so-called social role-and-stress model is being developed based on the data of an extensive employee survey that reflects the relationships between the identified employee types and the resulting occupational and non-occupational stress. This model is then integrated into the *OSim-GAM* simulation procedure. Thus, with the simulation-based organization of working time, in the future employee surveys will no longer be necessary, in principle, to determine the stress situation of employees and their occurring time conflicts between professional and private life. Finally, the effects of different working time models on work-life balance should be examined through a systematic simulation study, depending on the employee types.

## ACKNOWLEDGEMENT

The research presented here was carried out within the framework of the research project "Working time configuration considering the work-life balance using computer-supported simulation – *Ar*beitszeitgestaltung unter *B*erücksichtigung der *Wo*rk *L*ife Balance mit Hilfe der rechnerunterstützten Simulation *ARBWOL*", which is funded by the German Federal Ministry of Education and Research and the European Social Fund under grant number 01FH09046.

## REFERENCES

Bacher, J., Pöge, A. & Wenzig, K. 2010. *Clusteranalyse*. München: Oldenbourg.

Bogus, T. 2002. *Simulationsbasierte Gestaltung von Arbeitszeitmodellen in Dienstleistungsbetrieben mit kundenfrequenzabhängigem Arbeitszeitbedarf*. Aachen: Shaker.

Böker, K.-H. 2011. Spezielle Software für „Gute Schichtarbeit". *Computer und Arbeit* 1: 28–32.

Grzech-Šukalo, H., Hänecke, K., Jäger, C. & Nachreiner, F. 2003. *Software-gestützte Arbeitszeitgestaltung in der Praxis*, Bundesanstalt für Arbeitsschutz und Arbeitsmedizin (eds). Bautzen: Lausitzer Druck- und Verlagshaus.

Knauth, P. 2002. Arbeitszeitflexiblisierung aus arbeitswissenschaftlicher Sicht, In G. Zülch, P. Stock & T. Bogus, T. (eds), *Arbeitszeitflexibilisierung im Dienstleistungsbereich*: 51–74. Aachen: Shaker.

Kroll, L.E., Müters, S. & Dragano, N. 2011. Arbeitsbelastungen und Gesundheit. *GBE kompakt*: 2(5), Robert Koch-Institut (eds). http://www.rki.de/DE/Content/Gesundheitsmonitoring/Gesundheitsberichterstattung/GBEDownloadsK/2011_5_Arbeitsbelastungen.pdf?__blob=publicationFile, accessed 31.07.2014.

Leupold, M., Stock, P. & Zülch, G. 2013. Simulationsunterstützte Gestaltung flexibler Arbeitszeitmodelle. In M. Bornewasser & G. Zülch (eds), *Arbeitszeit – Zeitarbeit*: 221–240. Berlin: Springer Gabler.

Schmidt, D., Stock, P. & Zülch, G. 2012. Mitarbeitertypen für die Arbeitszeitgestaltung unter Berücksichtigung der privaten Belastungssituation. In Gesellschaft für Arbeitswissenschaft (eds), *Gestaltung nachhaltiger Arbeitssysteme*: 813–816. Dortmund: GfA-Press.

Stock, P., Schmidt, D., Leupold, M. & Zülch, G. 2013. Bewertung der Wirkung von Arbeitszeitsystemen auf die Work-Life-Balance mit Hilfe der personalorientierten Simulation. In Gesellschaft für Arbeitswissenschaft (eds), *Chancen durch Arbeits-, Produkt- und Systemgestaltung – Zukunftsfähigkeit für Produktions- und Dienstleistungsunternehmen*: 157–160 and 593–596. Dortmund: GfA-Press.

Stock, P. & Zülch, G. 2013. Belastungen und Beeinträchtigungen der Work-Life-Balance durch Arbeitszeitflexibilisierung. In M. Bornewasser & G. Zülch (eds), *Arbeitszeit – Zeitarbeit*: 203–220. Berlin: Springer Gabler.

Wedemeyer, H.-G. von 1989: *Entscheidungsunterstützung in der Fertigungssteuerung mit Hilfe der Simulation*. Düsseldorf: VDI.

Zülch, G. & T. Gamber, T. 2011. Agent-based planning and evaluation of working times in hospitals. In S. Albolino, S. Bagnara, T. Bellandi, J. Llaneza, G. Rosal & R. Tartaglia *Healthcare Systems Ergonomics and Patient Safety 2011*: 255–256. Boca Raton FL, London, New York NY, Leiden: CRC Press, Taylor & Francis Group.

Zülch, G., Stock, P., Schmidt, D. & Leupold, M. 2012. Herausforderungen für die Arbeitszeitgestaltung durch den demographischen Wandel. In E. Müller (eds), *Demographischer Wandel – Herausforderung für die Arbeits- und Betriebsorganisation der Zukunft*: 193–217. Berlin: GITO-Verlag.

*Bridging Research and Good Practices towards Patient Welfare – Shih & Liang (Eds)*
*© 2015 Taylor & Francis Group, London, ISBN 978-1-138-02716-9*

# Application of a modified SIR-based model-case of H1N1 in Taiwan

K.H. Yang
*Department of Industrial and Systems Engineering, Chung Yuan Christian University, Zhong-li, Taiwan, R.O.C.*

C.Y. Juan
*Department of Early Childhood Care and Education, Hsin Sheng College of Medical Care and Management, Longtan, Taiwan, R.O.C.*

J.Y. Hsu
*Institute for Information Industry, Taipei, Taiwan, R.O.C.*

Y. Chen
*Department of Industrial and Systems Engineering, Chung Yuan Christian University, Zhong-li, Taiwan, R.O.C.*

ABSTRACT: Several serious and large-scale influenza outbreaks happened in recent year in different countries, such as SARS in 2005 or H1N1 in 2009. These two influenza are world-wide spreading, which inflict heavy losses on people life and world economy. Another influenza, H7N9, doesn't outbreaks seriously, however, it continues from middle of 2013 to the beginning of 2014 without diminished in Taiwan. Consequently, influenza epidemics have greatly impacts not only on people's life and health, but medical systems in different countries. Although severe diseases continuously occur, they are not fully controlled. To face more and more severe disasters, governments in different countries have different measures to control influenza epidemics spreading. However, all measures have some impacts on people and society. Sometimes it is not easy to evaluate impacts. Therefore, it is necessary and essential to develop an appropriate model for measure evaluations. This study is to develop a SIR-based model for application to evaluate the candidate policies during the influenza outbreaks.

## 1 INTRODUCTION

Globalization makes distances and relationships between countries become close, which bring opportunities for people working or traveling around different countries. However, that also provides a channel for spread of infectious diseases. Every government seriously faces to the epidemic prevention issues and actively employs certain measures to prevent disease outbreak.Once diseases outbreak, a government has to try its best to stop negative impacts on economy and society. According to the statistics report of WHO in 2004, the infectious disease was the top 2 of the death causes. Influenza epidemic almost happen every several decades, which characteristics are not similar to seasonal influenza, i.e. the virus can be contagious among different species, and during the infection, the virus might be mutated to a new virus; examples are H5N1 in 2003 and H1N1 in 2009. To face those serious influenza epidemics, governments cautiously and continuously take certain prevention measures for potential outbreaks in the future.

For any prevention measure for influenza epidemic, a high level decision maker has to fully consider all aspects, such as social acceptance, economic impacts, resource limitations, and so on. However, future is hard to predict. It is not easy to evaluate what-if scenarios without any policy tool. Consequently, a suitable simulation model is useful for evaluations. The two popular models are compartmental models and agent based modes. Agent based models can model a complicated social system. However, they need large computer resources. There are not significantly breakthroughs until recently (Rakowski 2010). Compartmental models are modeled by a set of ordinary differential equations, which can analyze group behaviors (or status) of people and the spread of an influenza. Comparing to the agent based models, compartmental models are good for quick estimations and less requirements of computer resources, and however, they cannot predict individual's situations (Lant 2008).

The aim of this paper is to develop a new SIR-based model, which is named as SELMAHRD (see Section 3), for a policy maker to have a quick scan of some disease prevention measures during the outbreaks of an influenza epidemic. The primary measures include the pulse vaccination and the segregation. The embryo of the SIR model[1] was established by (Kermack & McKendrick 1927). The model defines 3 types of people groups, which includes susceptible people, infected people, and recovered people. In the model, there are some assumptions and limitations as follows; the total population is fix, people death due to diseases are not considered, and age and population structure is not included as well. Based on the prototype SIR model, a lot of researchers applied it to discuss the effectiveness of the policies or to analyze some important factors that will influence the behaviors of the epidemic. Yıldırım & Cherruault (2009) introduced a homotopy parameter, $p$, into SIR model and solved the model analytically to discuss the effects of the constant vaccination strategy. Dybiec (2009) modified the original SIR model to analyze susceptible agents during interactions with infectious neighbors and the mechanism of how those susceptible agents to become infectious. Others related reference are Zaman et al. (2009), Meng et al. (2010), Meng & Chen (2008), and etc. SEIR is also a SIR-based model. The difference between SIR and SEIR is a compartment, Exposed, added into the model. Wang et al. (2009), Chen et al. (2007) and Lekone & Finkenstädt (2006) applied SEIR model to analyze the control treatments for the epidemic. If we further think about deeply what are considerations that can be provided as necessary information for policy making, SIR, or SEIR model might not provide sufficient information, example, death number, hospitalized people, and so on.

Before this study, two different models, i.e., SEIRD and SEAHR, are developed by Lant (2008) and Hsu & Hsieh (2008). However, SEIRD considered the death compartment without the hospitalized compartment whiles SEAHR considered the hospitalized compartment but without the death compartment. Therefore, the goal of this study is to combine the death and hospitalized compartments into a SIR-based model so that the model will provide information as more as possible for a policy maker to evaluate the policies during the epidemic outbreaks. The following sections are organized as follows. Section 2 explains the briefs of SIR, SEIRD, and SEIAHR models. Section 3 describes the main model of this study, i.e., SELMAHRD model. Section 4 proceeds to the model validations. Section 5 demonstrates some results with or without different policies. Section 6 concludes this study in brief.

## 2 SIR-BASED MODELS

### 2.1 *Definitions of the symbols*

This section lists all the symbols that will be used for determine all models in Section 2 and Section 3.

$S(t)$: Numbers of susceptible people at time $t$
$E(t)$: Numbers of exposed people at time $t$
$I(t)$: Numbers of infected people at time $t$
$R(t)$: Numbers of recovered people at time $t$
$R_\infty$: The limiting recovered population size, that is, $\lim_{t\to\infty} R(t) = R_\infty$

[1] Another name of SIR model is Kermack-Mcermack model.

$D(t)$: Numbers of death people at time $t$
$N(t)$: Total number of people in the system at time $t$
$A(t)$: Numbers of asymptomatic people at time $t$
$H(t)$: Numbers of hospitalized people at time $t$
$L(t)$: Numbers of latent people at time $t$
$M(t)$: Numbers of symptomatic people at time $t$
$\alpha$: Transmission rate
$\alpha_I$: The ratio that latent people become symptomatic people
$\beta$: Contact rate
$\beta_I$: transmission probability
$\beta_{II}$: effective transmission probability
$a$: A control item to
$\delta_1$: A control item ratio to reduce latent people
$\delta_2$: A control item ratio to reduce symptomatic people
$\delta_3$: A control item ratio to reduce asymptomatic people
$\sigma$: The velocity that exposed people become latent people
$\sigma_2$: The velocity that symptomatic people become hospitalized people
$\gamma$: Recovery rate
$R_0$: Basic reproduction number
$\mu_I$: Death rate
$\mu_1$: The velocity that exposed people become latent people
$\mu_2$: The velocity that latent people become symptomatic people
$\upsilon$: The ratio that people gets a vaccination
$i$: Velocity of immunity after people gets a vaccination
$r_1$ Death rate of symptomatic people
$r_2$ Death rate of hospitalized people
$\gamma_2$: Removal rate of infected people become recovered people
$\gamma_3$: Removal rate of asymptomatic people become recovered people
$\gamma_4$: Removal rate of hospitalized people become recovered people
$\rho_1$: The ratio of symptomatic people become death people
$\rho_2$: The ratio of symptomatic people become hospitalized people
$p_1$: The ratio of hospitalized people become death people

## 2.2 *SIR*

SIR stands for Susceptible, Infected, and Recovered, which was developed by Kermack & McKendrick (1927). This model is the earliest mathematical compartmental model for epidemic researches. Equations are listed from Equations (1)–(5).

$$\frac{dS(t)}{dt} = -\beta I(t)S(t) \tag{1}$$

$$\frac{dI(t)}{dt} = \beta I(t)S(t) - \gamma I(t) \tag{2}$$

$$\frac{dR(t)}{dt} = \gamma I(t) \tag{3}$$

$$\frac{dS(t)}{dt} + \frac{dI(t)}{dt} + \frac{dR(t)}{dt} = 0 \tag{4}$$

$$N(t) = S(t) + I(t) + R(t) \tag{5}$$

## 2.3 *SEIRD*

SEIRD stands for Susceptible, Exposed, Infected, Recovered, and Death. The differences from SIR model are extra two compartments, namely, exposed and death compartments. Equations are listed from Equations (7)–(12).

$$\frac{dS(t)}{dt} = -\beta\alpha\frac{[E(t)+I(t)]S(t)}{N(t)} \tag{6}$$

$$\frac{dE(t)}{dt} = \beta\alpha\frac{[E(t)+I(t)]S(t)}{N(t)} - \sigma E(t) \tag{7}$$

$$\frac{dI(t)}{dt} = -\mu_I I(t) + \sigma E(t) \tag{8}$$

$$\frac{dR(t)}{dt} = \gamma I(t) \tag{9}$$

$$\frac{dD(t)}{dt} = \mu_I I(t) \tag{10}$$

$$\frac{dS(t)}{dt} + \frac{dE(t)}{dt} + \frac{dI(t)}{dt} + \frac{dR(t)}{dt} + \frac{dD(t)}{dt} = 0 \tag{11}$$

$$N(t) = S(t) + E(t) + I(t) + R(t) + D(t) \tag{12}$$

## 2.4 *SEIAHR*

SEIAHR is a model that includes an asymptomatic compartment and a hospitalized compartment, which provide policy intervention information, i.e. number of hospitalized people, for a policy maker to do the decision making to deal with epidemics. Equations (13)–(20).

$$\frac{dS(t)}{dt} = -\frac{\beta_I S(t)[I(t)+\delta_3 A(t)]}{S(t)+E(t)+I(t)+A(t)}\frac{\beta}{1+a[H(t)+R(t)]} \tag{13}$$

$$\frac{dE(t)}{dt} = \frac{\beta_I S(t)[I(t)+\delta_3 A(t)]}{S(t)+E(t)+I(t)+A(t)}\frac{\beta}{1+a[H(t)+R(t)]} - \mu E(t) \tag{14}$$

$$\frac{dI(t)}{dt} = \alpha_I \mu E(t) - (\sigma_2 + \gamma_2) I(t) \tag{15}$$

$$\frac{dA(t)}{dt} = (1-\alpha_I)\mu E(t) - \gamma_3 A(t) \tag{16}$$

$$\frac{dH(t)}{dt} = \sigma_2 I(t) - \gamma_4 H(t) \tag{17}$$

$$\frac{dR(t)}{dt} = \gamma_2 I(t) + \gamma_3 A(t) + \gamma_4 H(t) \tag{18}$$

$$\frac{dS(t)}{dt} + \frac{dE(t)}{dt} + \frac{dI(t)}{dt} + \frac{dA(t)}{dt} + \frac{dH(t)}{dt} + \frac{dR(t)}{dt} = 0 \tag{19}$$

$$N(t) = S(t) + E(t) + I(t) + A(t) + H(t) + R(t) \tag{20}$$

## 3 SIR-BASED MODELS

This study tried to extend the ideas of the SEIRD and SEIAHR to develop a model that includes hospitalized and death compartments which can provide "future" information for a policy maker. Some common ways to alleviate or prevent the influence are isolating susceptible people (adults or students) and letting them at home, doing pulse vaccination, or etc.

The assumptions are similar to those of SIR model, which include total population is fix, population structure is not considered in the model, latent period will not change with time, and recovered people will not get infected again. Based on those assumptions, the conceptual model is shown in Figure 1.

In the model, two new idea ideas are different from SEIRD and SEIAHR. First, Infected ($I(t)$) are subdivided into two compartments, that is, Latent ($L(t)$) and Symptomatic ($M(t)$) because nowadays, such as H1N1, some people may actually infect other people without any symptom, but later on their symptoms come out. Second, in order to include the "vaccination", an arrow from Susceptible ($S(t)$) to Recovered ($R(t)$) is put to represent the people who get vaccination will not get infection from epidemic.

$$\frac{dS(t)}{dt} = -\frac{\beta_{II}\beta S(t)(1-\upsilon)[\delta_1 L(t)+\delta_2 M(t)+\delta_3 A(t)]}{N(t)} - \upsilon i S(t) \tag{21}$$

$$\frac{dE(t)}{dt} = \frac{\beta_{II}\beta S(t)(1-\upsilon)[\delta_1 L(t)+\delta_2 M(t)+\delta_3 A(t)]}{N(t)} - \mu_1 E(t) \tag{22}$$

$$\frac{dL(t)}{dt} = \mu_1 E(t) - [\alpha\mu_2 + (1-\alpha)]L(t) \tag{23}$$

$$\frac{dM(t)}{dt} = \alpha\mu_2 L(t) - [\rho_1 r_1 + \rho_2 r_2 + (1-\rho_1-\rho_2)\gamma_2]M(t) \tag{24}$$

$$\frac{dA(t)}{dt} = (1-\alpha)L(t) - \gamma_3 A(t) \tag{25}$$

$$\frac{dH(t)}{dt} = \rho_2\sigma M(t) - [p_1 r_2 + (1-p_1)\gamma_4]H(t) \tag{26}$$

$$\frac{dD(t)}{dt} = \rho_1 r_1 M(t) + p_1 r_2 H(t) \tag{27}$$

$$\frac{dR(t)}{dt} = (1-\rho_1-\rho_2)\gamma_2 M(t) + \gamma_3 A(t) + (1-p_1)\gamma_4 H(t) \tag{28}$$

$$N(t) = S(t)+E(t)+L(t)+M(t)+A(t)+H(t)+D(t)+R(t) \tag{29}$$

$$\frac{dS(t)}{dt}+\frac{dE(t)}{dt}+\frac{dL(t)}{dt}+\frac{dM(t)}{dt}+\frac{dA(t)}{dt}+\frac{dD(t)}{dt}+\frac{dH(t)}{dt}+\frac{dR(t)}{dt}=0 \tag{30}$$

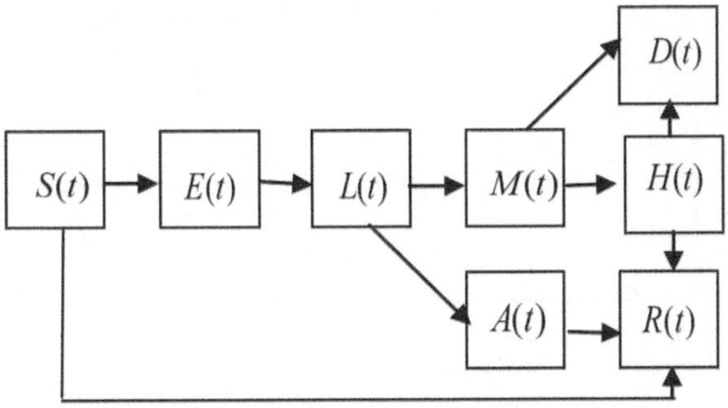


Figure 1. The structure of SELMAHRD model.

Followed by the deductions in [10], the basic reproduction number can be derived as the following equation, which can be a performance index of the model.

$$R_0 = \frac{\beta_{II}\beta(1-\upsilon)\delta_1}{\alpha\mu_2 + (1-\alpha)} \tag{31}$$

## 4 MODEL VALIDATION

In order to examine the validity of the SELMAHRD model, PowerSim is used for the development of the model and the symptomatic, hospitalized, and death of H1N1 of 2009 in Taiwan were used for the comparison with the model outputs. All initial values are listed in Table 1. All parameters are determined by try-and-error method, and the parameter values are within the range that previous researches suggested. Figure 2 to Figure 4 show the comparison results between the real data and simulation data.

Table 1. Initial parameter settings for SELMAHRD model.

| $S(0)$ | $E(0)$ | $L(0)$ | $M(0)$ | $A(0)$ | $H(0)$ | $R(0)$ | $D(0)$ | $\beta_{II}$ | $\beta$ | $\delta_1$ | $\delta_2$ | $\delta_3$ |
|---|---|---|---|---|---|---|---|---|---|---|---|---|
| 223998929 | 0 | 0 | 1 | 0 | 0 | 0 | 0 | 25% | 10 | 50% | 10% | 75% |
| $\upsilon$ | $i$ | $\mu_1$ | $\mu_2$ | $\alpha_{II}$ | $\rho_1$ | $\rho_2$ | $p_1$ | $r_1$ | $r_2$ | $\gamma_2$ | $\gamma_3$ | $\gamma_4$ |
| 0.21% | 14 | 1/3 | 1 | 67% | 0.2% | 3.3% | 0.1% | 0.1 | 0.1 | 0.1 | 1/3 | 0.2 |

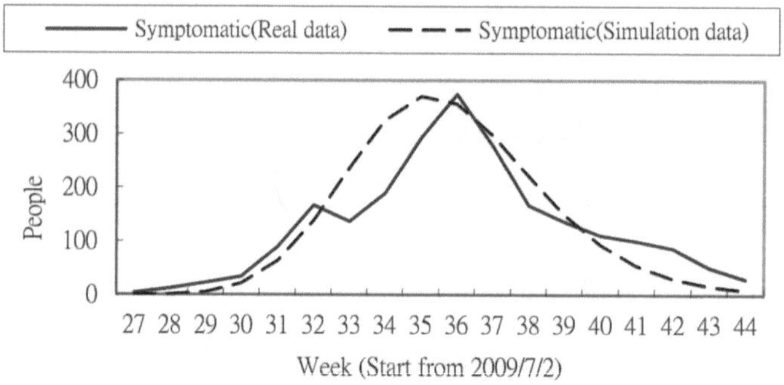


Figure 2. Symptomatic results comparison.

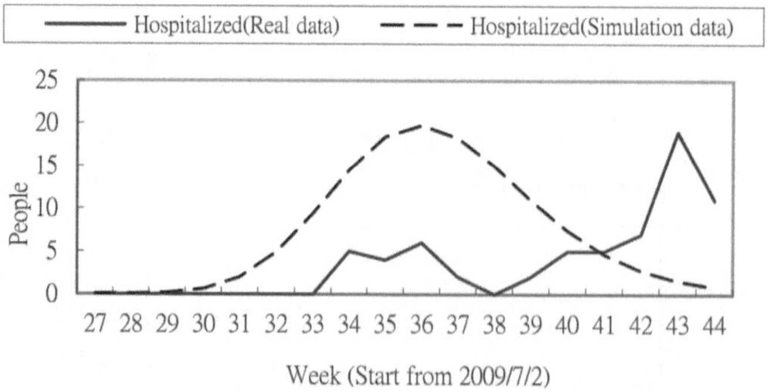


Figure 3. Hospitalized results comparison.

The comparison of symptomatic results between real data and model forecasts are very similar. The overall relative error percentage is 45.5%, the smallest error happens at week 36, only 5% difference, which is the peak value during the H1N1 epidemic period. From the Figure 6, one observation is that the real data seems to delay comparing to the simulation results. From the news history, when H1N1 happened in Mexico for a period of time, there was no H1N1 patent in Taiwan yet. Most people didn't pay H1N1 very much until few patents were reported. These might be a possible reason that the simulation data are ahead of the real data a little bit. But that inference needs more evidence. For the death comparisons, since the number is very small, therefore, it is hard to judge anything. The only thing that can be observed there is a tread explaining more death might be happen in the future.

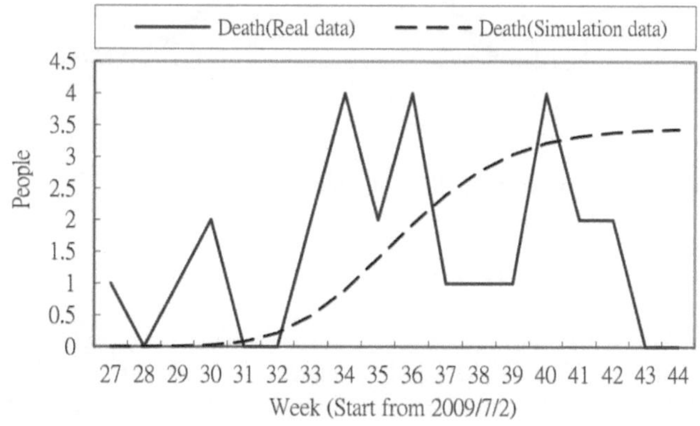


Figure 4. Death results comparison.

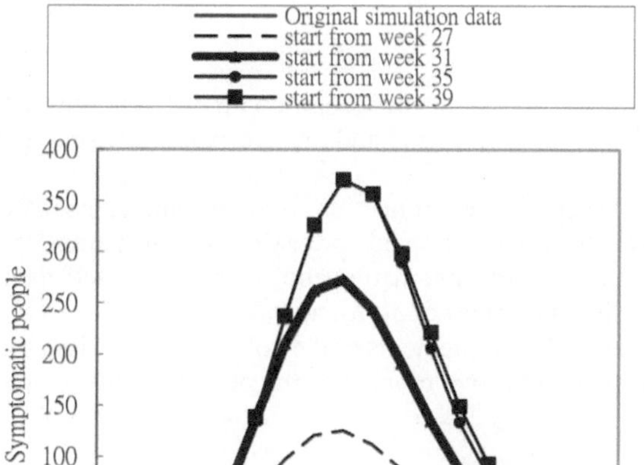


Figure 5. Self-management policy starting from different time.

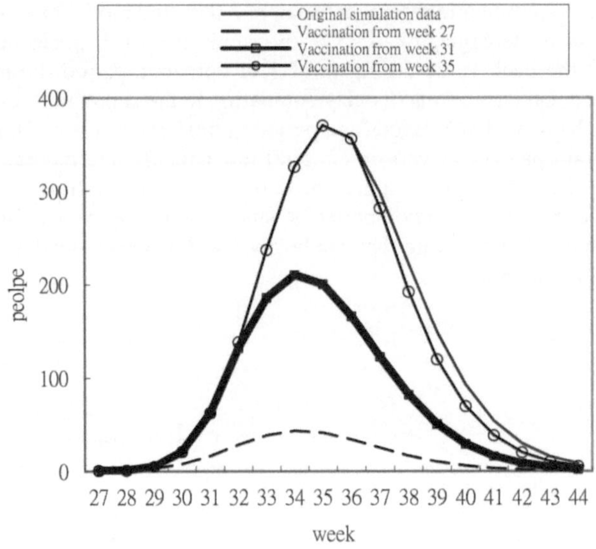


Figure 6. Vaccination policy starting from different time.

## 5 SCENARIO ANALYSIS FOR THE POLICIES

Since the model is still under development, the scenario results are preliminary just for the big picture of what-if analyses. Two different policy measures are considered. The first measure is to encourage susceptible people or enforce symptomatic people to stay at home until they are safe for other people, here; this scenario is named as "Self-management" scenario. And the second measure is to encourage people to be vaccinated. This study chose the parameters that related to the policies, and change the values to examine any possibility to decrease H1N1 epidemic. The parameters that relate to the first measure are $\delta_1$, $\delta_2$ and $\delta_3$ and the parameter that relates the second measure is $\upsilon$.

### 5.1 *Self-management scenario*

Figure 5 shows results of self-management scenario with different times. The original parameters settings are $\delta_1 = 50\%$, $\delta_2 = 10\%$, $\delta_3 = 75\%$ and the new parameter setting are $\delta_1 = 45\%$, $\delta_2 = 9\%$, $\delta_3 = 67.5\%$.

Equation (23) and (24) explain the implications of three parameters. Once a person has doubted himself infected or he or she has got infected, lower values of the parameters mean that less people become latent, symptomatic or asymptomatic people. Therefore, self-management policy can be implemented by decreasing the values of $\delta_1$, $\delta_2$, and $\delta_3$.

Figure 5 shows that the earlier implements of the policy cause the less symptomatic people. And if the policy is implemented after two months of the outbreak, this policy is not a good solution anymore.

### 5.2 *Vaccination scenario*

The parameter setting for the vaccination scenario is $\upsilon = 20\%$ instead of the original setting $\upsilon = 0.21\%$. Figure 6 shows vaccination starts from different time. Very similar to the self-management scenario, early intervention of the scenario will be more effective than late intervention of the measure.

## 6 CONCLUSIONS

Now a day, epidemic outbreaks seem to happen more frequently, and for each time, the situation is getting worse. People have to face the tough challenges each time to conquer the eco- and bio- disaster. This study established a SIR-based compartmental model and used it to discuss the effectiveness of two common policy measures. The model is still under development and has some flaws. Such as it doesn't cover age factor. Different ages of people have different immunity. Or the model doesn't link outer factors, that is, there is no relationship between influenza and economy or environment, which let this model not be a suitable policy tool. At the present model version, the analyzed results are might not be so strong, but the model has shown its potential that can be used for more complicated what-if analyses. In the following researches, the parameters have to be determined more strictly to get more convincing results. Different type of epidemics can be tested by this model to see the capability of this model.

## REFERENCES


Chen, S. C., C. F. Chang, L. J. Jou & C. M. Liao (2007). Modelling vaccination programmes against measles in Taiwan. *Epidemiology and infection*. Vol. 135, No. 5. 775–786.

Dybiec, B. (2009). SIR model of epidemic spread with accumulated exposure. *The European Physical Journal B*. Vol. 67. 377–383.

Hsu, S.B. & Y.H. Hsieh (2008). On the role of asymptomatic infection in transmission dynamics of infectious diseases.*Bulletin of Mathematical Biology*. Vol. 70, No. 1. 134–155.

Kermack, W. O. & A. G. McKendrick (1927). A contribution to the Mathematical Theory of Epidemics. *Proceedings of the Royal Society*. Vol. 115. 700–721.

Lant, T., M. Jehn, C. Christensen, O. M. Araz & J. W. Fowler (2008). Simulating pandemic influenza preparedness plans for a public university: a hierarchical system dynamics approach. *Winter Simulation Conference, Proceedings of the 2008 Winter*. 1305–1313.

Lekone, P. E. & B. F. Finkenstädt (2006). Statistical Inference in a Stochastic Epidemic SEIR Model and with Control Intervention: Ebola as a Case Study. *Biometrics*. Vol. 62. 1170–1177.

Meng X. & L. Chen (2008). The dynamics of a new SIR epidemic model concerning pulse vaccination strategy. *Journal of Mathematical Analysis and Applications*. Vol. 372. 162–180.

Meng, X., L. Chen & B. Wu (2010). A delay SIR epidemic model with pulse vaccination and incubation Times. *Nonlinear Analysis: Real World Applications*. Vol. 11. 88–98.

Rakowski, F., M. Gruziel, U. B-K-Jan & P. Radomski (2010). Influenza epidemic spread simulation for Poland – a large scale, individual based model study. *Physica A*. Vol. 389. 3149–3165.

Wang, X., Y. Tao & X. Song (2009). Pulse vaccination on SEIR epidemic model with nonlinear incidence rate. *Applied Mathematics and Computation*. Vol. 210. 398–404.

Yıldırım, A. & Y. Cherruault (2009). Analytical approximate solution of a SIR epidemic model with constant vaccination strategy by homotopy perturbation method. *Kybernetes*, Vol. 38, No. 9. 1566–1575.

Zaman, G., Y. H. Kang & I. H. (2009). Jung. Optimal treatment of an SIR epidemic model with time delay. *BioSystems*. Vol. 98. 43–50.

*Bridging Research and Good Practices towards Patient Welfare – Shih & Liang (Eds)*
 *ISBN 978-1-138-02716-9*

# A computer-aided system to categorize characteristics of children with special needs

C.Y. Juan
*Department of Early Childhood Care and Education,*
*Hsin Sheng College of Medical Care and Management, Longtan, Taiwan, R.O.C.*

J.W. Wu & K.H. Yang
*Department of Industrial and Systems Engineering,*
*Chung Yuan Christian University, Zhong-li, Taiwan, R.O.C.*

ABSTRACT: Nowadays, Taiwan government pays more and more attentions on the early-child health issues. Every year, it is around ten percent of 0~6 aged children who are identified with special needs in Taiwan. Early interventions are proved as effective ways to help for a child with special needs. The process of early intervention includes five phases: early detection, reporting, referral, identification, and treatment. The primary caregivers of the first three steps depend on parents or guardians, and the last two steps depend on formal organizations, such as hospitals, preschool, or special centers. In Taiwan, the families that raise children with special needs are mostly disadvantaged. It is not easy for them to get resources or information about early intervention for their child with special needs. Also in the current system of Taiwan, the identification process of the child with special needs of the hospital takes a lot of time to complete. It also probably delays the treatments for a child with special needs. Motivated by this situation, this study tries to develop a decision support system to facilitate initial diagnosis of a child with special needs. We establish a web-based system with PHP and MySql. In the current version of the system, it can help screen those three common symptoms of children with special needs, including autism, intellectual disabilities, and development delay. The objectives of this study focus on those parents can easily to use the system to acquire sufficient information about a suspected child; they will know what the next step is to help their child, and where they can ask for help. This system takes advantage of the convenience of the internet, and parents can easily and quickly preliminary get ideas about their children's symptoms. By the results of an initial screen, parents can more accurately and quickly to find the right doctor, and further it shortens the early treatment process.

## 1 INTRODUCTION

With a highly development of economic grow in Taiwan, the low fertility becomes a common situation. Small number of kids let presents pay more attentions on their children. A health and beautiful child is an expectation of most parents. However, there are still around 10 percent of children with special needs. When common parents suspect their children who might be children with special needs, they probably ask non-profession for help. Another situation is that maybe parents do not beware that their children have some problems. Adequate treatments on the children with special needs will be delayed. Because the issue is arousing, more and more people are aware of the importance of early intervention. Currently, the early intervention mainly includes five tasks;

early detection, reporting, referral, identification, and treatment, which determines structured and continuous services (Liao 1998, Huang & Chiu 1999). Although modern medical treatments or examinations are advanced, they sometimes take complicated and long processes, which cause a flaw for kids with special needs who need early intervention. Those kids might miss treatments of the best timing. Therefore, a short-term identification is important for bringing benefits for kids with special kids for further treatments.

Because preschool educations are important, researches on preschool educations are getting rich, especially a lot of studies focus on the subjects that under five years old. Child's brain grows with the fastest rate at a child under five years old. Without good care and education, development of a child's brain might get delayed. If a child's brain does not develop well, abilities of a child, such as recognition, language, sentiment, and so on, might be hindered. Special education specialists emphasize the early interventions and treatments are very important for children with failure to thrive. Early interventions and treatments not only alleviate degrees of children with failure to thrive, but prevent them from another handicap coming up (Chou et al. 2000).

The main purpose of this study is to establish a web-based computer system to help parents easily screen their kids if they are suspected to be kids with special needs. Since the system is still under development, the current version of the system can only screen autism, intellectual disabilities, and failure to thrive. The main difference between this study and a traditional identification process lies in parents can just answer the questionnaire with the questions they found in their daily life. Designed questions bases on the perspectives of families. According to parents' observations on their kids, they answer yes-no questions by checking a box item. It takes around 10 minutes to finish all questions. According to answers, the system will give scores of the three symptoms. The highest score might indicate suspected symptom of a kid. This can help parents have information to ask for further examination in the hospital. However, this system might run into some real difficulties, such as whether the screen results are legitimate. The limitations of this study might exists already, however, the preliminary results are still satisfied.

## 2 DELINIATION OF THE COMPUTER SYSTEM

The first step of this study is to establish a reliable questionnaire with collecting data. The data come from two sources. One are from parents' interviews carried by students of Hsin Sheng College of Medical Care and Management. Others are from references, such as (Paasche et al. 2004). We tried to determine the related statements corresponding to the symptoms of children with special needs. We also modified formal statements into oral-like sentences, which lets parents easily understand questions. Presents can easily answer questions just by daily observations on their children. Three categories of symptoms, autism, intellectual disabilities, and failure to thrive, are main targets because these types are very common in real life. After consulting experts and focus group discussions, we determined 155 questions in total, 78 items for autism, 49 items for intellectual disabilities, and 28 items from failure to thrive. The questions then fit in 5 development fields of a child, which include cognitive development, social and emotional development, speech and language development, fine motor skill development, and gross motor skill development. The advantage of this step is that parents easily observe all these five developmental behaviors in their daily life. Parents need not have advanced knowledge about children with special needs, and they can still fill in the questionnaire of this computer system. Because this computer is web-accessed, parents can just go to the webpage, login the system, finish the questionnaire, and then the system will give an initiative screen report of the child. In the future updated website, extra functions will be added, such as it can provide some suggestions on the next-step actions, for example, where are the suitable doctors and hospitals near-by.

The computer system is developed by PHP and MySql, which can define some internal rules to quickly screen out suspected children with special needs. Figure 1 shows the program structure and operational flow.

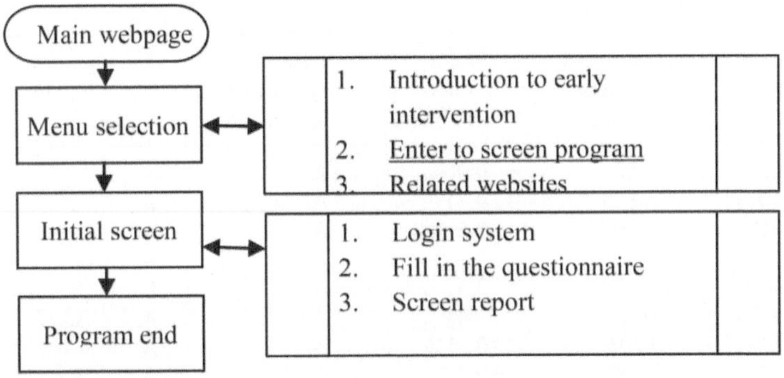


Figure 1. Computer system for initially screening children with special needs.

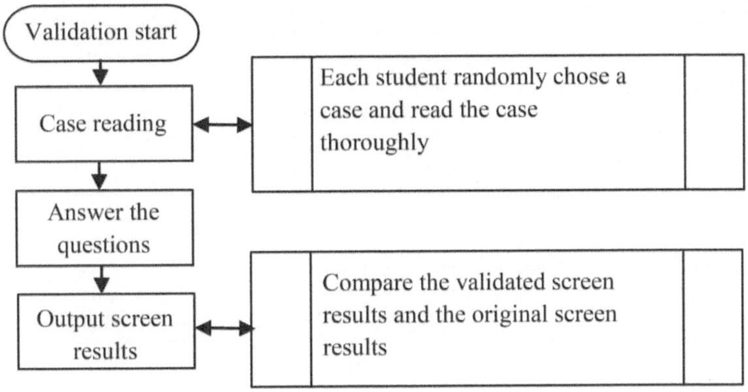


Figure 2. The validations of trial usage of the computer system.

## 3 STUDY RESULTS AND DISCUSSIONS

### 3.1 *Trial of the system*

When student interviewers collected data from parents, they asked parents' permission that they record their children with special needs in detail, including their typical behaviors and disability category. All the real cases are used as validated cases to test the reliability of the computer system. In the validation procedure, we recruited 18 students. Each student was responsible for a case, which hide disability category. He or she had to read one case in detail. According to the descriptions of a case, a student answer the questions in the computer system just by checking items. Figure 2 shows the validation procedures of this study.

Some results are from the validations. There were four cases related to failure to thrive. However, these four cases were misjudged, that is, the computer system judged other symptoms instead of failure to thrive. The other two symptoms have better results than failure to thrive does. The judgment rule is while the tester sent out his or her questionnaire, the system program start to count the item that matches the specific symptom. The highest score of a specific symptom is the suspected target symptom. For example, one tester, A; he checked the 10 items of autism questions, which represents 12.82% (10 out of 78); 1 item of intellectual disabilities questions, which represents 2.04% (1 out of 49); 0 items of failure to thrive questions; which represent 0.00% (0 out of 28). The percentage score of autism is the highest of all, and then the screen report shows the case is suspected as autism. The individual results of 18 testers are listed in Table 1 and summaries are listed in Table 2.

Table 1. System Validation results.

| Tester | Autism (A) Score | Intellectual disabilities (ID) Score | Failure to thrive (FT) Score | Disability type of an actual case | Judgment of the computer system | Judgment correctness |
|---|---|---|---|---|---|---|
| 1 | 12.82% | 2.04% | 0.00% | A | A | Y |
| 2 | 14.10% | 2.04% | 10.71% | A | A | Y |
| 3 | 15.38% | 2.04% | 3.57% | A | A | Y |
| 4 | 35.90% | 28.57% | 32.14% | FT | A | N |
| 5 | 6.41% | 16.33% | 3.57% | A | ID | N |
| 6 | 38.46% | 20.41% | 21.43% | A | A | Y |
| 7 | 20.51% | 10.20% | 10.71% | A | A | Y |
| 8 | 32.05% | 22.45% | 10.71% | A | A | Y |
| 9 | 19.23% | 12.24% | 14.29% | A | A | Y |
| 10 | 7.69% | 14.29% | 3.57% | ID | ID | Y |
| 11 | 15.38% | 8.16% | 14.29% | FT | A | N |
| 12 | 55.13% | 53.06% | 35.71% | A | A | Y |
| 13 | 28.21% | 32.65% | 21.43% | FT | ID | N |
| 14 | 20.51% | 38.78% | 21.43% | ID | ID | Y |
| 15 | 37.18% | 18.37% | 21.43% | A | A | Y |
| 16 | 28.21% | 18.37% | 0.00% | A | A | Y |
| 17 | 33.33% | 12.24% | 14.29% | ID | A | N |
| 18 | 20.51% | 24.49% | 10.71% | FT | ID | N |

Table 2. Overall performances of the computer system.

| Disability Type | Correctness Rate = number of correct judgments/total judgments |
|---|---|
| Autism | 10/11 = 90.91% |
| Intellectual disabilities | 2/3 = 66.67% |
| Failure to thrive | 0/4 = 0.00% |
| Overall Performance | (10 + 2 + 0)/(11 + 3 + 4) = 66.67% |

Table 3. Reliability analysis of the questionnaire of the computer system.

| Question Type | Number of questions | Cronbach's $\alpha$ |
|---|---|---|
| Autism | 78 | $0.867 > 0.8$ |
| Intellectual disabilities | 49 | $0.857 > 0.8$ |
| Failure to thrive | 28 | $0.64 < 0.8$ |
| Overall | 155 | $0.927 > 0.9$ |

From the analyzed results of Table 1 and Table 2, the correctness rate of Autism Type is higher than the other two types. Preliminary inferences might come from three aspects; 1. The questions are not reliable, 2. Contents of different questions are dependent, and 3. Number of representative questions is insufficient. The follows future discuss further by the above three points.

1. Comments on Reliability of questions

   In order to examine if the questions we developed are reliable, we analyze the results of Cronbach's $\alpha$ of questions. The results are shown in Table 3.

   Except for the questions of Failure to thrive, Cronbach's $\alpha$ of other question types are greater than the general suggested values. In the social science filed, some researchers (Nunnally 1978,

Table 4. Performance comparisons between this study and the traditional diagnosis.

| | This study | Traditional diagnosis |
|---|---|---|
| Assessment time | Minutes span | Days to weeks span |
| Cost of Assessment per Child | No cost | Relatively high |
| Preliminary screen report | Overall correctness is above 60%, especially good in Autism diagnosis | A child shown as Suspected has to go through further examinations |

Devellis 1991) considered that if Cronbach's $\alpha$ less than 0.6, the questionnaire is not acceptable and needs to be revised. From this point of view, the questionnaire is overall reliable and acceptable.

2. Comments on contents of different questions are dependent

According the previous discussions, reliability might not be an issue in this preliminary study. However, the point 2 might be an issue because different disabilities might have the same symptoms, or children with special needs have multi-disabilities. The other possibility might come from the unexperienced student interviewers, who cannot record the right behaviors or symptoms of children with special needs. This should be corrected by re-examining the whole questions and then deleting dependent questions in different categories of disabilities.

3. Comments on Number of representative questions is insufficient

This point is also an issue. Now in the system, more than 50% the questions are for Autism. Independent questions is difficult to find because normally Autism or Intellectual disabilities also have some symptoms of Failure to thrive. In the future, it is necessary to enlarge the field investigation and ask for more professionals' opinion to determine sufficient questions of different categories of children with special needs.

### 3.2 *Comparisons between this computer system and the traditional diagnosis for children with special needs*

Traditional diagnosis on a child with special needs often takes time and spends money. It might take around 4.5 to 156 hours or spends (Chen 2007) or spends more than NT 25000 dollars (Website: Outside the Square Psychology)[1]. The main reason that the traditional assessment takes time is a child has to be transited to the right place. Without professional and sufficient knowledge, parents don't know how to bring their child to the right assessment center efficiently and effectively. However, in this study, parents only need to fill in the questionnaire by daily observations on their child. They can get the preliminary screen report within 10 minutes. Also, the system will give some related information, such as the address or telephone number of an assessment center. Parents can have a basic idea what is the next step for their child. As for the cost, a child with special needs has to be diagnosed by professions. In different areas, a city or a suburb, the cost might different. However, it costs a lot no matter human resources or transportation. As for the web-based computer system, parents can easily login the system, and almost without any cost to get preliminary screen report of their child. From the technical point of views, PHP and MySql is not new. Even there is no need for a software engineer for developing this kind of web-based system. The setup cost of the system is related low comparing related facilities in the hospital. The maintenance of the system is also related low cost. The performance comparisons between this study and the traditional diagnosis are in the Table 4.

[1] Autism Spectrum Disorder Diagnostic Assessment.

## 4 CONCLUSIONS

The main purpose of this study is to establish a computer-aided system to categorize characteristics of children with special needs by PHP and MySql computer language. This system is presented by a web-based system. Users can easily assess this system and just fill in a questionnaire to get a quick report for a child. Although the system is still under development, it has already capabilities to provide sufficient information for parents to go to the next step actions for their child.

In the current system, it can screen three different symptoms, which includes autism, intellectual disabilities, and failure to thrive. After parents finish the questionnaire, the system will show the screen report with three scores corresponding to the three disability types. The highest scores of all represents a child is suspected to that disability type. However, the current system, due to questions has to be further revised, the overall screen correctness rate is 66.7%. It can screen autism very well, the correctness rate reaches around 90%. The worst screen result of all is failure to thrive. The potential issue might be no matter autism or intellectual disabilities, they accompany with failure to thrive at the same time, which makes the system difficult do the right judgment on a child with special needs.

In the future study, the correctness rates need to be improved so that the system can be applied in the real life and really help parents and children with special needs.

## REFERENCES


Chen, C. G. (2007). Accessments on special education students. Taiwan, Psychological Publishing Co., Ltd.

Chou, W. L., L. C. Cheng & H. Lin, F. 2000. Developemtns and future prospects of early intervention in Taiwan. *New Waves - Educational Research & Development* 5(4): 7–12.

Devellis, R. F. (1991). Scale Development: Theory and Application. London, Sage.

Huang, S. L. & Y. W. Chiu 1999. Nurses in Early Intervention for Children with Developmental Delays. *The Journal of Nursing* 46: 67–72.

Liao, F. H. 1998. The Team Model in the Early Intervention Program for Children with Developmental Delays. *Journal of the Physical Therapy Association of the R.O.C.* 23: 55–67.

Nunnally, J. C. (1978). Psychometric Theory, McGraw-Hill.

Paasche, C. L., L. Gorrill & B. Strom (2004). Children With Special Needs in Early Childhood Settings Wadsworth Publishing.

Website: Outside the Square Psychology. from http://www.outsidethesquarepsychology.com.au/main/page_assessment.html.

*Bridging Research and Good Practices towards Patient Welfare – Shih & Liang (Eds)*
*© 2015 Taylor & Francis Group, London, ISBN 978-1-138-02716-9*

# Simulation-based optimization method for operation scheduling using combined heuristics rules

Yuanyuan Guo, Feng Liang, Qian Wang & Jing Sui
*Department of Industrial Engineering, Nankai University, Tianjin, P.R. China*

ABSTRACT: Operation scheduling is one of the most crucial tasks in surgical center management. This paper focuses on operation scheduling optimization. A combined scheduling rule weighted composed of three simple scheduling rules is proposed to optimize operation scheduling performance. Considering plenty of real factors, a simulation model about operation scheduling system is built. According to two optimization objectives, tabu search algorithm is used to search for the optimal weight of simple rules in a combined scheduling rule based on simulation model. Finally, the performance comparison between the proposed combined scheduling rule and three simple rules are presented to indicate that the combined scheduling rule is more valid and efficient to sequence patients.

## 1 INTRODUCTION

It is well known that the surgical center is one of the most significant departments in a hospital. More than 40 percent of a hospital's total revenues come from the center and it also accounts for more than 40 percent of a hospital's total resource cost. Thus, it is regarded as the greatest revenue source as well as the largest cost unit of a hospital. Therefore, the surgical center has the largest potential in enhancing hospital profits, reducing resource cost and improving patient satisfaction.

Operation scheduling is one of the most crucial tasks in surgical center management. There are two factors making operation scheduling complex and difficult. First, the accurate prediction of operating duration is hard to obtain, because there are many uncertain factors and emergency conditions which cannot be predicted during operations. Second, an efficient operation scheduling rule is hard to be applied in every hospital's surgical center, because there are numerous differences between hospitals. For example, compared with private hospitals, managers in public hospitals pay less attention to hospitals' income. Instead, they are more concerned with fairness. So First Come First Service (FCFS), which is a relative fair scheduling rule, is always used in public hospitals. For the above reasons, a majority of hospitals schedule patients with empirical method. For example, the prediction of operating duration can be obtained by surgeons' estimation according to historical data of operations. In addition, several simple rules are usually applied to schedule patients, although these rules cannot meet every stakeholder's demand in operation scheduling system.

In the past several decades years, a large number of articles about operation room (OR) or operation scheduling optimization problems have been generated. Among these articles, plenty of methods are used popularly, such as mathematical programming, simulation and heuristic approach. Heuristic approach is one of the most effective research approaches study on OR and operation scheduling optimization. Beliën & Demeulemeester (2007) propose a cyclic master surgery schedule with leveled resulting bed occupancy, and use several different heuristic approaches to minimize the expected total bed shortage, which are repetitive MIP (a combined integer program) heuristic,

Quadratic MIP heuristic and simulated annealing. Hans et al. (2008) study the robust surgery loading problem and use the random exchange method (REM) and simulated annealing (SA) to minimize the total planned slack and maximize OR utilization. Hsu et al. (2003) use tabu search heuristic algorithm to deal with the operation scheduling problem from an ambulatory surgical center (ASC). Roland et al. (2006) study the surgeries planning and scheduling optimization problem which is similar to the resource-constrained project scheduling problem (RCPSP). Considering plenty of constraints on renewable and nonrenewable resources, they use genetic algorithm (GA) to minimize ORs opening costs and overtime pay. Denton et al. (2007) study how different schedules affect patient waiting time, OR idling time and overtime. They formulate a two-stage stochastic recourse model and propose several heuristics to form surgery sequencing, and compare actual schedules, optimal schedules with schedules formed by using the proposed heuristics. Blake et al. (2002) use the integer programming and an improvement heuristic to form a consistent schedule which can assign OR blocks to surgical groups based on objective of minimizing the shortfall between the target OR time of each surgical group and the actual allocation OR time. In addition, a majority of articles focus on a few similar objectives which have a profound effect on surgical center's benefit. For example, patient waiting time and patient throughput are two most crucial indexes in evaluating the performance of operation scheduling. Rohleder et al. (2011) use discrete event simulation modeling to identify optimized staffing levels and better operation scheduling in an effort to improve waiting time measures and reduce overall patient time in a clinic. Ahmed & Alkhamis (2009) present a method which combines system simulation with optimization to increase patient throughput and to reduce patient waiting time under budget restrictions by determining the optimal number of doctors, lab technicians and nurses required.

According to the literature review, there are numerous studies concentrating on outpatient scheduling optimization, while inpatient scheduling optimization gets less attention. However, the number of people who take operation is increasing dramatically and inpatient scheduling is mainly responsible for sequencing patients, so it becomes necessary to research inpatient-scheduling optimization. In reality, several simple scheduling rules are applied to sequence patients, such as First Come First Service (FCFS), Shortest Processing Time (SPT) and Critical Ratio (CR). However, every simple scheduling rule always sequences patients based on a specific purpose, it cannot meet every stakeholder's requirement in inpatient scheduling.

Hence, this paper focuses on inpatient scheduling optimization. A simulation model of operation scheduling system is built by the theory of discrete-event dynamic system simulation (DEDS), and a combined scheduling rule composed of three simple scheduling rules is proposed to optimize operation scheduling. Two optimization indexes including patient throughput and patient waiting time are chosen to evaluate the performance of operation scheduling. Finally, the performance comparison between the proposed combined scheduling rule and three simple rules is shown, which indicates that this combined scheduling rule is more valid and efficient to sequence inpatients.

The rest of this paper is organized as follows: section 2 proposes a combined scheduling rule which is composed of three simple scheduling rules by an initial weight. According to two optimization objectives, tabu search algorithm contained in Opt Quest optimizer is applied to search the optimal weight of every simple rule based on simulation model. Section 3 delivers the required information of a surgical center and provides major process of building operation scheduling simulation model. The performance comparison between the proposed combined scheduling rule and three simple rules are presented in the section 4. Finally, conclusions and future research contents are provided in the section 5.

## 2 A COMBINED SCHEDULING APPROACH

Sequencing patients reasonably is one of the most crucial tasks in surgical center management. If patients, ORs and surgeons are simply considered as work pieces, machines and materials, this problem can be regarded as a special job shop scheduling problem which can be stated as N jobs $\{J_1, J_2, \ldots, J_3\}$ are to be processed by M machines $\{M_1, M_2, \ldots, M_3\}$ within a given time period for several optimization objectives. Several simple scheduling rules are usually applied by managers to schedule patients. However, the combined scheduling rule based on several simple rules are superior to simple scheduling rules, so a combined scheduling rule which can improve the performance of operation scheduling is proposed in this paper. Three typical simple rules are selected and combined by an initial weight to compose this combined scheduling rule.

### 2.1 *Simple scheduling rules selection*

Scheduling rules are queue disciplines which determine the sequence of jobs in a job shop scheduling problem. Panwalkar & WafikIskander (1977) provide an exhaustive survey of scheduling rules which lists over 100 scheduling rules. These rules are divided into 3 main categories, including priority scheduling rules, heuristic scheduling rules and other rules. In addition, priority scheduling rules are divided into simple rules, combination of simple rules and weighted priority indexes. Among all, simple rules are used widely in manufacturing system. These rules are usually related to information about a specific job, such as arrival time, processing time, due date or slack (based on processing time and due date), etc. Among them, a arrival time based rule like First Come First Service (FCFS) is good for maintaining a fair between jobs, a processing time based rule like Shortest Processing Time (SPT) is beneficial to minimize flow time, and a due date based rule like critical ratio (CR) is usually applied for jobs which have a strict delivery time.

According to the analysis, three typical scheduling rules are chosen to compose a combined scheduling rule.

- First Come First Service (FCFS): it means the patient who has the longest patient waiting time is treated firstly. This rule is most widely and frequently used in operation scheduling. It is chosen from the perspective of fair.
- Shortest Processing Time (SPT): it means the operation which has the shortest operating duration is completed firstly. This rule is chosen from the perspective of hospital, because this rule can increase patient throughput and hospital's profits.
- Critical Ratio (CR): it means the patient who has the lowest critical ratio is treated firstly. The patient CR is the ratio between due date of patient and operating duration of patient. This rule is chosen from the perspective of patient. Because this rule determines that the patient who has shorter due date and longer operating duration is treated firstly.

### 2.2 *Simple scheduling rules quantification*

In order to get a combined scheduling rule, three selected simple rules must be quantified and dimensionless.

- First Come First Service (FCFS)

  Patient waiting time is applied to indicate patient arrival time, because the patient who has the longest waiting time is the one who arrives earliest. This rule stipulates that the patient who has the longest waiting time is to be treated firstly. After quantified, the value of this rule is represented by $C_{fcfs}$. The process of quantization is formulated as equation 1:

$$C_{fcfs} = \sqrt{\left(\frac{\frac{1}{WT_i} - \frac{1}{WT_{max}}}{\frac{1}{WT_{min}} - \frac{1}{WT_{max}}}\right)^2} \tag{1}$$

  where $C_{fcfs}$ represents the value of FCFS rule, $WT_i$ represents waiting time of patient i, $WT_{min}$ represents the shortest waiting time of all patients, $WT_{max}$ represents the longest waiting time of all patients.

- Shortest Processing Time (SPT)

  Operating duration of patient is applied to indicate processing time. This rule stipulates that the patient who has the shortest processing time is to be treated firstly. After quantified, the value of this rule is represented by $C_{spt}$. The process of quantization is formulated as equation 2:

$$C_{\text{spt}} = \sqrt{\left(\frac{OD_i - OD_{\min}}{OD_{\max} - OD_{\min}}\right)^2} \tag{2}$$

  where $C_{spt}$ represents the value of SPT rule, $OD_i$ represents operating duration of patient i, $OD_{\min}$ represents the shortest operating duration of all patients, $OD_{\max}$ represents the longest operating duration of all patients.

- Critical Ratio (CR)

  Critical ratio of patient is indicated by the ratio between due date of patient and operating duration of patient. Due date of patient which means remaining length of stay (LOS) is indicated by the difference between the maximum LOS of one patient and waiting time of this patient. This rule stipulates that the patient who has the lowest critical ratio is to be treated firstly. After quantified, the value of this rule is represented by $C_{cr}$. The process of quantization is formulated as equation 3, 4, 5:

$$DD_i = MaxLOS_i - WT_i \ (3) \quad CR_i = \frac{DD_i}{OD_i} \ (4) \quad C_{\text{cr}} = \sqrt{\left(\frac{CR_i - CR_{\min}}{CR_{\max} - CR_{\min}}\right)^2} \tag{3}$$

  where $C_{cr}$ represents the value of CR rule, $CR_i$ represents critical ratio of patient i, $CR_{\min}$ represents the lowest critical ratio of all patients, $CR_{\max}$ represents the highest critical ratio of all patients, $DD_i$ represents due date of patient i, $MaxLOS_i$ represents the maximum LOS of patient i, $WT_i$ represents waiting time of patient i.

  After quantified, three simple rules are combined by an initial weight to compose a combined scheduling rule, which is formulated as equation 6:

$$C_{\text{mix}} = k_{fcfs} \times C_{fcfs} + k_{spt} \times C_{spt} + k_{cr} \times C_{cr} \tag{4}$$

  where $C_{\text{mix}}$ represents the value of the combined rule, $k_{fcfs}$ represents weight of FCFS rule, $k_{spt}$ represents weight of SPT rule, $k_{cr}$ represents weight of CR rule.

In addition, if the weight of FCFS rule is greater than 0, it means the patient who has the longest waiting time is to be treated firstly, otherwise it means the patient who has the shortest waiting time is to be treated firstly. Similarly, if the weight of SPT rule is greater than 0, it means the patient who has the shortest processing time is to be treated firstly, otherwise it means the patient who has the longest processing time is to be treated firstly. If the weight of CR rule is greater than 0, it means the patient who has the lowest critical ratio is to be treated firstly, otherwise it means the patient who has the highest critical ratio is to be treated firstly.

## 3 SIMULATION MODEL

In this paper, the Arena Software is applied to build simulation model of operation scheduling system.

### 3.1 *Surgical center description*

#### 3.1.1 *Overview*

The example hospital is the first three level of first-class hospital in China and It was the largest cardiovascular disease specialist hospital in Asia at that time. The surgical center in this hospital mainly

Table 1. The resources and staffs of surgical center.

| Resources and staffs | Quantity |
|---|---|
| Sub-departments | 4 |
| Operating rooms (ORs) | 5 |
| Surgeons | 30 |
| Nurses | 10 |
| Technicians | 3 |

operates five kinds of cardiovascular interventional operations, including coronary angiography (CA), coronary artery stent implantation (CASI), radiofrequency ablation (RA), cardiac pacemaker implantation (CPI), coronary angiography and coronary artery stent implantation (CA+ CASI). These operations are all completed in the specific operating rooms (ORs) which have air purification devices. Generally, the ORs are open from 9 am to 5 pm Monday through Friday and are closed for holidays.

1. Resources and Staffs

   The resources and staffs of this surgical center are shown in Table 1. Every sub-department is independent and it contains its own surgeons who can complete one or more kinds of operations mentioned above. Every surgeon is capable of processing one or more types of operations. These sub-departments are responsible for their own patients. Four sub-departments share five ORs, ten nurses and three technicians. Every OR has one nurse assisting operations. Three technicians are accountable for apparatuses maintenance in five ORs. Nurses and technicians are scheduled by managers of surgical center based on their experience, and ORs and patients are scheduled according to scheduling rules.

#### 3.1.2 *Schedule stipulation and assumption*

- Generally, the length of the main operation scheduling plan is a week. Namely, this weekly plan will be made and completed before the start of every week, which contains patient sequence, surgeons of every patient, OR of every patient, operating duration and operating start time.
- There are five operating rooms (ORs). Each of them can be used by four sub-departments to perform five kinds of operations mentioned above. These ORs don't accept emergency operations which are usually completed in the specialized ORs. These five ORs are open from 9 am to 5 pm Monday through Friday and are closed for holidays.
- Generally, the preoperative examination needs two weeks (14 days), so patient waiting time is usual between 14 and 20 days. The longest Acceptable LOS of patient is 25 days.
- Every operation has two surgeons. Other resources and staffs needed in an operation are adequate.
- The preparation time of an operation is 15 minutes, including OR cleaning time, patient transfer time and resources preparation time.

### 3.2 *Data collection*

The data and information about operation scheduling in this surgical center from March 2011 to February 2012 are collected accurately, which are listed as follows:

1. the number of every type of operation in every sub-department

   After data processing, the percentage of every type of operation in every sub-department is shown in table 2.
2. the operating duration distribution of every type of operation in every sub-department

   The data are obtained from operation historical records from March 2011 to February 2012. The operating duration distribution of every type of operation is shown in table 3.

Table 2. The percentage of every type of operation.

| | Operating type | | | | |
|---|---|---|---|---|---|
| Sub-department | RA | CPI | CA+CASI | CASI | CA |
| A | 0.057 | 0.058 | 0.248 | 0.037 | 0.6 |
| B | 0.043 | 0.036 | 0.293 | 0.036 | 0.592 |
| C | 0.04 | 0.129 | 0.212 | 0.027 | 0.592 |
| D | 0.076 | 0.037 | 0.265 | 0.032 | 0.59 |

Table 3. The operating duration distribution of every type of operation.

| Sub-department | Operating type | Operating duration distribution (Parameters) |
|---|---|---|
| A | RA | NORM (99.1, 34.7) |
| | CPI | 40+WEIB (42.8, 1.12) |
| | CA+CASI | 15+ERLA (13.4, 5) |
| | CASI | 30+155 × BETA (1.52, 3.97) |
| | CA | 10+248 × BETA (2.28, 18.7) |
| B | RA | 42+238 × BETA (1.19, 4.38) |
| | CPI | 45+EXPO (57.8) |
| | CA+CASI | 20+GAMM (17.6, 3.24) |
| | CASI | TRIA (20, 51.2, 120) |
| | CA | 10+ERLA (12.8,2) |
| C | RA | NORM (91.9, 30.8) |
| | CPI | 65+EXPO (78.3) |
| | CA+CASI | 30+ERLA (29.1,2) |
| | CASI | 30+EXPO (54.5) |
| | CA | 10+GAMM (10.8, 2.29) |
| D | RA | TRIA (65, 78.9, 210) |
| | CPI | NORM (100, 42.2) |
| | CA+CASI | 40+135 × BETA (0.978, 2.66) |
| | CASI | 34.5+76 × BETA (0.437, 0.775) |
| | CA | NORM (32.7, 13.3) |

## 4 MODELING RESULTS AND ANALYSIS

The simulation model of operation scheduling system is built and two optimization objectives are chosen to evaluate the scheduling performance of every rule, which are maximizing patient throughput and minimizing patient waiting time. Tabu search algorithm contained in Opt Quest optimizer is used to determine the optimal combined scheduling rule according to two optimization objectives.

### 4.1 *The scheduling performance of three simple scheduling rules*

Three frequently used scheduling rules which are FCFS rule, SPT rule and CR rule are chosen to schedule patients for their operations. Two indexes which are patient throughput and patient waiting time are selected to evaluate the performance of these three scheduling rules. The simulation circle is 120 hours. Table 6 shows the performance of three simple scheduling rules.

Table 4 indicates that no matter managers in surgical center put emphasis on patient throughput or patient waiting time, SPT rule is the best choice among these three simple rules. From the perspective of patient throughput, SPT rule can achieve the largest patient throughput among three

Table 4. The scheduling performance of three simple scheduling rules.

| Scheduling rule | Patient throughput (case) | Patient waiting time (day) |
|---|---|---|
| FCFS rule | 171 | 20.322 |
| **SPT rule** | **206** | **18.976** |
| CR rule | 132 | 19.523 |

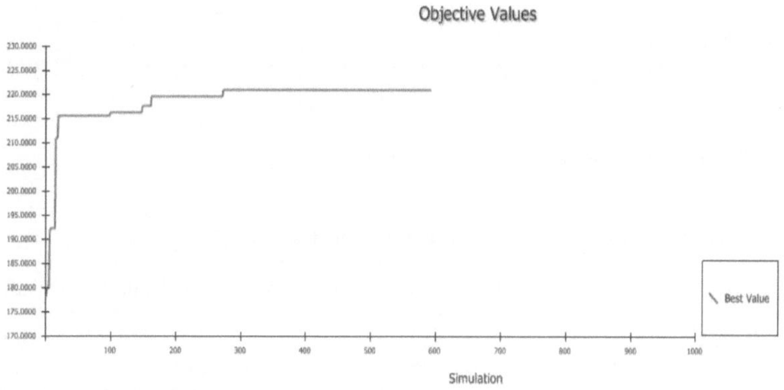


Figure 1. The optimization process of patient throughput.

rules, which is 206 operation cases completed in a week; and from the perspective of patient waiting time, SPT rule can also get the least patient average waiting time among the three rules, which is 18.976 days.

### 4.2 *The scheduling performance of the combined scheduling rule*

1. Setting patient throughput as optimization objective

   In this optimization process, patient throughput is served as the optimization objective. The optimization process of objective value is described in Fig. 1.

   According to the optimal weight of every simple rule, the optimal combined scheduling rule with the objective of maximizing patient throughput can be established. This optimal combined scheduling rule is called as TP rule, the representation of which is formulated as equation 7.

$$C_{tp} = -0.024117 \bullet C_{fcfs} + 0.990933 \bullet C_{spt} - 0.503051 \bullet C_{cr} \tag{5}$$

   where $C_{tp}$ represents the value of the combined scheduling rule with the objective of maximizing patient throughput, $C_{fcfs}$ represents the value of FCFS rule, $C_{spt}$ represents the value of SPT rule, $C_{cr}$ represents the value of CR rule.

2. Setting patient waiting time as optimization objective

   In this optimization process, patient average waiting time is served as optimization objective. The optimization process of objective value is described in Fig. 2.

   According to the optimal weight of every simple rule, the optimal combined scheduling rule with the objective of minimizing patient waiting time can be established. This optimal combined scheduling rule is called as WT rule, the representation of which is formulated as equation 8.

$$C_{wt} = -0.985033 \bullet C_{fcfs} - 0.438883 \bullet C_{spt} - 0.090595 \bullet C_{cr} \tag{6}$$

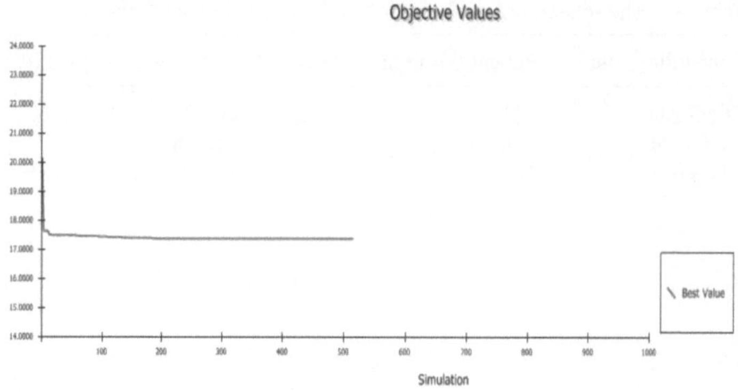


Figure 2. The optimization process of patient waiting time.

Table 5. The scheduling performance of two combined scheduling rules.

| Scheduling rule | Patient throughput (case) | Patient waiting time (day) | Total simulations | Best simulation |
|---|---|---|---|---|
| TP rule | **221** | 18.913 | 593 | 272 |
| WT rule | 148 | **17.388** | 516 | 184 |

Table 6. Comparative result between SPT rule and TP rule.

| Evaluation criterion | SPT rule | TP rule | Comparative result | |
|---|---|---|---|---|
| Patient throughput (case) | 206 | **221** | Adding 14 cases | Optimizing 7.28% |
| Patient waiting time (day) | 18.976 | **18.913** | Reducing 0.063 days | Optimizing 0.33% |

Table 7. Comparative result between SPT rule and WT rule.

| Evaluation criterion | SPT rule | WT rule | Comparative result | |
|---|---|---|---|---|
| Patient throughput (case) | 206 | 148 | Reducing 58 cases | Deteriorating 28.16% |
| Patient waiting time (day) | 18.976 | **17.388** | Reducing 1.588 days | Optimizing 8.37% |

where $C_{wt}$ represents the value of the combined scheduling rule with the objective of minimizing patient waiting time, $C_{fcfs}$ represents the value of FCFS rule, $C_{spt}$ represents the value of SPT rule, $C_{cr}$ represents the value of CR rule.

3. The performance comparison

In a summary, the scheduling performance of two combined scheduling rules is shown in Table 5, which indicates that each combined scheduling rule has its superiority. For instance TP performs well at maximizing patient throughput and WT rule does well in minimizing patient waiting time.

SPT rule which has the best scheduling performance among three simple rules is selected to compare with two combined scheduling rules. Patient throughput and patient waiting time are regarded as evaluation indexes. Table 6 describes comparative result between SPT rule and TP rule which indicates that the scheduling performance of TP rule is better than SPT rule in both patient throughput and patient waiting time. Table 7 shows comparative result between SPT rule and WT rule, which points that the scheduling performance of WT rule is better than SPT rule in patient waiting time.

Hence, if a hospital mainly aims at high-end patients and its managers put emphasis on patient satisfaction, reducing patient waiting time would become very important for them. In this circumstance, WT rule may be a good choice for managers to schedule patients. If a hospital mainly faces ordinary patients, hospital managers have to take account of patient throughput. In this situation, TP should be the best choice for managers to sequence patients.

## 5 CONCLUSION

In this study, heuristic optimization approach is applied to optimize operation scheduling. Firstly, a combined scheduling rule is proposed to optimize operation scheduling performance. This combined scheduling rule is composed of three simple scheduling rules by an initial weight; Secondly, a simulation model about operation scheduling system is built. Plenty of complex factors are considered in this model; Thirdly, based on two optimization objectives, tabu search algorithm is used to search for the optimal weight of simple rules in a combined scheduling rule. Finally, the performance of the combined scheduling rule compares with performances of three simple rules, which indicates that the combined scheduling rule is more valid and efficient to sequence patient. However, this paper only focuses on operation scheduling of elective patients, while it doesn't take non-elective (urgency or emergency) patients into account. And the kinds of simple scheduling rules contained in a combined scheduling rule are less. Only three frequently used simple rules are selected to compose a combined scheduling rule.

Thus, future research plan in operation scheduling optimization will consider non-elective patients in surgical center and establish a rule base which contains different kinds of scheduling rules.

## ACKNOWLEDGMENTS

The work described in this paper was substantially supported by Research Grants from National Natural Science Foundation of China (71271122).

## REFERENCES

B. Denton, J. Viapiano, A. Vogl. 2007. Optimization of surgery sequencing and scheduling decisions under uncertainty. Health Care Management Science 10 (1): 13–24.

B. Roland, C. Di Martinelly, F. Riane. 2006. Operating theatre optimization: A resource-constrained based solving approach. International Conference on Service Systems and Service Management. 1: 443–448.

Brecht Cardoen, Erik Demeulemeester, JeroenBeliën. 2010. Operating room planning and scheduling: A literature review. European Journal of Operational Research 201 (3): 921–932.

Denton B, Viapiano J, Vogl A. 2007. Optimization of surgery sequencing and scheduling decisions under uncertainty. Health Care Manage Science 10 (1): 13–24.

E. Hans, G. Wullink, M. Van Houdenhoven, et al. 2008. Robust surgery loading. European Journal of Operational Research 185 (3): 1038–1050.

Gordon T, Paul S, Lyles A, et al. 1988. Surgical unit time utilization review: resource utilization and management implications. Journal of Medical Systems 12 (3): 169–179.

Healthcare Financial Management Association. 2003. Achieving operating room efficiency through process integration. Health Care Financial Management 57 (3): suppl 1-7 following 112.

J. Beliën & E. Demeulemeester. 2007. Building cyclic master surgery schedules with leveled resulting bed occupancy. European Journal of Operational Research 176 (2): 1185–1204.

J.T. Blake, F. Dexter, J. Donald. 2002. Operating room manager's use of integer programming for assigning block time to surgical groups: A case study. Anesthesia and Analgesia 94 (1): 143–148.

Mohamed A. Ahmed & Talal M. Alkhamis, 2009. Simulation optimization for an emergency department healthcare unit in Kuwait. European Journal of Operational Research 198 (3): 936–942.
S. S. Panwalkar, WafikIskander. 1977. A survey of scheduling rules. Operations Research 25(1): 45–61.
Thomas R. Rohleder, Peter Lewkonia, Diane P. Bischak, et al. 2011. Using simulation modeling to improve patient flow at an outpatient orthopedic clinic. Health Care Management Science 14(2): 135–145.
V.N. Hsu, R. de Matta, C.-Y. Lee. 2003. Scheduling patients in an ambulatory surgical center. Naval Research Logistics. 50 (3): 218–238.

*Bridging Research and Good Practices towards Patient Welfare – Shih & Liang (Eds)*
*© 2015 Taylor & Francis Group, London, ISBN 978-1-138-02716-9*

# Finding significant factors for readiness of ISO 9001 implementation in a health care system

S.I. Chang
*Industrial and Manufacturing Systems Engineering, Kansas State University, Durland Hall, Manhattan*

K. Douglas-Mankin
*Vice-President, MentorNet, Stonegate Drive, Wellington*

L.A. Ang
*Clinical Quality Analyst, United Healthcare Community and State Hawaii*

B. Tavakkol
*Industrial and Systems Engineering Department, Rutgers, The State University of New Jersey, Piscataway*

ABSTRACT: Patient safety and quality improvement have been a driving force for many health care facilities to implement ISO 9001 in recent years. This study describes an effort to capture the ISO 9001-rollout experience of Sterile Processing Services (SPS) departments in a few pilot sites in the Veterans Health Administration (VHA) system. The objective of this work was to gauge the departments' readiness to implement ISO 9001. Possible factors were identified from the literature related to ISO 9001 implementation in healthcare. A series of four questionnaires for the Nurse Executive, and the Chief, Supervisor, and Staff (or Employee) of the SPS were developed and distributed to four sites in VHA. Inspired by Cameron and Quinn's competing values framework for diagnosing organizational culture, questionnaires were designed to contrast answers from management and staff. Spider Diagrams and statistical analysis were used to find emerging trends and factors.

## 1 INTRODUCTION

Organizational reliability is becoming a high priority for top VHA leadership. While efforts for continuous improvement have shown results, the maximum impact can only be achieved through a sustainable organization. The facility resources needed to sustain improvement efforts may be similar to those required for the improvements in the first place. However, they may be different. As ISO 9001 (American Society for Quality, 2000a, 2000b, 2000c) was rolled out to various Sterile Processing Services (SPS) departments, this study aimed to identify factors that help SPS (or other VHA organizations) achieve easier rollouts and to develop recommendations for readiness of rollouts. It provided: 1) an understanding of how successful implementations can serve to provide tools for sites that were not as effective; and 2) information about what to expect as the ISO implementation continues in other departments within Medical Centers.

The original ISO 9001 aims for manufacturing entities with operations in design/development, production, installation and servicing. The main assessment activities can be divided into internal audits and third-party external audits. Once an organization is certified, random periodic external audits will take place to maintain certification. This framework can be easily adopted into healthcare organizations such as VHA Medical Centers. However, VHA system is not seeking certification. The goal instead focuses on the use of ISO 9001 framework to streamline documentation and operation procedures. It is crucial that each facility within the VHA system is ready for this new paradigm change.

ISO 9001 certification has been proven to initiate and sustain lasting changes and quality improvements in a wide spectrum of industries. In the health care industry, it gives healthcare professionals a framework and methodology for monitoring and measuring conformity; provides accountability within the system; and ensures that the most important functions are carried out. ISO 9001 also starts the dialogue among departments and results in an increase in organizational knowledge. This increased information exchange brings a better appreciation of the healthcare system and can build relationships between departments. These changes will impact the organizational culture. The improvements made through ISO 9001 implementations often results in significant cost savings. (Levett and Burney, 2011)

VHA had decided to implement ISO 9001 in its Sterile Processing Services (SPS) departments first. The rollouts would be extended to the other parts of the VHA system in a later time. The objective of this research was to identify critical factors affecting the readiness of ISO rollouts in SPS's. In collaboration with the Office of Clinical Consultation & Compliance (OCCC) and Midwest Mountain Veterans Engineering Research Center (MWM VERC), specific characteristics of ISO 9001 implementation were distilled out that can be measured by an assessment. The VHA shareholders contributed to identifying VHA specific characteristics, such as those related to staff training in Lean or Analytics, which might also impact the organization's ability to function in a highly reliable manner using ISO 9001.

The first task was to identify critical factors known as reliability competency items that may affect the readiness for ISO 9001 implementation at VHA medical centers. The literature related to ISO 9001 implementation in healthcare was reviewed (Carson, 2003; Levett and Burney, 2011). Most work in the literature focuses on the measurement of outcomes after the implementation. In fact, no work had been found directly related to the assessment of readiness for ISO 9001 implementation. Therefore, we have drawn ideas from various literature related to health care organizational change and quality of care models (Donabedian, 1978; Gittel, 2009; Lukas, *et al.*, 2007; Hutton, 1994; Massaki, 1986; Stitt, 1990). Some of the literature was included in the VA TAMMCS Improvement Framework Guidebook (Department of Veteran Affairs, 2011).

## 2 REVIEW OF THE LITERATURE

### 2.1 *Reliability competency item list*

Reliability competency items are a set of characteristics that can be used to assess a SPS's readiness for ISO 9001 implementation. Under the framework of continuous improvement, such as Shewhart's PDCA (Plan, Do Check, Act) cycle, Six Sigma DMAIC (Define, Measure, Analyze, Improve, and Control) procedure, and VA-TAMMCS framework (Department of Veteran Affairs, 2011), the proposed list of characteristics measured a SPS's capability in planning, carrying out the change, studying and analyzing the results obtained, and the sustainment of this monitoring and improvement process through time. The generation of the reliability competency item list was based on the practical situations specifically for VHA medical facilities. First, a high-level hierarchy was identified as the critical category that would influence ISO readiness. The categories and the definitions used to guide the classification of each reliability competency item are as follows:

- Awareness – The Individual's Familiarity with ISO and Plans for Implementation.
- Culture – The organizational aspects that depend on the management and personnel within the SPS.
- Organizational Structure – The aspects that are independent of the managers or personnel within the SPS.
- Training – The processes for increasing the employees' skill sets.
- Policies and Procedures – The accepted methods for execution of work tasks within the SPS.

The identification of the top hierarchy category then dictated the second layer items as governed by a modified Donabedian's model (Donabedian, 1988). In assessing the quality of care, Donabedian's paradigm is based on a three-component approach – structures, process, and outcomes. Since the task at hand did not have the opportunity to obtain any outcomes before the ISO 9001 implementation, the focus was on the first two components. By structure, we mean the attributes of the settings in which a VHA facility (specifically a SPS) was ready for ISO 9001 implementation, including organizational structure, human resources, complexity of the facility, and policies/procedures. On the other hand, the process refers to the execution of healthcare tasks, awareness of the tasks at hand, training of nurses/technicians, and culture of the facility.

Lukas, Holmes, Restuccia, Shwartz and Charns (2007) propose another important work related to the study of transformational change in health care systems. Specially, an organizational model under this framework contains five critical elements or key drivers: (1) impetus to transform; (2) leadership commitment to quality; (3) improvement initiatives that actively engage staff in meaningful problem solving; (4) alignment to achieve consistency of organization-wide goals with resource allocation and actions at all levels of the organization; and (5) integration to bridge traditional intra-organization boundaries between individual components. Not only are these key drivers crucial to a health care organization in term of sustainable transformation, some of the interactions and iterations among these factors are significant as well. Note that VA-MAMMC (Department of Veteran Affairs, 2011) adopts this paradigm for elements of creating a culture of continuous improvement for all VHA medical facilities.

Before describing the category and the reliability competency items in detail, it is crucial to recognize that the goal of this list aimed to assess readiness rather than the outcomes of ISO 9001 implementation. An analogy is a farmer preparing for sowing before a planting season. The assessment of readiness in this case should focus on factors such as the ground condition, availability and quality of seeds, farmer's skill sets and workload, and farming style (corporate, co-op, or private) rather than on the yield (outcome) for the season. This assessment is very different from that of how much yield that this farmer will gain because the planting season has not arrived yet. In other words, the key assessment task is on readiness, not on outcome!

Table 1 contains the list of reliability competency items and their categories. These items were arranged in order so that the most important items are listed on top. Under the category of awareness, two items were identified as important. Item A1 is anticipation on ISO. This item measures employees' willingness, concern, and articulation of possible job changes due to ISO 9001 implementation while item A2 (Knowledge of ISO) measures the employees' initial understanding of ISO before an official ISO 9001 workshop is given. Regarding to the culture category, there are six items. Both items C1 (Leadership Style) and C2 (Organization Culture) are defined by Cameron (Cameron and Quinn, 2011). Cameron's survey tools were utilized to address these two items. Details of Cameron's competing Values framework are summarized in the next section. Item C3 is about organizational stability which can be measured by the turnover rates and the average number of years existing workers have held their current jobs. Items C4 and C5 are related to communication. The internal communication is related to the easiness and willingness of exchanging information within a department while the external communication is about inter-departmental exchange of information. Both C4 and C5 are related to Lukas et. al. (2007) model's key driver #5. Gittell (2009) also identifies communication as one of the crucial factors for high-performance health care organizations. Finally C6 is about empowerment which is the third key driver in the Lukas *et al.* (2007) model.

Under the category of organizational structure, five items listed include: O1 (Operational Complexity), O2 (Whether the SPS is Fully Staffed), O3 (Workload of Nurses and Technicians), O4 (SPS's Layout), and O5 (the Number of Initiatives the SPS is Currently Undertaking). This category is directly related to organizational structure within Donabedian's model. The category training includes four items: T1 (Training of Equipment), T2 (Training of Analytics), T3 (Impact of Training), and T4 (Tech/Computer Savviness). VHA employees have been through various analytics trainings. Item T2 is a measure of how much training SPS nurses and technicians have gone through while item T3 measures whether such training has translated into various process

Table 1. Categories and Reliability Competency Items for Readiness of ISO 9001 Implementation.

| Item No. | Reliability Competency Items | Category |
|---|---|---|
| A1 | Anticipation on ISO | Awareness |
| A2 | Knowledge on ISO | |
| C1 | Leadership style (Cameron) | Culture |
| C2 | Organizational culture (Cameron) | |
| C3 | Organizational stability | |
| C4 | Internal Communication | |
| C5 | External Communication | |
| C6 | Empowerment | |
| O1 | Operational complexity | Organizational Structure |
| O2 | Fully staffed | |
| O3 | Workload | |
| O4 | Facility layout | |
| O5 | Current initiatives | |
| T1 | Training of equipment | Training |
| T2 | Training of analytics | |
| T3 | Impact of training | |
| T4 | Tech Savviness | |
| P1 | Applicability of Policies & Procedures (P&P) | Policies and Procedures |
| P2 | Effectiveness of P&P | |
| P3 | P&P Updates | |

improvement projects. VA TAMMC (Department of Veteran Affairs, 2011) provides a very comprehensive list of non-statistical tools which may be available in yellow belt training. The green belt training may include additional statistical tools. Finally, the category policies and procedures (P&P) includes four items: P1 (applicability of P&P), P2 (effectiveness of P&P), and P3 (P&P updates).

The VHA All Employee Survey (AES), a survey that is distributed annually to all employees in the VHA system, dedicates one section directly related to P&P which is closely tied to ISO 9001 preparation. One of the most important steps for ISO 9001 implementation is the preparation of the quality manual that encompasses all processes, policies, and procedures. An organization with proper documentation of P&P and documentation control would definitely find ISO 9001 less intrusive in terms of possible changes in employees' daily work routines. On the other hand, employees of an organization with loose work standards and procedures will find the ISO 9001 implementation intrusive. The next section will share more detailed information on Cameron's framework which was used for comparing aspects of organizational culture.

### 2.2 *Cameron's organizational culture assessment instrument*

Cameron and Quinn (2011) suggest that in order for any systemic change to be effective and sustained it is critical for the organizational culture to change. They give historic examples of Total Quality Management Initiatives, Quality Circles, and Reengineering efforts, and his assertion was that the vast majority of these initiatives failed because the culture did not change. By utilizing the OCAI (to assess the organizational culture) and the MSAI (to assess the manager's skills related to cultural change), we were able to see if either of these characteristics align with the ease of implementation and the ability to sustain the change. OCAI stands for Organizational Culture Assessment Instrument while MSAI means Management Skills Assessment Instrument Specifically, MSAI can be used to assess C1 (Leadership Style) while OCAI can be used for C2 (Organizational Culture). Cameron's organization culture survey includes the following categories:

dominant characteristics, organizational leadership, and management of employees, organizational glue, strategic emphases and criteria of success.

If there is a pattern in the differences, Cameron and Quinn (2011) also propose a process for creating changes to organizational culture which should improve the ability to implement and sustain the ISO 9001 initiatives. Given the nature of the work done by SPSs, the expectation is that their culture would have a significant level of Cameron's hierarchy attribute, but it is the relative levels of clan, adhocracy, and market attributes in relation to the hierarchy attribute that will provide important insights.

## 3 LEARNING FROM THE PILOT SITES

Based on the detailed competency items listed in Table 1, we designed a series of four questionnaires for the Nurse Executive and the Chief, Supervisor, staff (or employees) of the SPS. The Chief of a SPS reports directly to the Nurse Executive while the Supervisors (including the Vice-Chief) reports to the Chief. Most questions in these questionnaires are the same or similar, but a few questions were tailored to specific positions. The questionnaire designs were first reviewed and vetted by MWM VERC and OCCC of VHA. This was followed by a pilot test before full implementation.

### 3.1 *Methods*

#### 3.1.1 *Data collection strategy*

Two methods were used for data collection – on-site interviews and teleconferencing. We physically visited two enterprise rollout sites selected to maximize the difference in medical center complexity, and conducted phone interviews with the remaining sites. Thus, physical visits were to Site 1 and Site 2.

With respect to interviews, they were voluntary, anonymous, and confidential; only the individual's position, not name, was recorded. When utilizing interviews as a data gathering technique, it is beneficial to allow individuals to review the transcript of their interview to insure accuracy. Each interviewee was provided with a unique number (selected by them) to enable distribution of the interview transcripts. The interviews were conducted on the first day of the visit, transcripts were prepared as soon as possible, and copies were distributed back to the interviewees utilizing the union representatives to insure confidentiality. This practice allowed each person to verify the accuracy of the data gathered during the interview. The local union representative was provided the numbered and sealed copies of the transcripts to be distributed to SPS staff who remembered their number. The communication accompanying the transcript specified that if changes were not provided by a specified date, the responses would be assumed accurately captured (and remaining transcripts can be destroyed by the union representative). The interviewee was provided a stamped self-addressed envelope to mail the transcript back as well as a fax number. This process was explained in the opening remarks during the interview.

#### 3.1.2 *Summary of numerical data gleaned from interviews*

The process of gleaning numerical results from the interview transcripts involved developing consensus among those who conducted either the site-visit interviews or phone interviews. After the interviews, answers were transcribed and returned to interview participants for validation. The answers were organized by positions and questions. For example, answers to the first question for the Nurse Executive were listed for Site 1, Site 2, Site 3, and Site 4 and were clustered in one group for scoring. A Microsoft Access Database was developed for managing and retrieving the answers.

The scoring scale used is from 1 to 5 where 1 is the worst score (indicating low readiness) and 5 (indicating high readiness). At least two interviewers were involved in each interview. Note that a blank cell represents either a missing data field or a field not to be answered by design. For example, the Nurse Executives were not asked the questions related organizational structure. Individuals were also given the option not to answer particular questions.

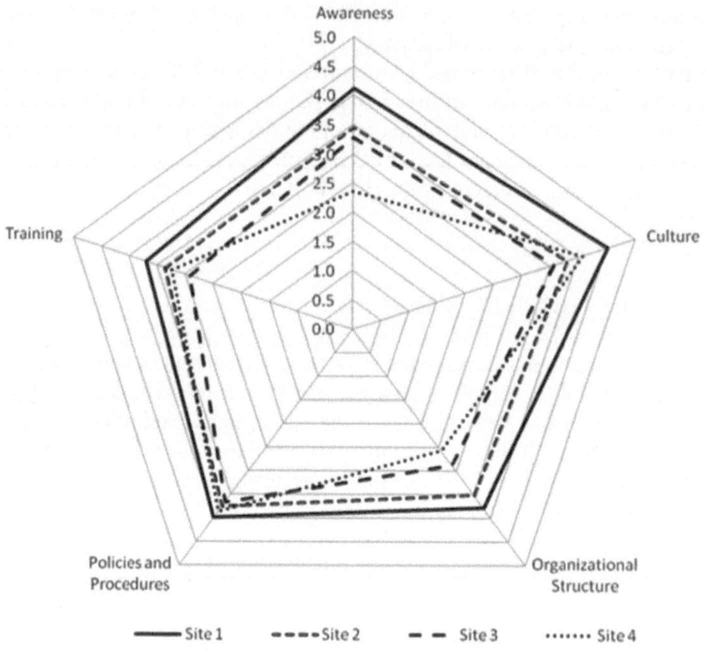


Figure 1. Spider Diagram of the Composite Score of Readiness Categories of Pilot Sites.

After the qualitative interview data was collected, each question was scored according to the scoring scale described above. With the readiness for ISO implementation in mind, a higher score (toward 5) was assigned to a facility based on answers indicating a higher level of readiness. A lower score (toward 1) reflects lower readiness. Some competency items had multiple questions associated with them. A composite score was then averaged based on the individual scores of these questions. The database created earlier for storing answers was used to record all scores as well. Various queries were developed to place all answers from all pilot sites for a specific question together. This practice prevents scoring bias among different sites because the scores were generated all at once for all sites. Therefore, all pilot sites were scored together for each question.

## 3.2 *Findings*

### 3.2.1 *Analyses of the numerical data gathered from the interviews*

The spider diagram shown in Figure 1 provides an overall comparison of four sites in terms of the readiness categories. Each vertex represents one category related to readiness for ISO implementation. The scoring scale is from 1 to 5 with 5 being the most desirable score. In other words, the site with the pentagon closest to all vertexes is the most desirable. Note that SPS averages were calculated from the average of the scores under each reliability competency item from the Chief, Supervisors, and Technicians only and do not include the scores from Nurse Executives. The reason is that the Nurse Executives were often very distant from the SPS and were not involved or familiar with the day-to-day operations of the department. Finally, sample sizes for each position were not considered in the average calculation because the ISO 9001 implementation will rely heavily on management. This average calculation will give more weight to the scores from management (i.e., Chief and Supervisor). The composite score is the average of the SPS averages of the reliability competency items under the same category. For example, the composite score of Awareness for Site 1 is 4.11. This is the average of the A1 (anticipation on ISO) and A2 (knowledge on ISO). A spider diagram based on the composite score was generated and is shown in Figure 1.

Table 2. Factors and Hypotheses that May Lead to Easiness of ISO Implementation.

| Factor | Name | Hypothesis | Category |
|---|---|---|---|
| 1 | Expectation | The more anticipation and expectation of potential benefits of ISO, the easier the ISO implementation. | Awareness |
| 2 | Support | The more support from the other departments, the easier the ISO rollout. | Culture |
| 3 | Cultural Tendency | The cultural perspective of the SPS Chief as compared to his/her subordinates facilitates ISO rollout. | Culture |
| 4 | Staffing | Having low turnover and being fully staffed leads to easier ISO implementation. | Organizational Structure |
| 5 | Project Coordinator Position | An SPS with existing positions related to quality leads to easier ISO implementation. | Organizational Structure |
| 6 | Computer Literacy | The better the computer skills are across the board the easier the ISO implementation. | Training |
| 7 | Task Variation | The more consistently the techs currently follow an SOP, the easier the ISO implementation. | Policies & Procedures |

Site 1 has the most desirable outcome among four sites interviewed. Site 2 follows as the second place but does not dominate the other two sites in the Culture or Training. The most significant category is Awareness because it has the largest spread in composite scores. The second most significant category is Organizational Structure followed by Culture and Training. There is no significant difference among pilot sites under the category Policies and Procedures.

Based on both quantitative and qualitative analyses of the interview data, the following factors emerge as important for SPS in term of readiness for ISO implementations. Table 2 summarizes the factors, the corresponding hypotheses and categories that each factor/hypotheses fall under. The identification of each factor will be discussed in detail in the following sections.

### 3.2.2 *Factor 1. Expectations*

Hypothesis: The more anticipation and expectation of potential benefits of ISO, the easier the ISO implementation. (Awareness)

This factor expectation can also be called anticipation. Site 1 was well prepared for ISO roll out because its Chief, Assistant Chief, supervisors, leads, and techs were knowledgeable about what ISO was and what it could do for them. During the site visit to Site 1, the quality related projects being pursued [e.g., PROQUIS (Process for Reliability Integrated System) software related to ISO] were illustrated on the white broad in their conference room. Most people interviewed at Site 1 were throughout the department's hierarchy were aware of the roadblocks and benefits of ISO which results in much higher scores in item A1 (Anticipation on ISO) than those from the other sites. This was in contrast to interviews at other sites where interviewees would response to the first question with "What is ISO?" Figure 1 portrays site scores for the Awareness category.

### 3.2.3 *Factor 2. Support*

Hypothesis: The more support from the other departments, the easier the ISO rollout. (Culture)

This factor was identified from responses primarily from SPS leadership. In the original interview design, support from departments other than SPS was not considered. It became apparent that the ISO rollout would not be smooth without the buy-in and cooperation from the other department related to SPS. This issue came up most in interviews that were close in time to the initial training. The sites were experiencing the first signs of lack of support in that their stakeholder departments were not willing to commit the time to participate in training.

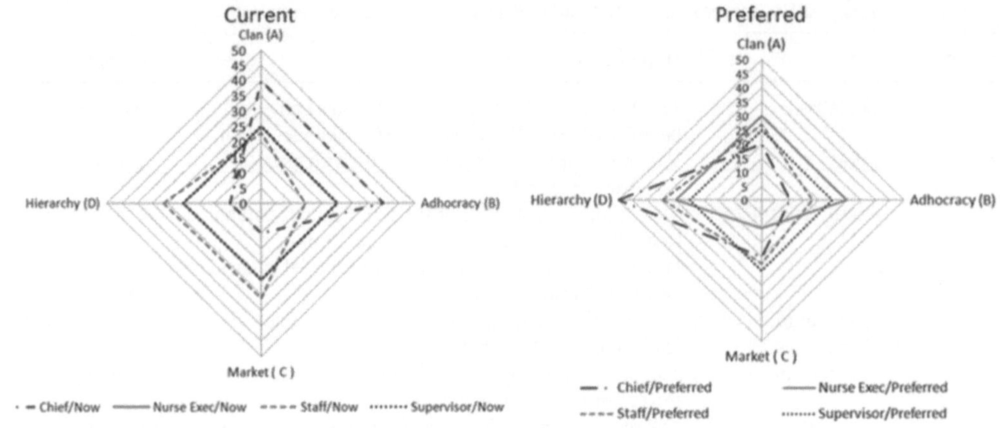


Figure 2. Site 1 OCAI Results.

### 3.2.4 *Factor 3. Cultural tendency*

Hypothesis: The cultural perspective of the SPS Chief as compared to his/her subordinates facilitates ISO rollout. (Culture)

Cameron and Quinn's Organizational Culture Assessment Instrument (OCAI) (2011) was used to identify this hypothesis. Cameron and Quinn's Competing Values Framework highlights four types of organizational culture: clan, adhocracy, market and hierarchy. Clan cultures emphasize empowerment, team building, employee involvement, human resource development, and open communication. In the clan culture, the leaders are facilitators, mentors, and team builders. An adhocracy culture focuses on the strategies of surprise and delight, creating new standards, anticipating needs, continuous improvement, and finding creative solutions. In an adhocracy culture the leaders are innovators, entrepreneurs, and visionaries. A market culture has the following traits: measuring customer preferences, improving productivity, creating external partnerships, enhancing competitiveness, and involving customers and suppliers. Leaders in a market culture are hard drivers, competitors, and producers. Finally, the hierarchy culture is about error detection, measurement, process control, systematic problem solving, and the use of set quality tools (e.g., fishbone diagrams, Pareto diagrams). Leaders in a hierarchy culture will be coordinators, monitors, and organizers.

According to the Cameron's OCAI data from Site 1 and Site 2, the staff in SPS see describe their culture as emphasizing hierarchy and market culture. The hierarchy result was expected as the VHA is a government organization and is very hierarchical in nature. However, what is interesting is the perspective of the SPS Chief at Site 1. The SPS Chief classified their current dominant culture as clan and adhocracy. Remember, clan would imply that the leader is a communicator, and adhocracy would imply that the leader is entrepreneurial. However, it is interesting to note that the Site 1 SPS Chief has a preferred culture of hierarchy which has leaders characterized as coordinators, monitors, and organizers. These latter characteristics are well aligned with ISO. Graphs of Site 1's OCAI results are provided in Figure 2.

### 3.2.5 *Factor 4. Staffing*

Hypothesis: Having low turnover and being fully staffed leads to easier ISO implementation. (Organizational Structure)

Stability in an organization is one of the most important factors in ISO roll out. The SPS at Site 1 was the largest with 56 full-time equivalents (FTE) while the one in Site 4 was the smallest with only 6 FTE. The SPS sites at site 3 and site 4 have 28 and 26 FTE, respectively. Figure 3 shows the comparison of the composite scores of all pilot sites in term of organizational structure. In particular, item O2 measures the degree to which an SPS is fully staff. An SPS with frequent

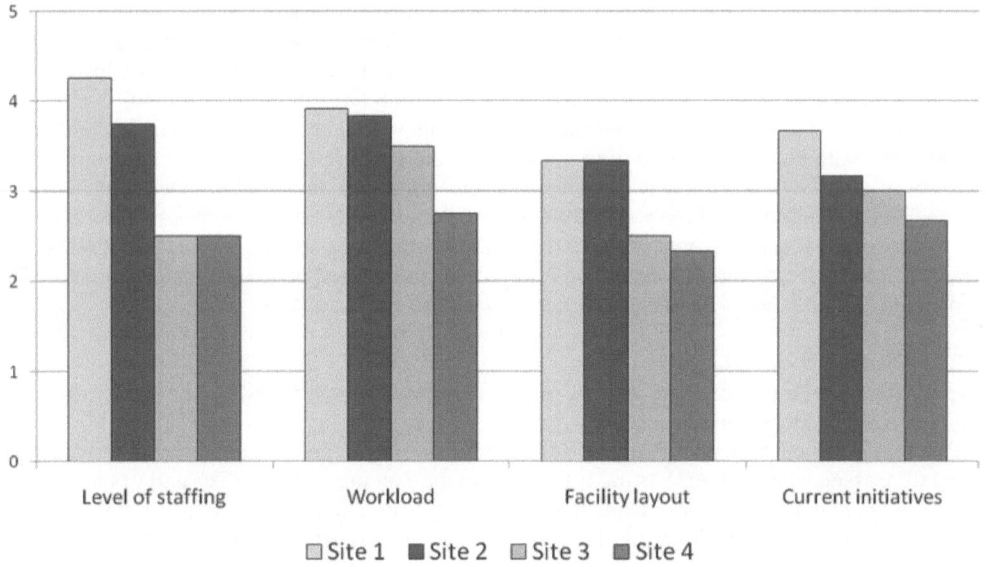


Figure 3. Site Composite Scores Related to the Organizational Structure Category.

turnover has a difficult time training new techs and meeting service demands at the same time. A new initiative such as ISO only makes this situation worse. Note that Site 1 scores the highest in term of O2 (less turnover or fully staffed) while Site 3 and Site 4 score lower due to either frequent turnover or not being fully staffed. It is interesting to note that several interviewees described the long-term benefits of ISO for dealing with turnover challenges. For example, if standard operating procedures (SOPs) are consistent across the VHA, an employee that moves from one medical center to another can be immediately productive. The other example given was that ISO would establish more consistency and documentation which would facilitate training new employees.

### 3.2.6 *Factor 5. Project coordinator position*

Hypothesis: An SPS with existing positions related to quality leads to easier ISO implementation. (Organizational Structure)

Figure 3 does not provide the scores for O1 which measure operational complexity. One way to measure organizational complexity is simply by comparing the number of FTEs in the SPS. The current staffing levels for Site 1, Site 2, Site 3, and Site 4 are 56, 25, 23, and 6 FTEs, respectively. It is not clear how organizational size will affect ISO rollout. A small SPS will face the difficulty of balancing existing tasks and ISO implementation but it is easier to communicate within a small SPS and provide training while a larger SPS may have more resources devoted to a special project such as ISO implementation. Site 2 had a Program Specialist Quality Assurance who is in charge of organizing the current SOP documentation and identifying changes needed (e.g., training, policy changes). He was very organized and disseminated changes to absent employees through e-mails and/or in paper form. Therefore, Site 2 was in a good position for ISO 9001 implementation. In a small SPS such as Site 4, the responsibility might fall on the shoulders of either the Chief or Assistant Chief who were already very busy. Item O3 (workload) scores also confirm this point.

### 3.2.7 *Factor 6. Computer literacy*

Hypothesis: The better the computer skills are across the board the easier the ISO implementation. (Training)

This factor may not be important in some organizations in term of ISO implementation. However, it is of upmost important for the VHA SPSs because the upcoming upgrades to its processing automation software PROQUIS. It was apparent from the interviews that some of the technicians were not very computer savvy. These employees were barely comfortable with email. In many of the interviews, the less computer savvy employees were described as "older" employees. In a couple of interviews, the suggestion was made that education level and socio-economic status were also contributors. It is anticipated that these technicians will have a more difficult time learning the new software that will be used with the ISO roll-out (e.g., PROQUIS). When conducting the first set of interviews at Site 1, it was clear that the employees they had selected for interviews were not a random sample of the population. Rather, they tended to be the more computer savvy of the employees. Thus, a question was added about the computer savviness of their coworkers. The score for T4 (Tech Savviness) was high, but the scores were based on the estimated averages. The wide range of computer skills within the department was really an issue.

### 3.2.8 *Factor 7. Task variation*

Hypothesis: The more consistently the techs currently follow an SOP, the easier the ISO implementation. (Policy and Procedure)

Reliability competency item P2 measures the effectiveness of policies and procedures. Respondents were asked whether there is a lot of variation in the way jobs are done as a part of the measure for P2. Note that a lower score means that this facility can benefit more from ISO implementation. A high score suggests that this facility will have an easier time to follow the spirit of ISO implementation.

## 4 CONCLUSIONS

We targeted four pilot sites. Five categories of factors studied in this project include awareness, culture, organizational structure, training, and policies and procedures. Note that culture is defined as the organizational aspects that depend on the management and personnel within an SPS. Based on these factors for assessing ISO-implementation readiness, several interview tools were developed for Nurse Executives, Chiefs, Assistant Chiefs, Supervisors, and Technicians. Cameron's competing values framework was used to diagnose organizational culture and leadership style of the four pilot SPS sites.

The research team visited the SPS sites at site 1 and site 2 to collect data using the developed interview tools. The other two sites were interviewed via telephone conferencing. Once all data was collected and validated (those who participated in either face-to-face or phone interviews had a chance to review their transcripts and to provide feedback), the research team provided numerical assessment of all factors for each site. The scores range from 1 to 5 with 1 representing the least ready and 5 the most ready in terms of ISO implementation. The spider diagram in Figure 1 summarizes the composite scores. It is clear that the SPS at site 1 dominated in all categories of ISO roll-out readiness. Site 2 ranked second among four pilot sites. Although Site 4 was ranked last in two categories (awareness and organizational structure), it had advantages over the others due to its small size.

Based on the analyses of both qualitative and quantitative data collected from the four pilot sites, the research team recommended seven factors that most significantly affect readiness of ISO rollouts. These factors include: expectation, support, culture tendency, staffing, project coordinator position, computer literacy, and task variation as shown in Table 2.

Although the above mentioned list was developed for SPS, these factors and findings can also be generally applied to VHA departments other than SPS or other health care organizations in terms of the preparation for an ISO implementation. Note that in the finding column there are a couple of places (specifically factors 3 and 5) where SPS is named. It can be easily replaced by any department other than SPS. In addition, the factors and their categories associated with readiness are generic rather than specific to SPSs. The organizational structure of employees and management is also universal for all organizations. The contrast in scores reflects not only in their knowledge of ISO but also reflect the level of communication.

ACKNOWLEDGEMENTS

We acknowledge the following organizations and persons who have made the completion of this research possible: Midwest Mountain Veterans Engineering Resource Center (MVM VERC) of the Veterans Health Administration for providing the funding the research for this paper; Dr. Jennifer Wagner, our project manager from the MVM VERC for guiding us through the VHA system, assisting us in contacting various people in the system and providing valuable feedback and support; Kimberly Moore of the Office of Clinical Consultation and Compliance, Dr. Peter Woodbridge, director of the MWM VERC and Peter Almanoff of the Veterans Health Administration for providing valuable feedback and support.

REFERENCES

American Society for Quality. (2000a). *ANSI/ISO/ASQ Q9000-2000. Quality management systems – fundamentals and vocabulary*. Milwaukee: American Society for Quality.

American Society for Quality. (2000b). *ANSI/ISO/ASQ Q9001-2000. Quality management systems – requirements*. Milwaukee: American Society for Quality.

American Society for Quality. (2000c). *ANSI/ISO/ASQ Q9004-2000. Quality management systems – guidelines for performance improvement*. Milwaukee: American Society for Quality.

Cameron, K., & Quinn, R. (2011). *Diagnosing and changing organizational culture based on the competing values framework* (3rd ed.). San Francisco, CA: Jossey-Bass.

Carson, B. (2003). *ISO 9001:2000 – A new paradigm for healthcare*. Milwaukee: ASQ Quality Press.

Department of Veteran Affairs. (2011). *VA TAMMCS improvement framework guidebook 2011, VA TAMMCS, vision analysis team aim map measure change sustain spread – It's how the VHA spells improvement.* (2nd ed.) VA National Office of Systems Redesign.

Donabedian, A. (1978). The quality of medical care. *Science, 200* (26), 856–865.

Drucker, P. (1967). *The practice of management*. New York: Harper and Row.

Gittel, J. H. (2009). *High performance healthcare – using the power of relationships to achieve quality, efficiency and resilience*. New York: McGraw Hills.

Hutton, D. (1994). *The change agents' handbook – A survival guide for quality improvement champions.* Milwaukee: ASQ Quality Press.

Levett, J., & Burney, R. (2011). *Using ISO 9001 in healthcare: Application for quality systems, performance improvement, clinical integration, and accreditatio*. Milwaukee: ASQ Quality Press.

Lukas, C., Holmes, S., Cohen, A., Restuccia, J., Cramer, I., Shwartz, M., & Charns, M. (2007). Transformational change in healthcare systems: An organizational model. *Health Care Management Review, 32*(4), 309–320.

Masaaki, I. (1986). *Kaizen: The key to Japan's competitive success*. New York: McGraw Hill/Irwin.

Stitt, J. (1990). *Managing for excellence – A systematic and holistic analysis of the process of quality and productivity improvement*. Milwaukee: ASQ Quality Press.

*Bridging Research and Good Practices towards Patient Welfare – Shih & Liang (Eds)*
*© 2015 Taylor & Francis Group, London, ISBN 978-1-138-02716-9*

# Optimizing surgery schedule with PICU nursing constraints

D. Ben-Arieh
*Industrial and Manufacturing Systems Engineering, Kansas State University, Manhattan, KS, USA*

ABSTRACT: Pediatric surgery, an important medical procedure in any children hospital, usually consists of scheduled and unscheduled surgeries. Many times surgeries have to be cancelled due to bottleneck in the patients' flow – specifically an overcrowded PICU (Pediatric Intensive Care Unit). This predominant controllable cause of cancellations is attributed to an unpredictable fluctuation in nursing demand at the pediatric intensive care unit.

This paper presents a new schedule designed based on an estimated patient length of stay and nursing profiles with the objective of minimizing day-by-day nursing level variation at the PICU unit. This variation is the main source of cancellation of surgeries which cause subsequent distress to patients and families as well as loss of revenue. In this work the expected nursing levels for each surgery type were calculated using cumulative distribution functions of patients' length of stay, multiplied by the required nursing profiles, predicted by nursing staff managers.

## 1 INTRODUCTION

The children hospital analyzed performs more than 14,500 surgeries every year utilizing 16 operating rooms. However, since some of the more complex surgeries require the patients to use the Pediatric Intensive Care Unit (PICU) after surgery, the unavailability of PICU beds can lead to cancellation of surgeries. Last year more than 1300 surgeries required the use of a PICU bed, with more than 160 surgeries cancelled due to shortage of those beds. The surgeries that utilize the lion-share of PICU beds include Cardiovascular, ENT, Ortho Spine, Plastics and Neurosurgery.

Currently, the PICU unit utilizes about 28 beds with an on-going effort to expand the unit up to 42 beds. The main constraint on expanding the PICU is nurse staffing shortage. This shortage combined with over-scheduling of the PICU beds lead to the cancellation of surgeries. Besides the monetary loss of a cancelled surgery, there are negative impacts on patient satisfaction and willingness to reschedule at the same hospital, and a risk of deteriorating patient's health. Recently, the hospital has taken more aggressive steps to remedy some of the scheduling problems between the surgical departments and the PICU producing noticeable improvement. However, there is a need for a more uniform process that will allow the surgical departments to operate more optimally considering the entire set of constraints they are facing.

Thus this project aims at developing a more optimal process and schedule of operations combined with an improved nurse staffing at the PICU that will allow higher throughput of patients through the surgical and the PICU departments.

In late 2012, this hospital experienced a sharp increase in surgery cancellations due to inadequate pediatric intensive care unit (PICU) nursing availability. Surgery cancellations are not uncommon within hospitals, and are a large source of revenue-loss and grief to patients and their families. Because PICU nursing staffing is within the control of hospital administrators, the issue was given more attention.

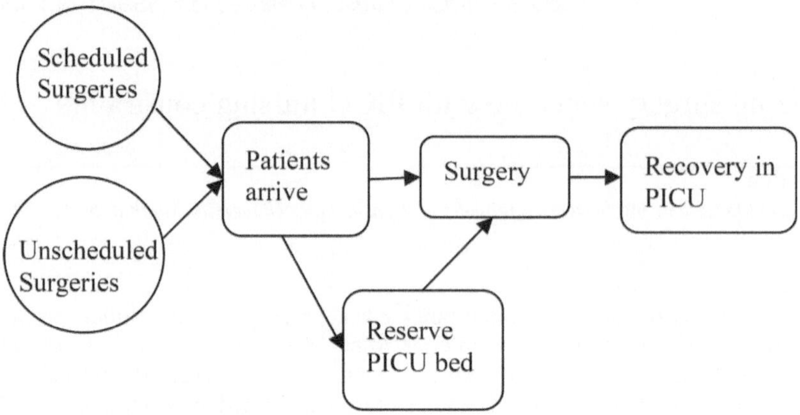


Figure 1. PICU bed reservation process.

Table 1. An example of various nursing profile.

| Surgery Type/Shift # | 1 | 2 | 3 | 4 | 5 | 6 | 7 | 8 | 9 | 10 | 11 |
|---|---|---|---|---|---|---|---|---|---|---|---|
| Cardiovascular Surgery/Cardiac Transplantation (1) | 1.0 | 1.0 | 0.5 | 0.5 | 0.5 | 0.5 | 0.5 | 0.5 | 0.5 | 0.5 | 0.5 |
| Cardiovascular Surgery/Cardiac Transplantation (2) | 1.0 | 1.0 | 0.5 | 0.5 | 0.5 | 0.5 | 0.5 | 0.5 | 0.5 | 0.5 | 0.5 |
| Cardiovascular Surgery/Cardiac Transplantation (3) | 1.0 | 1.0 | 1.0 | 1.0 | 1.0 | 1.0 | 0.5 | 0.5 | 0.5 | 0.5 | 0.5 |
| Cardiovascular Surgery/Cardiac Transplantation (4/5) | 1.0 | 1.0 | 1.0 | 1.0 | 1.0 | 1.0 | 1.0 | 1.0 | 1.0 | 1.0 | 0.5 |
| Other | 1.0 | 1.0 | 1.0 | 1.0 | 1.0 | 1.0 | 0.5 | 0.5 | 0.5 | 0.5 | 0.5 |

Upon analysis of surgery scheduling and PICU censuses leading up to the influx of surgery cancellations, it was realized that the number of nurses required during any given shift was highly volatile, with some shifts requiring significantly more nurses than others. This left some shifts overstaffed, and others understaffed. Nurse staffing for each shift was decided by the Nursing Staff Manager of the PICU, and based on current PICU census, patient surgery types (referred to as the Admitting Service), and nurses reserved to safeguard against the possibility of an inflow of unscheduled admits. The flowchart shown in Figure 1 below illustrates the PICU reservation process.

Specific nursing requirements for the admitting services ranged from a half-nurse per patient (meaning one nurse is assigned to two patients), to two nurses per patient for higher acuity surgeries. Nursing requirements for certain admitting services would often decrease after a patient's condition improved. The nursing requirements including the shift to fewer nurses is referred to as the nursing profile. Table 1 illustrates such nursing profiles for different surgery types showing the nursing demand changes over several shifts. The number under each shift represents the nursing requirement in that shift for each surgery type.

This project utilized two years of PICU patient data, which was filtered down to scheduled admissions alone. This means that scheduling unscheduled surgeries is obviously impractical and the main improvement to the system was due to better handling of scheduled surgeries. The past data shows that the main surgeries that required PICU bed belong to the Cardiovascular type surgeries which had 5 levels of complexity. Each such level in increasing order required more PICU resources. Figure 2 below shows the average load on the PICU due to cardiovascular surgeries.

Moreover, the number of surgeries varied every day of the week, due to surgeon availability, operating room assignment and similar factors. This daily load is presented in Figure 3.

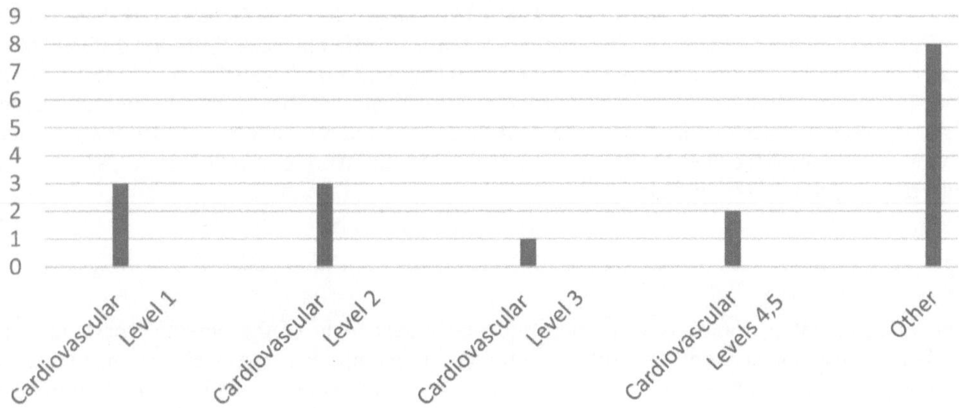


Figure 2. Weekly load on PICU due to various surgery types.

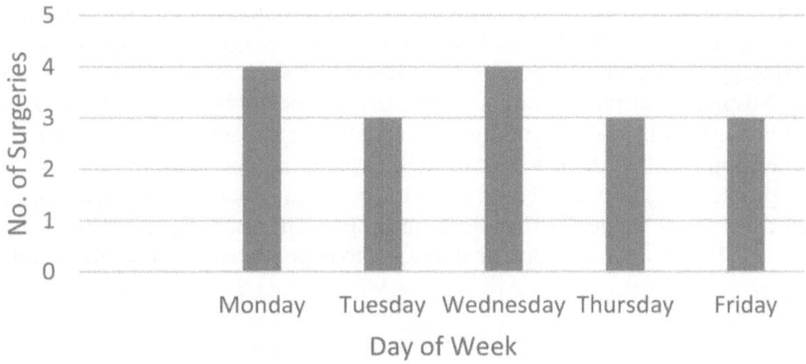


Figure 3. Distribution of surgeries during the week.

## 2 LITERATURE REVIEW

### 2.1 *Nurse workload analysis*

Since the problem on-hand attempts to schedule the surgeries so there is a minimal conflict with the nurses' schedule at the PICU, knowing how many nurses may be required is an important part of this schedule. Assessing the workload on nurses is a well-recognized problem with a variety of solution approaches (Edwardson and Giovannetti, 1994).

The literature reports ten methods used to calculate nurse workload related to acute care wards including emergency departments. According to *The nurse's workload in care situations*, a workload measurement system is necessary in order to staff nurses properly (Nuikka et al. 2001). Many hospitals are experiencing difficulty in retaining nurses because of a heavy workload, while other hospitals are interested in maintaining enough staff (Duffield et al. 2006). Additionally, intensity measures are needed in order to accurately calculate any type of workload for nurses (Meyer 1978). These reasons all explain why a workload measurement system is necessary for acute care department nurses.

The approaches to calculate the nurses' workload can be divided into three categories: based on patient classification, subjective measures and time based measurement approach. Nurse workload analysis based on patient classification, groups the patients based on the nursing resources required and creates a Nursing Intensity Index (Reitz 1985). Comparatively, many methods were based on the subjectivity of nurses to their jobs. The amounts of stress in different work shifts and different tasks relate directly to the workload assessed by each nurse (Nuikka 2001; Reitz 1985a). NASA developed a task load index called NASA-TLX that provides a workload scale based on the weighted average of six ergonomic scales that nurses rate (NASA Task Load Index). Time based load measurement system include the Modified Cooper-Harper scale, Overall Workload (OW) scale, NASA-TLX (Hart & Staveland, 1988), and other.

### 2.2 *Scheduling operating rooms*

Scheduling operating rooms is a complex problem independent of the nursing workload. This problem received much attention in the literature reporting optimization as well as implementation based approaches. The problem of operating room scheduling is recognized as a hard problem to solve usually employing variations of mix-integer programming formulations (for example Jebali et al., 2006).

More applied approaches include simulation as an instrument to improve scheduling of surgeries (Marjamaa et al., 2009; Santibanez et al., 2009). The difficulty in scheduling operating rooms is amplified by the uncertainty involved in which surgery duration is uncertain in addition to patients cancellations (e.g., Denton et al., 2007). Another approach for surgery scheduling involves decision support systems that can establish good solutions for a variety of conditions (e.g. Belien et al., 2009). Many hospitals run the operating rooms using block schedules that are decided by the surgeons and impose additional constraints. An analytical approach for using block scheduling is presented for example in Santibanez et al. (2007). Scheduling operating rooms is recognized as a very difficult problem to solve, especially since the problem may change frequently as operations get cancelled, delayed or change due to patients shifting health situation. Heuristic approaches have gain acceptance in this domain as a practical attempt to cope with this scheduling domain (e.g. Fei et al., 2009).

## 3 SURGERY SCHEDULING MODEL

### 3.1 *General model*

In order to reduce the number of surgeries canceled due to a shortage of PICU nurses, a mixed integer program was developed to find a weekly surgical template that would smooth the expected nurse staffing requirements throughout the week. The model was designed to find the optimal distribution of surgeries throughout the week from a PICU nurse staffing perspective and not to determine the optimal number of surgeries. Therefore the number of surgeries to be performed by each admitting service was an input to the model.

Because some patients stay beyond a week's duration, the template develop must consider the system performance at a steady state. This was accomplished by looking at a single week's worth of surgeries but over a horizon large enough that all of the patients leave the PICU. This was done for each of the weekly 14 shifts and then summed up across the entire horizon for each shift that represented the same weekday. This is illustrated below in Figure 4.

### 3.2 *Notations*

$i$: Weekday shift ranging from 1..14.
$j$: Admitting Service.
$k$: Absolute shift throughout entire model horizon ranging from 1..553.
$l$: Length of stay in PICU in shifts.

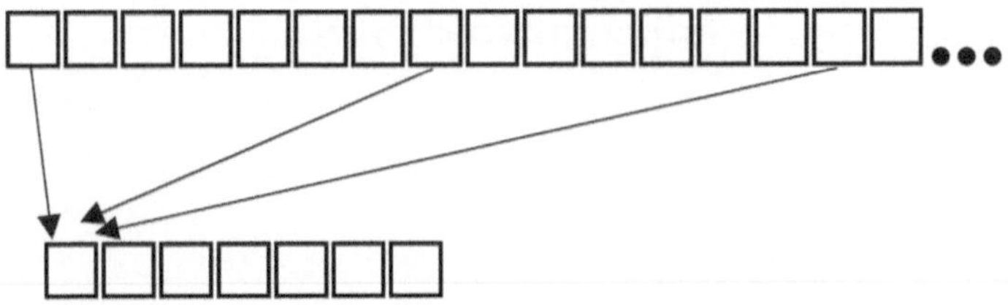

Figure 4. Assessing each shift for it full operational horizon.

### 3.3 *Parameters*

$N_{jl}$: Nursing requirement for a patient from admitting service $j$ after $l$ shifts in the PICU.
$P_{jl}$: Probability that a patient from admitting service $j$ will remain in the PICU after $l$ shifts.
$R_j$: Weekly surgery requirement for admitting service $j$.

### 3.4 *Variables*

$c_{ij}$: Count of surgeries on weekday shift $i$ performed by admitting service $j$.
$n_i$: Number of nurses staffed on weekday shift $i$.
$s_k$: Number of nurses staffed on shift $k$.

### 3.5 *Objective*

The objective (1) was to minimize the difference between the shifts with the maximum and minimum expected nurse staffing requirement. This effectively smoothes the number of PICU nurses staffed throughout the week as a result of scheduled surgeries.

$$Minimize\colon\ \max(n_i) - \min(n_i) \tag{1}$$

### 3.6 *Constraints*

The first constraint (2) that was added to the model requires that each shift be staffed to exactly meet the expected need. The reason that the staffing is required to be exact, and not greater than or equal is motivated by the objective function. If the number of nurses staffed $s_k$ during shift $k$ was allowed to be greater than the expected need then it is obvious that every shift would be staffed at the level of the shift with greatest expected need. This would result in the model being indifferent between the different feasible surgery schedules.

$$\sum_{i+l=k} \sum_{j} c_{ij} * N_{jl} * P_{jl} = s_k \qquad \forall\, k \tag{2}$$

Another constraint (3) required that each admitting service $j$ still performed the target weekly number of surgeries.

$$\sum_{i} c_{ij} = R_j \qquad \forall\, j \tag{3}$$

Constraint (4) sums the PICU nurse staffing throughout the entire horizon to the corresponding weekday shift $i$.

$$\sum_{k \equiv i\ (mod\ 14)} s_k = n_i \quad \forall i \tag{4}$$

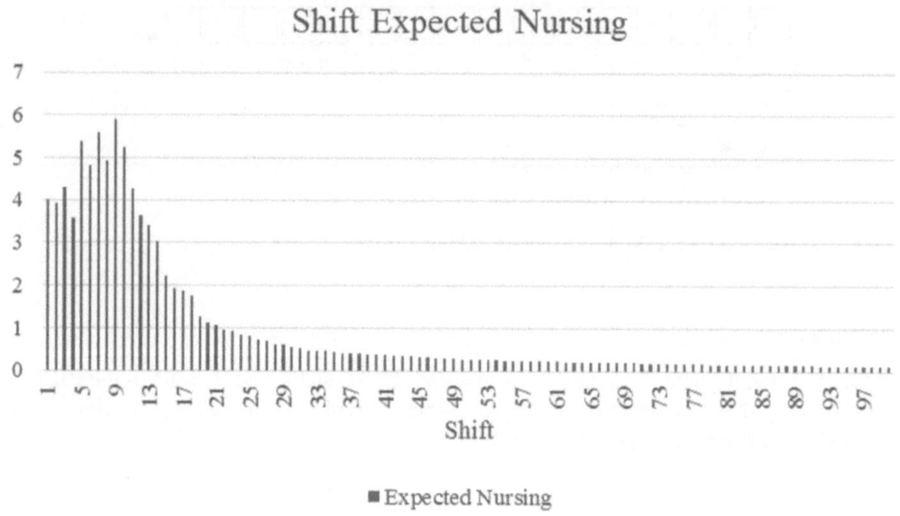


Figure 5. Model solution showing expected number of nurses in each shift.

The final constraints (5) and (6) simply prevent surgeries from being scheduled on a weekend or night shift.

$$c_{ij} \leq 0 \quad \forall\, i,j \;:\; i \equiv 0\ (mod\ 2) \tag{5}$$

$$c_{ij} \leq 0 \quad \forall\, i,j \;:\; i \in \{11,..14\} \tag{6}$$

An example of the total expected nursing level for the first 100 shifts (being on Monday morning) is shown in Figure 5. Note that this is not the expected nursing for the final weekly template, but before the expected nursing is consolidated into expected nursing in the weekly template.

As can be seen in the Figure, the expected number of nurses approaches zero as the shifts approach infinity. This is because it is less likely that patients stay in the PICU for extended periods of time, and therefore it is less likely that a nurse is expected to require scheduling.

The weekly template expected nursing is calculated by summing all Monday morning shifts with subsequent Monday morning shifts, summing all Monday evening shifts with all subsequent Monday evening shifts, and so on. After summing all shifts falling on the same day of the week, and time of day, we are left with expected nursing levels.

The hospital requested no surgeries be scheduled on the weekend or evening shifts. This was fulfilled by setting all $c_{ij}$ equal to zero for all shifts $i$ falling on the weekend and evenings. The model is highly customizable. For example, it can be customized so certain admitting services fall on certain shifts, or that no specific surgeries are scheduled certain shifts.

## 4 RESULTS

The optimized schedule tended to have lower acuity surgeries near the beginning of the week and higher acuity surgeries near the end of the week. Of course, the weekend has lower number of expected nurses due to no new scheduled arrivals in the PICU. Figure 6 shows the weekly expected nursing over a 14-shift weekly schedule, having a variance of approximately 1.22 expected nurses. Again, this represents the nursing load only due to scheduled surgeries. Unscheduled surgeries that take place daily are not mentioned in this paper.

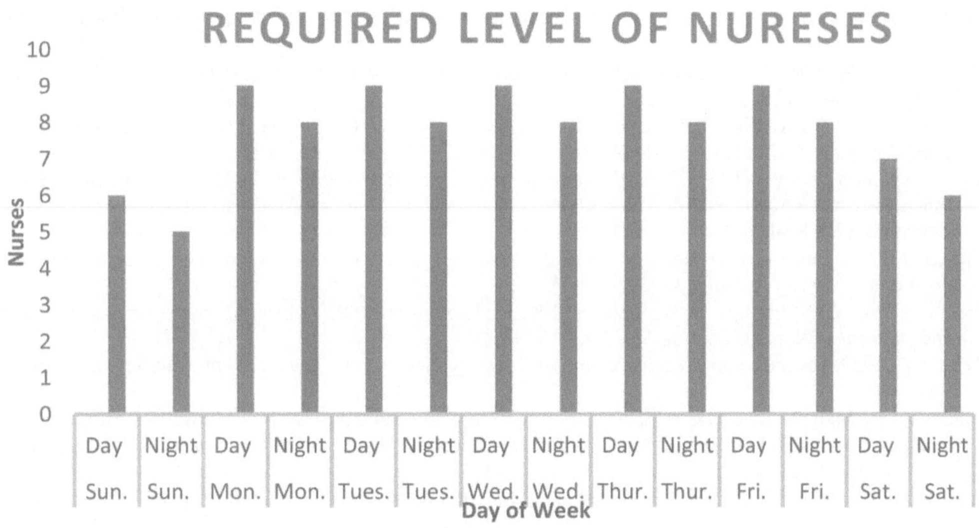


Figure 6. Consolidated weekly template displaying minimized expected nursing variation.

## 5 CONCLUSIONS

Scheduling operating rooms is a complicated task often accompanied by uncertainty, and many constraints such as the commonly used block schedule and surgeons availability. This work looks at scheduling operations with the objective of minimizing the variation of nurse staffing at the downstream PICU unit. The domain considered in this work is a children hospital with a variety of scheduled and unscheduled surgeries. Moreover, the data provided shows that the main surgeries that affect the PICU unit are limited to a few types.

This work uses a mixed integer programming to create a balanced schedule of PICU nurses so that with a proper and predictable staffing level a minimum number of surgeries will be cancelled. The tool was implemented using OPL and solved for any foreseeable horizon given the data from the last two years. The same model can be implemented whenever the data will change enough to justify a change in the model.

We expect the model to be implemented and to contribute to the actual operation of the surgical department.

## REFERENCES

Belien J., Demeulemeester and Cardoen B., 2009, A decision support systems for cyclic master surgery scheduling with multiple objectives, *J. Scheduling*, 12, 147–161.

Denton B., Viapiano J., and Vogl A., 2007, Optimization of surgery sequencing and scheduling decisions under uncertainty, *Health care Management Science*, 10, 13–24.

Duffield, C., M. Roche, and E. Merrick. 2006. Methods of measuring nursing workload in Australia. *Collegian* 13 (1): 16–22.

Edwardson S.R., and Giovannetti P.B., 1994, Nursing workload measurement systems, *Annual Review of Nursing Research*, 95–124.

Fei H., Chu C., and Meskens, 2009, Solving a tactical operating room planning problem by a column generation based heuristic procedure with four criteria, *Ann Oper. Res*, 166, 91–108.

Hart, S. G., and L. E. Staveland. 1988. *Development of NASA-TLX (task load index): Results of empirical and theoretical research.* In P. A. Hancock and N. Meshkati (Eds.) *Human Mental Workload*. Amsterdam: North Holland Press. http://human-factors.arc.nasa.gov/groups/TLX/.

Jebali A., Alouane A. H., and Ladet P., 2006, Operating rooms scheduling, Int. J. Production Economics, 99, 52–62.

Meyer, D. 1978. *GRASP: A patient information and workload management system*. Revised Edition ed. Morganton, NC: MCS.

Marjamaa R.A., Torkki P.M., Hirvensalo E.J., and Kervela O.A., 2009, What is the best workflow for an operating room? A simulation study of five scenarios, *Health care Management Science*, 12, 142–146.

*Nasa task load index (TLX) v.1.0: Paper and pencil package.* Moffett Field, CA: Human Performance Research Group, NASA Ames Research Center, http://humansystems.arc.nasa.gov/groups/TLX/downloads/TLX Scale.pdf.

Nuikka, ML, M. Paunonen, O. Hanninen, and E. Lansimies. 2001. The nurse's workload in care situations. *Journal of Advanced Nursing*, 33 (3): 406–15.

Prescott, P. A., C. Y. Phillips, J. W. Ryan, and K. O. Thompson. 1991. Changing how nurses spend their time. *Journal of Nursing Scholarship*, 23 (1): 23–8.

Reitz, J. A. 1985. Toward a comprehensive nursing intensity index: Part I, development. *Nursing Management* 16 (8): 21.

Santibanez P., Begen M., and Atkins D., 2007, Surgical block scheduling in a system of hospitals: an application to resource and wait list management in a British Columbia health authority, *Health Care Management Science*, 10, 269–282.

Santibanez P., Chow V.S., French J., Puterman M.L. and Tyldesley, 2009, Reducing patient wait times and improving resource utilization at British Columbia Cancer Agency's ambulatory care unit through simulation, *Health care Management Science*, 12, 392–407.

Verwey W.B., and Velrman, H.A., 1996, Detecting short periods of elevated workload: A comparison of nine workload assessment techniques, *J. of Experimental Psychology: Applied*, vol. 2, no. 3, 270–285.

*Bridging Research and Good Practices towards Patient Welfare – Shih & Liang (Eds)*
 *ISBN 978-1-138-02716-9*

# New mathematical models for vector-borne disease: Transmission of tomato spotted wilt virus

J. Wu, S. Zhao, Y. Kuang, D. Ben-Arieh, J. Nechols & D. Margolies
*Kansas State University, Manhattan, Kansas, KS, USA*

ABSTRACT: Tomato spotted wilt virus (TSWV) is a plant infecting virus and it impacts many food and ornamental crops, resulting in crop disease epidemics and huge economic loss. TSWV is transmitted exclusively by thrips, the most efficient of which is Western flower thrips, *Frankliniella occidentalis* (WFT). There is no single management method which can be adopted to control these insects and combinations of different management options properly should reduce the problem but still not easy to control them. In this paper, a mathematical model for vector-borne disease (WFT-TSWV) is constructed for the prediction of the transmission of TSWV by WFT. The purpose of the work is to provide mathematical analysis of the model and discuss some significant results arisen from the interplay of biological forces.

## 1 INTRODUCTION

Tomato spotted wilt virus (TSWV) is a plant infecting virus and impacts many food and ornamental crops, encompassing hundreds of plant species, resulting in crop disease epidemics of worldwide economic significance. TSWV is transmitted from plant-to-plant almost exclusively by thrips (Best 1968), the most efficient of which is the western flower thrips, *Frankliniella occidentalis* (Pergande) (WFT). Infected thrips, primarily adults, move into greenhouses from outdoor plants and, thus, introduce plant infection (Allen et al. 2000). In the North Central Region, and throughout the United States, greenhouse production of food and ornamental crops is at risk from this and other thrips-borne viruses. Risk of disease and associated crop losses from viruses result in expensive preventative treatments to exclude or attempt to eradicate thrips vectors.

TSWV was first described as 'spotted wilt' disease of tomato in Australia in 1915 (Brittlebank 1919). It was not until the early 1930's that the virus became known as tomato spotted wilt virus (Garden et al. 1935). The disease became prevalent in the floral crop industry and several field crops (e.g., peanut, tobacco and tomato) in the mid-1980's (Moyer et al. 1995). TSWV is the type member of the genus Tospovirus and is best-characterized with respect to vector interactions. Milne and Francki in (Milne & Francki 1984) were the first to recognize characteristic similarities between TSWV and a family of viruses that usually infect vertebrates, which later led to the inclusion of the Tospovirus genus into the family Bunyaviridae. Tospoviruses, the only plant-infecting genus in the virus family Bunyaviridae, are the cause of a number of significant plant diseases.

Tospoviruses are transmitted in a persistent propagative fashion (i.e., they replicate inside their thrips vector and are transstadially passed from molt-to-molt). In nature, TSWV is transmitted from plant-to-plant almost exclusively by thrips (Best 1968). While thrips are important as direct pests of agronomic and ornamental crops, it is the transmission of viruses that results in the most severe damage (Hunter & Ullman 1994). Thrips transmit plant viruses in the Tospovirus, Ilarvirus, Carmovirus, Sobemovirus and Machlomovirus genera (Jones 2005). TSWV has 7 known thrips vectors, all in the thysanopteran family Thripidae (Mound 1996, Webb et al. 1997). The western flower thrips is the most efficient vector of tospoviruses, transmitting five of the 14 currently described Tospovirus species (Whitfield et al. 2005).

Of the thrips species that transmit tospoviruses, WFT is the most economically important on greenhouse and ornamental crops worldwide (Ullman et al. 1997). The major threat from WFT is its ability to vector tospoviruses (Allen & Broadbent 1986, Jones 2005), which include, in addition to TSWV, impatiens necrotic spot virus, tomato chlorotic spot virus, chrysanthemum stem necrosis virus and groundnut ringspot virus (Wijkamp et al. 1996, Ullman et al. 2002). Infected thrips, primarily adults, move into greenhouses from outdoor plants and, thus, serve as the source of tomato infection (Allen et al. 2000). There are more than 147 known plant species that serve as hosts both for WFT and TSWV (Ullman et al. 2002, Sether & DeAngelis 1992). A number of weed species in the family Asteraceae serve as common reservoirs for infected WFT that invade greenhouses (Hobbs et al. 1996, Chatzivassiliou et al. 2001).

There is no single management method which can be adopted to control these insects and combinations of different management options properly should reduce the problem but still not easy to control them. As far as the authors know, no mathematical model is developed to study the transmission of TSWV through WFT. In this paper, we present mathematical models in both of homogenous population and heterogeneous meta-population to study the spatial-temporal disease spreading effects and the impacts of potential mitigation strategies. In addition, stability analysis using backward bifurcation theory revealed the stationary points for the nonlinear dynamic system which implies possible tipping points to control the vector borne disease.

The paper is organized as follows. In Section 2, we construct two mathematical models in homogeneous population without and with time-delay effects and age structures, respectively. In Section 3, we present the preliminary results on the prediction of transmission of TSWT and seek the significant parameters. In Section 4, we discuss the impacts of potential mitigation strategies and conclude our work.

## 2 MATHEMATICAL MODEL FOR HOMOGENEOUS POPULATIONS

Two mathematical models are constructed based on the transmission principle of TSWV by WFT, without and with time-delay and age-structure, respectively. The parameters used in the model are shown in Table 1.

The process of virus transmission is exclusively circulative-propagative (German et al. 1992). Plant virus transmission occurs during the adult thrips feeding process. In the process of feeding, saliva is injected into the plant creating the potential for the virus to be transmitted to the plant

Table 1. Parameters of the mathematical model.

| Parameter | Interpretation |
|---|---|
| $\nu_h$ | Birth rate of the host |
| $\beta_v$ | Bite rate of the thrip |
| $T_{hv}$ | Infection rate from vector to host |
| $\gamma_h$ | Age at harvest of host |
| $\mu_h$ | Death rate of host |
| $\tau_h$ | Incubation period for exposed host |
| $\nu_v$ | Birth rate of host |
| $K_v$ | Capacity of thrip population |
| $T_{vh}$ | Infection transmission rate from host to vector |
| $\mu_v$ | Death rate of thrip adults |
| $\mu_L$ | Death rate of thrip larvae |
| $\gamma_v$ | Age at maturity of WFT |
| $\tau_v$ | Incubation period for vector |

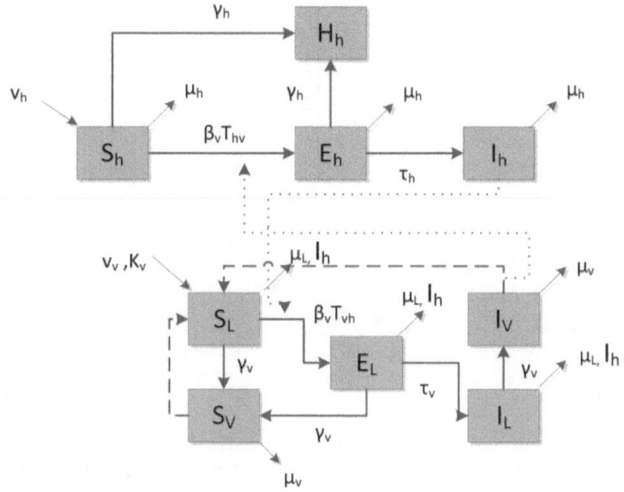


Figure 1. System Diagram of WFT-TSWV Model.

when it is present in the saliva of the thrips. Replication of the virion in the plant occurs upon entering the plant cells. WFT are only able to transmit TSWV if the virus is acquired in the earliest larval stage (Van de Wetering et al. 1996, Assis Filho et al. 2004). There are 6 stages for WFT development: the egg, two larval stages, two transformation stages, and adult. WFT acquire viruses from infected plants during their first and second nymphal instars (two larval stages) and virions propagate within the thrips during thrips development (Ullman et al. 2002). Thrips can acquire the virus at later immature and adult stages but are then not able to transmit the virus (Ohnishi et al. 2001). Adult thrips that feed on infected plants do not become viruliferous even if they are allowed lengthy feeding on tospovirus-infected plants (Ullman et al. 1989).

The system diagram of TSWV transmission between the host and WFT is shown in Figure 1. In this section, we assume the population is homogeneous. A modified susceptible-exposed-infected (SEI) model is constructed for both of host and WFT.

In our models, WFT are classified into two groups: larvae and adults. WFT larvae are distributed among susceptible $S_L$, exposed $E_L$, and infected $I_L$ compartments and adults are distributed between non-infective $S_v$ and infective $I_v$. Specifically, the susceptible larvae can either develop to non-infective adults or be transmitted to exposed larvae. The exposed larvae can either develop to non-infective adults successfully or be transferred to infected larvae. The infected larvae develop to infective adults. The hosts are distributed among susceptible $S_h$, exposed $E_h$, infected $I_h$, and harvest $H_h$. The susceptible host may either harvest or be infected by infective WFT adults. The exposed hosts may either harvest before becoming infected or transfer to infected hosts. The infected hosts cannot recover or harvest.

## 2.1 *Homogeneous population without time-delay effects and age structures*

Based on this principle of TSWV transmission process, the mathematical model is shown below:

### 2.1.1 *Host population*

$$\frac{dS_h}{dt} = v_h - (\beta_v T_{hv} I_v S_h) / N_h - (1/\gamma_h) S_h - \mu_h S_h \quad (1)$$

$$\frac{dE_h}{dt} = (\beta_v T_{hv} I_v S_h) / N_h - (1/\tau_h) E_h - (1/\gamma_h) E_h - \mu_h E_h \quad (2)$$

$$\frac{dI_h}{dt} = (1/\tau_h)E_h - \mu_h I_h \tag{3}$$

$$\frac{dH_h}{dt} = (1/\gamma_h)S_h + (1/\gamma_h)E_h \tag{4}$$

$$\frac{dN_h}{dt} = v_h - \mu_h S_h - \mu_h E_h - \mu_h I_h \tag{5}$$

where
$S_h$ = the number of susceptible hosts,
$E_h$ = the number of exposed hosts,
$I_h$ = the number of infected hosts,
$H_h$ = the number harvest hosts,
$N_h$ = the total number of hosts.

The above model is a modified SEI model with compartment $H_h$, which represents the number of harvest hosts. On each day, there is a constant birth rate $v_h$ for the susceptible hosts. There are $\beta_v T_{hv} I_v S_h / N_h$ susceptible hosts moved from compartment $S_h$ to $E_h$ due to TSWV transmission, $\gamma_h S_h$ susceptible hosts transferred to $H_h$ resulting from harvest and $\mu_h S_h$ susceptible hosts removed because of natural death. Similarly, $\sigma_h E_h$ exposed hosts becomes infected hosts, $\gamma_h E_h$ exposed hosts transfer to $H_h$, and $\mu_h E_h$ exposed hosts are removed due to natural death. Infected hosts cannot harvest and $\mu_h I_h$ infected hosts are removed from the model.

2.1.2 *WFT population*

$$\frac{dS_L}{dt} = v_v(S_v + I_v)(1 - N_v / K_v) - (\beta_v T_{vh} I_h S_L)/N_h - (1/\gamma_v)S_L - \mu_L S_L \tag{6}$$

$$\frac{dE_L}{dt} = (\beta_v T_{vh} I_h S_L)/N_h - (1/\tau_v)E_L - (1/\gamma_v)E_L - \mu_L E_L \tag{7}$$

$$\frac{dI_L}{dt} = (1/\tau_v)E_L - (1/\gamma_v)I_L - \mu_L I_L \tag{8}$$

$$\frac{dS_v}{dt} = (1/\gamma_v)S_L + (1/\gamma_v)E_L - \mu_v S_v \tag{9}$$

$$\frac{dI_v}{dt} = (1/\gamma_v)I_L - \mu_v I_v \tag{10}$$

$$\frac{dN_v}{dt} = v_v(S_v + I_v)(1 - N_v / K_v) - \mu_v N_v \tag{11}$$

where
$S_L$ = the number of susceptible larva,
$E_L$ = the number of exposed larva,
$I_L$ = the number of infected larva,
$S_v$ = the number of non-infective adult,
$I_v$ = the number of infective adult,
$N_v$ = total number of WFT.

There are two stages in the above model for WFT: larvae and adults. Specifically, larvae stage includes the first 5 stages during development and adults' stage only represents the last stage, adults. Only susceptible larvae equation includes birth rate because the female WFT do not transmit TSWV vertically. $\beta_v T_{vh} I_h S_L / N_h$ susceptible larvae are infected by the infected host and move to compartment $E_L$, $(1/\gamma_v)S_L$ susceptible larvae are developed to susceptible adults, and $\mu_L S_L$ susceptible larvae are removed due to natural death. For the exposed larvae, $(1/\tau_v)E_L$ of them transmit to the compartment $I_L$ after incubation period, $(1/\gamma_v)E_L$ develop to non-infective adults before they become infected larvae and $\mu_L S_L$ are dead. $(1/\gamma_v)I_L$ infected larvae transfer to infective adults and $\mu_L I_L$ of them are removed because of natural death. Non-infective adults do not become infective adults even though they are infected by the infected hosts.

### 2.2 *Homogeneous population with time-delay and age structures*

In this section, we add two more characteristics: time-delay effect of incubation period and age structures for development period.

In Section 2.1, the reciprocal of incubation period is adopted as the transfer rate from exposed compartment to infected compartment. However, more realistic models with time-delay effects are studied in SEIRS epidemic models (Cooke 1996, Gao 2008). Considering the incubation periods $\tau$h and $\tau$v for the host and WFT, respectively, the new equations for Eh, Ih, EL, IL are shown below:

$$\frac{dE_h}{dt} = (\beta_v T_{hv} I_v S_h)/N_h - e^{-\mu_h \tau_h}(\beta_v T_{hv} I_v(t-\tau_h)S_h(t-\tau_h))/N_h - (1/\gamma_h)E_h - \mu_h E_h \tag{12}$$

$$\frac{dI_h}{dt} = e^{-\mu_h \tau_h}(\beta_v T_{hv} I_v(t-\tau_h)S_h(t-\tau_h))/N_h - \mu_h I_h \tag{13}$$

$$\frac{dE_L}{dt} = (\beta_v T_{vh} I_h S_L)/N_h - e^{-\mu_L \tau_v}(\beta_v T_{vh} I_h(t-\tau_v)S_L(t-\tau_v))/N_h - (1/\gamma_v)E_L - \mu_L E_L \tag{14}$$

$$\frac{dI_L}{dt} = e^{-\mu_L \tau_v}(\beta_v T_{vh} I_h(t-\tau_v)S_L(t-\tau_v))/N_h - (1/\gamma_v)I_L - \mu_L I_L \tag{15}$$

Before discussing age-structure model, we briefly recall Lotka-Sharpe model which is the classical age-structure model in an isolated population (Sharpe 1911). The model is shown below:

$$\frac{\partial}{\partial t}n(a,t) + \frac{\partial}{\partial a}n(a,t) = -\mu(a)n(a,t) \tag{16}$$

$$n(0,t) = \int_0^\infty b(a)n(a,t)da \tag{17}$$

$$n(a,0) = n_0(a) \tag{18}$$

Here $\mu(a)$ and $b(a)$ are age-dependent mortality and fertility rate, respectively.

If we give a threshold age $\tau > 0$, $a < \tau$ represents juvenile and $a > \tau$ represents adults. Let $n(a, t)$ denote the number of individuals of age $a$ at time t, and thus,

$$n(a,t) = \begin{cases} n_1(a,t), & a \in [0,\tau) \\ n_2(a,t), & a \in (\tau,\infty) \end{cases} \tag{19}$$

$$n_1(t) = \int_0^\tau n_1(a,t)da \tag{20}$$

$$n_2(t) = \int_\tau^\infty n_2(a,t)da \tag{21}$$

This idea was extended into a prey-predator model with age-structure (Mohr et al. 2013) and can be adopted by our WTF-TSWV model.

In the host model, the transmission from susceptible hosts and exposed hosts to harvest hosts is a process based on the age threshold $\gamma_h$. Assume there is no mortality at the moment becoming age $\gamma_h$, the total number of susceptible host $S_h$, exposed hosts $E_h$ and harvest hosts $H_h$ are given by

$$S_h = \int_0^{\gamma_h} S_h(a,t)da \tag{22}$$

$$E_h = \int_0^{\gamma_h} E_h(a,t)da \tag{23}$$

$$H_h = \int_{\gamma_h^+}^{\infty} H_h(a,t)da \tag{24}$$

where $\gamma_h^- = \sup\{a : a < \gamma_h\}$, $\gamma_h^+ = \inf\{a : a > \gamma_h\}$

Considering the interaction with WFT, we have

$$\frac{\partial}{\partial t} S_h(a,t) + \frac{\partial}{\partial a} S_h(a,t) = -(\beta_v T_{hv} I_v S_h(a,t)) / N_h - \mu_h S_h(a,t) \tag{25}$$

$$\frac{\partial}{\partial t} E_h(a,t) + \frac{\partial}{\partial a} E_h(a,t) = \beta_v T_{hv} I_v S_h(a,t) / N_h - e^{-\mu_h \tau_h} \beta_v T_{hv} I_v (t-\tau_h) S_h(a, t-\tau_h) / N_h - \mu_h E_h(a,t) \tag{26}$$

$$S_h(0,t) = v_h \ (t > 0) \tag{27}$$

$$E_h(0,t) = 0 \ (t > 0) \tag{28}$$

As such, the total number of susceptible hosts satisfies:

$$\begin{aligned} \frac{dS_h}{dt} &= \int_0^{\gamma_h^-} \frac{\partial S_h}{\partial t}(a,t)da = \int_0^{\gamma_h^-} (-\frac{\partial S_h}{\partial a}(a,t) - (\beta_v T_{hv} I_v S_h(a,t)) / N_h - \mu_h S_h(a,t))da \\ &= v_h - S_h(\gamma_h^-, t) - (\beta_v T_{hv} I_v S_h) / N_h - \mu_h S_h \end{aligned} \tag{29}$$

Similarly,

$$\begin{aligned} \frac{dE_h}{dt} &= \int_0^{\gamma_h^-} \frac{\partial E_h}{\partial t}(a,t)da \\ &= \int_0^{\gamma_h^-} (-\frac{\partial E_h}{\partial a}(a,t) + \beta_v T_{hv} I_v S_h(a,t) / N_h - e^{-\mu_h \tau_h} \beta_v T_{hv} I_v (t-\tau_h) S_h(a, t-\tau_h) / N_h - \mu_h E_h(a,t))da \\ &= -E_h(\gamma_h^-, t) + \beta_v T_{hv} I_v S_h / N_h - e^{-\mu_h \tau_h} \beta_v T_{hv} I_v (t-\tau_h) S_h / N_h - \mu_h E_h \end{aligned} \tag{30}$$

$$\frac{dI_h}{dt} = e^{-\mu_h \tau_h} (\beta_v T_{hv} I_v (t-\tau_h) S_h (t-\tau_h)) / N_h - \mu_h I_h \tag{31}$$

$$\frac{dH_h}{dt} = S_h(\gamma_h^-, t) + E_h(\gamma_h^-, t) \tag{32}$$

$$\frac{dN_h}{dt} = v_h - \mu_h S_h - \mu_h E_h \tag{33}$$

A general solution for the count $n(a, t)$ in any compartment at age $a$ and at time $t$ is shown below in which $n(a, t)$ satisfies the equation:

$$\frac{\partial}{\partial t} n(a,t) + \frac{\partial}{\partial a} n(a,t) = b(a,t) - \mu(a,t) n(a,t) \tag{34}$$

where $b(a, t)$ is a function of $a$ and $t$.

$$n(a,t) = e^{\int_0^a \mu(x,-a+t+x)dx} \left(\int_0^a e^{-\int_0^y \mu(x,-a+t+x)dx} b(y,-a+t+y)dy + C(-a+t)\right) \tag{35}$$

where $C(-a+t)$ is a function related to the initial condition. As such, we have

$$S_h(\gamma_h^-, t) = e^{\int_0^{\gamma_h^-} (-\beta_v T_{hv} I_v (-\gamma_h^- + t + x) / N_h (-\gamma_h^- + t + x) - \mu_h) dx} C_1(-\gamma_h^- + t) \tag{36}$$

$$\begin{aligned} E_h(\gamma_h^-, t) = e^{-\mu_h \gamma_h^-} (\int_0^{\gamma_h^-} & e^{\mu_h y} (\beta_v T_{hv} I_v (-\gamma_h^- + t + y) S_h (y, -\gamma_h^- + t + y) / N_h (-\gamma_h^- + t + y) \\ & - e^{-\mu_h \tau_h} \beta_v T_{hv} I_v (-\gamma_h^- + t + y - \tau_h) S_h (y, -\gamma_h^- + t + y - \tau_h) / N_h (-\gamma_h^- + t + y)) dy + C_2(-\gamma_h^- + t)) \end{aligned} \tag{37}$$

where:

$$C_1(t)=\begin{cases} S_h(-t,0) & if\ t\le 0 \\ v_h & if\ t>0 \end{cases} \tag{38}$$

$$C_2(t)=\begin{cases} E_h(-t,0) & if\ t\le 0 \\ 0 & if\ t>0 \end{cases} \tag{39}$$

Adopting the same method, a new WFT population model is constructed below:

$$\frac{dS_L}{dt}=v_v(S_v+I_v)(1-N_v/K_v)-(\beta_v T_{vh} I_h S_L)/N_h - S_L(\gamma_v^-,t)-\mu_L S_L \tag{40}$$

$$\frac{dE_L}{dt}=(\beta_v T_{vh} I_h S_L)/N_h - e^{-\mu_L\tau_v}(\beta_v T_{vh} I_h(t-\tau_v)S_L(t-\tau_v))/N_h - E_L(\gamma_v^-,t)-\mu_L E_L \tag{41}$$

$$\frac{dI_L}{dt}=e^{-\mu_L\tau_v}(\beta_v T_{vh} I_h(t-\tau_v)S_L(t-\tau_v))/N_h - I_L(\gamma_v^-,t)-\mu_L I_L \tag{42}$$

$$\frac{dS_v}{dt}=S_L(\gamma_v^-,t)+E_L(\gamma_v^-,t)-\mu_v S_v \tag{43}$$

$$\frac{dI_v}{dt}=I_L(\lambda_v^-,t)-\mu_v I_v \tag{44}$$

$$\frac{dN_v}{dt}=v_v(S_v+I_v)(1-N_v/K_v)-\mu_v N_v \tag{45}$$

where:

$$S_L(\gamma_v^-,t)=e^{\int_0^{\gamma_v^-}(-\beta_v T_{vh} I_h(-\gamma_v^-+t+x)/N_h(-\gamma_v^-+t+x)-\mu_L)dx}\,C_3(-\gamma_v^-+t) \tag{46}$$

$$\begin{aligned}E_L(\gamma_v^-,t)=e^{-\mu_L\gamma_v^-}\Big(&\int_0^{\gamma_v^-} e^{\mu_L y}(\beta_v T_{vh} I_h(-\gamma_v^-+t+y)S_L(y,-\gamma_v^-+t+y)/N_h(-\gamma_v^-+t+y)\\ &-e^{-\mu_h\tau_h}\beta_v T_{vh} I_h(-\gamma_v^-+t+y-\tau_v)S_L(y,-\gamma_v^-+t+y-\tau_v)/N_h(-\gamma_v^-+t+y-\tau_v))dy+C_4(-\gamma_v^-+t)\Big)\end{aligned} \tag{47}$$

$$\begin{aligned}I_L(\gamma_v^-,t)=e^{-\mu_L\gamma_v^-}\Big(&\int_0^{\gamma_v^-} e^{\mu_L y}e^{-\mu_h\tau_h}\beta_v T_{vh} I_h(-\gamma_v^-+t+y-\tau_v)S_L(y,-\gamma_v^-+t+y-\tau_v)/N_h(-\gamma_v^-+t+y-\tau_v))dy\\ &+C_5(-\gamma_v^-+t)\Big)\end{aligned} \tag{48}$$

$$C_3(t)=\begin{cases} S_L(-t,0) & if\ t\le 0 \\ v_v(S_v+I_v)(1-N_v/K_v) & if\ t>0 \end{cases} \tag{49}$$

$$C_4(t)=\begin{cases} E_L(-t,0) & if\ t\le 0 \\ 0 & if\ t>0 \end{cases} \tag{50}$$

$$C_5(t)=\begin{cases} I_L(-t,0) & if\ t\le 0 \\ 0 & if\ t>0 \end{cases} \tag{51}$$

Assume $S_h$, $E_h$, $I_h$, $H_h$,$N_h$, $S_L$, $E_L$, $I_L$, $S_v$, $I_v$, and $N_v$ are equal to 0 at any time $t < 0$.

## 3 PRELIMINARY RESULTS

In this section, we compare the two models without and with time delay and age structures, respectively, and discuss the significant parameter, particularly for the second model.

Figure 2 shows the dynamic transmission of TSWV with the initial configuration: $v_h = 5$, $\beta_v = 0.6$, $T_{hv} = 0.4$, $\gamma_h = 40$, $\mu_h = 0.015$, $\tau_h = 4$, $K_v = 10000$, $v_v = 5$, $T_{vh} = 0.2$, $\mu_v = 0.03$, $\mu_L = 0.4$, $\gamma_v = 10$, $\tau_v = 4$. On the first day, there are 80 susceptible hosts, 9 non-infective WFT, and 1 infected WFT. The simulation time is 180 days. Figure 2a depicts the four different stages

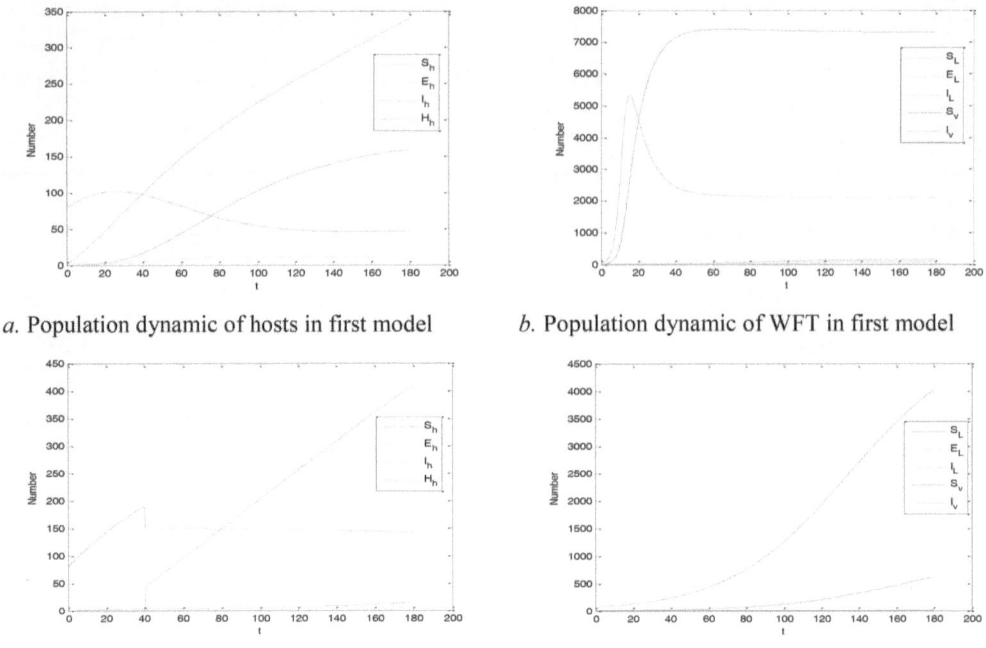


*a.* Population dynamic of hosts in first model

*b.* Population dynamic of WFT in first model

*c.* Population dynamic of hosts in second model

*d.* Population dynamic of WFT in second model

Figure 2. Number of Each Compartment on Each Day in Two Models.

of the host (tomato plants), the susceptible, exposed, infective, and harvested in the first model. Figure 2b shows the population dynamics between lava and adult thrips for susceptible, exposed and infective stages in the first model, respectively. Figure 2c shows the population dynamics of hosts in the second model and Figure 2d shows the population dynamics of WFT in the second model. We can find that the infected hosts and infected WFT are controlled in a lower level with time delay and age structures compared to the first model in this specific case. If we have more infected hosts or infected WFT in the beginning or we adjust the parameter which improves the transmission process, these two models would have very similar plots after first few days. However, the virus transmission on the first few days has significant effects on the spread of disease and we are only interested in some certain periods before host. We do not need to care about disease transmission after harvest time And therefore, we prefer to adopt the second model which is more realistic instead of first one, so we cannot simplify the development process and incubation periods to be a rate by using the reciprocal of corresponding periods.

Figure 3 shows the sensitivity analysis for $\gamma_h$ and $\mu_L$, which are age at maturity of WFT and mortality rate of larvae, respectively. If it takes 12 or more days from eggs to adults for WFT, the transmission of TSWV can be controlled and all infected WFT are dead eventually in our initial configuration from the left figure. Similarly, if the mortality rate of larvae is greater than 0.42, the disease cannot spread out.

In Figure 4, incubation periods for both of hosts and WFT are considered. Although $\tau_v$ and $\tau_h$ are very significant and the number of infected hosts and infected WFT will decrease as incubation periods increase, it is very difficult to control these two parameters in the real world. Only when $\tau_v$ is greater than 10 days, transmission of TSWV can be controlled. For $\tau_h$, we cannot stop the spread of disease even the incubation period of hosts are greater than 20 days.

For other parameters, most of them are very significant; however, it is very hard to change those parameters. For the mitigation strategies in reality, combination of different management methods may impact multiple parameters simultaneously and reduce the problem.

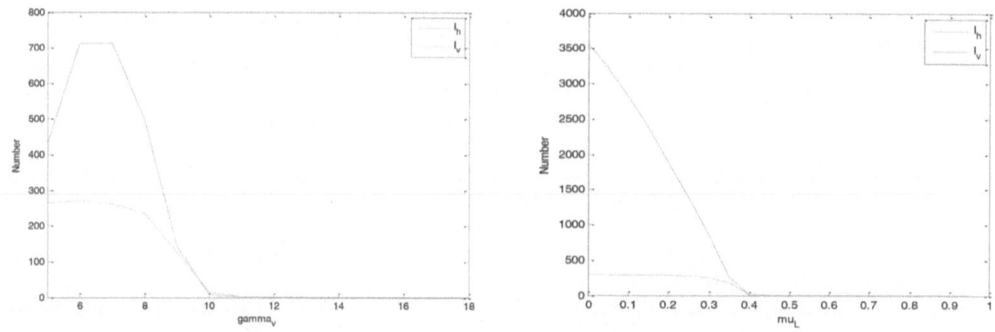


*a*. Change of infected hosts and WFT with $\gamma_h$ *b*. Change of infected hosts and WFT with $\mu_L$

Figure 3. Sensitivity Analysis for $\gamma_h$ and $\mu_L$.

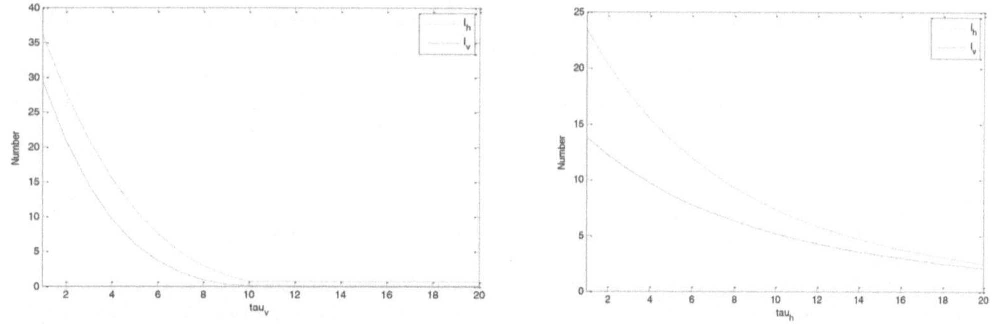


*a*. Change of infected hosts and WFT with $\tau$ *b*. Change of infected hosts and WFT with $\tau_h$

Figure 4. Sensitivity Analysis for $\tau_v$ and $\tau_h$.

## 4 DISCUSSION

In this paper, two mathematical models for WFT-TSWV transmission are constructed. From the preliminary results, a time delay and age structures model is recommended to describe the dynamic transmission of TSWV. So far there is no single management method which can be to control these insects and combinations of different management options properly should reduce the problem but still not easy to control them. Through the model, wefind the development period of WFT and the mortality rate of larvae are significant and adjustable parameters. The thrips life cycle takes about 5–20 days from egg to adult depending on temperature. The proper control of temperature in greenhouse may impact the disease spread as well as the plants grow. Effectively killing the larvae is another option to control the transmission of TSWV. The preliminary results provide more information which may be referred when considering the combination of management methods. However, the models proposed in this paper are very basic models. More properties are needed to consider, such as spatial structures, the mobility of WFT, etc. Biological experiments are also required to calibrate this model.

## REFERENCES

Allen, W.R. & Broadbent, A.B. 1986. Transmission of Tomato Spotted Wilt Virus in Ontario Greenhouses by Frankliniella occidentalis, *Canadian Journal of Plant Pathology 8*, 33–38.

Allen, C.T., Kharboutli, M.S., McAllister, C.D., Coker, C.M. & Cooper, P.E. 2000. Thrips, Weeds, and Tomato Spotted Wilt Virus, *Arkansas Agricultural Experiment Station* 475, 11–16.

Assis Filho, F.M. de, Deom, C.M. & Sherwood, J.L. 2004. Acquisition of Tomato Spotted Wilt Virus by Adults of Two Thrips Species, *Phytopathology* 94, 333–336.

Best, R.J. 1968. Tomato Spotted Wilt Virus. *Advances in Virus Research. Academic* 13, 65–146.

Brittlebank, C.C. 1919. Tomato Diseases. *Journal of the Department of Agriculture, Victoria* 17, 213–235.

Chatzivassiliou, E.K., Boubourakas, I., Drossos, E., Eleftherohorinos, I., Jenser, G., Peters, D. & Katis, N.I. 2001. Weeds in Greenhouses and Tobacco Fields are Differentially Infected by Tomato Spotted Wilt Virus and Infested by Its Vector Species. *Plant Disease* 85, 40–46.

Cooke, K.L. & van Den Driessche, P. 1996. Analysis of an SEIRS Epidemic Model with Two Delays. *Journal of Mathematical Biology* 35, 240–260.

Gardner, M.W., Tomkins, C.M. & Whipple, O.C. 1935. Spotted Wilt of Truck Crops and Ornamental Plants. *Phytopathology* 25, 17.

Gao, S., Chen, L. & Teng Z. 2008. Pulse Vaccination of an SEIR Epidemic Model with Time Delay. *Nonlinear Analysis: Real World Application* 9(2), 599–607.

German, T. L., Ullman, D. E.& Moyer, J. W. 1992. Tospoviruses: Diagnosis, Molecular Biology, Phylogeny and Vector Relationships. *Annual Review of Phytopathology* 30, 315–348.

Hobbs, H.A, Johnson, R.R., Story, R.N., Black, L.L. & Kuo, C.G. 1996. Weed Hosts and Thrips Transmission of Tomato Spotted Wilt Virus in Louisiana. *International Symposium on Tospoviruses and Thrips of Floral and Vegetable Crops, ISHS Acta Horticulturae* 431, 291–297.

Hunter, W.B.& Ullman, D.E.1994. Precibarial and Cibarial Chemosensilla in the Western Flower Thrips, Frankliniella occidentalis (Pergande) (Thysanoptera: Thripidae). *International Journal of Insect Morphology and Embryology* 23, 69–83.

Jones, D.R.2005. Plant Viruses Transmitted by Thrips.*European Journal of Plant Pathology* 113, 119–157.

Milne, R.G.& Francki, R.I.B. 1984. Should Tomato Spotted Wilt Virus be Considered as a Possible Member of the Family Bunyaviridae. *Intervirology* 22, 72–76.

Mohr, M., Barbarossa, M.V. & Kuttler, C. 2013. Predator-Prey Interactions, Age Structures and Delay Equations. *Arxiv*, (submitted).

Mound, L. 1996. The Thysanopteran Vector Species of Tospoviruses. appears. in Kuo, G. (ed), Tospoviruses and Thrips of Floral and Vegetable Crops: An International Symposium, Taichung, Taiwan, Acta Horticulturae, Technical Communications of International Society for Horticultural Science. *Asian Vegetable Research and Development Center*, Taiwan, 298–309.

Moyer, J.W., Daub, M.E. & Robb, K.L.1995. Thrips and TSWV/INSV," appears inNew Guinea Impatiens: A Ball Guide. eds. W. Banner and M. Klopmeyer. *Ball Publishing Co. Batavia Illinois*, 119–131.

Ohnishi, J., Knight, L.M., Hosokawa, D., Fujisawa, I. & Shinya, T. 2001. Replication of Tomato Spotted Wilt Virus after Ingestion by Adult Thrips setosus is Restricted to Midgut Epithelial Cells. *Phytopathology* 91, 1149–1155.

Sether, D.M. & DeAngelis, J.D. 1992. Tomato Spotted Wilt Virus Host List and Bibliography. *Special Report no. 888, Agricultural Experiment Station, Oregon State.*

Sharpe, F. R. & Lotka, A. J.1911. A Problem in Age Distribution. *Philosophical Magazine* 6(21), 435–438.

Ullman, D.E., Wescot, D.M., Hunter, W.B. & Mau, R.F.L. 1989. Internal Anatomy and Morphology of Frankliniella occidentalis (Pergande) (Thysanoptera:Thripidae) with Special Reference to Interactions between Thrips and Tomato Spotted Wilt Virus. *International Journal of Insect Morphology and Embryology* 18, 289–310.

Ullman, D.E., Sherwood, J.L. & German, T.L. 1997. Thrips as Vectors of Plant Pathogens. appears in T.S. Lewis (ed.), *Thrips as Crop Pests, CAB International*, 539–565.

Ullman, D.E., Meideros, R., Campbell, L.R., Whitfield, A.E., Sherwood, J.L. & German, T. 2002. Thrips as Vectors of Tospoviruses. *Advance in Botanical Research* 36, 113–140.

Van de Wetering, F., Goldbach, R. & Peters, D. 1996. Tomato Spotted Wilt Tospovirus Ingestion by First Instar Larvae of Frankliniella occidentalis is a Prerequisite for Transmission. *Phytopathology* 86, 900–905.

Webb, S.E., Kok-Yokomi, M.L. & Tsai, J.H. 1997. Evaluation of Frankliniella bispinosa as a Potential Vector of Tomato Spotted Wilt Virus (Abstr.). *Phytopathology* 87, S102.

Whitfield, A.E., Ullman, D.E. & German, T.L. 2005. Tospovirus-Thrips Interactions. *Annual Review of Phytopathology* 43, 459–489.

Wijkamp, I., Almarza, N., Goldbach, R. & Peters, D. 1996. Distinct Levels of Specificity in Thrips Transmission of Tospoviruses. *Phytopathology* 85, 1069–1074.

*Bridging Research and Good Practices towards Patient Welfare – Shih & Liang (Eds)*
*© 2015 Taylor & Francis Group, London, ISBN 978-1-138-02716-9*

# Medical device training classification system

M. Rys, M. Prieb & C. Taylor
*Kansas State University, Manhattan, Kansas, KS, USA*

ABSTRACT: With the growing number of advancements in medical technologies, the task of effectively incorporating a new device into a hospital is becoming an overwhelming task for hospital administration. With such a wide range of medical devices, planning for the procurement of a new technology through training and actual implementation can be a complex process. If appropriate steps aren't taken, patient safety can be compromised. This paper presents a systematic way for hospitals to classify medical technologies and recommend an appropriate level of training. To develop this process, a classification system, scoring system, validation plan, and training matrix was designed. A classification methodology was created with two main steps. First, medical devices are placed into a class (Infrastructure, Clinical Support, Direct Medical). The second step places the device into a level within the class (Low, Medium, High). In order to classify devices into their corresponding level, a scoring system was developed. This system was statistically validated using medical device experts. Also, a comprehensive training and preparation matrix was created. Once a device is placed in a class and level, the training and preparation matrix will recommend an appropriate level of training and additional preparation considerations.

## 1 INTRODUCTION

For healthcare organizations to remain competitive and profitable, continuous improvement of processes and technologies is a must. Emerging medical technologies are continuing to transform the way surgeons, doctors, and nurses deliver patient care. With the growing number of advancements in medical technologies, the task of effectively incorporating a new device into a hospital is becoming an overwhelming effort for hospital administration. Planning for the procurement of a new technology through training and actual implementation can be a timely process. This process becomes very complex with multiple individuals involved in various steps. Those that will actually use the devices can feel an added amount of stress with the addition of new technologies.

The first step in the process of technology procurement is determining what devices should be purchased. Oftentimes, the devices are identified by each department based on need. When purchasing, the process can be drawn out with negotiations between administration and device sales personnel on what services will be provided before and after implementation. Once the item is received by the hospital, it will undergo some type of assessment to determine the amount of training required. After training is selected, it must be delivered to nurses, clinicians, administrative personnel, and anyone that may come in contact with the new technology. Since hospital staff is always pressed for time, it can be difficult to get everyone through the training in an efficient manner. Additionally, there is great variation in the complexity of technologies that are acquired by a hospital; therefore, there is great variation in the amount of training needed. A standard procedure for determining how much and what type of training would best fit the technology would help reduce this variability.

## 2 PROJECT DESCRIPTION

The following section presents the project objectives as well as the project scope. Additionally, the project deliverables are outlined.

### 2.1 *Objective*

Currently there is not a standard process for determining the necessary training and preparation for medical devices being implemented into hospitals. This can lead to variability, ambiguity, excessive use of time and resources, and most importantly medical errors. The objective of this project was to create a systematic way for hospitals to classify medical technologies and recommend an appropriate level of training and preparation. This system was designed to incorporate any technologies procured by the hospitals ranging from employee timekeeping software to complex robotic surgery systems.

### 2.2 *Deliverables*

To achieve this objective, a classification system and corresponding training matrix to aid in implementation was designed. The deliverables of this project include the following:

- Classification system
- Scoring system for classification
- Validation plan
- Training and preparation matrix

These four tasks were developed after extensive review of relevant literature. These sources included professional journals, scholarly articles, and current device manufacturer literature.

## 3 CLASSIFICATION SYSTEM

It is important to find an appropriate classification system to manage the growing number of medical technologies. Currently, few classifications have been developed for the wide range of medical technologies that exist. After reviewing existing classification methodologies for medical devices and general technologies, a two step classification system was developed. The first step of classification divides the medical devices into a class based on its primary use. The second step further expands the previous classification into three additional levels. These two classification steps are detailed below.

### 3.1 *Classes*

After reviewing a number of classifications for medical devices (Geisler & Heller, 1997, Geisler, et al., 2010, Bunger & Tummler, 1994, McGuinness & Noy, 2011, FDA Medical Device Classification Procedures, 2012, Schnoll, 1997), it was found that devices were commonly grouped together based on similar use or functions. The number of classes for the existing medical device classifications systems range from two to six classes (Geisler & Heller, 1997, Geisler, et al., 2010). Since the purpose of the new classification system is to be used in conjunction with a training matrix, a simple three category classification would be most effective. With three categories, there is a decreased likelihood that classes will overlap and less confusion when placing devices into the appropriate category.

The three classes created for this classification system are Infrastructure, Clinical Support, and Direct Medical.

Definitions for these three classes were derived to serve two main purposes. First, they must encompass any medical technology that may be procured by a hospital. Secondly, the three classes

| | Infrastructure | Clinical Support | Direct Medical |
|---|---|---|---|
| Low | | | |
| Medium | | | |
| High | | | |

Figure 1. Classification Matrix.

should have fundamental differences in the necessary amount and type of training. The first class, Infrastructure, is defined as devices that support day-to-day tasks of staff and clinicians. Devices in this category may or may not be medical in nature. Additionally, they may be used for medical storage, sterilization, or administrative purposes. The second class, Clinical Support, is defined as devices that are indirectly involved in patient diagnosis and treatment. They may provide daily patient support or may be used as supplementary devices during treatment, diagnosis, or surgical procedures. This class may also include software or monitoring devices. The third class, Direct Medical, is defined as devices that are task-specific and are the primary tools necessary for treatment, diagnosis, or surgical procedures. These devices often come in direct contact with the patient during use.

### 3.2 *Levels*

Although the three classes effectively categorize medical technologies, devices in each class have a wide range of training needs. For example, the Direct Medical class includes devices as simple as a scalpel and as complex as the da Vinci Robotic Surgery System. To further develop the classification system, each class was divided into three levels: low, medium, and high. These levels more accurately classify the devices into groups with similar training needs. This methodology is similar to the classification system created by Paul Shrivastava and William E. Souder when they classified a broad range of technologies in their research (1984). Using the classes and levels, a visual classification matrix was created. This matrix can be seen in Figure 1.

## 4 SCORING SYSTEM

With the classification matrix outlined, the next step was to create a systematic way of placing devices in their appropriate level. To assist with this placement, a scoring system was created. This system is detailed in the following section.

### 4.1 *Characteristics*

Historically, classifications systems for general technologies have been based on their fundamental characteristics. From extensive research (Geisler, et al., 2010, Calantone & Garcia, 2002, Gemuednden, et al., 2006, Anderson, et al., 2000, Shrivastava & Souder, 1984, Rosenthal & Tatikonda, 2000, Pavlas, et al., 2010, Nielsen 1993, ISO, 1993, Ginsburg, 2004, Sharp & Rogers, 2002), a number of technology characteristics were compiled. These characteristics included simplicity, error handling, ease of integration into existing technologies, innovativeness, technological novelty, sophistication, consistency and degree of feedback. Many of the characteristics reviewed were combined and adapted to fit technologies found in the medical field. From this, five characteristics were finalized to aid in the placement of devices into their relevant levels. These characteristics include: complexity of interface, usability, learnability, newness, and riskiness (Geisler, et al., 2010, Shrivastava & Souder, 1984, Anderson, et al., 2000, Rosenthal & Tatakonda, 2000, Pavlas, et al.,

2010). Complexity of interface focuses on the simplicity of a devices' interface and controls. Usability focuses on the procedures of the devices and its dependence on additional components or subsystems. Learnability focuses on the ease of use and degree of memorability which can be affected by the number of steps, complexity of steps, and the frequency or use. Newness focuses on the novelty of a device compared to its predecessors at the hospital or to the industry as a whole. Finally, riskiness focuses on the potential risk to the user and the patient if used improperly. The riskiness category is heavily based on the FDA's risk classification ("Medical Device Classification," 2012).

### 4.2 *Scale and scoring guidelines*

To define the level of a device, a scoring system was designed using the aforementioned five characteristics. Each characteristic is scored on a zero to six point scale. This scoring system is similar to the general technology classification instrument used by Shrivasta and Souder (1984). Once a device is evaluated on each characteristic, the scores are summed. This total score will correspond to the level the device should be placed. The levels range from zero to ten points, 11 to twenty points, and 21 to thirty points for low, medium, and high respectively.

To aid in the scoring process, a scoring guideline was created. This guideline breaks down the zero to six scale for each characteristics by including descriptive statements. If a device matches the majority of the statements, it should be scored with the corresponding number of points. Devices may fit multiple scores' descriptions; however, one should select the score with the descriptions that most accurately match the device. One exception to this exists; when scoring the riskiness characteristic, if a device fits multiple score definitions, the highest score should apply. Additionally, devices in the Infrastructure class should not exceed two points for riskiness. When completing scoring for devices, it should be assumed that clinicians and specialist that will be using device have completed all formal education in the area of their practice. The complete scoring guideline can be seen in Table 1.

## 5 VALIDATION

With the completion of this classification system, it was necessary to ensure that it would provide reliable and consistent results. To validate this system, a two step validation was put in place. The first step was a qualitative expert review of the system. The second step was involved a quantitative statistical validation. Each step is detailed in the following section.

### 5.1 *Expert review and feedback*

The first step in the validation process involved five medical device experts. Each expert was given the classification system and a sample list of 24 technologies. The experts were given two weeks to complete the classification and scoring of each device. Additionally, they were asked to provide feedback on areas of confusion or areas for improvement. This step in the validation provided two important aspects to the validation. First, the definitions of classes and scoring guidelines were refined to clarify any misunderstandings. Additionally, the scores provided by the experts were used to create the base scores for each technology. The five experts' scores and the KSU's team scores for each device were averaged to create a base score. These base scores were used as a standard to compare other individuals' scores during the second validation step.

### 5.2 *Statistical validation*

The second step in the validation process was a statistical review. To start, professionals working in the health care industry were asked to complete the classification and scoring of the same sample 24 technologies. Five validations that categorized the class appropriately were received while only

Table 1. Characteristics Scoring Guideline.

| 0–2 points | 3–4 points | 5–6 points |
|---|---|---|
| **Complexity of Interface [2, 3, 6]** | | |
| – No user interface<br>– Non-electronic/ simple electronic<br>– Controls' actions are visibly defined (device does exactly what user expects)<br>– Descriptive labels and feedback cues utilized throughout | – Simple user interface<br>– More than five different functioning buttons/controls<br>– Descriptive labels are used where needed<br>– Feedback cues are sufficient | – Interactive user interface<br>– Complex control panel (digital or physical)<br>– Numerous icons & features without clear labels<br>– Lack sufficient feedback cues |
| **Usability [4]** | | |
| – One intended purpose<br>– One fixed procedure<br>– Technology does not require subsystems<br>– Device performed individually | – Few purposes/differing procedures<br>– Technology requires few sub-systems, attachments, or components<br>– Procedure may require few supporting staff | – Technology requires substantial subsystems, components, or attachments<br>– Requires substantial decision making<br>– Varying procedures based on scenarios/decision making<br>– Procedure requires participation from multiple individuals |
| **Learnability [2, 6]** | | |
| – Task can be easily accomplished during first encounter<br>– Few/simple steps in process (less than 5)<br>– Very frequent use for intended user<br>– Users already have inherent or learned skills required for use | – Steps can be memorized after number of iterations<br>– Number of steps is between 6–15<br>– Steps are more complex<br>– Procedure may require hand eye coordination | – Large number of steps (would require list or manual)<br>– Different scenarios/ decisions may require different steps<br>– Infrequent use or high degree of memorability<br>– Procedure requires extreme hand eye coordination and precision |
| **Newness [2, 3, 5, 8]** | | |
| – Very similar to previous device<br>– Not new to industry or<br>– hospital | – Includes significant new features from previous model<br>– May be new for hospital, but not necessarily to industry | – Technology and/or Process is novel/has no predecessors<br>– New for hospital and industry |
| **Riskiness [3, 5, 7]** | | |
| – No direct contact with patient<br>– Strong error prevention (users are not allowed to make a mistake)<br>– Listed as FDA Class I | – General controls are insufficient to assure safety<br>– Possibility of indirect harm to patient care upon misuse or failure of device<br>– Technology is listed as FDA Class II | – Supports or maintains human life<br>– Technology is listed as FDA Class III (invasive, long duration of use, critical location of use, hazardous mode of application) |

three validations with the scoring of all items were completed. The purpose of this validation was to test inter-rater reliability. The Kappa statistic was used to compare the rater's scores and classes against the base scores and classes that were developed from the expert review. The Kappa Statistic is shown in Equation 1:

$$K=(Po-Pe)/(1-Pe) \quad (1)$$

| | Poor | Slight | Fair | Moderate | Substantial | Almost perfect |
|---|---|---|---|---|---|---|
| Kappa | 0.0 | .20 | .40 | .60 | .80 | 1.0 |

| Kappa | Agreement |
|---|---|
| < 0 | Less than chance agreement |
| 0.01–0.20 | Slight agreement |
| 0.21–0.40 | Fair agreement |
| 0.41–0.60 | Moderate agreement |
| 0.61–0.80 | Substantial agreement |
| 0.81–0.99 | Almost perfect agreement |

Figure 2. Kappa Statistic Interpretation (Landis and Koch, 1977).

Figure 2 shows how the results of the Kappa Statistic should be interpreted (Landis & Koch, 1977). The three experts that completed the scoring showed fair agreement between the experts and the base scores with an average kappa statistic of 0.29. The agreement when classifying devices was also fair with a kappa of 0.29. These moderately low agreement levels can be attributed to the raters being from different hospitals or unfamiliar with some of the devices. Additionally, more participants for validation would be ideal. More recommendations concerning validation are detailed in the recommendation section to follow.

## 6 TRAINING AND PREPARATION MATRIX

The last project step was to complete a matrix to recommend training and preparation tasks for devices in each class and level. After conducting literature review and reviewing manufacturer recommendations, training and preparation considerations were defined for each level in the classification system. These recommendations are outlines in the following sections.

### 6.1 *Infrastructure*

To start, the low level of infrastructure devices are inherently straight-forward devices. No formal training should be required; however, it would be beneficial for a biomedical engineer or procurement specialist to create a quick reference guide using the user manual (Gardner-Bonneau, et al., 2010). This reference sheet should be visible where the device is in use. For devices in the medium level, quick references or job aids with step by step instructions should be used ("Kronos," 2012). Additionally, completion of self-paced tutorials or a competency test should be used (Prendergast, 2008). In addition to the requirements for low and medium, high devices should also incorporate a follow-up Q&A session with internal expert for those that have additional education needs (Pillar & Jacox, 1991).

### 6.2 *Clinical support*

Similar to devices in the low level of infrastructure, low clinical support devices should have a quick reference guide with instructions on use and common errors posted near or with the device (Gardner-Bonneau, et al., 2010). Devices in the medium level of clinical support, by nature, will require more formal training. Recommendations for this level of device include in-service education delivered from the manufacturer ("Inova," 2012, Ingram, 2012). These in-service training sessions should be followed with competency tests for all individuals that may come in contact

*Selecting super-users:* Choose well regarded faculty (i.e., nurse managers, unit secretaries, nurses who are comfortable with computers) that will be able to effectively teach their peers. Choose at least one representative for each unit during each shift. Try to achieve a 1:13 ratio of super-users to trainees.

*Amount of super-user training*: 8 hours of total training staggered over 3 training sessions that lasted 2-3 hours each. The sessions occurred over a 6 week period. In order to maintain an effective learning environment, class sizes should be limited to 15 users.

*Type of training*: To accommodate all types of learning styles, a diverse mix of training strategies are recommended. This can include: hands-on opportunities, lectures, handouts, online computer simulations, and group discussions/participation. Overall, an emphasis should be placed on "using the system".

*Final steps*: Super-users should complete competency tests that measure their knowledge of the system. These final steps should be completed at least 3 weeks prior to system go-live to allow sufficient time for general user training.

Figure 3. Recommendations for Super-User (Karuppan, 2000).

with the device. It will also be helpful to appoint a clinical champion in the organization for when the manufacturer has finished training (Ingram, 2012). This champion should receive extra training from the manufacturer and be available to give extra training or answer questions for the organization. It is recommended for devices in the high level to use super-users for training.

Since devices at that level often require a larger time commitment for training, super-users is the most practical method to deliver training to multiple people simultaneously (Karuppan, 2000). This is especially useful in larger hospitals where the device may affect multiple departments. Additionally, super-users will be responsible for assessing performance and continuing education efforts after initial training and go-live is complete ("Training for Electronic," 2012).

There are many aspects to consider when using super-uses to train hospital staff on a new technology. One study based on a large scale patient care management system implementation outlined these important considerations including how to select super-users, as well as, the amount and type of training delivered to hospital staff. These recommendations are summarized in Figure 3.

### 6.3 *Direct medical*

Devices in the low direct medical level should have quick reference material developed that emphasizes common errors, alarms, and any operational changes from a previous device. This can be developed by procurement specialists or the training department by consulting with the manufacturer sales representative (Gardner-Bonneau, et al., 2010). In addition, for some new devices in this category, a low tech simulator, such as a mannequin may be useful. A mannequin can be used before a device is in full use to give the user a better feel for the device as they practice physical maneuvers or procedures (Eder-Van Hook, 2004).

Medium level devices will also be very similar to those in the indirect medical class in that in-service training is recommended as the primary source for training. This training event should be delivered by a manufacturer representative and should include hands-on practice in a mock setting (Karuppan, 2000). Since devices in this class are often used during procedures or are in direct contact with the patient, extra precautions should be taken. It is recommended to have all users complete a self or peer assessment using a competency checklist (Karuppan, 2000, Paull & Mazzia, 2011). These competency reviews should be repeated periodically to ensure users stay up-to date. Additionally, partial task trainers may be appropriate for devices that fall in this level. Partial task trainers are a type of simulator that has visual, audio, and touch cues integrated with computers to replicate a clinical setting (Eder-Van Hook, 2004).

Table 2. Training and Preparation Matrix.

| Infrastructure | Clinical Support | Direct Medical |
|---|---|---|
| **Low** | | |
| – Internal expert review user manual<br>– Create quick reference guide for basic, critical, or confusing aspects of device operation and troubleshooting. Post quick reference with or near the device. | – Internal expert review user manual and provide information packet and/or quick reference of devices controls, operation, etc. | – Provide user manuals<br>– Optional training using manikins or other training devices |
| **Medium** | | |
| – Internal expert review user manual<br>– Create quick reference guide for basic, critical, or confusing aspects of device operation and troubleshooting. Post quick reference with or near the device. | – Provide intended users with web-based or video training. This training can be created internally or purchased from device manufacturer.<br>– Post-test should be used at completion of training to ensure competency. If post-test is failed, user should meet with administration or education team to determine further steps. | – In-service training event delivered by device manufacture representative. Training should include hands-on practice in mock setting.<br>– Supervised use until supervising physician signs off that user is competent.<br>– Continuing education or periodic competency assessments. |
| **High** | | |
| – Internal expert review user manual<br>– Create quick reference guide for basic, critical, or confusing aspects of device operation and troubleshooting. Post quick reference with or near the device.<br>– Optional Q&A session for intended users with internal expert | 1. *Select Super Users** to attend manufacturer training sessions.<br>2. *In-Service Training for Groups of Users:* Super user will give training to small groups of users in modules. Users will complete post quiz after each module. Hands-on training will be part of training modules when applicable.<br>3. *Quick Reference Guide:* For use after training. | 1. *Select Super Users** to attend manufacturer training sessions.<br>2. *In-Service Training:* All potential users should attend in-service training sessions given by device manufacturing representatives. Follow up sessions should be given by super users and all users should attend. Training sessions should involve hands-on training in mock settings<br>3. Supervised use until supervising physician signs off that user is competent.<br>4. Users should be provided with information packets and/or videos for review during their own free time<br>5. Continuing education or periodic competency assessments for all users |
| **Additional Preparation** | | |
| – Address any physical workplace requirements for device (electrical, drainage, lighting) | – Review post-implementation support included in Manufacturer's contract. Revise or add to contract if needed.<br>– Address any physical workplace requirements for device (electrical, drainage, lighting)<br>– Create and communicate new protocols (cleaning, maintenance, safety)<br>– Determine if any team workflows or individual responsibilities need to be reviewed | – Review post-implementation support included in Manufacturer's contract. Revise or add to contract if needed.<br>– Address any physical workplace requirements for device (electrical, drainage, lighting)<br>– Create and communicate new protocols (cleaning, maintenance, safety)<br>– Determine if any team workflows or individual responsibilities need to be reviewed |

Finally, devices in the highest level of direct medical have the most stringent training recommendations. To start, initial training should be completed using super-users. The recommendations for super-user training outlined in Figure 2 still apply for this level of training. This super-user training should rely heavily on hands-on activities (Karuppan, 2000). It has been shown that hands-on training can help users become proficient be enhancing their psychomotor skills (Hilbig, 1998). Additional literature or videos should be provided to all trainees for review during free time. This may be useful to help struggling users feel more confident or for users unable to attend a training session. Another key aspect of training should include some type of simulator training to enhance the user's visio-spatial, perceptual, and motor skills (Basdogan, et al., 2007). The most appropriate type of simulation tool for this level of device is a realistic mannequin-based simulator. These full length mannequins are completely computer driven. This type of training allows for team practice in lifelike settings (Eder-Van Hook, 2004). Once the initial training is completed, all users should have supervised use until a supervising physician signs off that user is competent (Prendergast, 2008). Again, periodic competency reviews should be recorded and kept up-to-date on all users (Phoenix, et al., 2008).

### 6.4 *General preparation considerations*

In addition to general training recommendations, supplementary preparations must also be considered. For all devices, keep in mind and address any physical workplace requirements necessary for the device. This could be any electrical, heating, drainage, or lighting requirements (Mytton, et al., 2010). Additionally, all devices should be reviewed to determine if any new protocols need to be created and communicated. These protocols could be related to manufacturer recommendations for cleaning, maintenance, or safety of the device.

For devices that are classified as Clinical Support or Direct Medical, it is critical to consider any team workflows or individual responsibilities that may need to be revised or established. Often times, as new devices are implemented, individual or team responsibilities may be unclear or even unknown. For devices that are classified in the medium or high categories of Clinical Support or Direct Medical, it may also be important to review any post-implementation support that is included in the manufacturer's contract. The contract should be revised if post-implementation support appears to be insufficient. Devices that are classified in the high category of Direct Medical often times require the substantial collaboration between many personnel and subsystems. One study found that pre and post-operative briefing checklists had many related benefits. These benefits included: undesirable events caught, increased "on-time" starts to surgeries, and also better use of the equipment (Paull & Mazzia, 2011). Table 2 summarizes the training recommendations and the additional preparation considerations.

## 7 RECOMMENDATIONS

One limitation of this study is that the system cannot be completely validated without some form of variance. With the system being validated from a variety of medical procurement specialists, there is some discrepancy in each expert's opinion on the complexity, usability, learnability, newness, and riskiness of the example devices used in the validation process. This is due to the fact that the experts were from different hospitals that have different degrees of technology advancements. Further validations should be completed with experts from the same hospital to see how much the scores vary within one hospital. Furthermore, this validation can be completed at multiple hospitals to ensure validity. Another limitation regarding the validation process is that the characteristic scoring process requires a sufficient level of knowledge related to the given devices. Some experts noted concern that they did not feel informed enough on certain devices to effectively rank them. Finally, it is recommended that this system is tested in a hospital in order for more enhancements to be made.

## REFERENCES


Anderson, P., Gatignon, H., Smith, W., & Tushman, M. L., 2000, "A Structural Approach to Assessing Innovation: Construct Development of Innovation Locus, Type, and Characteristics," Management Science, 48(9), 1103–1122.

Basdogan, C., Harders, M., Sedef, M. & Wesarg, S., 2007, "VR-Based Simulators for Training in Minimally Invasive Surgery," IEEE Computer Graphics and Appplication, 27(2), 54–66.

Calantone, R. & Garcia, R., 2002, "A Critical Look at Technological Innovation Typology and Innovativeness Terminology: A Literature Review," 19, 110–132.

Eder-Van Hook, J., 2004, "Building a National Agenda for Simulation-based Medical Education," Advanced Initiatives in Medical Simulation, 1–40.

Gardner-Bonneau, D. J., Weinger, M. B. & Wiklund, M. E., 2010, "Handbook of Human Factors in Medical Device Design," CRC Press, Boca Raton, FL.

Geisler E. & Heller O., 1997, "Management of Medical Technology: Theory, Practices, and Cases," $1^{st}$ Edition, Kluwer Academic Professionals, Norwell, MA.

Geisler E., Spadoni, E. & Tuchetti, G., 2010, "Health Technology Assessment: Evaluation of Biomedical Innovative Technologies," IEEE Engineering in Medicine and Biology Magazine, 29(3), 70–76.

Hilbig, J. 1998. "The Operating Room Registered Nurse: Medical Device Traning Needs and Current Practice in Australia," Biomedical Instrumentation & Technology, 32(1), 25–31.

Ingram, J., Oct 23 2012, "In-service training for Phillips Intellivue Patient Monitor," Interview by C. Taylor [Personal Interview].

"Inova Introduces Accuvein AV300 Devices," Inova Health Systems. Retrieved from http://inovanursing.org/2012/10/ Oct 19, 2012.

Karuppan, C., 2000, "Training Super Users in Large Health Care Facilities," Journal of Information Technology Management, 11(3–4), 21–28.

"Kronos: Education Services," 2012, Retrieved from http://www.kronos.com/download/thank-you.aspx?did=11747&rr=0&LangType=1033.

Landis, J.R. & Koch, G.G., 1977, "The Measurement of Observer Agreement for Categorical Data," Biometrics, 33, 159–74.

"Medical Device Classification Procedures," 21 CFR 860.3, 2012.

Mytton, O., Velazquez, A., Banken, R., Mathew, J., Ikonen, T., Taylor, K., Painter, F., Jean-Baptiste, R., Poon, A. & Ruelas, E., 2010, "Introducing new technology safely," Quality & Safety in Health Care Journal, 19(2).

Paull, D. D. & Mazzia, L. M., 2011, "Medical Team Training in the Veterans Health Administration: Checklist – Guided Preoperative Briefings and Postoperative Debriefings," Advanaces in Human Factors and Ergonomics in Healthcare, 147–152.

Pavlas, D., Salas, E., Sharit, J., & Wilson, K. A., 2010, "Eight Tips for Implementing New Technologies in Health Care," Ergonomics in Design: The Quarterly of Human Factors Applications, 18(20), 20–25.

Phoenix, P., Clarke, H., Young, J., Gillespie, A., Smith, A. & Hodgkinson, R. NHS Bolton, Risk Management, 2008, Medical devices & equipment management policy (RM032), 1–39.

Pillar, N. & Jacox, A., 1991, "The Introduction of New Technology on the Nursing Unit," Nursing Economics, 9(1), 51–63.

Prendergast, A. Clinical Strategy and Performance Committee, (2008), Medical devices policy (POL/CL/WCN/654), 1–31.

Prendergast, A., 2008, "Medical Device Policy," The Walton Centre NHS Foundation Trust, 1–31.

Rosenthal, S. R. & Tatikonda, M. V., 2000, "Technology Novelty, Project Complexity, and Product Development Project Execution Success: A Deeper Look at Task Uncertainty in Product Innovation," IEEE Transaction on Engineering Management, 47(1), 74–87.

Shrivastava, P. & Souder, W. E., 1984, "Towards a Scale for Measuring Technology in New Product Innovations," Research Policy, 14, 151–160.

2012.,"Training for Electronic Medical Records Software," Retrieved from http://www.micromd.com/emr/training.html

# *Part E: Environment and system design*

*Bridging Research and Good Practices towards Patient Welfare – Shih & Liang (Eds)*
*© 2015 Taylor & Francis Group, London, ISBN 978-1-138-02716-9*

# An experience and evaluation of objective structured clinical examination in nursing education at Kaohsiung Medical University

Shu-Yuan Jian
*College of Nursing, Kaohsiung Medical University, Kaohsiung, Taiwan (R.O.C.)*

Fan-Hao Chou
*School of Nursing, Kaohsiung Medical University, Kaohsiung, Taiwan (R.O.C.)*

ABSTRACT: Objective Structured Clinical Examination (OSCE) is consistently recognized as a reliable and valid strategy to assess students' competence and clinical performance. Based on the OSCE principles, the School of Nursing, Kaohsiung Medical University (KMU) has carried out reforms progressively each year in the following dimensions: "instructor training," "reform of teaching environment and equipment," "curriculum planning," "more standardized checklist," "design of clinical situations and adoption of standardized patients (SPs)," and "immediate and multidimensional evaluation." The effects of reforms included the following: there was a significant promotion of hardware and software capabilities; nine cross-disciplinary integrated cases have been completed over the past two years; and the overall satisfaction of instructors and students reached over 85%. We will continuously improve the OSCE in the following directions: (1) building the checklist of nurse-patient communication and interaction; (2) adding evidence-based data of the application of the OSCE in nursing to enhance or improve its use in the nursing field.

## 1 INTRODUCTION

Nursing is a science requiring synthesis of "academic theories" and "skills." The goal of nursing education is to integrate professional nursing knowledge with skills to educate students as professional nursing staff with ample knowledge and sophisticated skills. It has been a challenge for nursing education to provide skills instruction and examinations that conform to changing clinical needs. Having a reliable and valid method to assess students' competence and clinical performance fairly, is especially challenging to all nursing faculty. The Objective Structured Clinical Examination (OSCE) is an examination where "students demonstrate their competence under a variety of simulated conditions" (p. 424) and originally developed in Dundee in the mid-seventies (Harden & Gleeson, 1979). It uses a simulated and standardized format to examine the synthesis of knowledge and clinical skills. The OSCE provides an innovative learning experience for students as compared to the experience offered by traditional learning. Standardized patients (SPs), trained from individuals based on a controlled clinical situation that is realistic and nonthreatening, are important for the OSCE (O'Neill & Mccall, 1996). The OSCE has been used to assess medical students since the mid-1970s, and it has been described in abundance in the literature (Wilkinson, Newble, & Frampton, 2001). The OSCE has been also utilized increasingly to assess nursing and allied health professional students during the past decade. A review article explores that OSCE can be used most effectively to assess the clinical performance in nurse undergraduate curricula (Mitchell, Henderson, Groves, Dalton, & Nulty, 2008).

## 2 THE EXPERIENCE OF OSCE IN NURSING EDUCATION AT KMU

Based on the OSCE benefits and principles, the School of Nursing, Kaohsiung Medical University (KMU) dedicated its efforts to six dimensions changed in order to enhance and examine students' clinical ability.

### 2.1 *Instructor training*

Nursing education emphasize the instruction and examination of nursing skills. Therefore, instructor training is undoubtedly the primary task. In 2002, the School of Nursing sent two instructors to receive an eight-week training program including problem-based learning (PBL), OSCE, and application of SPs in the OSCE at McMaster University, Canada. Since then, we have conducted a series of seed instructor's on-campus training, and attended programs conducted by universities in Canada, America, and Hong Kong. In 2005, the seed instructors conducted a demonstration lecture regarding OSCE in the course of Physical Examination and Assessment to all instructors of the School, which aimed at obtaining a consensus of the teaching staff and reviewing the feasibility of the OSCE in the School of Nursing and its associated problems.

### 2.2 *Curriculum planning*

Although the utility of the OSCE is beyond doubt, a few study findings show that major gaps exist in the examination of the psychometric properties and communication performances (Mitchell et al., 2008; Walsh, Bailey, & Koren, 2009). Based on the spirit of the OSCE, we adopted a progressive improvement project. We initially conducted micro-adjustment on courses of "Fundamental Nursing" and "Physical Examination and Assessment," in which the School of Nursing already had current clinical examinations. Another, the original OSCE, comprised a series of 16 to 20 stations (Harden & Gleeson, 1979). Considering our courses and students' competence, we modified it to be one station including assessment, examination, and communication from the original OSCE. As for objectivity, we standardized skills, achieving the consistency among instructors.

### 2.3 *Checklist standardized*

With the consensus of the OSCE, the first step of reforms was to standardize skills: the main instructor of a certain skill must incorporate academic theories and current clinical skills, standardize the step-by-step skill, demonstrate the process to their peers, modify it after careful peer review and discussion, edit the checklist according to the modified standardized steps, and assign a scoring weight for each step. The processing time for each skill is set to take 10-15 minutes to complete. The checklist had to include four parts: preparation of materials, operation process of skills, patient preparation and communication, and attitude toward the operation process. Finally, the checklist was reviewed and tested again to complete the standardization of a skill.

### 2.4 *Reform of teaching environment and equipment*

Effective implementation of the OSCE requires both equipment and locations. Based on OSCE principle, we modified existing spaces, and improved equipment of the students' self-learning, instructors' teaching, and clinical skill examination. As for software, we standardized a variety of clinical skills and videotaped them. All these video files were saved in the computers of the self-learning computer labs to allow students to learn by themselves. We also improved hardware components year by year according to fund-raising status. We installed a central computer controller in the basic skills learning center and placed a bedside computer beside each practice bed, allowing students to watch videos and immediately modify their skills during practice. In addition, we also installed a 360-degree comprehensive video recorder on the demonstrations bed, allowing students to watch the process and contents of instructors' demonstration from each side. The process of students' clinical examinations can also be videotaped, serving as the basis of immediate review and evaluation for both instructors and students after the test.

## 2.5 *Design of clinical situations and adoption of SPs*

Another important concept of the OSCE is to assess students' clinical competence. Unlike the focus of traditional classroom-based learning, clinical competence is the ability that nursing students and staff should possess to calmly apply critical thinking and solutions and to accurately evaluate and cope with patients' problems in facing clinical situations and actual patients (Wass, Van der Vleuten, Shatzer, & Jones, 2001).

To assess students' competence and clinical performance in clinical situations, the School of Nursing abolished the conventional method by which students were tested with informed skill items and replaced this with clinical scenarios. We edited the situation of clinical patients according to students' learning level and needs. Students must read the scenario and decide which physical examination and assessment or nursing skill should be applied (see Sample 1).

***Sample 1***

***Applicable participants:** the second-year students of the School of Nursing*

***Applicable courses:** Fundamental Nursing Skills*

*The case has an insertion of central venous catheter into the right subclavian vein. As the day-shift nurse assigned to care for this case, you find a pressure ulcer over the case's sacrum due to prolonged bed rest when you visit the case in earlier shift exchange. Now it is time for daily routine care, and please provide the case with appropriate nursing interventions.*

The School of Nursing also introduced SPs into the students' technique examination in 2010. Before then, nursing students practiced and examined clinical skills between peers. The SP method is an innovative addition to improve and assess the clinical competence of communication among nurse learners (Luctkar-Flude, Wilson-Keates, & Larocque, 2012; Robinson-Smith, Bradley, & Meakim, 2009). These SPs had been well training from the Clinical Skill Center of Kaohsiung Medical University.

We conducted a satisfaction survey on utilization of SPs in clinical examination over the past two years among instructors and students, and found that instructors' satisfaction on SPs reached over 95% (see Table 1) in the first survey of the 2010 academic year. In the second survey, we found a downward trend in satisfaction after modifying certain item questions. However, the overall satisfaction remained at 90%. The qualitative results showed that most instructors supported the role-play of SPs and their effects on clinical examination. Instructors had more needs on "giving students appropriate responded results and feelings" and they were more unsatisfied with "by order of the test, without giving reminders," which was consistent with the quantitative results.

From the students' perspective, SPs' professional attitude, communication and interaction, and assistance in understanding clinical problems were assessed. There was a downward trend in students' satisfaction on SPs year by year, but the overall satisfaction level remained at around 90% (see Table 2). The qualitative results showed that most students attributed their eased sense of fear to SPs' attitude; that the participation of SPs gave a greater sense of reality; and that students expected SPs to provide concrete suggestions after tests.

Table 1. Satisfaction of instructors on SPs' performance.

| Item | Level of satisfaction (%) 2010[a] n1 = 9 1st | 2nd | 2011[b] n2 = 9 |
|---|---|---|---|
| The overall satisfaction of SPs' performance | 95.6 | 90.5 | 46.9 |
| 1. Does not arrive late or leave early, showing earnest and professional attitude | 96.7 | — | — |
| 2. Can fully understand and remember the script and contents of the teaching plan | 94.5 | 82.5 | 96.9 |
| 3. Acts according to instructions or requirements of examiners | 95.6 | 81.1 | 96.9 |
| 4. Plays the role realistically and shows the signs accurately | 96.7 | 82.8 | 95.4 |
| 5. Concentrates on the role, not distracted | 95.6 | 80.3 | 96.9 |
| 6. Shows emotion, attitude and responses of the patient in the script appropriately | 95.6 | 81.4 | 98.5 |
| 7. Maintains consistency of the role-play | 95.6 | 82.1 | 98.5 |
| 8. Possesses good interaction and communication skills | 94.5 | — | — |
| 9. Does not leak out a clue for tests or give incorrect information | 94.5 | 60.9 | 25.0 |
| 10. Does not transmit any issue unrelated to the test | 95.6 | 80.3 | 95.3 |
| 11. Can answer questions naturally | — | 82.8 | 95.4 |
| 12. Only talks with the examinee during the test | — | 82.8 | 95.3 |

Instructors' opinions
1. SPs could act according to the script.
2. SPs have good cooperation and coordination.
3. SPs assist in conducting tests smoothly.
4. I hope that SPs can provide students concrete suggestions and tell them "how to do it" for a better performance next time.
5. SPs could follow the test rules, not giving reminders

[a]The 2nd semester, Academic Year 2010; [b]The 1st semester, Academic Year 2011

Table 2. Satisfaction of students on SPs' performance.

| Item | Satisfaction (%) 2010[a] n1 = 82 | 2011[b] n2 = 73 |
|---|---|---|
| The overall satisfaction of SPs' performance | 93.3 | 93.5 |
| 1. Shows appropriate professional attitude | 91.3 | 89.9 |
| 2. Is good at communication and interaction | 92.2 | 89.3 |
| 3. Shows real emotion and response of a patient | 90.6 | 86.5 |
| 4. Helps examinees understand realistic clinical problems | 93.3 | 89.3 |
| 5. Promotes examinees' clinical responses | 93.0 | 89.9 |
| 6. Elevates examinees' communication skills | 92.2 | 88.3 |
| 7. Does not transmit any issue unrelated to this test | 92.3 | 88.8 |

Students' opinions
1. SPs are kind, giving me confidence.
2. SPs' suggestions are valuable and helpful.
3. SPs are able to react true feelings of a patient
4. Use of SPs is helpful in hospital learning experience
5. Use of SPs gave much more sense of immediacy than utilization of peer nursing students
6. Great experience
7. It is difficult to explain to the patient in a limited time. Patients do not understand how we expect them to act, so sometimes we have to take time to explain further.
8. I hope that SPs do not give me hints because I won't be scored on it in this case. It is appropriate to let me know which assessment should be done on my own.

[a]The 2nd semester, Academic Year 2010; [b]The 1st semester, Academic Year 2011

### 2.6 *Immediate and multidimensional evaluation*

The immediate feedback to students during the OSCE has been found to improve competencies and quality of the learning experience for students (Cushing, Abbott, Lothian, Hall, & Westwood, 2011). Historically, clinical examination in the School of Nursing has maintained such evaluation methods. After the utilization of SPs in simulated clinical situations, we also requested SPs to give feedback on communication, attitude, and skills to students.

## 3 EVALUATION AND DISCUSSION

Many researchers reported that the OSCE–having more advantages than those of traditional assessment methods–is a reliable and valid method of assessing the students' nursing competency skills (McWilliam & Botwinski, 2012; Selim, Ramadan, El-Gueneidy, & Gaafer, 2012; Walsh et al., 2009). In this reform program, the OSCE does make the technical teaching and evaluation more consistent and standardized than in the past in teachers' opinions. On the other hand, teachers also consented to immediate and multidimensional evaluation in light of reviewing and discussing performance with students following the standardized examination contents. This content and feedback allowed students to ascertain their "blind spots" and help instructors understand their teaching effects so that they could improve the content and methods of teaching.

We performed student evaluations on the OSCE using questions such as "I think that application of the OSCE can elevate clinical learning experience;" "I think that the OSCE can help me understand what I need in future learning;" "I think that the OSCE will be helpful in future clinical work;" "I think that the OSCE can help me understand my professional competence;" and "I think that the OSCE can help me understand the attitude and reaction in coping with clinical situations." Most students gave positive responses with an affirmation degree more than 85%, which is the same as the results of several previous studies, and even higher than some other studies (Brosnan, Evans, Brosnan, & Brown, 2006; Kurz, Mahoney, Martin-Plank, & Lidicker, 2009; Rentschler, Eaton, Cappiello, McNally, & McWilliam, 2007).

We also conducted a satisfaction survey three times over the past two years on utilization of SPs in clinical examination, among instructors and students. From the view of instructors, we used 12 items, which included SPs' professional performance, role-play performance, concentration, communication, and interaction as an overall satisfaction on SPs. We found that instructors' overall satisfaction on SPs was 95.6% and 90.5% respectively in the first and second survey of the 2010 and 2011 academic year. However, the overall satisfaction dropped to 46.9% in the third survey. The discriminate downward trend appeared in the sub-item "do not leak out a clue for tests or giving incorrect information" with a satisfaction level from 94.5% and 60.9% dropping to 25.0% from the first to the third evaluation. The qualitative results showed that most instructors supported the role-play of SPs and their effects on clinical examination. Instructors had more needs on "giving students appropriate responded results and feelings" and they were more unsatisfied with "by order of the test, without giving reminders," which was consistent with the quantitative results.

From the students' perspective, SPs' professional attitude, communication and interaction, and assistance in understanding clinical problems were assessed. The levels of overall satisfaction and sub-items both remained over 90%. The qualitative results showed that most students attributed their eased sense of fear to SPs' attitude; that the participation of SPs gave a greater sense of reality; and that students expected SPs to provide concrete suggestions after tests. These quantitative and qualitative results are consistent with previous research findings (Kurz et al., 2009; Lin, Chen, Chao, & Chen, 2012). The only single negative statement, however, was "I hope that SPs will not give me further hints because I won't get scored on it in this case. It is appropriate to let me know which evaluation should be done on my own." Overall, students' opinion was in line with that of instructors. In brief, both quantitative and qualitative data demonstrated that instructors and students were more unsatisfied with the hints and reminders given by SPs.

Many reports or studies have shown some limitations of application of the OSCE in nursing. Compared to traditional methods, the OSCE cannot necessarily assess students' clinical competence, especially in the psychometric properties and high cost (Palese et al., 2012; Walsh et al., 2009). In this reform program, we are confronted with some problems when implementing reforms of the OSCE. Our students are mainly junior university students, and these two courses of study focused on technology. In addition, the professional communication with patients cannot be assessed easily at the junior stage. Therefore, we kept the essence of the OSCE and modified a few steps to replace typical implementation. It will be a formidable task in the future to design a checklist incorporating nurse-patient communication and patients' psychological aspects into the examination. We also faced the biggest problem of "high cost." In the current examination, only one skill was assessed at a time for every student. Therefore, the examination of one class involved 9–12 instructors, 9–12 SPs and several working staff, and took a whole working day. Each course is required to have 2–4 times of examination on every course during each semester. The personnel and financial costs of using the typical OSCE to access clinical skills and competence will quickly become prohibitive.

Moreover, sufficient clinical scenarios are required. We found that students and peers or students a year before or behind them would discuss the scenarios on which they were tested with each other. It might result in a failure of assessment of students' clinical competence in facing clinical situations and a failure of the essence of the OSCE.

## 4 CONCLUSION

In addition to satisfaction, this reform did not use a control group to show any significance results. In addition, we know that the many barriers of implementing the OSCE may appear overwhelming. We still strongly anticipate its good reception and positive effects. Therefore, based on the essence of the OSCE, we will conduct continuous reforms according to the current needs. These reforms include: increasing the editing of clinical scenarios to offer diverse test contents to inspire students' judgment in clinical situations; continuously communicating with SPs to avoid unduly assisting or confusing students; building a checklist of nurse-patient communication and interaction to incorporate nurse-patient communication competence into the OSCE; completing typical OSCE for senior students as early as possible to assess their real clinical competence; and adding evidence-based data to enhance and improve the application of the OSCE in nursing.

## REFERENCES


Brosnan, M., Evans, W., Brosnan, E. & Brown, G. 2006. Implementing objective structured clinical skills evaluation (OSCE) in nurse registration programs in a centre in Ireland: A utilization focused evaluation. Nurse Education Today 26(2): 115–22.

Cushing, A., Abbott, S., Lothian, D., Hall, A. & Westwood, O.M. 2011. Peer feedback as an aid to learning–what do we want? Feedback. When do we want it? Now! Medical Teacher 33(2): e105–12. doi: 10.3109/0142159X.2011.542522.

Harden, R.M. & Gleeson, F.A. 1979. Assessment of clinical competence using an objective structured clinical examination (OSCE). Medical Education 13(1): 39–54.

Kurz, J.M., Mahoney, K., Martin-Plank, L. & Lidicker, J. 2009. Objectives Structured Clinical Examination and examination and advanced practice nursing students. Journal of Professional Nursing 25(3): 186–191. doi: 10.1016/j.profnurs.2009.01.005.

Lin, E.C., Chen, S.L., Chao, S.Y. & Chen, Y.C. 2012. Using standardized patient with immediate feedback and group discussion to teach interpersonal and communication skills to advanced practice nursing students. Nurse Education Today. Advance online publication. doi:10.1016/j.nedt.2012.07.002.

Luctkar-Flude, M., Wilson-Keates, B. & Larocque, M. 2012. Evaluating high-fidelity human simulators and standardized patients in an undergraduate nursing health assessment course. Nurse Education Today 32(4): 448–452. doi: 10.1016/j.nedt.2011.04.011.

McWilliam, P.L. & Botwinski, C.A. 2012. Identifying strengths and weaknesses in the utilization of Objective Structured Clinical Examination (OSCE) in a nursing program. Nursing Education Perspectives 33(1): 35–39.
Mitchell, M.L., Henderson, A., Groves, M., Dalton, M. & Nulty, D. 2008. The objective structured clinical examination (OSCE): Optimising its value in the undergraduate nursing curriculum. Nurse Education Today 29(4): 398–404. doi: 10.1016/j.nedt.2008.10.007.
O'Neill, A. & Mccall, J.M. 1996. Objectively assessing nursing practices: A curricular development. Nurse Education Today 16(2): 121–126.
Palese, A., Bulfone, G., Venturato, E., Urli, N., Bulfone, T., Zanini, A., ...Dante, A. 2012. The cost of the objective structured clinical examination on an Italian nursing bachelor's degree course. Nurse Education Today 32(4): 422–426. doi:10.1016/j.nedt.2011.03.003
Rentschler, D.D., Eaton, J., Cappiello, J., McNally, S.F. & McWilliam, P. 2007. Evaluation of undergraduate students using Objective Structured Clinical Evaluation. Journal of Nursing Education 46(3): 135–139.
Robinson-Smith, G., Bradley, P.K. & Meakim, C. 2009. Evaluating the use of standardized patients in undergraduate psychiatric nursing experiences. Clinical Simulation in Nursing 5(6): e203–e211.
Selim, A.A., Ramadan, F.H., El-Gueneidy, M.M. & Gaafer, M.M. 2012. Using Objective Structured Clinical Examination (OSCE) in undergraduate psychiatric nursing education: Is it reliable and valid? Nurse Education Today 32(3): 283–288. doi:10.1016/j.nedt.2011.04.006.
Walsh, M., Bailey, P.H. & Koren, I. 2009. Objective Structured Clinical Evaluation of clinical competence: an integrative review. Journal of Advanced Nursing 65(8): 1584–1595. doi: 10.1111/j.1365-2648.2009.05054.x
Wass, V., Van der Vleuten, C., Shatzer, J. & Jones, R. 2001. Assessment of clinical competence. Lancet 357(9260): 945–949.
Wilkinson, T.J., Newble, D.I. & Frampton, C.M. 2001. Standard setting in an objective structured clinical examination: use of global ratings of borderline performance to determine the passing score. Medical Education 35(11): 1043–1049.

*Bridging Research and Good Practices towards Patient Welfare – Shih & Liang (Eds)*
*© 2015 Taylor & Francis Group, London, ISBN 978-1-138-02716-9*

# Retained Surgical Instruments: Using technology for prevention and detection

Y.-C. Teng
*School of Clinical Medicine, University of Cambridge, Cambridge, Cambridgeshire, UK*

J. Ward, T. Horberry & P.J. Clarkson
*Engineering Design Centre, University of Cambridge, Cambridge, Cambridgeshire, UK*

V. Patil
*Cambridge University Hospitals NHS Foundation Trust, UK*

ABSTRACT: Retained surgical instruments (RSI) are preventable "Never Events", yet our UK Hospital Trust experienced five retentions between 2011 and 2012. To reduce the retention risk and to aid rapid detection, we propose the deployment of additional technology-based controls: Surgical Data Recorders (SDR) and Radio-Frequency Identification (RFID) systems. In our research, the five relevant incident reports were reviewed, and the risk factors and the report recommendations were analysed. Furthermore, a comprehensive literature search on SDR and RFID was conducted. The current strategy for RSI prevention is manual countingof surgical instruments, but breakdowns in the process is a risk factor for RSI. Our work found that the technological systems proposed are a potentially promising means for accident prevention, monitoring, detection and post-incident learning for the five RSI cases. It is recommended that a systems approach, incorporating solutions like SDR and RFID should be adopted to reduce human errors in "Never Events".

## 1 INTRODUCTION

### 1.1 *Patient safety*

As healthcare systems evolve to unprecedented complexities in the 21st century, delivering safe and efficient care is becoming increasingly challenging. Patient safety has emerged as an important healthcare concern globally in response to the huge impact of medical errors. Medical errors and iatrogenesis present risks, harming as many as 850,000 patients in England per year. Similarly, an estimated 400,000 deaths are caused by preventable mistakes in the US every year (National Patient Safety Agency 2005; James 2013). Despite all the efforts on improving safety in healthcare, the number of patient safety incidents reported to the National Reporting and Learning System (NRLS) in the UK rose steadily between 2003 and 2012 to 340,000 in January to March of 2012 (National Reporting and Learning System 2013). While this increase is in part because local risk management organisations became better connected to NRLS during the time period, the numbers alone make a compelling statement that patient safety can be, and should be, improved.

A key area concerning patient safety that needs to be addressed urgently is "Never Events", defined as "serious, largely preventable patient safety incidents that should not occur if the available preventative measures have been implemented by healthcare providers" (National Reporting and Learning System 2010). Retained Surgical Instruments (RSI) is an example of a preventable Never Event, yet between 2011 and 2012 the UK Hospital Trust where this research was undertaken saw at least five retentions. Serious Incident reports on RSIs typically put forward recommendations that focus on the individuals directly involved in the error. This attempts to reduce behaviours that deviate from the protocol and limit recurrences. The advice often includes carrying out competency assessments on staff, training and re-educating the operators, filling in checklists and enforcing the WHO Surgical Safety Checklist. Yet the "solutions" implemented generally tackle overt operator errors and fail to recognise that such errors largely stem from the underlying design of the healthcare system.

### 1.2 *Systems approach to human error*

Rather than focusing on the fallibility of humans and our predisposition to committing errors,the systems approach accepts that human errors are inevitable even in ideal conditions (Reason 2000). Errors are not deemed to be solely caused by carelessness, forgetfulness or negligence in the individuals. Instead, they are believed to be the result of weaknesses in the systems, such as "recurrent error traps" (Reason 2000). The systems approach calls for re-examination of the healthcare organisation, the conditions surrounding the medical error, and the system defences/controls that were intact and perhaps more importantly, inadequate at the time of the incident. Measures targeting the operator, the procedure, the hospital environment and the healthcare system may be implemented together to provide a more holistic solution.

Information technology offers promising alternative systems solutions to medical errors. In the context of RSI, Surgical Data Recorders and Radio-Frequency Identification (RFID) tagging systems are two emerging technologies that hold encouraging potentials to tackling the problems.

### 1.3 *Surgical data recorder*

A Surgical Data Recorder (SDR) is a concept which is adapted from Flight Data Recorders in aircraft aimed at capturing the complex circumstantial information in patient safety incidents (Nascimento & Calil 2010). The system has had some uptake in the surgical setting in the past, examples of which include a BUPA Foundation research grant in 2001 on the potential of surgical 'black box' in improving clinical standards (Imperial College London 2001). However its implementation has been limited especially for SDRs that record for the purpose of accident investigation, and debates about what it could and should record remain. Information recorded in previous trials that can be incorporated in SDR include patients' physiological data, video, equipment settings and status, and audio recordings (Nascimento & Calil 2010). The captured recordings may serve to prevent serious incidents and improve patient safety in four ways: accident prevention, real-time monitoring, detection and post-incident learning. One may either record the whole surgical procedure or only the "predefined, crucial steps of the procedure, e.g. initial findings, dissection, hemostasis, suture, final findings" (Gambadauro & Magos 2011). The extensive possible combinations of recording medium and method necessitate the identification of how SDR should be aimed at reducing RSI.

### 1.4 *Radio-frequency identification*

Radio-Frequency Identification (RFID) and Data Matrix Code (DMC) are two emerging tagging systems that have been used in medical trials on RSI. A RFID system consists of a tag attached to the surgical item and a "reader" that receives unique radio wave signals from the tags. DMC works on a similar principle – a unique Data Matrix tag on each surgical item is scanned by a barcode reader as it enters and leaves the sterile field (Asiyanbola et al. 2012). While both systems automate the counting process, several differences exist between them. DMC readers are typically laser-based scanners that require a direct line of sight to the whole barcodes. RFID does not need such specific orientations and unlike DMC scanners, RFID readers may detect partially or completely concealed tags. The system is well-suited for preventing and detecting RSI since the items can be scanned more quickly and they have a lower chance of possessing defective tags compared to DMC. And not only can the items be counted before and after the operation, RFID can also scan them during the surgery while they are within the patient's body cavity. This feature does not exist with DMC technology, and thus DMC will not be discussed further for the purpose of RSI prevention.

Objects labelled with RFID tags can be detected by a handheld scanner to automatically count before, during and after the surgery to improve prevention and detection. RFID systems can minimise human error in counting and shorten the detection time. Counting errors have been shown to happen in 1.53% of patients undergoing surgery (Rupp et al. 2012). With a hypothetical detection rate of 100%, RFID may be a far more effective method than manual counting and at the same time promotes prevention and rapid detection of RSI (Macario et al. 2006).

## 2 METHOD

The study was subdivided into two components. First, the Never Event reports at the UK Hospital Trust where this research was conducted were analysed.The selection criteria were retained surgical instruments within the timeframe of 2011 to 2012. The reports follow the general format of: chronology of event, post-incident findings, root cause of retention, contributory factors and recommendations. The analysis identified common themes in the root cause, contributory factors and recommendations.

The second part of the study consisted of a worldwide literature review on how SDR and RFID systems could have prevented the selected RSIs. Papers from 2004 to 2014 indexed in leading databases including Google Scholar, Pubmed and Scopus were selected and reviewed systematically. The review focused on four prevention strategies:

- Error prevention: How technology can prevent the RSI altogether
- Error detection: How technology can provide a faster response when errors do occur
- Real-time monitoring: How technology can monitor surgery in real-time and request for backup assistance when theatres are understaffed
- Post-incidental learning: How technology can record and store crucial information on RSI events to be analysed post-operatively

After the case studies and wider literature were analysed, an inventory of the ways SDR and RFID systems could have helped to prevent the RSIs was generated systematically. In particular, the ways SDR and RFID could have tackled the factors contributing to the five RSIs were inspected.Finally, we reviewed the technologies' potentials in complementing the RSI reports' recommended solutions.

## 3 RESULTS

### 3.1 *Case studies*

Five case studies qualified for the study's selection criteria. Of these, two involve the retention of tampons, one of a swab, one of an Acland clamp, and one of a lapromat. While tampons and swabs generally consist of absorbent cotton, Acland clamp is a small (generally less than 5 mm) clamp that grips blood vessels tightly to occlude blood flow during surgery, and a lapromat is a silicone mat inserted into the abdomen that assists wound closure and is normally removed before the final stitches are completed. As seen in Table 1, six (1.1 to 1.6) risk factors were identified under 'post-incident findings' and 'contributory factors' in the RSI reports in two or more of the incidents, classifying themselves as potential common causes of RSI. Other risk factors (1.7 to 1.11) were identified in single case studies.

After the reports discuss each incident's causes and findings, they propose future recommendations aiming to reduce RSIs, included in Table 2. Some suggestions are identified intwo or more case reviews (2.1 to 2.5), while others are in single cases (2.6 to 2.11).

Table 1. Risk factors in the five incident reports.

| # | Risk factors |
|---|---|
| 1.1 | Staff were aware of the normal practice, but could not remember what happened during the surgery in question |
| 1.2 | The checks were documented as completed when in reality it was not conducted adequately |
| 1.3 | Inconsistencies in counting and use of a checklist in different theatres. In a delivery room, the swab count only involves the operating surgeon, whereas counting requires two persons check in the theatre. |
| 1.4 | Inconsistencies in the instrument sets used between different theatres for the same procedures |
| 1.5 | Time pressure, leading to less than robust counts |
| 1.6 | Failure to communicate and document the full counts on the whiteboard |
| 1.7 | Theatre staff did not escalate staff shortage and call for further assistance |
| 1.8 | Breakdown in communications between theatre staff |
| 1.9 | The Count Policy at the Trust states, "When being counted out swabs should be placed in a clear plastic bag singly or in fives". However, counting swabs (or any other item) in a plastic bag is difficult and more susceptible to counting errors |
| 1.10 | Failure to conduct a visual check on the presence of item |
| 1.11 | Two instruments that are normally used together are counterintuitively packaged in separate different sets. This complicates the checking and counting process |
| 1.12 | The retained item is not radio-opaque, rendering it difficult to detect on routine post-operative radiographs |

Table 2. Recommendations given by the reports to reduce future RSIs.

| # | Recommendations |
|---|---|
| 2.1 | Standardise the counting procedure across all environments |
| 2.2 | Standardise the instrument set used across all environments |
| 2.3 | Implement competency checks and include them in the induction programme for all new staff |
| 2.4 | Decide on whether tampon classifies as a type of swab or not, and whether it should be included separately from the swabs count in the surgical record |
| 2.5 | The surgeon should not rush the team under non-emergency situations |
| 2.6 | Instead of counting used swabs inside a plastic bag, devise a less error-prone method |
| 2.7 | Explore system solutions to minimise the risk of retained swabs |
| 2.8 | Details on the key timings of an operation, operating surgeons and clinical staff at any stage of the procedure should be documented in the theatre operating notes. |
| 2.9 | Switch items to those that are radio-opaque and hence more detectable on radiographs |
| 2.10 | Modify the packaging of surgical items to simplify the counting process |
| 2.11 | Establish clear leadership in the theatre at all times |

### 3.2 *Surgical Data Recorder*

The surgical data recorder system in the operating theatre is a developing concept with great potential in improving patient safety. Numerous SDRs have been devised in the literature to monitor surgeries and the performance of the surgical team in the theatre. Yet few trials to date have been conducted specifically for the purpose of detecting and preventing patient safety incidents. While every operating theatre and surgical procedure is slightly different from each other, SDRs commonly encompass video and audio documentation (Guerlain et al. 2005; Cone et al. 2006; de Korne et al. 2010). Video recordings may target the technical equipment and the surgical staff, while audio recordings may be directed at the theatre in general or be done via lightweight, wireless microphones attached to individual theatre staff (Rockstroh et al. 2014).

#### 3.2.1 *Error prevention: Could SDR have prevent the five RSI cases altogether?*

The concept of SDR as a surgical "black box" implies that it does not actively prevent RSIs during surgical interventions. Unless the feeds are analysed by an operator in real-time, SDRs cannot identify retention before the incisions are closed. On the other hand, periodic reviews of the records can pinpoint the systemic and environmental factors responsible for "near misses" and never events. This provides a platform where surgical staff can learn when and how medical errors are prone to occur. This is especially useful in "near misses" which are nearly impossible to detect post-operatively from traditional surgical documentations alone. The increase in functional awareness amongst the team minimises adverse events and prepares them for unexpected complications (Gambadauro & Magos 2011). In the case examples, periodic reviews of surgery recordings ensures that the checks are completed (Table 1, number 1.2), conducted systematically (1.3) and confirmed visually where appropriate (1.10). Post-operative analysis can identify inappropriate time pressure (1.5), staff shortage (1.7) and breakdown in communications on counts and between theatre staff (1.6 and 1.8).

3.2.2 *Error detection: SDR enables a faster response pathway when errors do occur*

Investigations into RSIs often rely on data from medical logging and subjective personal accounts from staff in interviews to pinpoint the root cause of the event. Yet often the data logging is incomplete (Table 1, number 1.1). And even when data exist, they may be insufficient to piece together the whole chronology surrounding the incident. It is often hard to determine how and when a complication arises, and it is even more difficult to detect and evaluate "near misses" – incidents that could have potentially caused harm to patients but are prevented. The SDR system presents objective documentation of the surgery in the event of a RSI detected post-operatively. Video and audio-recordings of the surgical staff provide an unbiased and exact sequence of events, provided that they are arranged to have minimal blind spots but maintaining patient confidentiality. SDR has the potential to reduce the time spent on inquiries into the individuals involved and surgical documentations, and in turn facilitate the investigations leading to Serious Incident Reports.

3.2.3 *Real-time monitoring: SDR may monitor surgeries and request backup assistance when theatres are understaffed*

It is conceivable that the video and audio recordings by SDR may be monitored in real time by a central operator. Live streaming video of surgeries has been tested in the past, including for teaching and monitoring purposes (Schneider et al. 2007; Santomauro et al. 2013). In theory it should be effective at reducing medical errors and RSIs, providing the monitoring staff are vigilant at all times. A major challenge however would be to gather enough skilled staff to monitor a high volume of surgical procedures at any given time. Equally, operator acceptance of such potentially intrusive technology is uncertain. Until these can be overcome, real-time monitoring may first be implemented on a small scale in random operations to evaluate the values of SDR in monitoring and detecting staff shortage in theatres.

3.2.4 *Post-incidental learning: SDR can record and store crucial information on RSI events to be analysed post-operatively*

Choosing the modes of information to be recorded systematically and the method of such implementation are critical to establishing a SDR that can pinpoint the root cause of an event. Provided that SDRs have a minimal composition of audio and video-recording, surgical procedures may be documented in two ways alongside conventional surgical documentation. One may either record the whole surgical procedure or only the "predefined, crucial steps of the procedure" (Gambadauro & Magos 2011). Whole procedure recordings may be reviewed at random for quality control or systematically in the event of a RSI. Selective recordings may be analysed according to the key moments of a procedure. As the critical moments become statistically connected to patient safety incidents and "near misses", SDR may be continuously refined to optimise the times it selects to record. Data from SDR may be compared with the theatre's records and the surgical outcome.

An effective SDR system should complement the recommended actions set out in Table 2. In the event of a RSI, SDR provides detailed documentation on whether the counting performed conforms to the protocol (this facilitates the solution from Table 2, number 2.1) and whether the theatre is under time-constraints in non-emergency operation (2.5). It records the roles of the operating surgeons and clinical staff at any stage of the procedure (2.8) and helps determine whether leadership is clearly established in the event (2.11).

## 3.3 *RFID*

RFID tags can be attached to surgical items and in turn detected by a handheld scanner that automatically counts the tags before, during and after the surgery. Systems approach to RSIs recognises that human error in counting surgical items is largely unavoidable. A wide range of factors have been shown to contribute to RSI, including incorrect surgical count, emergency surgeries,lack of reminders in place for counting and absence of teamwork and communication amongst the surgeons and nurses (Gawande et al. 2003; Agrawal 2012; Stawicki et al. 2013).

### 3.3.1 *Error prevention: RFID could have prevented the RSIs altogether*

The incidents were all triggered by breakdowns in the routine checking procedures, including failure to complete the count and an inconsistent approach to the check. By implementing RFID alongside standard checking protocols, it minimises the chances of error arising from breakdowns in the counting procedure and it further validates the final count. Egorova et al retrospectively analysed the data stored from 153,263 operations in 2000 to 2004. It was found that count discrepancies occurred at a rate of 0.69%, or 1 in 145 surgical procedures, and the odds of a RSI occurring grew to more than 100 times in the event of mismatched counts (Egorova et al. 2008). The propensity of manual checks to human errors calls for more reliable methods. An experimental clinical trial was conducted at Stanford University Medical Centre, in which a handheld RFID scanner successfully detected all RFID-tagged sponges with 100% detection rate, specificity and sensitivity (Macario et al. 2006). Furthermore, RFID system features the ability to count not only before and after, but also during the surgery while the items are within the patient's body cavity, something not available in traditional counting protocol.

### 3.3.2 *Error detection: RFID enables a faster response pathway when errors do occur*

Traditionally RSIs are detected when the post-operative count mismatches the number counted before the operation, when the item is identified on routine post-operative imaging or when the patient presents with RSI-induced complications. Unfortunately the act of counting multiple times during an operation lends itself to higher chance of errors. In a prospective study enrolling 2285 patients, the RFID system detected one near miss when the manual counting protocols falsely showed all surgical items had been accounted for. The technology further reduced detection time for 35 miscounts, while "no known RSI occurred during the study period" (Rupp et al. 2012). RFID offers the capacity to count the instruments before, during and after an operation. The ability to count the items just before the incisions are closed allows any retained item to be detected and removed, resolving the otherwise Never Event into a "near miss" immediately. This can prevent RSIs from being incidental findings on routine post-operative radiographs, a phenomenon that occurred in 59% of all detected retained foreign objects in a study at the Mayo Clinic (Cima et al. 2008). RFID has been shown to count tagged sponges at 100% detection rate in less than 3 seconds on average (Macario et al. 2006). The speed and accuracy of this technology potentially renders it an efficient detection method supplementing standard manual counting protocols. Equally, it may be a less intrusive technology than SDR, so operator acceptance of it may be higher.

After the items are retained, RSI technology can mitigate the risk factors identified in the five cases. When the checks are documented as completed, a handheld RFID scanner can provide an accurate count rapidly that can then be compared against the manual count (reducing the risk of point 1.2). RFID can minimise the chance of retention in the presence of risk factors that facilitate RSI, including inconsistencies in the counting and use of checklist across different theatres (1.3), time pressure (1.5), absence of visual check (1.10), counterintuitive packaging of surgical items (1.11) and use of radio-transparent items difficult to detect on radiographs (1.12).

#### 3.3.3 *Real-time monitoring: How can RFID monitor surgery and request backup assistance when theatres are understaffed*

Unlike SDR, RFID tagging technology on surgical items does not offer the capacity to detect and assist theatres that are understaffed. An alternative, although intrusive, application may be to tag surgical staff instead and install longer-range RFID detectors in the theatre. As each RFID tag is unique, it can convey the identification of the staff present to a central data processor. These data can be monitored and compared against a pre-set minimum staff requirements in the theatre, such that understaffed surgeries can be detected by the system immediately.

#### 3.3.4 *Post-incidental learning: RFID can record and store crucial information on RSI to be analysed post-operatively*

It would be simple and inexpensive to record the counts performed by the RFID reader next to those done manually. As the literature suggests that RFID has a "detection rate, specificity and sensitivity of 100% each", in theory RSIs should not occur under RFID system and thus there should not be post-accident analysis (Macario et al. 2006). However, there is potential for humanerror in the operation of the handheld detection device. This poses a very small chance of the error coinciding with an inaccurate manual count during an RSI, leaving the object undetected. But appropriate training on the use of the scanner can easily minimise this possibility, and the benefits of RFID would likely outweigh the potential of committing this error.

## 4 DISCUSSION

### 4.1 *Main risk factors in RSI events*

Retained surgical instruments persist at a rate of 1 in 5500 to 1 in 18,760 in-patient operations (Gawande et al. 2003; Cima et al. 2008; Egorova et al. 2008).In this research, we selected five case studies on RSI and analysed the factors and causes that contribute to their occurrences. The five RSIs include retention of two tampons, one swab, one Acland clamp, and one lapromat. The risk factors underlying the RSIs found in this analysis included inconsistencies in the checking procedure and instrument sets leading to incorrect count, time pressure and breakdowns in communication between theatre staff. These factors are in broad agreement with those identified in the literature. Two observational studies by Gawande et al. and Lincourt et al. uncovered several risk factors for RSI. Time pressure arising from emergency operations, the number of major procedures performed, operation theatre time and procedures conducted after 5pm are predisposed to RSI. Incorrect counts of sponges and instruments facilitate RSIs to remain undetected.Changes during a planned procedure, such as handovers during procedure and unexpected changes to the operation also contribute (Gawande et al. 2003; Lincourt et al. 2007).

Controls targeting individuals to eliminate deviations in their execution of the procedures often have short-lived impact before the general awareness of risks of RSIs diminishes again. In an ideal world, medical professionals should adhere to the recommendations completely; however, theatres are highly complex and dynamic in nature, where interruptions, time constraints, unexpected complications and staff handovers are commonplace. SDR and RFID provide automatic and reliable technological alternatives that can complement the recommendations implemented by the RSI reports.

### 4.2 *Benefits of implementing a surgical data recorder*

An encouraging finding from the analysis is that all five cases could have benefitted from a SDR system that contains a minimum of audio and video recordings. SDR can provide easily retrievable, objective documentation of the incident that can be analysed post-operatively in the investigation. It can reduce the time spent on inquiries into the staff and surgical documentation, and in turn facilitate the completion of the Serious Incident Reports for the RSI. Furthermore, whole surgical procedure recordings can be reviewed to identify and subsequently used to reduce inappropriate clinical practice. In our case study analysis, the risk factors identified contain several examples of poor clinical practice. These include deviations from standard counting protocol, breakdowns in communicating the count numbers and surgeons unduly rushing the team, resulting in inadequate checks.

As SDR is still relatively new, several challenges must first be overcome. Since raw video and audio files are simply direct recordings of incidents, the next task is to determine how the data should be captured and processed to aid post-operative reviews (Guerlain et al. 2004). And while it may be efficient to analyse recordings of RSIs, if SDR data is to be reviewed at random for quality control then the analysis may be more time- and labour-consuming. Optimisation of the review process would be ideal if SDR data is to be reviewed periodically to identify risk factors and feedback to the surgical team. Another immediate issue with SDR is the perceived risk and threat to the recorded subjects in the event of litigation, and it is understandable that some may refuse to be recorded altogether. Similarly, the ethics of preserving the privacy of the recorded individuals need to be weighed against the ethics of using the data to reduce the incidence of RSIs (Gelbart et al. 2009).

### 4.3 *Benefits of RFID technology in tackling RSI*

Retained surgical instruments have been estimated to occur at a rate of 1 in 5500 to 1 in 18,760 inpatient operations (Gawande et al. 2003; Cima et al. 2008; Egorova et al. 2008). By obtaining a detailed understanding of the failure mechanisms in the case studies reviewed, our analysis found that RFID technology could have theoretically prevented all five RSIs had it been in place at the time of the incidents. If a RSI does occur, the RFID reader can rapidly detect retention via post-operative scanning and facilitate RSI removal. The benefits of RFID are based on a notionally perfect RFID system, and numerous past studies have indeed shown the high accuracy of such technology (Macario et al. 2006; Rupp et al. 2012). This allows RFID to mitigate the human errors that lead to RSIs, complementing the operator-specific solutions given the RSI reports.

Potential issues in RFID include errors in the operation of the handheld reader that could lead to a missed RSI. The cost-effectiveness of the technology is another concern. Regenbogen et al. estimated that using RFID may cost USD 720,000 for every additional RSI prevented, compared to manual counting. This is cheaper and more effective than adopting universal and selective radiographs for high-risk operations, which they estimated to cost USD 1.1 to 1.4 million for every RSI averted. And as medical and liability often cost more than USD 200,000 per RSI incident (at least within the United States),technologies such as RFID may have a comparatively acceptable cost (Regenbogen et al. 2009).

## 5 CONCLUSIONS

With 234.2 million major surgeries performed globally per year, human error in the healthcare system needs to be carefully managed through a range of effective controls (Weiser et al. 2008). Based on detailed incident case study reviews and a comprehensive review of the worldwide literature, our research has shown that both surgical data recorder and RFID technology could provide an additional layer of "information technology" control. Such controls can be used for accident prevention, monitoring, detection and post-incident learning. Successful implementation of such technologies would not only reduce RSI but also dramatically speed up counting and facilitate investigations into patient safety incidents. However, this is not to say that technology is the perfect solution to RSI. It provides an extra layer of control that complements the solutions implemented by the RSI reports. Appropriate measures targeting human and system factors remain indispensable for resolving the persisting problem of retained surgical instruments.

ACKNOWLEDGEMENTS

The authors wish to thank Cambridge University Hospitals NHS Foundation Trust for providing the funding, support and access to the Serious Incident Reports that made this research possible. Dr Horberry acknowledges the support of an EC Marie Curie Fellowship 'Safety in Design Ergonomics' (project number 268162) he holds at the University of Cambridge, UK.

REFERENCES

Agrawal, Abha. 2012. "Counting Matters: Lessons from the Root Cause Analysis of a Retained Surgical Item." *Joint Commission Journal on Quality and Patient Safety / Joint Commission Resources* 38 (12): 566–74.

Asiyanbola, B, R Etienne-Cummings, and Jonathan S Lewi. 2012. "Prevention and Diagnosis of Retained Foreign Bodies through the Years: Past, Present, and Future Technologies." *Technology and Health Care: Official Journal of the European Society for Engineering and Medicine* 20 (5): 379–86. doi:10.3233/THC-2012-0687.

Cima, Robert R, Anantha Kollengode, Janice Garnatz, Amy Storsveen, Cheryl Weisbrod, and Claude Deschamps. 2008. "Incidence and Characteristics of Potential and Actual Retained Foreign Object Events in Surgical Patients." *Journal of the American College of Surgeons* 207 (1): 80–87. doi:10.1016/j.jamcollsurg.2007.12.047.

Cone, Stephen W, Anna Leung, Francisco Mora, Azhar Rafiq, and Ronald C Merrell. 2006. "Multimedia Data Capture and Management for Surgical Events: Evaluation of a System." *Telemedicine Journal and E-Health: The Official Journal of the American Telemedicine Association* 12 (3): 351–58. doi:10.1089/tmj.2006.12.351.

De Korne, Dirk F, Jeroen D H van Wijngaarden, U Frans Hiddema, Fred G Bleeker, Peter J Pronovost, and Niek S Klazinga. 2010. "Diffusing Aviation Innovations in a Hospital in The Netherlands." *Joint Commission Journal on Quality and Patient Safety/Joint Commission Resources* 36 (8): 339–47.

Egorova, Natalia N, Alan Moskowitz, Annetine Gelijns, Alan Weinberg, James Curty, Barbara Rabin-Fastman, Harold Kaplan, et al. 2008. "Managing the Prevention of Retained Surgical Instruments: What Is the Value of Counting?" *Annals of Surgery* 247 (1): 13–18. doi:10.1097/SLA.0b013e3180f633be.

Gambadauro, P., and A. Magos. 2011. "Surgical Videos for Accident Analysis, Performance Improvement, and Complication Prevention: Time for a Surgical Black Box?" *Surgical Innovation* 19 (1): 76–80. doi:10.1177/1553350611415424.

Gawande, Atul A, David M Studdert, E John Orav, Troyen A Brennan, and Michael J Zinner. 2003. "Risk Factors for Retained Instruments and Sponges after Surgery." *The New England Journal of Medicine* 348 (3): 229–35. doi:10.1056/NEJMsa021721.

Gelbart, B, C Barfield, and A Watkins. 2009. "Ethical and Legal Considerations in Video Recording Neonatal Resuscitations." *Journal of Medical Ethics* 35 (2): 120–24. doi:10.1136/jme.2008.024612.

Guerlain, Stephanie, Reid B Adams, F Beth Turrentine, Thomas Shin, Hui Guo, Stephen R Collins, and J Forrest Calland. 2005. "Assessing Team Performance in the Operating Room: Development and Use of a 'Black-Box' Recorder and Other Tools for the Intraoperative Environment." *Journal of the American College of Surgeons* 200 (1): 29–37. doi:10.1016/j.jamcollsurg.2004.08.029.

Guerlain, Stephanie, Beth Turrentine, Reid Adams, and J.Forrest Calland. 2004. "Using Video Data for the Analysis and Training of Medical Personnel." *Cognition, Technology & Work* 6 (3). doi:10.1007/s10111-004-0154-2. http://link.springer.com/10.1007/s10111-004-0154-2.

Imperial College London. 2001. "'Black Box' for Ops, and Safer Drug Prescribing to Be Investigated by Local Doctors." *Imperial College London*. August 24. http://www.imperial.ac.uk/college.asp?P=2871.

James, John T. 2013. "A New, Evidence-Based Estimate of Patient Harms Associated with Hospital Care:" *Journal of Patient Safety* 9 (3): 122–28. doi:10.1097/PTS.0b013e3182948a69.

Lincourt, Amy E, Andrew Harrell, Joseph Cristiano, Cathy Sechrist, Kent Kercher, and B Todd Heniford. 2007. "Retained Foreign Bodies after Surgery." *The Journal of Surgical Research* 138 (2): 170–74. doi:10.1016/j.jss.2006.08.001.

Macario, Alex, Dean Morris, and Sharon Morris. 2006. "Initial Clinical Evaluation of a Handheld Device for Detecting Retained Surgical Gauze Sponges Using Radiofrequency Identification Technology." *Archives of Surgery (Chicago, Ill.: 1960)* 141 (7): 659–62. doi:10.1001/archsurg.141.7.659.

Nascimento, L. N., and S. J. Calil. 2010. "The Clinical Data Recorder: What Shall Be Monitored?" In *XII Mediterranean Conference on Medical and Biological Engineering and Computing 2010*, edited by

Panagiotis D. Bamidis and Nicolas Pallikarakis, 29:995–98. Berlin, Heidelberg: Springer Berlin Heidelberg. http://www.springerlink.com/index/10.1007/978-3-642-13039-7_252.
National Patient Safety Agency. 2005. "Medical Error". National Patient Safety Agency. http://www.nrls.npsa.nhs.uk/resources/?EntryId45=61579.
National Reporting and Learning System. 2010. "Never Events - Framework: Update for 2010-11." *National Reporting and Learning System*. March 24. http://www.nrls.npsa.nhs.uk/resources/?entryid45=68518.
National Reporting and Learning System. 2013. "NRLS Quarterly Data Workbook up to March 2012". Data report CB_PS_0031. http://www.nrls.npsa.nhs.uk/resources/collections/quarterly-data-summaries/?entryid45=135152&p=1.
Reason, J. 2000. "Human Error: Models and Management." *BMJ* 320 (7237): 768–70. doi:10.1136/bmj.320.7237.768.
Regenbogen, Scott E, Caprice C Greenberg, Stephen C Resch, Anantha Kollengode, Robert R Cima, Michael J Zinner, and Atul A Gawande. 2009. "Prevention of Retained Surgical Sponges: A Decision-Analytic Model Predicting Relative Cost-Effectiveness." *Surgery* 145 (5): 527–35. doi:10.1016/j.surg.2009.01.011.
Rockstroh, Max, Stefan Franke, and Thomas Neumuth. 2014. "Requirements for the Structured Recording of Surgical Device Data in the Digital Operating Room." *International Journal of Computer Assisted Radiology and Surgery* 9 (1): 49–57. doi:10.1007/s11548-013-0909-4.
Rupp, Christopher C, Mary J Kagarise, Stella M Nelson, Allison M Deal, Susan Phillips, Janet Chadwick, Tamara Petty, Anthony A Meyer, and Hong Jin Kim. 2012. "Effectiveness of a Radiofrequency Detection System as an Adjunct to Manual Counting Protocols for Tracking Surgical Sponges: A Prospective Trial of 2,285 Patients." *Journal of the American College of Surgeons* 215 (4): 524–33. doi:10.1016/j.jamcollsurg.2012.06.014.
Santomauro, Michael, G Anthony Reina, Sean P Stroup, and James O L'Esperance. 2013. "Telementoring in Robotic Surgery." *Current Opinion in Urology* 23 (2): 141–45. doi:10.1097/MOU.0b013e32835d4cc2.
Schneider, Armin, Dirk Wilhelm, Dietrich Doll, Uwe Rauschenbach, Michael Finkenzeller, Helga Wirnhier, Klaus Illgner, and Hubertus Feussner. 2007. "Wireless Live Streaming Video of Surgical Operations: An Evaluation of Communication Quality." *Journal of Telemedicine and Telecare* 13 (8): 391–96. doi:10.1258/135763307783064386.
Stawicki, Stanislaw P A, Susan D Moffatt-Bruce, Hesham M Ahmed, Harry L Anderson 3rd, Tara M Balija, Irina Bernescu, Liza Chan, et al. 2013. "Retained Surgical Items: A Problem yet to Be Solved." *Journal of the American College of Surgeons* 216 (1): 15–22. doi:10.1016/j.jamcollsurg.2012.08.026.
Weiser, Thomas G, Scott E Regenbogen, Katherine D Thompson, Alex B Haynes, Stuart R Lipsitz, William R Berry, and Atul A Gawande. 2008. "An Estimation of the Global Volume of Surgery: A Modelling Strategy Based on Available Data." *Lancet* 372 (9633): 139–44. doi:10.1016/S0140-6736(08)60878-8.

*Bridging Research and Good Practices towards Patient Welfare – Shih & Liang (Eds)*
*© 2015 Taylor & Francis Group, London, ISBN 978-1-138-02716-9*

# Holistic enhancement in hip surgery by design

A. Albayrak & R.H.M. Goossens
*Delft University of Technology, Faculty of Industrial Design Engineering, Delft, The Netherlands*

S.B.W. Vehmeijer
*Reinier de Graaf Hospital Delft, Delft, The Netherlands*

H. Bodewes, T. Sunderman & J. Lancaster
*Biomet, Dordrecht, The Netherlands*

ABSTRACT: There is a discrepancy between the patient empowerment trend and patients' physical dependency during rehabilitation. On the one hand, patient empowerment means patients expect to be able to take a more active part in their own healthcare but on the other hand, patients become highly dependent upon other people and feel insecure during the rehabilitation period after a total hip replacement (THR). These people need to re-organize their life during this period, which is often a very hard and isolated process. The aim of this study is to explore patient centered design opportunities to facilitate the patient in a convenient way during their rehabilitation. For this paper, three design cases are selected and discussed.

## 1 INTRODUCTION

Total hip replacement (THR) surgery involves removing a diseased hip joint and replacing it with an artificial joint, called a prosthesis. Hip replacement is typically used for patients with hip joint damage from arthritis, avascular necrosis, loosening of components from a previous THR surgery or an injury (Wright et al., 1994). The main goal of THR is to relieve pain and increase the mobility and function of a damaged hip joint. Followed by rehabilitation, hip replacement can relieve pain and restore range of motion and function of the hip joint. In 2008, almost 30000 Total Hip Arthroplasties were performed and it is expected that the demand for a hip replacement will triple by 2030 (Kurtz et al., 2009). In the Netherlands, there has been an increase of 68% in the number of primary total hip replacements from 1986 to 1997 (Ostendorf & Johnell, 2002). According to the National Institute for Public Health and the Environment (Ministry of Health, Welfare and Sport – RIVM) this number will increase up to 50.000 hip surgeries in the year 2030 (Nationaal Kompas Volksgezondheid, 2009).

In the past developments in THR mostly aim to optimize technical solutions such as improvement of performance of the prosthesis or required surgical tools. However, rehabilitation and treatment after hip surgery has recently seen many developments focusing on a shorter in-hospital stay (Berend, Lombardi et al. 2004; Berger, Jacobs et al. 2004; MacDonald, Arthur et al. 2005). One of the advancements is the Rapid Recovery Program (RRP). As described by Husted and Kristensen, patients stay in hospital for 2–4 days, this is less than the 5–11 days required for a regular procedure (Jorgensen & Kehlet, 2012).

With the RRP patients can recover in their home environment, which has shown to be beneficial for progress of rehabilitation (Mallinson, Bateman et al. 2011), and can support healthcare facilities in achieving an increased efficiency of the perioperative process and cost reduction. Previous research argues that early discharge and guided rehabilitation in the home environment improves function and quality of life (Siggeirsdottir, Olafsson et al. 2005).

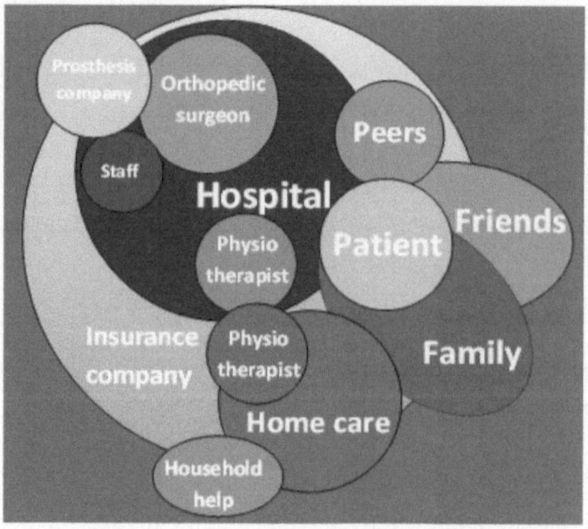


Figure 1. Different stakeholders involved in a THR surgery (adopted from Benito Junquera et al. 2014).

However, the patients that recover at home, experience an elongated and less efficient way of rehabilitation (Jan et al. 2004). A patient needs to feel respected and cared for, which is less noticeable after hospital discharge. Nevertheless, a strong therapeutic alliance between all the stakeholders (Fig. 1), including the patient, increases "patient satisfaction with decisions and patient commitment to treatment plans" (Street et al. 2009). Patient satisfaction is an area, which has been gaining attention (Husted et al. 2008), and the expectation is that in the coming years the healthcare will become more patient centered. Patient empowerment, which means that patients expect to be involved more actively in decision making, will be embedded more and more in the healthcare system (Forbes Healthcare Summit 2013, 2013).

The aim of this study is to explore patient centered design opportunities to facilitate the patient in a convenient way during their rehabilitation. For this paper, three student projects from the master course Advanced Concept Design is selected based on the diversity of their solution.

## 2 MATERIALS AND METHODS

### 2.1 *Approach*

The course Advanced Concept Design is compulsory for the students who are following the Integrated Product Design master program at the Faculty of Industrial Design Engineering at Delft University of Technology in the Netherlands. The first ten weeks of this course 4-5 students are working together in a design team and go through the analysis phase of the design process. During this phase, the design team does design related research on the user, context and technology and finish this part with a report and a personal design vision. This vision describes on an abstract, non-manifest level the intended user experience and the interaction between the user-product-context. In the second part of this course (ten weeks) the students work individually on the conceptualization of their design vision and come-up with innovative concepts at the end.

### 2.2 *Cases*

At the end of the course, Advanced Concept Design there were in total fourteen individual design projects which is clustered in three domains; Website/App (2 projects), Product (4 projects), Product & service (8 projects). For each domain, one case is selected and will be discussed in this paper. Each of the cases addresses different design opportunities to facilitate the patient in a convenient

Figure 2. Interaction 'hearty encouragement'.

Figure 3. Interaction 'support & motivate'.

way during their rehabilitation. The text and drawings used to describe the cases are derived from the original reports of the selected student projects.

2.2.1 *Case I – 'Activities' by A. Weiss (Benito Junquera, A. et al., 2013)*
Design Vision
Helping hip replacement patients to keep their social life going during rehabilitation and involve their family and friends, both through hearty encouragement. Preventing social isolation and making it natural for people close to help.

The interaction of hearty encouragement is a little like having some warm soup in a cold with family and friends around (Fig. 2). Re-energizing, strengthening and providing some intimacy.

2.2.2 *Case II – 'Transformo' by N. Paricharak (Kleissen, K. et al., 2013)*
Design Vision
Support & motivate hip patients during their rehabilitation by providing them feedback on their body posture.

The interaction of involving the patient actively in the progress of their rehabilitation place the patient in such a situation in which he sees the product as a personal trainer who is helping and motivating him/her to get well (Fig. 3). The patient should experience the rehabilitation procedure as a positive challenge. Visualization of the progress is hereby the key to motivation.

2.2.3 *Case III – 'Biomet You' by A. Benito Junquera (Benito Junquera, A. et al., 2013)*
Design Vision
'Everyone is a different world'. Monitor patient's lifestyle before surgery to help them set limits, goals and expectations during rehabilitation, satisfying their need of individualization. At the same time, give the medical specialist a tool to obtain information and establish a more personalized treatment.

Figure 4. Interaction 'everyone is a different world'.

The interaction of personalized treatment enables the patients to set clear and realistic goals to achieve, satisfying their need of individualization, and allows the medical specialist to approach them in such a way that their expectations can be met.

## 3 RESULTS

### 3.1 *Case I – 'Activities' by A. Weiss (Weiss, 2014)*

During rehabilitation, patients experience several problems. One of these is social isolation. They have a hard time keeping up with their social lives and become somewhat isolated, as they are physically impaired and cannot leave home. Additionally, they will become highly dependent on other people and their help even for basic tasks. Both will frustrate them, potentially making them unhappy and reducing their satisfaction with the rehabilitation. This can be prevented by allowing patients to more easily and enjoyably involve people close to them. Both in order to tell them how they are doing and feeling; sharing their rehabilitation and getting motivation out of it. As well as getting relatives, who become more involved, to more frequently provide with chores or impossible tasks.

The concept Activities is a small online platform, with a user-facing app and website and a service connecting the users. In short, Activities allow its users to plan activities together (Fig. 5). Hip replacement patients can invite their family and friends through the application. It also allows them to automatically invite peers who live nearby. Once the patient has invited people, they can plan activities and invite them. The invitees can also plan activities with the patient and invite each other.

The screens share many visual and interaction elements. The most important of which are the cards used for people and activities. These cards allow the user to build or compose an activity. They were inspired by the idea of composing something using physical blocks or cards. As a tablet allows direct manipulation using touch, this digital form retains much of that feel, resulting in a simple, playful, clear and very visual interaction. A toggle allows the user to switch the library to showing people, from which the user can pick the people they would like to invite (Fig. 6).

Both can simply be dragged to the composing area marked by a dashed line (Fig. 7). Dragging them back to the library removes them again (Fig. 7). Some activities allow the user to pick a suggestion for that activity, like a recipe for cooking. When the user adds such an activity, the activities are replaced with the suggestions for it. These can be recipes for cooking, locations for outings etc. Once an activity is selected, the additional information area shows either a motivating description or information about the selected suggestion such as a recipe and allows the user to ask invitees to do the things required for the activity, such as getting groceries or driving. Once the user is done, they tap the send button. Emails are sent out to the invitees and the plan is added to the plans screen, where it can be adjusted.

Users of Activities can interact with the application either using a tablet application or through a web browser. Both provide a very similar interface, using the same visual elements and mostly the same interactions, but optimized for their respective screens. Although the tablet application is

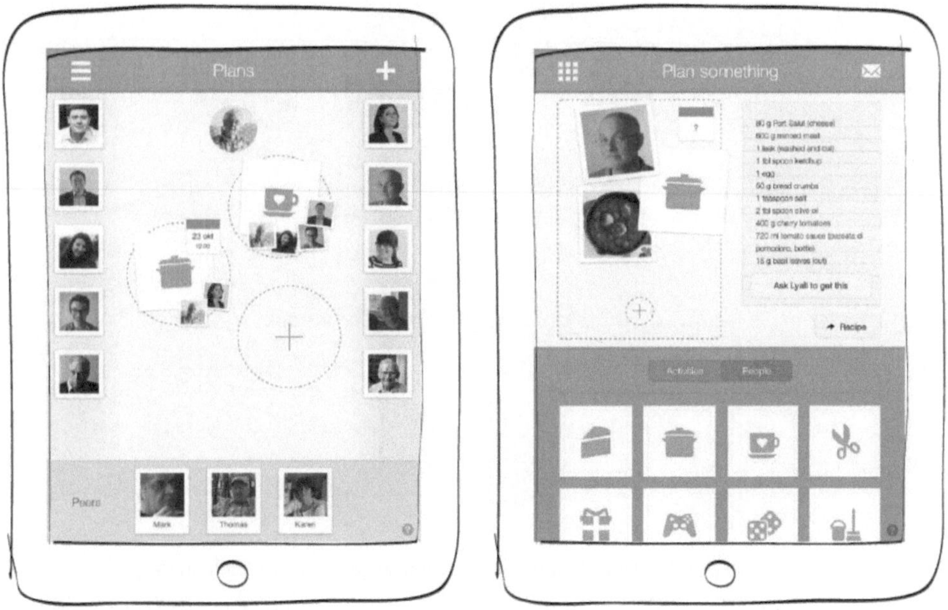


Figure 5. Two main screens of Activities.

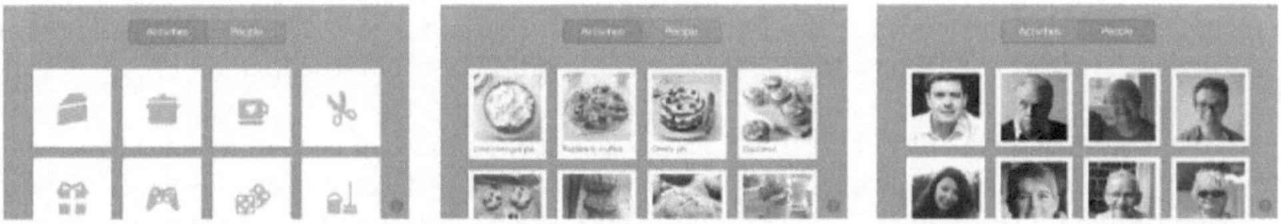

Figure 6. Switching between activities, suggestions and people.

Figure 7. Intuitive usage steps of the Activities.

preferred, due to the touch interaction, which greatly enhances the feeling, the web version makes is easier for family and friends to participate. Requiring only a mouse click to visit it.

### 3.2 *Case II – Transformo' by N. Paricharak (Paricharak, 2014)*

A hip replacement journey is a complex procedure in the life of the patient and there are many factors, which influence the experience of the patient. One of these factors is the gait abnormality of the patient. After surgery almost everyone, tends to demonstrate some degree of walking abnormalities. These walking impairments are often the cause of a referral for physical therapy care to help correct these problems after surgery (Scott, 2011). A common problem, seen in physical therapy departments, is progressing from a walker to no assistive device at all. Discontinued use of walking

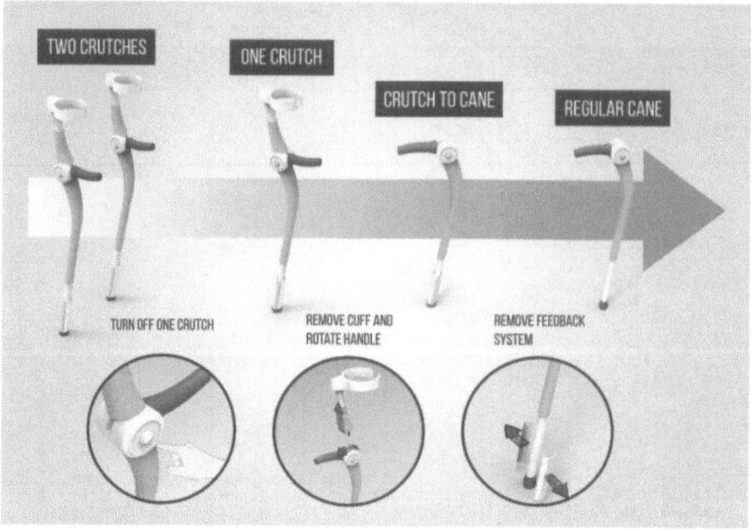


Figure 8. Transformation process depending on the different stages of the rehabilitation.

aids after walker is not recommended and the patient should get support and feedback during this process on their gait to rehabilitate on a convenient way.

The concept 'Transformo' is a smart crutch, which is designed to gradually support a hip replacement patient in all the different stages of the rehabilitation phase. Since healing is a gradual process, the crutches transform of shape allowing the patient to gradually put more load on his leg muscles (Fig. 8).

In consultation with the physiotherapist, the patient can remove elements of the crutches, which are no longer, required due to progress. To control the amount of load on the leg muscles, a vibration feedback system is also attached to the product. This system vibrates and informs the patient if too much or too less support is taken from the crutches while using (Fig. 9).

The stages of the transformation are based on strength improvements of muscles around the hip joint which patient should undergo during the rehabilitation process. The milestones are directly translated to transformations of the crutches. The change in use and shape ensures the specific purpose of each stage.

Stage 1: The patient should be able to put more pressure on their operated leg, in order to use the muscles around the hip joint. By decreasing the support to one crutch, the load on the leg muscles automatically increases.

Stage 2: Once the patient achieves stability in walking, he can advance to a cane. Hereby the load on the patient's leg muscles will increase even more, since the patient cannot use much of his upper body for support.

Stage 3: In order to remove the feedback system, which is meant to correct the walking pattern of the patient, the patient has to be able to walk without a limp when not using an assistive device. In order to train that, the patient has to walk by taking long and regular strides with both legs. If the strides are too small, or irregular, the patient still puts pressure on the supporting tool. The patient can then use the regular cane as a mild support.

### 3.3 *Case III – Biomet You by A. Benito Junquera (Benito Junquera, A. 2014)*

Previous research to the target group made clear how different people are, and how different their lifestyles, mindsets, environments, needs and habits are (Benito Junquera, A., et al. 2013). This evidences the difficulty of generalizing, and urges the need of creating tailor made solutions so patients see themselves reflected and attended in a special way.

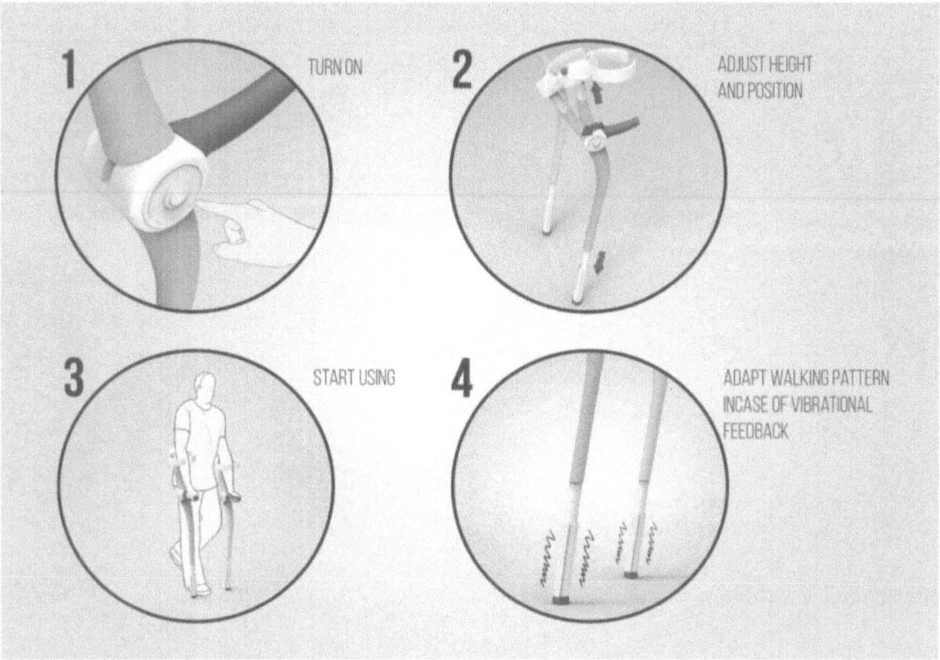


Figure 9. 'Transformo' in use.

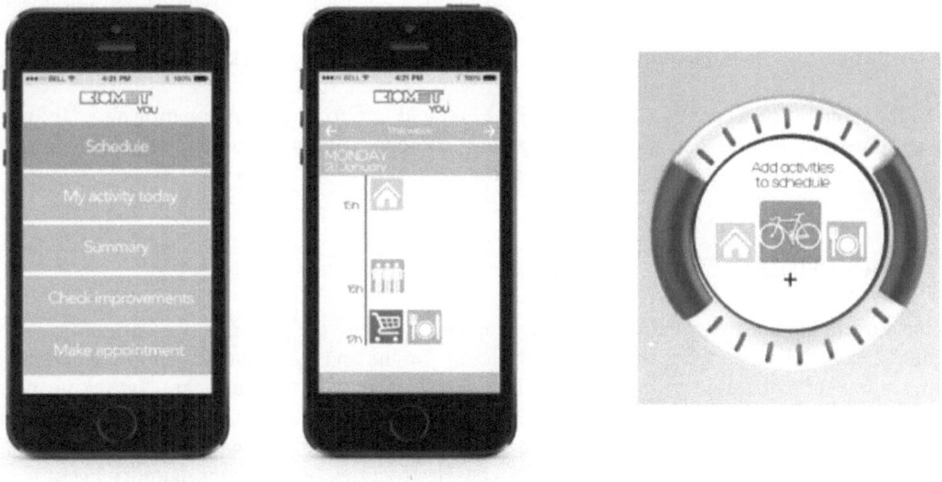


Figure 10. Biomet You, a Smartphone application and a handheld device.

The concept 'Biomet YOU' consist of a handheld device and a Smartphone application and is a solution created to give a personal and adaptable rehabilitation to the patients. It also helps them perceive an impersonal procedure as more personal and individualized. It registers patient's routines, habits and levels of activity before the surgery and this input is used to create a custom-made rehabilitation program after the surgery. It is helping the patients to have realistic expectations and set more familiar and understandable goals during rehabilitation (Fig. 10).

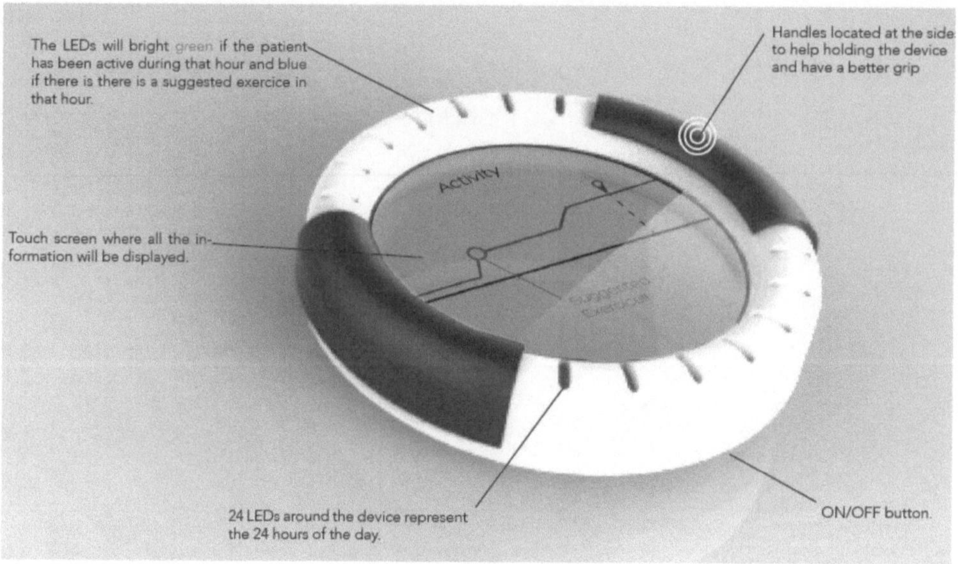


Figure 11. Handheld device showing the activities.

After surgery, a medical specialist and the patient set a custom-made rehabilitation program where the habits and activities of the patient before surgery is included and combined with necessary exercises during rehabilitation. The Smartphone application allows the patient to enter the activities per day/weeks and to make changes if needed. The data stored in the Smartphone application can be synchronized with the handheld device, which shows, and keep track the daily activities of the patient (Fig. 11).

At the same time Biomet YOU, is a tool for the medical staff to gather information about the different patients and increase their understanding how different patients will undergo their rehabilitation and which problems may occur and can be avoided.

## 4 CONCLUSION

In the current healthcare system, there is discrepancy between the patient empowerment trend and patients' physical dependency during rehabilitation. On the one hand, patient empowerment means patients expect to be able to take a more active part in their own healthcare, by taking part in decision making for example (Forbes Healthcare Summit 2013, 2013). On the other hand, patients become highly dependent upon other people and very insecure during the rehabilitation period after hip replacement. Even simple activities like getting groceries or doing the laundry become very difficult for a while. During the rehabilitation phase patients, feels themselves insecure because of lack of information on their activities and exercises.

We believe that patient centered design solutions can empower the patient, facilitate them in a convenient way and improve the communication between the different stakeholders during the rehabilitation phase. The three cases discussed here show how different these solutions can be. The concept 'Activities' addresses the social isolation of the patient due to their physical condition after hip surgery and describes an intuitive interaction which brings different stakeholders together and share the joy of participation in a joint activity. The concept 'Transformo' is a solo product solution, which supports the patient in a convenient way and gives feedback on the gait to improve the walk pattern. The transformation that Transformo undergoes provides the outside world with information on the progress of the patient. The final concept 'Biomet You' is a solution, which combines a handheld device and a Smartphone application. The interaction is based on customization of the rehabilitation program by including the daily life activities of the patient before surgery into this program.

Considering all the fourteen projects from the course Advanced Concept Design it can be concluded that more and more integrated design solutions are created where an application the usage of a physical product extends to the digital world. The shift towards a more product & service design solutions is not surprising seeing the developments in the digital and social media.

## ACKNOWLEDGMENT

The authors would like to acknowledge the contribution of the students: Backus, M., Benito Junquera, A., Cleveringa, E.H., Hoeksema, J., Kersteman, M.P.C., Kleissen. K.J., Leeuwen van, J., Linhardt, A.L.K., Ljungh, H, Paricharak, N.A., Vreede, M.M.A., Waard de, S.T., Weiss, A. J., Winters, M., & Zonneveld van, P., and would like to thank the employees of Reinier de Graaf hospital in Delft for their participation.

## REFERENCES

Berend K.R., Lombardi, A.V. & Mallory, T.H. 2004. Rapid recovery protocol for peri-operative care of total hip and total knee arthroplasty patients. *Surg Technol Int.* 13:239–47.

Berger, R.A., Jacobs, J.J., Meneghini, R.M., Della Valle, C., Paprosky, W. & Rosenberg, A.G. 2004. Rapid, Rehabilitation and Recovery with Minimally Invasive Total Hip Arthroplasty. *Clin Orthop Relat Res.* 429:239–47.

Forbes Healthcare Summit 2013. 2013. Empowering The Patient Revolution. Retrieved January 15, 2014 from http://www.forbes.com/conferences/2013/forbes-healthcare-2013.html

Husted, H., Holm, G. & Jacobsen, S. 2008. Predictors of length of stay and patient satisfaction after hip and knee replacement surgery: fast-track experience in 712 patients. Acta Orthop. 79(2): 168–173.

Jan, M.H., Hung, J.Y., Lin, J.C., Wang, S.F., Liu, T.K. & Tang, P.F. 2004. Effects of a home program on strength, walking speed, and function after total hip replacement. *Arch Phys Med Rehabil.* 85(12):1943–51.

Jorgensen, C.C. & Kehlet, H. 2012. Role of patient characteristics for fast-track hip and knee arthroplasty, *Br J Anaesth.* 110(6):972–80

Kurtz, S.M., Lau, E., Ong, K. Zhao, K., Kelly, M., & Bozic, K.J. 2009. Future Young Patient Demand for Primary and Revision Joint Replacement: National Projections from 2010 to 2030. *Clin Orthop Relat Res.* 467(10): 2606–2612

MacDonald, V., Arthur, B. & Parent, S. 2005. The Vancouver General Hospital joint replacement rapid recovery program: Optimizing outcomes through focused pathways. *Journal of Orthopaedic Nursing* 9(2): 95–102.

Mallinson, T.R., Bateman, J., Tseng, H.Y., Manheim, L., Almagor, O., Deutsch, A. & Heinemann, A.W. 2011. A Comparison of Discharge Functional Status After Rehabilitation in Skilled Nursing, Home Health, and Medical Rehabilitation Settings for Patients After Lower-Extremity Joint Replacement Surgery. *Arch Phys Med Rehabil.* 92(5): 712–720.

Ostendorf, M. & Johnell, O. 2002. The epidemiology of total hip replacement in the Netherlands and Sweden, present status and future needs, *Acta Orthop Scand* 73(3):282–6.

Kleissen. K.J., Paricharak, N.A, Linhardt, A.L.K., Hoeksema, J. & Waard de, S.T. 2013 Midterm report course Advanced Concept Design (ID4170)

Paricharak, N.A. 2014. Transformo – Hip Replacement. Final report Course Advanced Concept Design (ID4170)

Scott, D.F, 2011. Total Hip Joint Replacement Gait Instructions. Spokane Joint Replacement Center. Revised September 6, 2011. Retrieved January 10, 2014 from http://www.spokanejointreplacement.com/

Siggeirsdottir, K., Olafsson, O., Jonsson, H., Iwarsson, S., Gudnason, V. & Jonsson, B.Y. 2005. Short hospital stay augmented with education and home-based rehabilitation improves function and quality of life after hip replacement: randomized study of 50 patients with 6 months of follow-up Acta Orthop. 76(4):555–62.

Street, R.L., Makoul, G., Arora, N.K. & Epstein, R.M. 2009. How does communication heal? Pathways linking clinician–patient communication to health outcomes. *Patient Educ Couns.* 74(3): 295–301.

Volksgezondheid Toekomst Verkenning, Nationaal Kompas Volksgezondheid. 2009 Bilthoven: RIVM, <http://www.nationaalkompas.nl>, Nationaal Kompas Volksgezondheid, versie 3.19, 24 september 2009

Benito Junquera, A., Cleveringa, E.H., Kersteman, M.P.C., Vreede, M.M.A. & Weiss, A. J. 2013. A study to define opportunities for improving rehabilitation of THR patients. Midterm report course Advanced Concept Design (ID4170)

Weiss, A.J. 2014. Activities – A product concept for Biomet. Final report Course Advanced Concept Design (ID4170)

Wright, J.G. Rudicel, S. & Feinstein, A.R. 1994. Ask Patients What They Want. Evaluation of Individual Complaints Before Total Hip Replacement. *British Journal of Bone & Joint Surgery,* 76-B (2): 229–234.

*Bridging Research and Good Practices towards Patient Welfare – Shih & Liang (Eds)*
*© 2015 Taylor & Francis Group, London, ISBN 978-1-138-02716-9*

# Establishment of a checklist for triaging STEMI patients in the emergency department

R.F. Lin & R. Pang
*Department of Industrial Engineering & Management, Yuan Ze University, Chung-Li, Taiwan*

K.-C. Tsai
*Department of Emergency, Far Eastern Memorial Hospital, New Taipei City, Taiwan*

A.-H. Li
*Department of Cardiology, Far Eastern Memorial Hospital, New Taipei City, Taiwan*

ABSTRACT: A checklist, Heart Broken Index (HBI), is current used by the department of emergency, the Far Eastern Memorial Hospital to triage potential STEMI patients. However, the performance of HBI was not superior because it was developed based on physicians' subjective experience. Hence, this study aimed at developing an effective checklist using the lens model approach. Two hundred and fifty-eight clinical cases, randomly selected from 3991 STMEI-suspected cases occurred between 2009 to 2010, were used to develop and test models. Despite the original HBI, a new HBI and an ECG model were developed based on the collected data. The comparison of these three models showed that the ECG model developed based on the lens model approach had the best performance in triaging patients who need the implementation of electrocardiography (ECG). Several suggestions were discussed to enhance the checklist for future practical usage.

## 1 INTRODUCTION

Since 2007, the Far Eastern Memorial Hospital, a 1000-bed regional teaching hospital in northern Taiwan, has engaged in employing methods to reduce the Door-to-Balloon (D2B) time for patients with ST segment elevation myocardial infarction (STEMI). These methods include introducing an audit program (Lai *et al.* 2009), redesigning patient processing flow, modifying bottle-neck processes using six-sigma method, and triaging STEMI patients utilizing the Heart Broken Index (HBI) by Chen *et al.* (2012). However, according to Chen *et al.* (2012), although the employment of these methods helped reduce the D2B time, there were still 25% of STEMI patients required more than 90 minutes of D2B time, a benchmark suggested by the clinical practice guidelines by the United States (Antman *et al.* 2004) and Europe (Van de Werf *et al.* 2003). The development of HBI was an excellent idea to speed the triaging process before sending these patients for electrocardiography (ECG) inspection. However, because the HBI was developed based on physicians' subjective experience, the checklist only obtained a low hit rate, helping little for reducing the D2B time. To help the Far Eastern Memorial Hospital reduce the D2B time, it is necessary to develop an effective checklist using a systematic method.

## 2 LITERATURE REVIEW

### 2.1 *ST-elevation myocardial infarction*

A ST segment elevation myocardial infarction (STEMI), caused by a prolonged period of blocked blood supply, is one type of acute coronary syndromes (ACS), often result from disruption of mildly stenotic vulnerable plaques leading to thrombotic complication (Ambrose 1996). Based on the spectrum and severity of clinical conditions, ACS could be categorized into unstable angina (UA), non-ST-segment elevation myocardial infarction (NSTEMI), and ST-segment elevation myocardial infarction (STEMI) (Kumar and Cannon 2009). Each year in the United States, the hospitalizations of these three types of ACS are approximately 0.55, 0.57, and 0.24 million (Lloyd-Jones *et al.* 2009). Because of differences in the pathophysiology of UA/NSTEMI and STEMI, different therapeutic goals and approaches are applied to patients. For patients with UA/NSTEMI, the goal of antithrombotic therapy is to prevent further thrombosis and to allow endogenous fibrinolysis to dissolve the thrombus and reduce the degree of coronary stenosis (Lewis *et al.* 1983). Hence, revascularization is frequently used to increase blood flow and prevent reocclusion or recurrent ischemia. Since patients with STEMI are more severe than those with UA/STEMI, the focus of this study was placed on STEMI and more details about the therapy and procedures for patients with STEMI will be detailed in the following sections.

### 2.2 *STEMI treatment*

For patients with STEMI, unlike UA/NSTEMI, because their infarct-related artery is usually totally occluded, they should be immediately treated for obtaining normal coronary blood flow (Cannon and Braunwald 1996). Two types of approaches are currently available for establishing timely coronary reperfusion: pharmacological (fibrinolysis) and mechanical (primary percutaneous coronary intervention, PCI) (Boersma *et al.* 2003, Keeley *et al.* 2003). Although the clinical superiority of primary PCI is clear and widely used (Cannon *et al.* 2000, McNamara *et al.* 2006b, Moscucci and Eagle 2006), the main challenge lies in the ability to implement it promptly. Cannon *et al.* (2000) conducted a multivariate adjusted analysis and found that, for patients undergoing primary PCI, increases in the D2B time, especially by more than 2 hours, were associated with higher mortality rates. Evidences (Berger *et al.* 1999, Cannon *et al.* 2000) showed that the D2B time is highly associated with the mortality rate. To his end, the American College of Cardiology and the American Heart Association (ACC/AHA) recommended that all patients with STEMI undergo rapid assessment for reperfusion therapy and that a reperfusion strategy be implemented promptly after the patient's contact with the medical system (Antman *et al.* 2007, Kushner *et al.* 2009). More specifically, the fibrinolytic therapy should be implemented within 30 minutes (door-to-needle time or first medical contact-to-needle time) and the intracoronary balloon inflation should be implemented within 90 minutes (door-to-balloon time or first medical contact-to-balloon time) of the patient's arrival at the hospital or first contact with the medical system.

### 2.3 *Methods for reducing D2B time*

To reduce mortality rates in patients with STEMI, medical centers around the world are making efforts to reduce D2B Time. Bradley *et al.* (2006) surveyed 365 hospitals in the United States and categorized six strategies that were effective for reducing the door-to-balloon time. These strategies included (1) having emergency medicine physicians activate the catheterization laboratory, (2) having a single call to a central page operator activate the laboratory, (3) having the emergency department activate the catheterization laboratory while the patient is en route to the hospital, (4) expecting staff to arrive in the catheterization laboratory within 20 minutes after being paged, (5) having an attending cardiologist always on site, and (6) having staff in the emergency department and the catheterization laboratory use real-time data feedback. However, according to Bradley *et al.* (2006), despite of the advantages of these strategies, few hospitals were using them due to various reasons.

In fact, only a minority of hospitals treat patients with STEMI within 90 minutes and the issue has not be improved substantially in recent years (McNamara *et al.* 2006a).One of these difficulties is making quick and correct judgments when the potential patient arrived at the emergency department. Because the exhibiting signs and clinical symptoms are complicated and varied among patients with a variety of diseases, making a quick and correct diagnosis in the triage stage is

Table 1. Heart broken index (Hsu *et al.* 2011).

| Item | Score | Assessment |
|---|---|---|
| Chest pain (More than 30 years-old with typical chest pain) | 2 | |
| Epigastralgia or severe epigastric pain (not acute pain) | 1 | |
| Cold sweating | 1 | |
| Dyspnea | 1 | |
| | Total Score: | |

difficult. The accurate diagnosis of STEMI can be based on clinical characteristics and persistent ST-segment elevation as demonstrated by 12 lead electrocardiography (ECG). However, due to cost constraint, not every hospital is able to implement the ECG inspection for every STEMI-suspected patient. Hence, triaging patients effectively and efficiently can solve the cost issue and speed the D2B time.

### 2.4 *Far-Eastern Memory Hospital*

The difficulty of speed the D2B time also occurs in the Far Eastern Memorial Hospital. Since 2007, the Far Eastern Memorial Hospital has been applying several methods to reduce the D2B time. One of these methods is to implement a self-developed checklist for triaging STEMI patients. The checklist, called the "Heart Broken Index", was developed by Hsu *et al.* (2011) using four critical symptoms, comprising chest pain, epigastralgia or epigastric pain, cold sweating, and dyspnea (Table 1). Paramedics in the emergency department (ED) use the HBI to triage the patient whom is suspected with STEMI. If the symptom of chest pain or any two of the other three symptoms occur on the patient, he/she would be considered as an highly potential STEMI patient and sent to implement the ECG inspection. The idea of applying the HBI was fascinating and aimed at utilizing effectively and efficiently medical resources on STEMI patients. However, an investigation showed that the implementation of HBI contributed little helps, and the potential reason was that the cues and the corresponding weightings of HBI were determined based on physicians' subjective experiences, lacking of a systematic development method. To this end, a decision-making modeling method, called the "lens model", is introduced in the following section.

### 2.5 *Method of lens model*

The lens model framework was developed by Brunswik (1952) who studied the human behavior of judgment (Guo 2003). As shown in Figure 1, the three basic elements of the lens model are the cues ($x_1, x_2, \ldots, x_n$) available in the environment, actual outcomes ($Y_e$), and decisions ($Y_s$). Two linear regression models (Equations 1 and 2) were created to represent two linear relationships between the cues and actual outcomes ($Y_e$) and linear predictions of human outcomes ($Y_t$), and between the cues and human decisions ($Y_s$). The model generates linear predictions ($Y_f$) of human decisions. The four sets of $Y$ values obtained can be further used to obtain five correlation indices. Each index can be used to evaluate certain aspects of a decision maker's judgment performance (see Tucker 1964 for more details).

$$Y_e = a + \sum_{i=1}^{n} b_i X_i \tag{1}$$

$$Y_s = a + \sum_{i=1}^{n} b_i X_i \tag{2}$$

where, Equations 1 and 2 are ordinary multiple regressions.

The lens model have been widely applied to a variety of domains, such as social policy making (Hammond and Stewart 1975, Brehmer 1986, Dalgleish 1988, Rohrbaugh 1988, Hammond 1996),

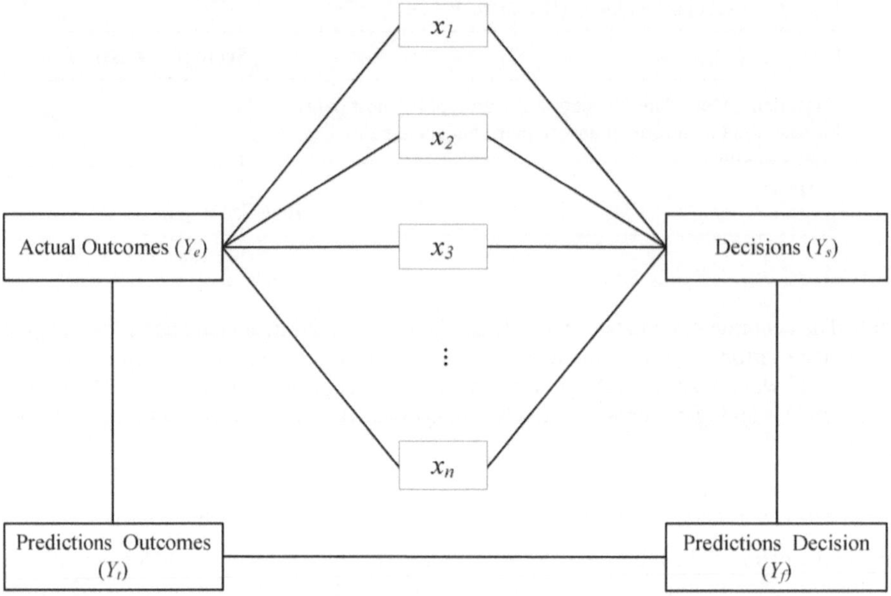


Figure 1. The lens model framework (Kim and McLeod 1999).

reservoir control (Jha and Bisantz 2001), weather forecasting (Stewart *et al.* 1992), and educational judgments (Cooksey *et al.* 1986). More specifically, the lens model approach has been successfully applied to several clinical fields, such as assessing how medical staff use patient characteristics and facial expressions to make pain judgment (Hirsh *et al.* 2011), conferring the relationship between clinical data and physiological measures of hemodynamic status (Speroff *et al.* 1989), and analyzing the judgments of 30 psychologists and psychiatrists assessing imminent danger of psychiatric patient (Werner *et al.* 1983).

### 2.6 *Research objectives*

To increase the accuracy of triaging STEMI patients so as to decrease the D2B time, this study aimed at utilizing the lens model approach (Brunswik 1956) to establish an effective check list for triaging STEMI patients.

## 3 METHOD

### 3.1 *Data collection*

To develop the checklist,clinical data were first collected and analyzed. From 2009 to 2010, the department of emergency of the Far Eastern Memorial Hospital applied the HBI by Hsu *et al.* (2011) to 3991 patients whom were suspected with STEMI. Due to the limitations of time and costs, 258 cases were randomly selected for data analyses. For each case, clinical symptoms and the final diagnosis of disease were recorded. For those clinical cases with diagnoses that required the implement of ECG were treated as signal cases (i.e., $Y_e = 1$); whereas, the remaining clinical cases were treated as noise cases (i.e., $Y_e = 0$). Note that due to limited number of STEMI patient (only one out of 258 cases), the targeted diagnosis was changed from the STEMI patient to the patient who need the implementation of ECG.

Table 2. Twenty-one symptoms reported in the 258 clinical cases.

| Symptom | Frequency | Symptom | Frequency | Symptom | Frequency |
|---|---|---|---|---|---|
| Chest Pain | 107/41.47% | Dyspnea | 28/10.85% | Syncope | 3/1.16% |
| Chest Tightness | 93/36.05% | Radiation Pain | 19/7.36% | Diarrhea | 3/1.16% |
| Nausea/Vomiting | 46/17.83% | Fever | 11/4.26% | Dyspnea on Exertion | 2/0.78% |
| Dizziness | 41/15.89% | Weakness | 8/3.10% | Chest Compression | 1/0.39% |
| Short of Breath | 40/15.50% | Abdominal Pain | 7/2.71% | Chillness | 1/0.39% |
| Cold Sweating | 32/12.40% | Dry Throat | 5/1.94% | Pallor | 1/0.39% |
| Epigastric Pain | 31/12.02% | Left Arm Soreness | 4/1.55% | Difficult to Walk | 1/0.39% |

### 3.2 *Models development*

To demonstrate the effectiveness of the new developed checklist,two models were developed. The first model, called the "new HBI", was developed based on the four cues used by the original HBI, comprising chest pain, epigastralgia, cold sweating, and dyspnea. After statistically testing whether or not the four cues are statistically influential cues for judging patients who needed the implementation of ECG, the influential cues were calculated with appropriate weights using a linear regression model (Equation 1). The second model, called the "ECG model", was developed starting by statistically determining the influential symptoms from the all symptoms reported in the 258 clinical cases. Only statistically influential cues were used to calculate corresponding cue weights for building the model using a linear regression model as well (Equation 1). After the two models were developed, the original 258 clinical cases were used for testing their performance of triaging patients and their performance were compared with that of the HBI using the signal detection theory (Egan 1975, Wickens 1992).

## 4 RESULTS

### 4.1 *Clinical data analyses*

Twenty-one symptoms (Table 1) were reported in the 258 clinical cases.These symptoms listed in descending order of frequency are chest pain, chest tightness, nausea/vomiting, dizziness, short of breath, cold sweating, epigastric pain, dyspnea, radiation pain, fever, weakness, abdominal pain, dry throat, left arm soreness, syncope, diarrhea, dyspnea on exertion, chest compression, chillness, pallor, and difficult to walk.

Among the 258 clinical cases, 124 types of diagnoses were made by physicians, in which 10 diagnoses occurred twice and 91 diagnoses only occurred once. Table 3 summaries 23 diagnoses that occurred more than twice. Among all the 124 diagnoses, those that need the implementation of ECG when the patient arrived at the department of emergency are summarized in Table 4.

### 4.2 *Developing the new HBI and the ECG model*

A statistical test of the four cues used by the original HBI (Hsu *et al.* 2011) showed that the symptoms of chest pain ($p < 0.001$), cold sweating ($p < 0.01$), and dyspnea ($p < 0.001$) were statistically influential cues, but epigastric pain ($p > 0.05$) was not. Thus, the new HBI was assigned with three influentialcues, comprising chest pain, cold sweating, and dyspnea. Their corresponding cue weights are 0.475, 0.095, and 0.283, respectively.

Another regression test showed that among the 21 reported symptoms only chest pain ($p < 0.001$), chest tightness ($p < 0.001$), chest compression ($p < 0.05$), and short of breath ($p = 0.001$) were statistically influential cues for triaging patients who need to apply the ECG. Thus, the STEMI

Table 3. Twenty-three common diagnoses diagnosed in the 258 clinical cases.

| Diagnosis | Frequency | Diagnosis | Frequency |
|---|---|---|---|
| Chest Pain | 29/11.24% | Acute Gastritis | 4/1.55% |
| r/o* Coronary Artery Disease | 11/4.26% | Acute Pancreatitis | 4/1.55% |
| Hyperventilation (HTN) | 10/3.88% | Arrhythmia | 4/1.55% |
| Atypical Chest Pain | 8/3.10% | Bronchitis | 4/1.55% |
| Coronary Artery Disease (CAD) | 8/3.10% | Chest Pain, HTN | 4/1.55% |
| r/o Gastro Esophageal Reflux Disease | 8/3.10% | NSTEMI | 4/1.55% |
| Acute Bronchitis | 6/2.33% | Palpitation | 4/1.55% |
| Atrial Fibrillation (AF) | 6/2.33% | r/o Anxiety | 4/1.55% |
| Chest Pain, r/o CAD | 6/2.33% | Acute Myocardial Infarction (AMI) | 3/1.16% |
| Dizziness | 5/1.94% | Congestive Heart Failure (CHF) | 3/1.16% |
| Vertigo | 5/1.94% | Ventricular premature contraction (VPC) | 3/1.16% |
| Acute Coronary Syndrome (ACS) | 4/1.55% | | |

*Note: r/o means "rule out".

Table 4. Forty-eight diagnoses that need the implementation of ECG.

| | Diagnosis | |
|---|---|---|
| Acute Coronary Syndrome (ACS) | Chest Pain rule out Anxiety | r/o* CHF, AF with Rapid Ventricular Response |
| Acute Bronchitis HCVD | Chest Pain rule out CAD | |
| Acute Bronchitis HTN | Chest Tightness | r/o Chronic Kidney Disease with Acute Exacerbate CHF |
| Acute Lung Edema Uremia | Congestive Heart Failure (CHF) | |
| Atrial Fibrillation | Dizziness | r/o Mitral Valve Prolapse |
| Acute Myocardial Infarction | Dizziness HTN | r/o NSTEMI, Paroxysmal Supraventricular Tachycardia (PSVT) |
| Angina | Dysrhythmia | |
| Aortic Aneurysm | End-Stage Renal Disease CHF | r/o Unstable Angina |
| Atrial Premature Contractions | Fever Atrial Fibrillation | r/o VPC, r/o ACS |
| Arrhythmia | HCVD | Short of Breath |
| Atypical Angina | Hypertension (HTN) | STEMI |
| Atypical Chest Pain | Myocardial Infarction | Suspected Angina |
| Coronary Artery Disease (CAD) | NSTEMI | Suspected NSTEMI |
| Chest Pain | Palpitation | Unstable Angina |
| Chest Pain Headache | Pulmonary Embolism | Upper Respiratory Infection HTN |
| Chest Pain HTN | rule out ACS HTN | Valvular Heart Disease Status Post |
| Chest Pain LBP | rule out CAD | Ventricular premature contraction (VPC) |

*Note: r/o means "rule out".

model was assigned with these four cues and their corresponding cue weights were 0.48, 0.54, 1, and 0.3.

Different cues weights were determined using linear regression so that the extent to which each cue can contribute various importance while triaging patients. After tests, however, the judgment criteria of both models of new HBI and lens STEMI model were set at zero. That is, if one or more than one symptom occurs on a patient, both model triage the patient as a patient who needed the implement of ECG.

Table 5. Model performance.

| Model | Hit (%) | False Alarm (%) | Miss (%) | Correct Rejection (%) | *d'* |
|---|---|---|---|---|---|
| HBI | 46.62 | 39.20 | 53.38 | 60.80 | 0.189 |
| New HBI | 55.64 | 52.00 | 44.36 | 48.00 | 0.092 |
| ECG Model | 91.73 | 66.40 | 8.27 | 33.60 | 0.964 |

### 4.3 *Performance comparison of the three models*

The performance of the three models, comprising the original HBI, new HBI, and ECG model are summarized in Table 5 using the signal detection theory. As shown in the table, the new HBI had a higher rate of hit and a lower rate of miss than the original HBI. However, the discriminability index (*d*') of the new HBI was lower than that of the original HBI. The ECG model had the best performance of rates of hit, false alarm, miss, correct rejection, and had the highest value of discriminability index (*d*').

## 5 DISCUSSION

The comparison of the three models using the signal detection theory showed that the HBI, the new HBI, and the ECG model had the discriminability index (*d*') of 0.189, 0.092, and 0.964 (the higher value, the better prediction performance), respectively. Although the discriminability index of the new HBI was lower than that of the original HBI, the new HBI had better performance of hit (55.64% vs. 46.62%) and miss (44.36% vs. 53.38%). When making decisions for determining whether or not the patients should implement ECG, it would be better to behave in a risky manner. That is, increasing hit rate and reducing miss rate because the cost of miss a STEMI patient is much higher than that of mistakenly alarming a patient without STEMI. Even though, the performance of the new HBI was still poor.Based on the clinical experience, the symptoms of chest pain, epigastralgia or epigastric pain, cold sweating, and dyspnea were used by Hsu *et al.* (2011) to develop the HBI, in which the weight of chest pain was 2, showing the importance. After statistical test of all the 21 symptoms reported in the 258 cases, we found that only chest pain was the statistically influential symptom for triaging patients who should implement ECG, and the cue of chest pain was the third important cue. The reason may be that the HBI employed the symptoms that occur most frequently. However, this is a mistake because the most frequent symptoms may not be the critical cue for making a decision. By applying a systematic approach to determine influential cues and cue weights, the ECG model was formed with the four influential symptoms of chest pain, chest tightness, chest compression, and short of breath. Although these cues were assigned with various weights of 0.48, 0.54, 1, and 0.3, respectively, because the judgment criterion was set at zero, every symptom played a critical role for making decision. Model test results showed that the ECG model had much better performance than the two HBI models did. All the decision outcomes obtained by the ECG model were much better. More specifically, the ECG model had a hit rate at 91.73% and a miss rate at 8.27%.

Although the ECG model had the best performance, there are suggestions for enhance the checklist. First of all, sufficient clinical data would be beneficial to improve the model performance. As found in this study, STEMI was one of 124 diagnoses in the 258 clinical cases and it only occurred once, which is not one of 24 common diagnosis summarized in Table 3. The low occurrence of STEMI made this study develop the checklist for triaging patients who need the implementation of ECG, but not for patients with STEMI. Although triaging ECG patients can reduce D2B time as well, directly triaging STEMI patient can speed D2B time more effectively. To develop the checklist for triaging STEMI patient, adequate clinical data are necessary and enable model be trained and tested with different data sets. Second, the lens model approach can be further applied to model and evaluate human decision-making processes. In this study, we only used clinical data to train the model (i.e., environmental model, $Y_e$). Future study can use physicians' decision to build a new model (i.e., human model, $Y_s$) for further model enhancement.

## 6 CONCLUSIONS

Based on 258 clinical cases that applied a checklist, called the "Heart Broken Index (HBI)", in the Far Eastern Memory Hospital, this study used the lens model approach to develop a new HBI and a ECG model for triaging patients who need the implementation of ECG. The new HBI was developed based on the four symptoms used by the HBI; whereas, the ECG model was developed based on all the 21 symptoms reported in the 258 clinical cases. The comparison of these three models showed that the ECG model had the best performance, demonstrating the potential application of the lens model approach in developing the checklist for helping paramedics in the department of emergency to effectively triage ECG patients. In future research, the checklist can be further enhanced with adequate clinical data and by modeling physicians' decision-making processes. The use of the enhanced checklist should help triage patients who need ECG inspection and thus reduce the D2B time.

## REFERENCES


Ambrose, J.A., 1996. Angiographic correlations of advanced coronary lesions in acute coronary syndromes. *In* Fuster, V. ed. *Syndromes of Atherosclerosis: Correlations of Clinical Imaging and Pathology.* Armonk, NY: Futura Publishing Company, Inc., 105–122.

Antman, E.M., Anbe, D.T., Armstrong, P.W., Bates, E.R., Green, L.A., Hand, M., et al., 2004. Acc/aha guidelines for the management of patients with st-elevation myocardial infarction—executive summary: A report of the american college of cardiology/american heart association task force on practice guidelines (writing committee to revise the 1999 guidelines for the management of patients with acute myocardial infarction). *Journal of the American College of Cardiology,* **44** (3), 671–719.

Antman, E.M., Hand, M., Armstrong, P.W., Bates, E.R., Green, L.A., Halasyamani, L.K., et al., 2007. 2007 focused update of the acc/aha 2004 guidelines for the management of patients with st-elevation myocardial infarction. *Circulation,* **117** (2), 296–329.

Berger, P.B., Ellis, S.G., Holmes, D.R., Granger, C.B., Criger, D.A., Betriu, A., et al., 1999. Relationship between delay in performing direct coronary angioplasty and early clinical outcome in patients with acute myocardial infarction: Results from the global use of strategies to open occluded arteries in acute coronary syndromes. *Circulation,* **100** (1), 14–20.

Boersma, E., Mercado, N., Poldermans, D., Gardien, M., Vos, J. and Simoons, M.L., 2003. Acute myocardial infarction. *The Lancet,* **361** (9360), 847–858.

Bradley, E.H., Herrin, J., Wang, Y., Barton, B.A., Webster, T.R., Mattera, J.A., et al., 2006. Strategies for reducing the door-toballoon time on acute myocardial infarction. *The New England Journal of Medicine,* **355** (22), 2308–2320.

Brehmer, B., 1986. Social judgment theory and the analysis of interpersonal conflict. *Psychological Bulletin,* **83**, 985–1003.

Brunswik, E., 1952. *The conceptual framework of psychology* Chicago: University of Chicago Press.

Brunswik, E., 1956. Historical and thematic relations of psychology to other sciences. *Scientific Monthly,* **83**, 151–161.

Cannon, C.P. and Braunwald, E., 1996. Time to reperfusion: The critical modulator in thrombolysis and primary angioplasty. *Journal of Thrombosis and Thrombolysis,* **3** (2), 117–125.

Cannon, C.P., Gibson, C.M., Lambrew, C.T., Shoultz, D.A., Levy, D., French, W.J., et al., 2000. Relationship of symptom-onset-to-balloon time and door-to-balloon time with mortality in patients undergoing angioplasty for acute myocardial infarction. *The Journal of the American Medical Association,* **283** (22), 2941–2947.

Chen, K.C., Hsu, J.C., Cheng, I.N., Lai, C.L., Fan, C.M., Chen, K.C., et al., 2012. Using presenting symptoms to revise the "heart-broken index" for the triage of acute myocardial infarction. *APSC Subspecialty Congress.*

*Intervention and Imaging 2012.* Taipei International Convention Center: Taiwan Society of Cardiology, 1–78.

Cooksey, R.W., Freebody, P. and Davidson, G.R., 1986. Teacher's prediction of children's early reading achievement *American Educational Research Journal,* **23** (1), 41–64.

Dalgleish, L.I., 1988. Decision making in child abuse cases: Application of social judgment theory and signal detection theory. *Human Judgment: The SJT View, B. Brehmer and C. R. B. Joyce, Eds. Amsterdam, The Netherlands: North-Holland,* 317–360.

Egan, J.P., 1975. *Signal detection theory and roc analysis* New York: Academic Press.

Guo, J.M., 2003. *Policy capturing on budget management performance judgment – an application study of lens model.* ROC National Defense University.

Hammond, K.R., 1996. *Human judgment and social policy* New York: Oxford Univ. Press.

Hammond, K.R. and Stewart, T.R., 1975. *Social judgment theory* New York: Academic Press.

Hirsh, A.T., Callander, S.B. and Robinson, M.E., 2011. Patient demographic characteristics and facial expressions influence nurses' assessment of mood in the context of pain: A virtual human and lens model investigation. *International Journal of Nursing Studies,* **48** (11), 1330–1338.

Hsu, J.-C., Chen, K.-C., Cheng, I.-N. and Li, A.-H., 2011. Using heart broken index to improve the diagnostic accuracy of patient with stemi and shorten door-to-balloon time on emergency department. *American Heart Association 2011 Scientific Sessions.* Orlando, Florida.

Jha, P. and Bisantz, A.M., 2001. Modeling fault diagnosis in a dynamic process control task using a multivariate lens model. *the Human Factors and Ergonomics Society 45th Annual Meeting.* Minneapolis/St. Paul.

Keeley, E.C., Boura, J.A. and Grines, C.L., 2003. Primary angioplasty versus intravenous thrombolytic therapy for acute myocardial infarction: A quantitative review of 23 randomised trials. *The Lancet,* **361** (9351), 13–20.

Kim, C.N. and Mcleod, R., 1999. Expert, linear models, and nonlinear models of expert decision making in bankruptcy prediction: A lens model analysis. *Journal of Management Information System,* **16** (1), 189–206.

Kumar, A. and Cannon, C.P., 2009. Acute coronary syndromes: Diagnosis and management, part. *Mayo Clinic Proceedings,* **84** (10), 917–938.

Kushner, F.G., Hand, M., Smith, S.C., King, S.B., Anderson, J.L., Antman, E.M., et al., 2009. 2009 focused updates: Acc/aha guidelines for the management of patients with st-elevation myocardial infarction (updating the 2004 guideline and 2007 focused update) and acc/aha/scai guidelines on percutaneous coronary intervention (updating the 2005 guideline and 2007 focused update): A report of the american college of cardiology foundation/american heart association task force on practice guidelines. *Journal of the American College of Cardiology,* **54** (23), 2205–2241.

Lai, C.-L., Fan, C.-M., Liao, P.-C., Tsai, K.-C., Yang, C.-Y., Chu, S.-H., et al., 2009. Impact of an audit program and other factors on door-to-balloon times in acute st-elevation myocardial infarction patients destined for primary coronary intervention. *Academic Emergency Medicine,* **16** (4), 333–342.

Lewis, H.D., Davis, J.W., Archibald, D.G., Phil, M., Steinke, W.E., Smitherman, T.C., et al., 1983. Protective effects of aspirin against acute myocardial infarction and death in men with unstable angina — results of a veterans administraction cooperative study. *The New England Journal of Medicine,* **309**, 396–403.

Lloyd-Jones, D., Adams, R., Carnethon, M., Simone, G.D., Ferguson, T.B., Flegal, K., et al., 2009. Heart disease and stroke – 2009 update: A report from the american heart association statistics committee and stroke statistics subcommittee. *Circulation,* **119** (3), 480–486.

Mcnamara, R.L., Herrin, J., Bradley, E.H., Portnay, E.L., Curtis, J.P., Wang, Y., et al., 2006a. Hospital improvement in time to reperfusion in patients with acute myocardial infarction, 1999 to 2002. *Journal of the American College of Cardiology,* **47** (1), 45–51.

Mcnamara, R.L., Wang, Y., Herrin, J., Curtis, J.P., Bradley, E.H., Magid, D.J., et al., 2006b. Effect of door-to-balloon time on mortality in patients with st-segement elevation myocardial infarction. *Journal of American College of Cardiology,* **47** (11), 2180–2186.

Moscucci, M. and Eagle, K.A., 2006. Reducing the door-to-balloon time for myocardial infarction with st-segement elevation. *The New England Journal of Medicine,* **355** (22), 2364–2365.

Rohrbaugh, J., 1988. Cognitive conflict tasks and small group processes. *Human Judgment: The SJT View, B. Brehmer and C. R. B. Joyce, Eds. Amsterdam, The Netherlands: North-Holland,* 199–226.

Speroff, T., Connors, A.F. and Dawson, N.V., 1989. Lens model analysis of hemodynamic status in the critically iii. *Med Decis Making,* **9** (4), 243–252.

Stewart, T.R., Heideman, K.F., Moninger, W.R. and Reagan-Cirincione, P., 1992. Effects of improvedinformation on the components of skill in weatherforecasting. *Organizational Behavior and Human Decision Processes,* **53**, 107–134.

Tucker, L.R., 1964. A suggested alternative formulation in the developments by hursch, hammond, and hursch, and by hammond, hursch, and todd. *Psychological Review,* **71** (6), 528–530.

Van De Werf, F., Ardissino, D., Betriu, A., Cokkinos, D.V., Falk, E., Fox, K.A., et al., 2003. Task force on the management of acute myocardial infarction of the european society of cardiology. Management of acute myocardial infarction in patients presenting with st-segment elevation. *European Heart Journal,* **24** (1), 28–66.

Werner, P.D., Rose, T.L. and Yesavage, J.A., 1983. Reliability, accuracy, and decision-making strategy in clinical predictions of imminent dangerousness. *Journal of Consulting and Clinical Psychology,* **51** (6), 815–825.

Wickens, C.D., 1992. *Engineering psychology and human performance,* second ed. New York: Harper Collins Publishers.

*Bridging Research and Good Practices towards Patient Welfare – Shih & Liang (Eds)*
*© 2015 Taylor & Francis Group, London, ISBN 978-1-138-02716-9*

# Examined the CCT 4000K lighting environment on visual fatigue during working

Chinmei Chou & Tsu-Yu Chen
*Department of Industrial Engineering & Management, Yuan Ze University, Taiwan*

Chih-Yong Chen
*Occupational Hygiene Division, Institute of Occupational Safety and Health, Ministry of Labor*

ABSTRACT: The aim of this study was used two lighting Correlated Color Temperature (CCT) to design a new lighting condition to discuss participant's visual and psychological performance during the simulate office job. This study were applied Eye trackers (View Point Eye Tracker®) and Visual Accommodative Microfluctuations Instruments (Speedy-K; Nikon, Tokyo, Japan) to measure participant's visual performance which included eye's blink, fixation and ciliary Muscle Accommodative. To examined whether the new lighting design could decrease participant's visual fatigue effectively. The results showed that blink duration and time had significance effect under lighting conditions on tasks.Through the results of blink paper-sheet and tablet work looks suitable under the lighting which base on 4000K. Additionally the performance of participant's fixation would be affected by the different lighting conditions and tasks. In summary,we hope these results can improve human's working place lighting, and even enhance the comfort of ward for patient as well.

*Keywords*: CCT, Lighting Condition Design, Visual performance, Eyestrain

## 1 INTRODUCTION

Through the evolution of technology, people are spending more time working indoors. However, we have noticed that spending considerable time working indoors might cause people to experience visual fatigue and discomfort when lighting is insufficient. Focus on the lighting devices which applied widely in present was equipped in single Correlated Color Temperature (CCT). There were few or even none of lighting devices would contain two different lighting CCT conditions in one device and discussed human's working performance and eyestrain at the same time. The definition of term correlated color temperature was introduced when the chromaticity of a selective radiator (e.g., fluorescent lighting) is not exactly equal to any of the chromaticties of a blackbody radiator (Wyszecki 1967). In the view of this issue, this study tried to develop a new lighting characteristic which utilized the fittest lighting for human where applied in the working places to improve human's working performance and physiological uncomforted. To determine the fittest lighting for human, this study focus on the Correlated Color Temperature (CCT) which was the most directive to affect human's perception. In view of this, this study is going to discuss the specific Correlated Color Temperature (CCT) of lighting environments which are not only proven to offer better eye comfort to people, but also proceed to enhance working performance. According to relevant literatures reviewed (Navvab 2002, Manav 2007, Hoffmann G 2008, Sarah Laxhmi Chellappa and Cajochen* 2011) investigated and indicated that CCT would affect human's alert, subjective psychological, and work performance. Among of these literatures, CCT 4000K was recommended by its higher psychological values which conduct by the experimental and assessment by participants. On the other hand, CCT 6500K were evidenced to enhanced participants' work performance and alertness, which measured by the ratio of task corrective and objective human's physiological responses. Another issue which has been known that visual fatigue was associated with blink duration. According to the (Morris and Miller 1996, Häkkänen, Summala et al. 1999, Sirevaag and Stern J.A 2000, Van Orden, Jung et al. 2000, Verwey and Zaidel 2000, Schleicher, Galley et al. 2008) investigated and indicated blink duration was related to the visual fatigue. The part of fixation (Schleicher, Galley et al. 2008) was developed a new concept to evaluate the visual fatigue by three different fixation time, which the time were set less than 150 ms, between 150 ms to 900 ms, greater than 900 ms. And other relevant researches (Velichkovsky, Dornhoefer et al. 2000, Pannasch, Dornhoefer et al. 2001, Phillips and Edelman 2008, Maria Laura Mele 2012) proved that fixation could be an indices to evaluate visual fatigue, and different length of fixation time might caused different visual performance. The High Frequency Component (HFC) value which measured by the ciliary muscle accommodative micro fluctuations, Suzuki et al. (SUZUKI SETSUKO 2001). Due to most of relevant literatures which concerned about the lighting environment evaluation or lighting environment design were conduct in a laboratory and evaluated and determined by the results of experiment which were optimal lighting conditions. In the view of this,focusing on two lighting conditions based on CCT 4000K, as a means of evaluating human's visual fatigue while working in this environment. To interpreted the relationship between lighting CCT and visual performance, this study followed this methodology to design the experimental tasks and protocol to simulate the working status where conduct in a laboratory as well. The purpose of this study is to discuss whether the two lighting environments as examined by this paper affect human's visual fatigue during the working and used three different fixation intervals to figure the change of fixation out and test this theory which developed by Schleicher.

## 2 METHODOLOGY

### 2.1 *Participants*

Eight participants, aged between 20 to 24 were recruited for this study. All participants possessed basic computer and tablet skills necessary for the experiment. Further, all participants had normal or correct to normal vision.

### 2.2 *Experimental setup and equipment*

This study was conducted in an office-like laboratory. The study measured participant's eye movement as a means of evaluating participant's eye fatigue. The instruments used in this study as Figure 1 and 2 shown below which included eye trackers (View Point Eye Tracker®) and visual accommodative microfluctuations instruments (Speedy-K; Nikon, Tokyo, Japan). During the experiment, the

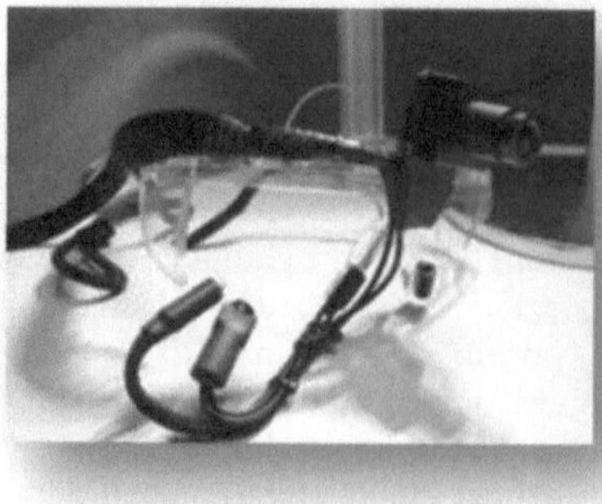

Figure 1. Eye trackers (View Point Eye Tracker®).

desktop illumination was set at 500 lx, the controlled room temperature was 25°C, and the relative humidity was 50% RH.

### 2.3 *Parameter design*

Two lighting conditions as Figure 3 shown for participants were established as a means of evaluating their visual fatigue while working. The lighting conditions included CCT 4000K lighting sources and 6500K lighting sources. The main lighting region was equipped with CCT 4000K lighting sources in the center of the lighting device, and the part of edge side in the devices would set another lighting sources included CCT 4000K, and 6500K; therefore the two lighting sources were combined to construct two different types of lighting conditions which we defined as CCT 4000K–4000K and CCT4000K–6500K.

### 2.4 *Experimental variables*

The dependent variables were blink counts, blink duration, three fixation time interval counts. These variables were measured and recorded by equipped the Eye Tracker to participants,and the HFC value were measured and recorded by the visual accommodative microfluctuations instruments (Speedy-K; Nikon, Tokyo, Japan). Participants were asked to watch the icon (as Figure 4 showed below) which inside the instrument, then the instrument was start to recorded HFC value of participants'. Subjective questionnaires which evaluated eye fatigue contained the subjective eye fatigue, environment feeling, and task satisfaction. The independent variables were the two lighting conditions which this study designed.

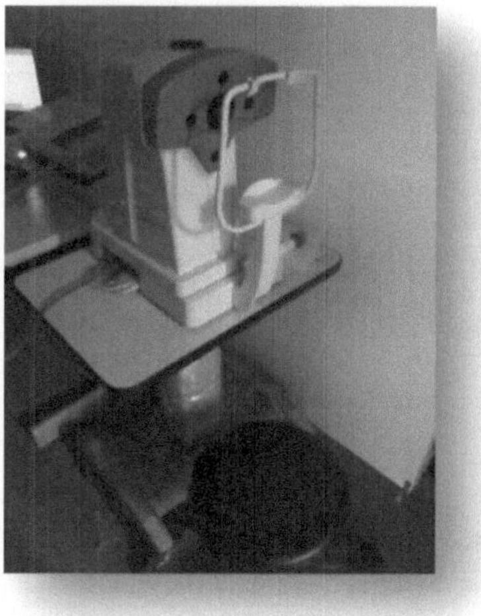

Figure 2. Visual Accommodative Microfluctuations Instruments (Speedy-K; Nikon, Tokyo, Japan).

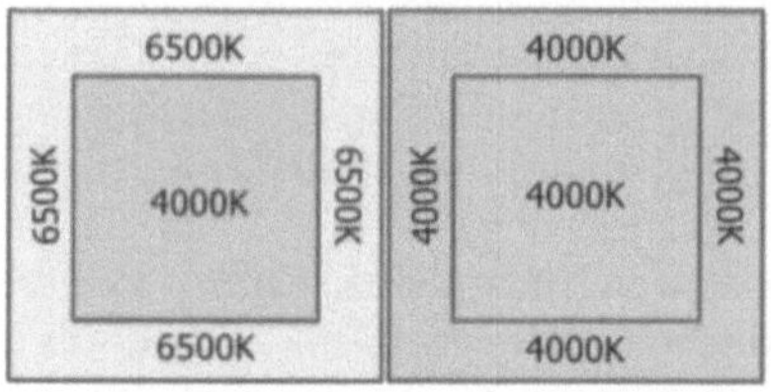


Figure 3. Lighting conditions which this study determined, it contained Two Correlated Color Temperature 4000K and 6500K.

Figure 4. The icon which participants were asked to stare in the visual accommodative microfluctuations instruments (Speedy-K; Nikon, Tokyo, Japan) for measured the performance of Ciliary muscle.

Figure 5. Three primary tasks which this study conduct, the order from left to right were Landolt-Rings paper-sheet task, Three-Digit questions laptop task, and Listening and Wrong Identified tablet task.

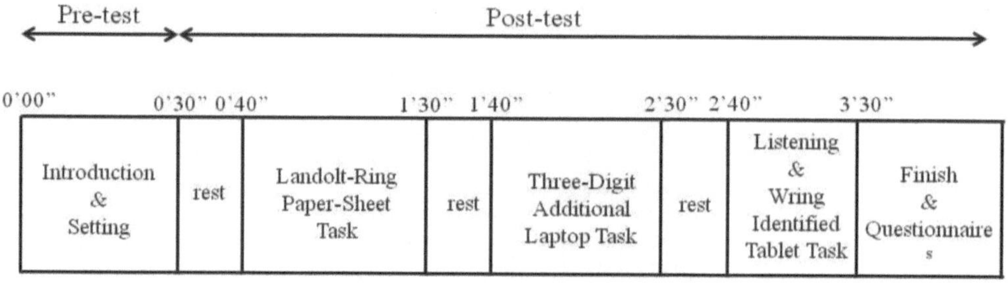


Figure 6. The protocol of this study, which contained three primary tasks and rest during the whole experimental procedure.

## 2.5 *Experimental protocol*

This study required participants to undertake three set tasks under the respective lighting conditions (As shown in Figure 5 and 6). These tasks included paper-sheet, laptop, and tablet task. Paper-sheet task were Landolt-Ring, participants were gave an A4 size paper which contained 240 different directions Landolt-Rings. Participants were asked to count the direction of Landolt-Rings. Laptop task were three-digit additional math questions, participants were asked to use laptop to count the three-digit math questions, the questions of this task were showed up randomly, and the question would change to another until the participant enter the correct answer. Tablet task were Listening and Wrong Identified task, participants were gave a Chinese literature manuscript and asked to listening the literature which the content of literature were similar to the manuscript, and participants needed to correct the different words which would hide in the manuscript. Each task was fifty minutes in duration and the order of tasks were fixed. Participants were given a ten minute break between each task. At the end of this protocol, Participants' HFC value was measured by microfluctuations instruments at the beginning and the end of this protocol. When after the tasks of experiment, participants were gave a questionnaires to state their subjective perception. The figures of tasks, and protocol were shown below.

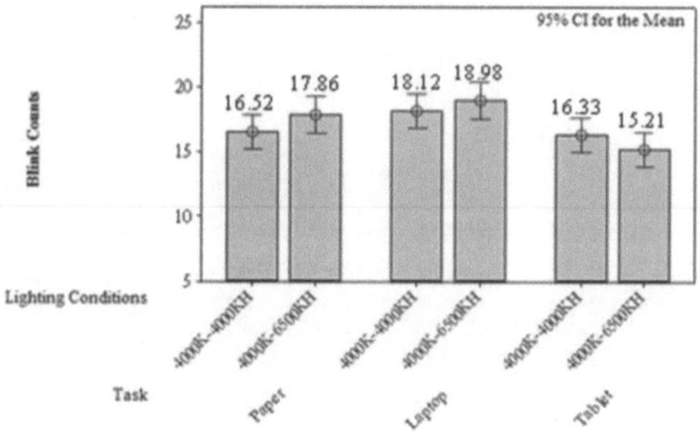


Figure 7. The result of blink counts which reached the significance by three primary tasks designed in this study.

### 2.6 *Statistical methodology*

In this study Analysis of Variance (ANOVA) were applied to test whether the independent variables were existed difference to dependent variables. Post-analysis were also applied by Tukey method for test which levels of dependent variables were truly affected by independent variables.

## 3 RESULTS

### 3.1 *Blink*

Due to the previous literatures indicated Blink Duration and Frequency were an index to evaluate human's visual performance. According to the data analysis which used ANOVA indicated that Blink Duration and Frequency didn't get the significance difference (p-value less than 0.05), however the factor "Task" which means three tasks which this study conduct were got the significance difference. The results were described in detail in follow sections.

#### 3.1.1 *Blink counts and blink duration*

The results of blink counts indicated that through the figure could observe that the new lighting condition had higher blink duration while conducting the task except for tablet task.Blink counts didn't reach the significance different in these two lighting conditions; however different tasks were existed significantly affected by blink counts. According the post-analysis the tablet task scored the lowest blink counts which is 15.77 times/min (as shown in Figure 7). On the other hand, the results of blink duration indicated different tasks were significantly affected by blink duration. The tablet task scored the highest blink duration 37.49 ms (as shown in Figure 8). Further, there were no significant differences between the lighting conditions. In summary, the duration of blink were not be affect by lighting conditions but affected by different tasks.

### 3.2 *Fixation duration counts*

Consistent with the literatures which concern to discuss the fixation related to visual fatigue. Therefore this study referenced the methodology that indicated by R. Schleicher who divided the fixation durations into short (fixation time less than 150 ms), middle (between 150 ms to 900 ms), and overlong fixations (greater than 900 ms) and calculated the counts expected to figure the trend which related to visual fatigue (Schleicher, Galley et al. 2008). Once the fatigue was showed up

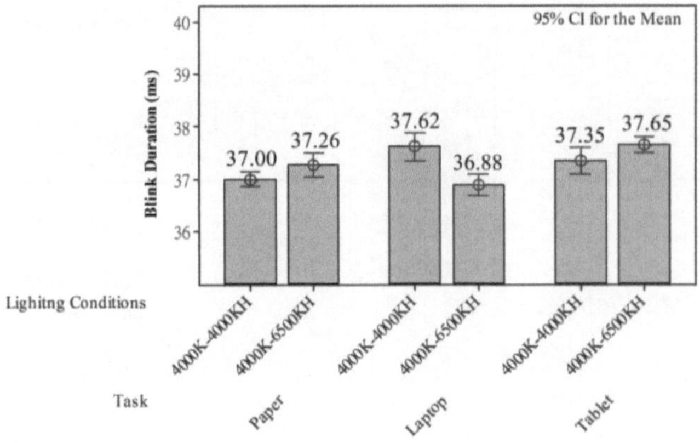


Figure 8. The result of blink duration which reached the significance by three primary tasks designed in this study.

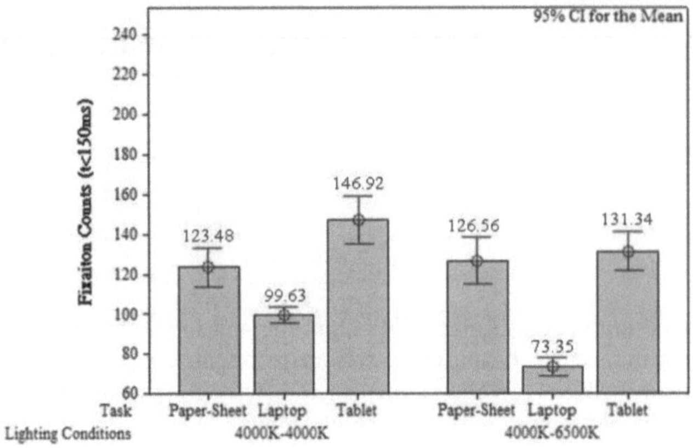


Figure 9. The changed of three tasks fixation counts (t < 150 ms) under these two lighting conditions.

the trend of fixation duration were increased in the middle interval and decreased in the short and overlong interval. And the results would discuss in following section.

### 3.2.1 *Fixation time less than 150 ms*

The results showed that different lighting condition had the significance effect fixation time when the fixation time less than 150 ms. According to the Tukey comparison, as shown in Figure 9. The tablet task scored the highest fixation counts 146.92 times under 4000K–4000K and 131.34 times under 4000K–6500K. in summary, there were no obvious increased trend of fatigue showed in this time interval under these two lighting conditions.

### 3.2.2 *Fixation time between 150 ms to 900 ms*

The results showed that different lighting conditions had a significance effect; through the Tukey comparison as Figure 10 shown, CCT 4000K–4000K scored the higher fixation counts than 4000K–6500K, however in this time interval the fixation counts didn't reach the significance decreased trend to represent the visual fatigue. In summary, the results of post-comparison under these two

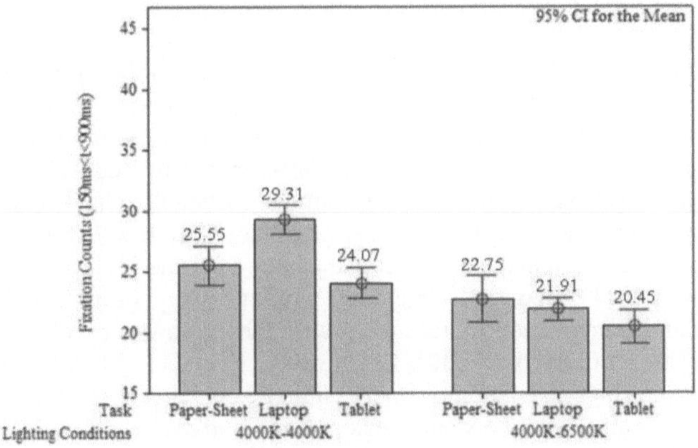


Figure 10. The changed of three tasks fixation counts (150 ms < t < 900 ms) under these two lighting conditions.

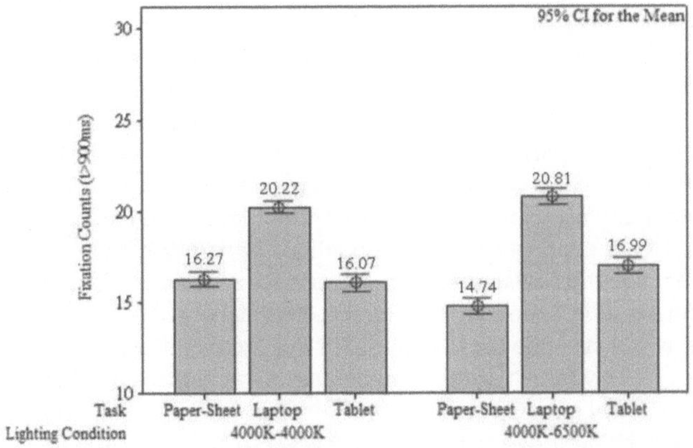


Figure 11. The changed of three tasks fixation counts (t > 900 ms) under these two lighting conditions.

lighting conditions indicated that the laptop tasks scored the highest fixation counts 29.31 times under 4000K–4000K condition.

#### 3.2.3 *Fixation time greater than 900 ms*

In this fixation time interval, as shown in Figure 11 the laptop task scored the highest fixation count 20.22 times and 20.81 times while under 4000K–4000K and 4000K–6500K lighting conditions, the comparison of different lighting conditions had no significance effect. In summary, there was no significance increased trend to represent the visual fatigue in this fixation time interval.

### 3.3 *HFC value and subjective eye fatigue questionnaires*

The results of the HFC value showed that different lighting conditions had no significance effect, however the delta of the HFC had an obvious trend of decreasing under the CCT 4000K–6500K lighting condition. The result of the subjective questionnaires measurement showed that different lighting conditions had no significance effect on participants' subjective eye fatigue, environment feeling, and task satisfaction.

## 4 DISCUSSION

The aims of this study were utilized eyes evaluative indices blink and fixation to examine the lighting conditions which this study designed were suitable for participants'working status and tracked participants' eyes statement by eye-tracking systems and Visual Accommodative Microfluctuations Instruments. According to the previous literatures, Different Correlated Color Temperature (CCT) would affect human's emotion, cognitive of environment, psychological status, and even working performance was be proved and evidenced by many similar researches (Billmeyer 1981, Sato 1992, Linhart and Scartezzini 2011). On the other hand, the issue of visual performance has been research over 25 years (Kowler 2011), among of these much similar researches indicated that blink and fixation were typical and fundamental indices to discuss the eyestrains by observed eye movement (Gooding, Miller et al. 2000, Salvucci 2001, Lavine 2002, Fukuda K 2005, Ulf Ahlstrom 2006). As previous literatures indicated that eye movement could be evidenced its validity and wide applied in present issue which concern to the visual performance study and eyestrains detect. Through the experiment for this study, the results between these two lighting conditions which didn't reach the significance affect on indices that associated with visual fatigue might due to the varieties of these two lighting conditions were not that sensitively; therefore our eyes couldn't reorganized and perception the difference between these conditions. Another reason which we can't detect the eye fatigue by HFC value which performed by ciliary muscle might caused by the protocol and experimental design, due to the nature mechanism of eyes accommodation, the fatigue couldn't detect the fatigue by Visual Accommodative Microfluctuations Instruments. On the other hand, from the aspect of visual performance, previous visual fatigue researches were applied EOG (Electrooculographic) to assess the blink and fixation performance, eye-tracker system applied to collect the data which concern to blink and fixation performance was rarely and develop recently. The counts of three fixation time intervals included express ($t < 150$ ms), cognitive ($150 \text{ ms} < t < 900$ ms), and overlong ($t > 900$ ms) fixation intervals which these interval were developed and named by Schleicher et al in 2008 was used EOG to conduct (Schleicher, Galley et al. 2008). This study expected to follow the same method to judge the visual fatigue by observed the trends of fixation counts; however the trend of fixation counts didn't show up the fatigue obviously. Although the literature's methodology was evidenced could evaluate the visual, the in this study still existing some difference to the original methodology. Therefore this study might reassessment the way of data transformer and the calculated which this study conduct was suitable for this variety of data in the future. The part of HFC value which was the performance of ciliary muscle visual accommodation was a new indicator to judge the visual fatigue,however due to the layout of experimental environment which had a distance from experimental conducting room to HFC vale measure point and the nature mechanism of human's eyes, the recovery rate of eyes fatigue were too rapidly to detect the fatigue phenomena, examiner might miss the period to detect the fatigue. Therefore the performance of ciliary muscle (HFC value) didn't have any difference in this study. The research direction of future work might be able to apply in the medical area. For instance, the patient who lived in ward were needed the comfortable lighting to relax their emotion and psychological statement. Therefore, constructing a good lighting environment for patients to rest or to recuperate is quite benefit. In spite of the literatures which associated with hospital or medical place were not that sufficient, the information of this indoor lighting study may be able to be a referenced prototype for medical place to construct suitable lighting environment to patient, not only enhance the comfort of indoor environment, but also moving forward to change the lighting to make it more suitable, appropriate to human.

## 5 CONCLUSION

The results of this paper show that although most of these visual fatigue indicators were not affected by different lighting conditions as set during the experiments of this study. In another words, maybe this lighting condition setting was suitable for human's eye while conducting the office tasks. In the part of three fixation time interval fixation counts had a lower fatigue tendency under the lighting conditions which this paper set. Thus, these two lighting conditions might suitable for short fixation tablet tasks, long fixation paper and laptop tasks, and in the cognitive interval 4000K–4000K was more suitable than 4000K–6500K. In summarized, this study concluded that these two lighting conditions had its benefit to human's visual perceptions, thus this study hope to apply the lighting not only in the work place, but also to utilize this lighting environment to ward. Hope it could improve patients' physiological comfort and constructed a better lighting environment for patients.

## REFERENCES


Billmeyer, F. W. (1981). "IES Lighting Handbooks, 1981 Reference Volume and 1981 Applications Volume, John E. Kaufman, Ed. Illuminating Engineering Society of North America, New York, 1981, 500 pp. each. Price: $50.00 each, $90.00 set." *Color Research & Application* **6**(4): 253–253.

Fukuda K, S. J., Brown TB, Russo MB. (2005). "Cognition, blinks, eye-movements, and pupillary movements during performance of a running memory task." *Aviat Space Environ Med* **76**: C75–85.

Gooding, D. C., M. D. Miller and T. R. Kwapil (2000). "Smooth pursuit eye tracking and visual fixation in psychosis-prone individuals." Psychiatry Research **93**(1): 41–54.

Häkkänen, H., H. Summala, M. Partinen, M. Tiihonen and J. Silvo (1999). "Blink duration as an indicator of driver sleepiness in professional bus drivers." *Sleep* **22**(6): 798–802.

Hoffmann G, G. V., Griesmacher A, Bartenbach C, Canazei M, Staggl S, Schobersberger W. (2008). "Effects of variable lighting intensities and colour temperatures on sulphatoxymelatonin and subjective mood in an experimental office workplace." *Applied Ergonomics* **39**(6): 719–728.

Kowler, E. (2011). "Eye movements: The past 25 years." *Vision Research* **51**(13): 1457–1483.

Lavine, R. A., et al. (2002). "Eye-Tracking measure and human performance in a vigilance task." *Aviation,Space,and Environmental Medicine* **73**: 367–372.

Linhart, F. and J.-L. Scartezzini (2011). "Evening office lighting e visual comfort vs. energy efficiency vs. performance?" *Building and Environment* **46**: 981–989.

Manav, B. (2007). "An experimental study on the appraisal of the visual environment at offices in relation to colour temperature and illuminance." *Building and Environment* **42**(2): 979–983.

Maria Laura Mele, S. F. (2012). "Gaze and eye-tracking solutions for psychological research." *Cogn Process* **13** (1): 261–265.

Morris, T. L. and J. C. Miller (1996). "Electrooculographic and performance indices of fatigue during simulated flight." *Biol Psychol* **42**(3): 343–360.

Navvab, M. (2002). "Visual Acuity Depends on the Color Temperature of the Surround Lighting." *Illuminating Engineering Society* **31**: 70–84.

Pannasch, S., S. M. Dornhoefer, P. J. A. Unema and B. M. Velichkovsky (2001). "The omnipresent prolongation of visual fixations: saccades are inhibited by changes in situation and in subject's activity." *Vision Research* **41**(25–26): 3345–3351.

Phillips, M. H. and J. A. Edelman (2008). "The dependence of visual scanning performance on saccade, fixation, and perceptual metrics." *Vision Research* **48**(7): 926–936.

Salvucci, D. D. (2001). "An integrated model of eye movements and visual encoding." *Cognitive Systems Research* **1**(4): 201–220.

Sarah Laxhmi Chellappa, R. S., Peter Blattner, Peter Oelhafen, Thomas Götz, Christian and Cajochen* (2011). "Non-Visual Effects of Light on Melatonin, Alertness and Cognitive Performance: Can Blue-Enriched Light Keep Us Alert?" *PLoS ONE* **6**(1).

Sato, H. M. M. (1992). "The Effect of Color Temperature of Lighting Sources on the Autonomic Nervous Functions." *Physiol, Anthrop* **11**(5): 533–538.

Schleicher, R., N. Galley, S. Briest and L. Galley (2008). "Blinks and saccades as indicators of fatigue in sleepiness warnings: looking tired?" *Ergonomics* **51**(7): 982–1010.

Sirevaag, E. J. and Stern J.A (2000). "Ocular measures of fatigue and cognitive factors" *In: R.W. Backs and W. Boucsein,eds. Engineering psychophysiology. Mahwah, NJ: Lawrence Erlbaum Associates*: 269–288.

Suzuki Setsuko, K. M., Kato Keiichiro (2001). "Evaluation of Accommodative Function by High Frequency Component of Accommodative Microfluctuation." *Japanese Journal of Visual Science* **22**: 93–97.

Ulf Ahlstrom, F. J. F.-B. (2006). "Using eye movement activity as a correlate of cognitive workload." *Industrial Ergonomics* **36**: 623–636.

Van Orden, K. F., T.-P. Jung and S. Makeig (2000). "Combined eye activity measures accurately estimate changes in sustained visual task performance." *Biological Psychology* **52**(3): 221–240.

Velichkovsky, B. M., S. M. Dornhoefer, S. Pannasch and P. J. A. Unema (2000). Visual fixations and level of attentional processing. *Proceedings of the 2000 symposium on Eye tracking research & applications*. Palm Beach Gardens, Florida, USA, ACM: 79–85.

Verwey, W. B. and D. M. Zaidel (2000). "Predicting drowsiness accidents from personal attribute,eye blinks and ongoing visual behavior." *Personality and Individual Differneces* **28**: 123–142.

Wyszecki, G. and Stiles, W.S. (1967). *Color science: Wiley New York.*

*Bridging Research and Good Practices towards Patient Welfare – Shih & Liang (Eds)*
*© 2015 Taylor & Francis Group, London, ISBN 978-1-138-02716-9*

# Facilitating teamwork in orthopaedic surgery by design

M. Melles, G. Anastasiadis, C.C.M. Moes & R.H.M. Goossens
*Faculty of Industrial Design Engineering, Delft University of Technology, Delft, The Netherlands*

J. Lancaster
*Biomet Europe BV, Dordrecht, The Netherlands*

S.B.W. Vehmeijer
*Orthopaedics Group, Reinier de Graaf Hospital, Delft, The Netherlands*

ABSTRACT: The tools and products medical specialists use affect how they communicate and collaborate. Still, most medical products are designed from a clinical perspective only. The aim of this investigation is to explore the phenomenon of design and teamwork in orthopaedic surgery, focusing on how the built environment and the tools that are used impact teamwork. Twelve orthopaedic surgeries were observed and staff involved was interviewed. The results include different work models of orthopaedic surgery (i.e. physical, sequence and flow models) yielding different interpretations of the surgical work situation and a set of insights and preliminary design guidelines for facilitating teamwork during surgery by design.

## 1 INTRODUCTION

The impact of teamwork during surgery on patient outcome is increasingly acknowledged and scientifically proven (eg. Hull et al. 2011, Lingard et al. 2004, Reader et al. 2006). Multiple factors are known to influence teamwork during surgery. Next to the expertise of each team member and previous experiences with each other, examples include the design of the artifacts that are used. For example, during certain types of orthopaedic surgery the scrub staff wears helmets, complicating the visual and oral communication between team members. Or the required handling of specific tools in a confined incision area might block the surgeon's view from the incision. Still, most medical products are designed from a clinical perspective only. This investigation explores the topic of designing products that also take into account the impact products have on teamwork. Starting point in this paper is to investigate teamwork in the orthopaedic surgical environment and how the built environment, including the products that are used, affect teamwork. Insights and a set of preliminary design guidelines for facilitating teamwork during orthopaedic surgery will be presented.

### 1.1 *Teamwork in the operating room*

The teamwork skills of healthcare professionals are part of their set of so-called non-technical skills. Non-technical skills are skills that are not directly related to the technical or clinical expertise of healthcare professionals but do influence their performance (Yule et al. 2006). Different categorisations of non-technical skills exist, but in general four categories can be distinguished (based on Flin et al. 2007 and Reader et al. 2006): (1) communication and teamwork – about exchanging information, establishing a shared understanding, coordinating team activities and supporting others; (2) situation awareness – about gathering and understanding information and anticipating future situations; (3) decision making – about considering options and implementing and reviewing decisions; and (4) task management – about planning and prioritising tasks and utilising resources.

The importance of non-technical skills in healthcare has been more and more acknowledged over the last years. Teamwork in particular has gained in attention due to societal trends as aging – leading to increased comorbidity, patient empowerment – resulting in patients and relatives having a more active role in their treatment, and advanced technology making multiple specialists needed. As a consequence, teamwork manifests itself in many domains in healthcare today.

In this paper the focus is on teamwork in the orthopaedic operating room (OR). The required collaboration between the multiple teams during surgery make the OR one of the most complex working environments in health care (Christian K. et al. 2006). For each operation, four to ten medical staff members need to work in the same OR. These professionals have different trainings and backgrounds. They do not always know each other and they have different needs while working in the same room. The ORs have to fit all these different needs and provide an ideal environment that facilitates collaboration among the members of the OR team.

Orthopaedic OR teams consist of professionals from at least three different specialties; surgeons, anaesthesiologists, and nurses (Sevdalis et al. 2008). Additionally, a radiologist can be present during certain types of surgeries. These professionals formulate the following four sub-teams;

- *Surgical team*
  The surgical team consists of senior surgeons (SS) and resident surgeons (RS). More complicated surgeries can involve up to four surgeons. The senior (attending) surgeon is the leader of the whole OR team and has the responsibility to make sure the surgery proceeds according to plan without any errors. The senior surgeon guides the sub-teams through the different steps of the operation. He works closely with the anaesthesiologist team to ensure the patient's current condition. The resident surgeon works under the supervision of the senior surgeon.
- *Nurses team*
  The nurses team typically consists of two to four nurses that help the operation to proceed smoothly. The registered nurse (RN) works closely with the surgical team by assisting them with all the necessary tools and equipment. The RN makes sure that the equipment has been successfully sterilized, passes the required tools to the surgical team and, ideally, anticipates what the surgeon will need. The circulating nurse (CN) moves the trolleys with the equipment from the sterilization room to the operation room, unwraps the non-sterilized wrappings for the scrub nurse and assists all sub-teams during the operation. In general, the main goal of the nurses team is to minimize any distractions that may lead the surgical team loose their focus (Hull et al. 2011).
- *Anaesthesiologist team*
  The anesthesiologist team usually consists of one anesthesiologist and one assistant. The team is in charge of the patient's level of consciousness, medication, heart rate, blood and urine loss, and respiration. Typically, in the team's briefing at the beginning of the operation, the anaesthesiologist asks the patient about the medication options and any allergic history. The team needs to communicate with the surgeons many times during an operation in order to be aware of the moments that a change in the patient's medication is needed.
- *Radiologist team*
  During certain types of surgeries, a radiologist needs to be present. This sub-team typically consists of only one radiologist who is specialized in the diagnosis of diseases and injuries using two or three dimensional imaging (e.g. X-ray, Computer Tomography – CT, ultrasound and Magnetic Resonance Imagining – MRI). When the radiologist is needed during surgery, he or she communicates with the surgical team in order to ensure the patient's condition.

All aforementioned professionals need to perform their responsibilities at the same time, while they are working at the same workplace. While each user has individual tasks, there are many moments that communication and collaboration between the sub-teams are crucial. Poor or inefficient teamwork might lead to misunderstandings affecting situation awareness and patient safety (Hull et al. 2011, Reader et al. 2006), possible delays in the surgical procedure (Lingard et al. 2004), and repetition of certain tasks and usage of equipment that is not required, again leading to delays in the procedure (Lingard et al. 2004). The aim of this investigation is to explore the phenomenon of teamwork and design in orthopaedic surgery (see also Anastasiadis, 2013). The result will be a set of preliminary insights and guidelines for facilitating teamwork during surgery by design.

## 2 METHOD

### 2.1 *Participants*

In total 12 orthopaedic surgeries were observed, followed by semi-structured interviews with the OR staff members involved, in four different hospitals in the Netherlands. The observations included four Total Hip Arthroplasty (THA) operations using the Anterior Supine Intermuscular (ASI) approach, five THA operations using the conventional approach and three ankle screw removal surgeries.

### 2.2 *Procedure*

Each surgery was attended from the pre-operative phase until the post-operative phase. The interaction between the members of the orthopaedic surgical teams was observed with a focus on communication and collaboration. A special focus was on how the working environment, including all artifacts, influenced teamwork. Due to privacy reasons, the operations couldn't be recorded with any audiovisual mean. Instead, sketches were made illustrating the working environment and the performed actions. Relevant behavior and actions of the team members were written down. The interviews were performed after each operation, while the team was having its break between operations. Overall, the goal of the interviews was to understand communication difficulties from the different team members' perspective. Each observation lasted between 1.5 and 2 hours. The interviews took about 20 minutes.

### 2.3 *Analysis*

Work modeling techniques of Beyer & Holtzblatt (1998) were used to analyse the observations and interviews. Work modeling consists of five different perspectives on a work situation. These different interpretations are expected to lead to a better understanding of needs and work context, eventually leading to more suitable products. So far, we used three of the five models: the physical model, the sequence model and the flow model. Physical models visualise the physical environment, in our case showing how the sub-teams organise their working environment and how the environment supports or constraints their work. Sequence models capture the steps to accomplish a task, including goals and problems. Our sequence models were created in order to identify the different strategies that OR professionals use and to identify problems. And finally, flow models were created to capture the communication and coordination between the OR sub-teams, their formal and informal communication patterns, roles and responsibilities. Physical models, sequence models and flow models were made of each operation. One final flow model was made incorporating all findings from the different observations. Remarkable observations related to teamwork and design were clustered into six themes. Eventually, based on all findings, six design guidelines for facilitating teamwork in orthopaedic surgery were formulated. For a more detailed description of the procedure and analysis see Anastasiadis (2013).

## 3 RESULTS

### 3.1 *Physical models*

In order to gain an impression of the working area of each sub-team, the spatial usage of the environment per sub-team per surgery was mapped. Figure 1 shows an example of such a physical model made of a THA surgery using the ASI approach. As illustrated in this model the working environment of the surgical sub-team is confined around the incision. The other sub-teams are using the remainder of the space in order to attend to the needs of the surgeons. In general, the observations made clear that the senior surgeon is leading the surgery and consequentially also leading in how and when the members of the other sub-teams carry out their own tasks. Hence, awareness among all sub-team members of what the surgical team is doing is crucial for the efficiency and continuity of the surgical procedure.

During this particular type of THA surgery, plastic translucent sheets are used hanging in the sagittal plane of the patient. As a consequence, the registered nurse (RN) needs to stand at the side of the incision in order to properly see the incision. The RN therefore stands behind the surgical team, which complicates communication and tool handling. The sheets also block the direct view between surgical team and radiology team and (partly) anaesthesiology team, this way hindering the communication between these teams. Since the anaesthesiologists need to know the exact phase of the operation in order to control analgesia, the assistant anaesthesiologist is regularly walking close to the operating table (see figure 1, AA). This way they can see and communicate orally with the surgical team if and when they have to increase dosages. The cooperation of the radiologist was needed three times during this surgery: at the start of the operation in order to check the patient's damaged hip and twice after the placement of the prosthesis to ensure that its location is optimal. At these moments, there is oral communication between the radiologist and the senior surgeon while analysing the imaging results. The circulating nurse (CN) is walking around and collaborating with all sub-teams to answer to their needs.

### 3.2 *Sequence models*

Sequence models were made of all 12 observed operations for each phase (pre-operative, intra-operative and post-operative). Overall all procedures occurred with success without any serious irregularities. Figure 2 shows an example of a sequence model made of a THA surgery using the ASI approach. Attention can be paid to the fact that equipment was missing during the intra-operative phase. This led the procedure to pause for a few minutes until the team found the equipment in the OR. The team informed the observer that it's not a rare phenomenon to miss equipment from the storage room. Another point of attention concerns the importance of the responsibilities of the RN. The RN has to be alert during the whole procedure about the current state of the operation, anticipating the equipment needed by the surgeon and this way collaborating with the surgeon in the most optimal way. It was observed that a few times the SS had to place the equipment used on the surgical table instead of giving it back to the RN (which is procedure), because the RN was busy with other tasks.

### 3.3 *Flow model*

Figure 3 shows the flow model that is made based on all observations and interviews. All the team members have their own predetermined tasks, but also need to communicate with the other team members on a regular base. It was observed that all collaborations have as final recipient the surgical team. Additionally, the CN always stands as the middle person between the surgical team and the rest of the professionals. In figure 3, one can see that the whole team communicates multiple times in order to assure the current task of the surgical team. According to the particular phase the surgical team is in, the rest of the OR team needs to start performing other tasks.

### 3.4 *Preliminary insights*

- Task management
  All participants indicated the importance of time during an operation. Delays and time consuming actions should be minimized or even eliminated. It was observed that many small delays were found on delivering the right instrument at the right moment.

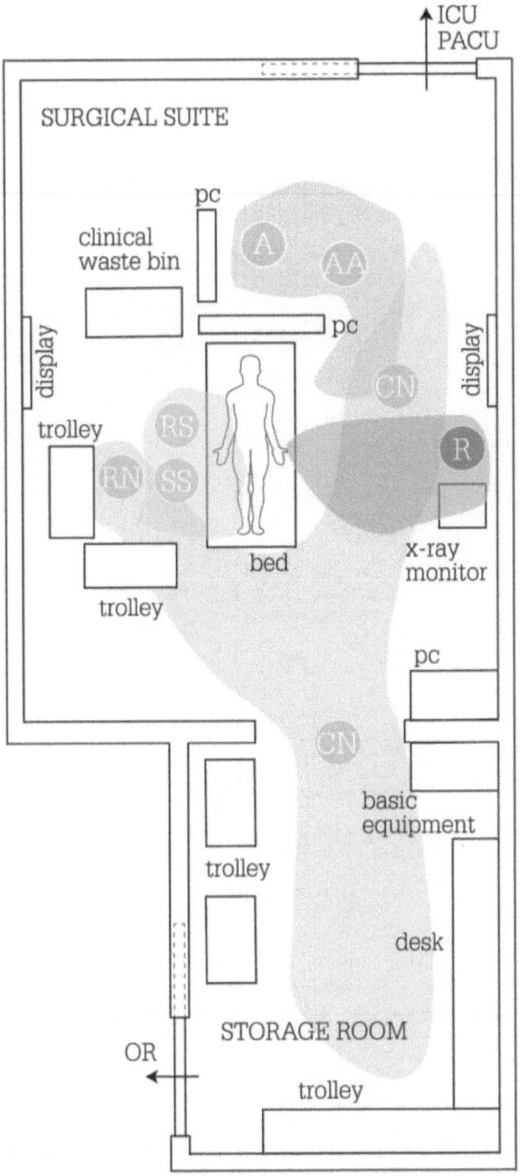


Figure 1. Physical model of a THA surgery using the ASI approach, based on work modeling techniques of Beyer & Holtzblatt (1998). SS = senior surgeon, RS = resident surgeon, RN=registered nurse, CN = circulating nurse, A = anaesthesiologist, AA = assistant anaesthesiologist, R = radiologist.

- Awareness of the operation phase
  At various moments during surgery, the sub-teams seem not fully aware of the current operation phase. This leads to small delays in operation time, since the sub-teams are not fully aware when to perform their (pre-determined) tasks.
- Visual communication
  The effectiveness of the visual and oral communication among the sub-teams seemed questionable since various used artifacts blocked their way, such as translucent plastic sheets, trolleys, helmets and sterilized suits.

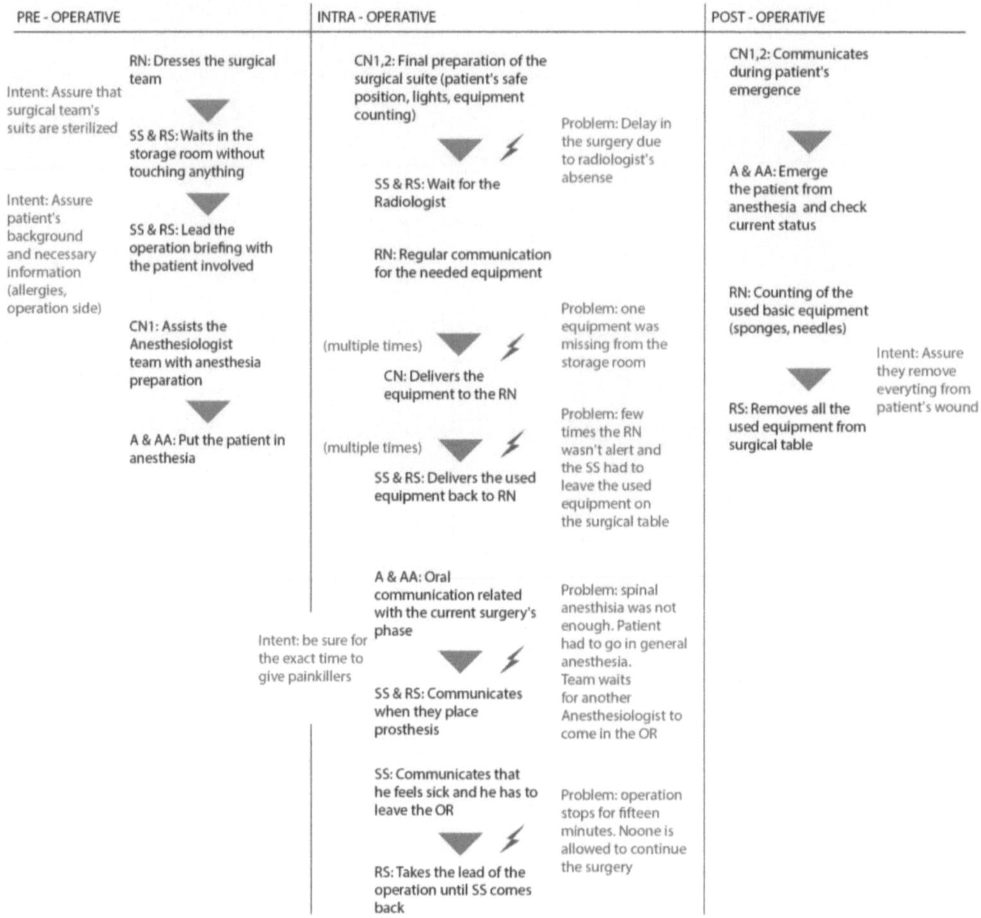


Figure 2. Sequence model of a THA surgery using the ASI approach, based on work modeling techniques of Beyer & Holtzblatt (1998). SS = senior surgeon, RS = resident surgeon, RN = registered nurse, CN = circulating nurse, A = anaesthesiologist, AA = assistant anaesthesiologist, R = radiologist.

- Handling of tools
  Another point of attention is the way the current tools are handled. For example, the resident surgeon (RS) needs to hold the retractors at certain positions for a long period of time. This makes the collaboration among the surgical team members more complicated since the RS's hands block the surgeon's view on the incision. Additionally, the RS spends a lot of the operation time handling the retractors while their main role should be to learn the procedure by constantly communicating with the surgeon.
- Position of the sub-teams in the OR
  Both surgeons who performed the ASI approach indicated that it is preferable to work with the scrub staff opposite to each other. This is not possible though because the person who stands on the other side of the table cannot fully see the incision due to the plastic sheets used.
- Uncertainty
  From the field research done, it seemed that orthopaedic operations do not pose serious uncertainty errors. Most of the errors that occur during hip replacement surgery are predictable. The only uncertainty that was identified is on the level of bone defection. This is because the surgeons

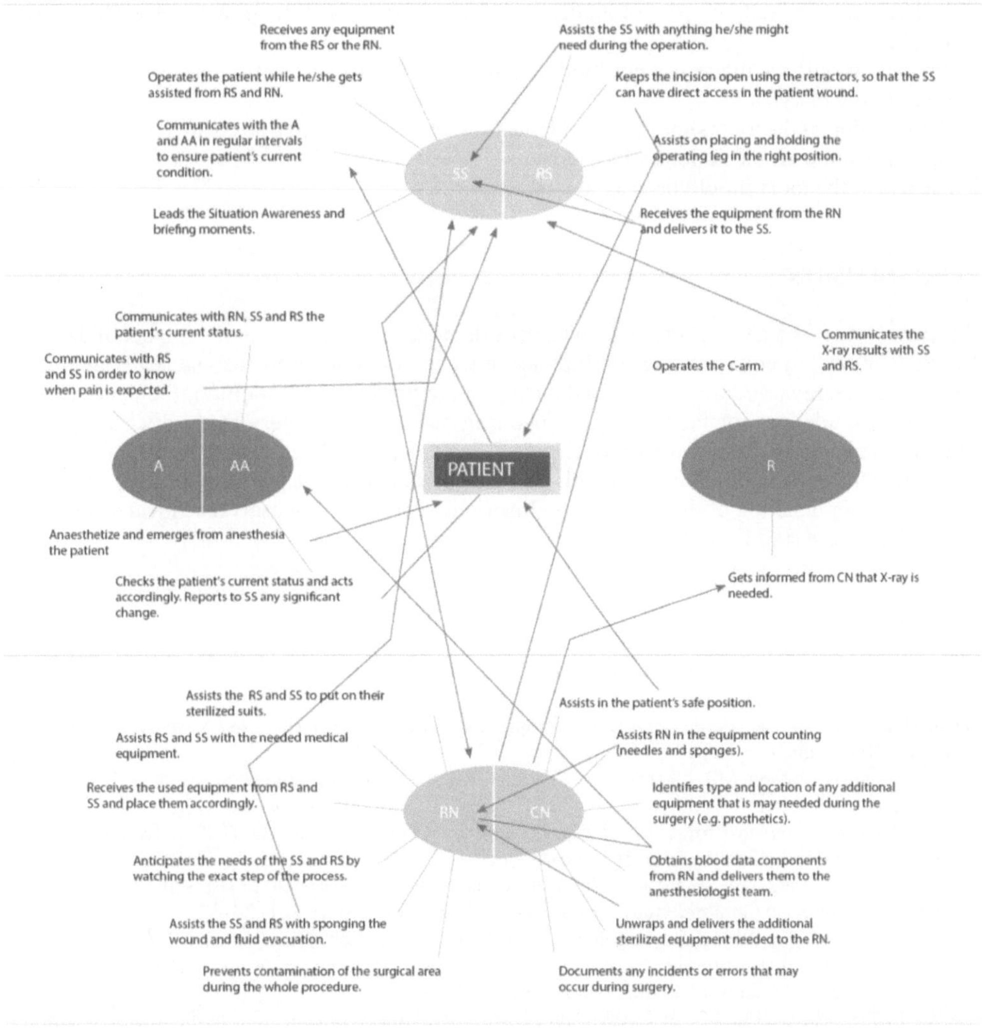


Figure 3. Flow model of the intra-operative phase in orthopaedic surgeries, based on work modeling techniques of Beyer & Holtzblatt (1998). SS = senior surgeon, RS = resident surgeon, RN = registered nurse, CN = circulating nurse, A = anaesthesiologist, AA = assistant anaesthesiologist, R = radiologist.

see it for the first time when the incision is made and they have reached the acetabulum and the femur.

## 4 PRELIMINARY DESIGN GUIDELINES FOR FACILITATING TEAMWORK DURING ORTHOPAEDIC SURGERY

Our preliminary findings led to six guidelines for teamwork-facilitating products:

- Products should facilitate a shared awareness among all OR team members about the current phase of the surgery.
- Products should facilitate the registered nurse to work in a proactive manner, fully anticipating the surgical needs.

- Products should facilitate (or not hinder) a proper view on the patient/the incision for all team members.
- Products should facilitate (or not hinder) direct communication between the team members, either orally of visually.
- The handling of the tools should not hinder teamwork: tools should be easy to handle and require the minimum possible actions while its in use.
- The use of the tools should be easy and intuitive.

## 5 CONCLUSION

This paper presents a new perspective on teamwork in the operating room; the impact of design on communication and collaboration. Preliminary insights into how products and working environment impact teamwork during orthopaedic surgery are presented. In addition, the paper presents preliminary guidelines for facilitating teamwork in orthopaedic surgery by product design. Currently we are working on design proposals embodying these guidelines (Anastasiadis, 2013), which are being developed in direct contact with all stakeholders involved. This approach is expected to result in further detailing of the design guidelines and in product proposals that advance teamwork among healthcare staff in orthopaedics.

## REFERENCES


Anastasiadis, G. 2013. *Facilitating teamwork in orthopaedic surgery by product design.* MSc thesis. Delft: Delft University of Technology.

Beyer, H. & Holtzblatt, K. 1998. *Contextual Design – Defining Costumer Centered Systems*. San Francisco: Morgan Kaufmann.

Christian, C.K., Gustafson, M.L., Roth, E.M., Sheridan, T.B., Gandhi, T.K., Dwyer, K., Zinner, M.J., & Dierks, M.M. 2006. A prospective study of patient safety in the operating room. *Surgery* 139(2): 159–173.

Flin, R., Yule, S., Paterson-Brown, S., Maran, N., Rowley, D., & Youngson, G. (2007). Teaching surgeons about non-technical skills. *Surgeon* 5(2): 86–89.

Hull, L., Arora, S., Kassab, E., Kneebone, R., & Sevdalis, N. 2011. Assessment of stress and teamwork in the operating room; an exploratory study. *The American Journal of Surgery* 201: 24–30.

Lingard, L., Espin, S., Whyte, S., Regehr, G., Baker, G.R., Reznick, R., Bohnen, J., Orser, B., Doran, D., &Grober, E. 2004. Communication failures in the operating room: an observational classification of recurrent types and effects. *Qual Saf Health Care* 13: 330–334.

Reader, T., Flin, R., Lauche, K., & Cuthbertson, B.H. 2006. Non-technical skills in the intensive care unit. *British Journal of Anaesthesia* 96: 551–559.

Sevdalis, N., Forrest, D., Undre, S., Darzi, A., & Vincent, C. 2008. Annoyances, disruptions, and interruptions in surgery: the disruptions in surgery index (DiSi). *World Journal of Surgery* 32: 1643–1650.

Yule, S., Flin, R., Paterson-Brown, S., & Maran, N. 2006. Non-technical skills for surgeons in the operating room: a review of the literature. *Surgery* 139:140–149.

*Bridging Research and Good Practices towards Patient Welfare – Shih & Liang (Eds)*
 *ISBN 978-1-138-02716-9*

# Classification of soothing music using Fuzzy C-Means clustering algorithm

Ya-Wen Hsu, Hong-Pin Tsai, Ming-Chuan Chiu & Sheue-Ling Hwang
*Department of Industrial Engineering and Engineering Management, National Tsing Hua University, Taiwan*

Hsiang-Lan Shih, Fang-Ting Huang & Chun-Ting Lee
*Industrial Technology Research Institute, Taiwan*

ABSTRACT: Life of modern people becomes more convenient and rich in material side but worse in mental side due to life stress. This results in bloom of some diseases such as insomnia. Listening to music could be one way to make people feel smooth. Some previous literature had advocated the efficiency of music therapy, however, only a few previous studies discussed and connected subjective indicators (personal cognition) with objective indicators (music features). Therefore, the aim of the study is to investigate what kind of music characteristics can spiritually relax people and obtain the therapeutic music from above results. Firstly, this study collected 25 different styles of music as samples. These songs were classified with Fuzzy C-Means clustering algorithm. According to our experimental result, music with mild amplitude, slow speed, and subjectively positive feeling can enable soothing in mind. The findings would also fit in with physiological signals (Heart Rate Variability) to ensure the consistency in psychology and physiology. This finding can provide suggestions on selection of therapeutic music. In addition, musicians can compose appropriate therapeutic music for patients of different mental illness.

## 1 INTRODUCTION

As the growth of economics and development of technologies, people's life became more adequate and convenient. However, pursuing material benefits often makes our mental side being under pressure. Psychologist had pointed out that regularly low quality sleep has high probability to cause chronic fatigue and results in many physical or mental illness such as constipation, headache and hypertension. In order to effectively relieve psychological stress, music is considered as one of effective ways to cease the problem. In early 1800s, music has been applied to improve the patient's sleep, but research also showed some unique melodies and rhythms of music can release blood pressure, and slow down both basal metabolism and breathe rate. The physiological response becomes more moderate when subjects are under pressure. In recent years, music therapists also adopted music in the medical procedure as psychological and physiological treatment because therapeutic music has the efficiency on release of pressure and achieves the relax pressure.

According to Two-dimensional emotion representation in Thayer's model as Fig. 1 shown, the emotion of human can be divided into two dimensions from silent to energetic and from negative emotions to positive emotions (Taylor, 2001). There are various styles of music, and the purpose of this study is to search for music characteristics that can guide a peace and harmonious status on people. Therefore, this study will categorize the "positive" and "silent" music by clustering music styles. After clustering music styles, the psychological and physical experiment will be conducted to examine if these music indeed have effectiveness, and this research expects to get the result for the steady state on psychology and physiology.

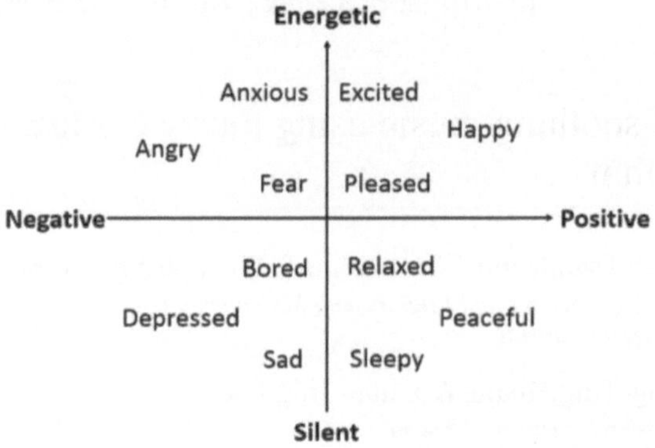


Figure 1. Two-dimensional emotion representation in Thayer's model (Taylor, 2001).

## 2 LITERATURE REVIEW

### 2.1 *Music and emotion*

The emotion of music can be mainly expressed as adjective such as happy, angry, sad, calm and so on. Thayer had proposed the concept of emotion plane, and classifying the emotion in accordance with two features: emotion stress and emotion energy. According to the result of the classification, the emotion plane can be built up. As shown in Fig. 1, the model used "Energetic" and "Calm" to indicate the degree of emotional energy, and it used "Happy" and "Anxious" to indicate the degree of emotional stress (Han, 2010; Gaston, 1951).

### 2.2 *Physiological signals and emotion*

There are many physiological indicators to reflect the various emotions, and the most direct and obvious index is heart rate variability (HRV). HRV are normally reflecting the works of autonomic nervous system (ANS) in individuals (Ellis & Thayer, 2010). The ANS is divided into two subsystems: the parasympathetic nervous system (PNS) and sympathetic nervous system (SNS) (Dorland, 2011). In general, the PNS indicate the decreasing of heart rate, blood pressure and relate to smooth emotion; the SNS indicate the increasing of heart rate, blood pressure and relate to energetic emotion (Kim & Kim, 2004).

### 2.3 *Music therapy*

The aim of music therapy is to establish a health service that is similar to occupational therapy and physical therapy using music therapeutically to address physical, psychological, cognitive and/or social functioning for patients of all ages (Bruscia, 1989). The impact of Music therapy on psychiatric patients has been widely studied. Witkin, (1954) has proposed the psychological effects of music on patients as follows:

(1) The music can attract attention and increase the sustained attention of the patients.
(2) Music can replace the grievous emotion of patients generated by others.
(3) Music can relieve the patient inner tension and conflict.
(4) The rhythmic stimulation of the music can increase the times of physical activity.

According to the differences features in the music, Gaston (1951) classified music into exciting music and calm music. The former refers to the passion and fast-paced music; the latter means gentle and lyrical music, which can make people feel calm and steady effect. Kerr (1942) had

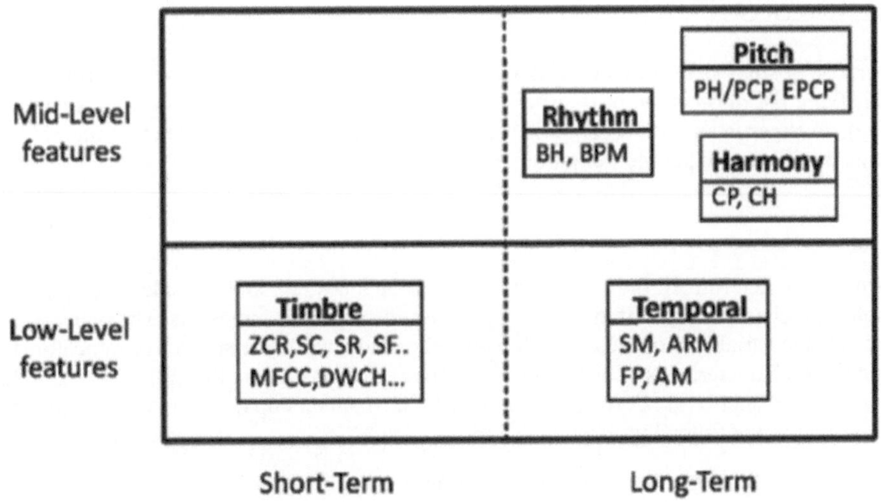


Figure 2. Characterization of audio features (Fu, 2011).

conducted sample surveys for the patient who is ongoing music therapy, in which 90% of the sample believes that music can reduce fatigue, and 88% of the sample believes that music will ease the mood. There are also many studies indicated that music therapy have a significant impact on the operation for people's autonomic nervous system, and the autonomic nervous system plays an important role in physiological functions such as heart rate variability, respiration, blood pressure, digestion and metabolism .

Based on the above review, music therapy is indeed helpful for people's psychology and physiology. Accordingly, some specific styles of music can soothe the soul and physiology reaching to a steady state.

### 2.4 *Music clustering methods and clustering index*

The current music classifications can be divided into two directions: one is based on music content (music content-based) to analyze the classification, including the melody, rhythm, chords and other music characteristics as a basis for classification (Lin, 2004; Brecheisen, 2006); another classification is based on learning machine, such as linear analysis, Bayesian analysis and artificial neural network approach to music classification (Loh, 2006; Mandel, 2008). Levitin (2011) has proposed that the basic elements of music are pitch, rhythm, tempo, contour, timbre, loudness (amplitude) and response. Fu (2011) organized the four categories of audio features into the two-directional diagram based on the short and long term, and music-oriented features. Fig. 2 shows that music features can be divided into long-term characteristics with moderate tone, harmony and rhythm; beats with long-term and low level of eigenvalues; timbre with Short-term and low level of eigenvalues.

### 2.5 *Summary*

In the following music clustering indicators, this study mainly adopted as clustering based on the physical properties. However, "Emotion" is a word difficult to objectively defined and only analysis in the physical properties might cause the error. Therefore, this study will consider the classified characteristics physical characteristics (amplitude, velocity) at the same time. Fuzzy algorithm would be applied to translate the uncertainty of emotion in order to conduct the analysis.

## 3 METHODOLOGY

The purpose of study is to find the soothing music by using Fuzzy C-Means Clustering Algorithm by classifying the music style in first priority. After finding the target group of music (silent and positive), an experiment with ergonomic principle is conducted to verify whether the positive and silent music can soothe people's emotion effectively.

### 3.1 *Data collection*

This study first determined the number of music to be analyzed, and considered of physical and psychological characteristics of music as objective and subjective indicators. The objective indicators include the variability of music rhythm and amplitude, while the subjective indicators include psychological characteristics of positive and negative emotions. Then, the indicators that related to the membership function are defined and analyzed. The membership function of the three indicators is represented as the following formula:

(1) Membership functions with low level of amplitude variability

$$\mu_1(x) = e^{-\frac{1}{220}x^2}, \; x > 0 \tag{1}$$

(2) Membership function of slow music

$$\mu_2(x) = \begin{cases} 1, & 0 \le x \le 1 \\ \frac{1}{2} - \frac{\sin\left(\frac{\pi}{4}*(x-3)\right)}{2}, & 1 \le x \le 5 \\ 0\;, & x \ge 5 \end{cases} \tag{2}$$

(3) Membership functions of music to give people positive emotions

$$\mu_3(x) = \begin{cases} 0\;, & 0 \le x \le 1 \\ \frac{(x-1)^2}{16}, & 1 \le x \le 5 \\ 1, & x \ge 5 \end{cases} \tag{3}$$

(4) Define high comfort level music by the following membership function ($\mu_r$) as follows:

$$\mu_4(x) = \begin{cases} 0\;, & 0 \le x \le 1 \\ \frac{(x-1)^3}{64}, & 1 \le x \le 5 \\ 1, & x \ge 5 \end{cases} \tag{4}$$

The indicators of collecting music are as follows:

#### 3.1.1 *Average amplitude variation*

Using well-known music software "Max" to analyze music, and after calculating amplitude for every second, the amplitude variance of whole song will be find and considered as indicators.

#### 3.1.2 *Tempo*

Our research set five kinds of major changes for music tempo, including Lento, Andante, Moderato, Allegro and Presto (Sachs, 1988), and rating it with the range of value 1 to 5. Some fine-tuning are needed when the same as the adagio song but not exactly the same speed. For example: two songs are belongs to Adagio, but one is faster, then the speed evaluation may be1.25, 1.5, 1.75 ... and so on. These indicators are measured by physical characteristics, and the analysis focused on the music itself.

### 3.1.3 *Forward degree of comfort*

Since that the positive mood belongs to the subjective feelings of listeners, this study e conducts experimental design with questionnaire. Detailed description of the experiment is shown as follows:

(1) The subjects entered the laboratory and listened 30 seconds of natural sounds first to avoid interference due to different emotions generated.
(2) Play music that each song is about 1 to 3 minutes.
(3) After playing music, the participants had 30 seconds to answer two questions:
    Does the song make you positive emotions?
    Does the song make you feel comfortable after listening?
(4) Play 30 seconds of natural sounds.
(5) Play another song.
(6) Go back to step 3 until the end of the whole songs finished.
(7) Finish songs playback and end the experiment.

## 3.2 *Fuzzy c-Means clustering*

We converted the rhythm of the music and the forward degrees of music into appropriate data by using the fuzzy relations function and the original data of the degree of amplitude variability that mentioned in the previous sections. Fuzzy c-Means clustering would be used to deal with these data as follows:

$$\text{Min } z = (\tilde{U}) = \sum_{i=1}^{c} \sum_{k=1}^{n} (\mu_{ik})^m \|x_k - v_i\|^2$$

$$\text{s.t. } v_i = \frac{\sum_{k=1}^{n} (\mu_{ik})^m x_k}{\sum_{k=1}^{n} (\mu_{ik})^m}, \forall i$$

According to Two-dimensional emotion representation in Thayer's model, this study expects to find the positive and silent music. The process of applying Fuzzy C-Means Clustering Algorithm in this research is shown as follows:

Step 1. According to Thayer's model, this study expects to find the positive and silent music. So, the number of clustering c ($2 \leq c \leq n$) were defined to be 4, which is the number of four clusters, and the weight m ($1 < m < \infty$) was taken the general values: 2. Next, these raw data will be transferred to define as $x_k = [\mu_1(x_k), \mu_2(x_k), \mu_3(x_k)]$, $\mu_{ik} \in [0, 1]$ in the matrix $\tilde{U}^1$ will be established. Further, since for any point $x_k$, the sum of the degrees in individual clusters should be exactly equal to 1, the following equation must be satisfied:

$$\sum_{i=1}^{c} \mu_{ik} = 1, \forall k \tag{5}$$

Step 2. Using the following formula to calculate for each center of clusters $v_i$.

$$v_i = \frac{\sum_{k=1}^{n} (\mu_{ik})^m x_k}{\sum_{k=1}^{n} (\mu_{ik})^m} \tag{6}$$

Step 3. Calculate the new matrix $\tilde{U}^{l+1}$ in accordance with the following equation with $v_i$.

$$\mu_{ik} = \frac{\left[\frac{1}{\|x_k - v_i\|_G^2}\right]^{1/(m-1)}}{\sum_{j=1}^{c} \left[\frac{1}{\|x_k - v_i\|_G^2}\right]^{1/(m-1)}} \tag{7}$$

where, $i = 1,\ldots,c$; $k = 1,\ldots,n$.

Step 4. Calculating $\Delta = \|\tilde{U}^{l+1} - \tilde{U}^l\|_G$, if it is greater than a certain standard, then l is set to l +1 and return to Step 2. If it is below a certain standard, the algorithm will end.

### 3.3 *Comparative analysis*

In this phase, we compare and analyze the different clusters according to the soothing music of subjective decision and the result of fuzzy clustering and verify whether subjective ratings and objective data analysis is consistent or not.

### 3.4 *Identify the characteristics of soothing music*

According to the result of above result, we compile the same group of soothing music, and the common characteristics of this music rhythm, amplitude variability and positive emotion and relationships.

### 3.5 *Validation of physiological signals*

After finding the relationship between physical indicators (music rhythm, amplitude variability) and positive emotions, we begin to find the relationship between physical indicators and physiological signals (HRV will be used as an indicator here). Also, how the different music styles affect the physiological signals is interested in this research. As the result, we conduct an experiment with ergonomic principles to verify the relationship.

## 4 CASE STUDY

### 4.1 *Data collecting and experiment*

The whole experiment of the study can be divided into two parts. The first part is to classify the target music style and compare the clustering results with psychological result. According to the previous section, the study selected 25 different styles of music (including five homemade songs) for analysis, and there are four subjects in this experiment. Experimental procedures are described as Section 3.1, each subjects would first restore the original state of nature before listening to the experimental music and then continue the experiment until every songs are completed. All subjects were asked to answer “How comfortable this song brings you?” and “How positive this song brings you?” within 30 seconds. The two questions are in the form of a five-point scale, the greater the value representing and the more comfortable/possible the song being. Finally, the results of the questionnaires and physical signals would be converted into relative functions.

The second experiment were cooperated with Industrial Technology Research Institute (ITRI), and the procedures were shown as Section 3.5. Starting with the five homemade songs, 10 subjects (which have been working in ITRI, five men and five female and they are between the ages of 50 to 60) joined this experiment. All subjects would hear the five songs and the nature sound in interval to avoid unnecessary interruption. Besides, the subjects worn on the wrist strap with a sensing device to secure the HRV related signal during the sound play and they would be requested to answer a questionnaire to pair the suitable emotion themselves toward different music. After the experiment, the indicators of music and the five different music styles would be compared with the HRV of each subject.

### 4.2 *Data clustering and analysis*

#### 4.2.1 *Find the soothing music*

This study used the software Matlab 2012a published by The MathWorks to analyze above data, and the clustering results are as follows in Fig. 3.

Fig. 3 show the results of fuzzy clustering, and Fig. 4 is the result of questionnaires by converting the soothing relationship function. As shown in Fig. 3, Group 2 is largely fitting in with soothing song (membership degree greater than or equal 0.8). Thus, we had analyzed the three indicators of Group 2, it had a lower level of amplitude variability and the tempo was slow. However, the tracks

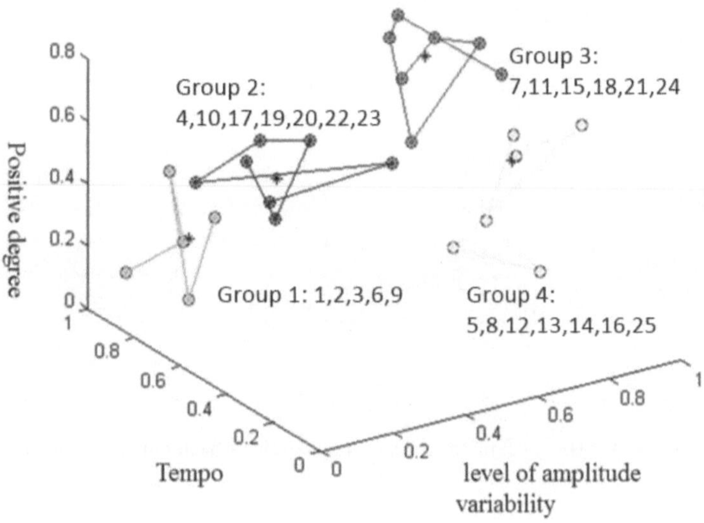


Figure 3. Fuzzy clustering results.

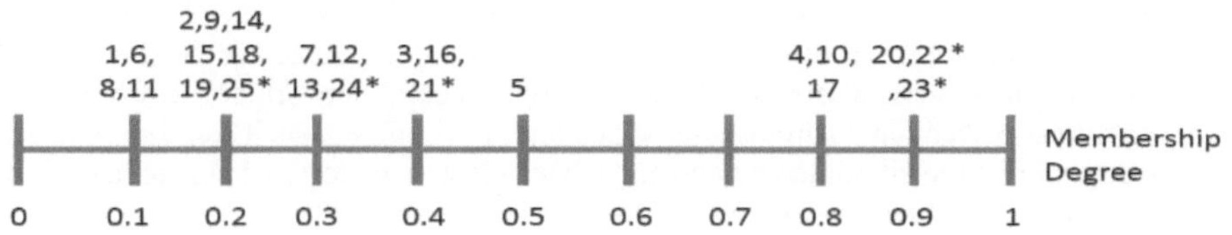


Figure 4. The results of subjective questionnaires.

within Group 2 are belonging to high positive degree. This result could actually fit in the reality of the true feelings. In general, people feel comfortable in a song whose melody wouldn't have high level of amplitude variability, slow tempo and bring people more positive feelings.

### 4.2.2 *HRV related signals in music*

After finding the soothing group of music, this study selected five homemade songs from track 21 to 25, and examining the HRV related data of all subjects to verify if the physiological signals correspond with psychological response and physical properties of music. Each subject's HRV related data and pattern are shown as Fig. 5 due to error and missing data occurred, we exclude the data of some subjects and only have effective sample: Subject 1, 2, 5, 7 and 8. Here, this study uses LF/HF value to measure the ratio of SNS and PNS. When the LF/HF ratio value more than 1, the SNS take part more; the ratio value equals to 1, SNS is same as PNS; the ration value less than 1, the PNS take parts more (Godleski et al, 2000).

Fig. 5 shows LF/HF value of each subject in the five songs which have been minus the initial value. The track order for higher proportion of parasympathetic is: 22, 21, 23, 25 and 24, however, the results just show a little difference in the patterns. Returning to the result of Fuzzy clustering method, Track 22 and 23 are belonging to the best soothing group and the two songs also take top three in the higher proportion of parasympathetic order. That means the clustering result is not only matching up the psychological response but also roughly corresponds to the operation of physiological signals.

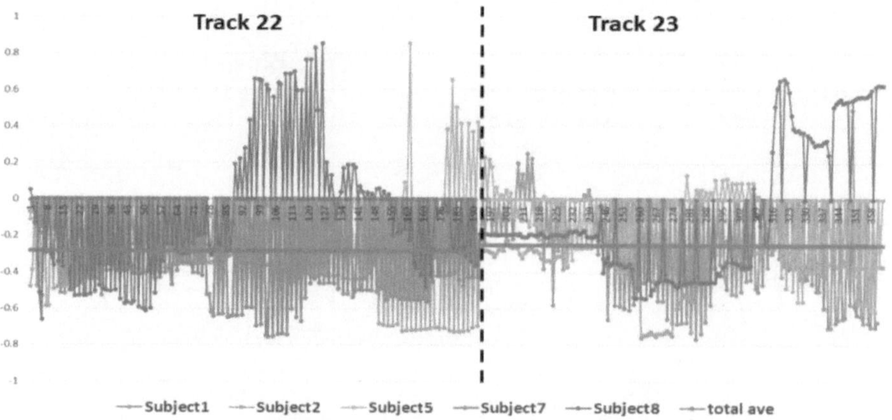


Figure 5. Each subject LF/HF value in Track 22 & Track 23 (have been minus the initial value).

The result of questionnaires and fuzzy clustering algorithm are roughly matched up, even with the physiological consequence. As the second group shown in Fig. 3, it fitted the expectations of this study: To identify appropriate music style as a therapeutic music which can make people stay in a soothing state. In the previous studies, most of them only focus on the classification of the music in physical properties such as: Signal wave, rhythm and frequency as the clustering basis. And the clustering methods are often used linear analysis, neural network and probability model, however, emotional vocabularies are usually vague and more subjective. Therefore, this study in addition to considering the amplitude, tempo and other objective factors, also using fuzzy relations function to convert the subjective feelings.

## 5 CONCLUSION

Using fuzzy c-mean clustering algorithm can make the result more fit in with the expected emotional model, and the HRV related data could be a verification to test if subjects' mentality and physiology are consistent. This study initially understood the soothing music style has features with lower amplitude, slower tempo and is characterized by positive emotions.

In order to simplify the process of experiment, the study excluded diversified timbre in different musical instrument and let all songs being piano-based. However, in reality, timbre may be an important factor to affect human's mood changes, and it must be carefully considered. The presenting way of music is diversified, in addition to the music features and the influence of external environment; it might also be associated with the player's performance skills or current mood. If above factors are taken into account, more resources would be incorporated in the analysis.

Besides, in the ergonomics experiment, the outliers of the samples are too much, so we can just tell roughly situation of LF/HF data from the average value in different songs. Therefore, the output signal of sensitive device need to be monitored carefully and the sample could be added more in future. In that way, the changing of physiological signals could be recognized clearly. Furthermore, the weight of each indicator of music is identical, however, some indicators may affect us more and some may affect less. As a result, the weight analysis or Analytic Hierarchy Process (AHP) could be added before clustering music styles in future research to get a more precise consequence. In conclusion, this study gives a direction for composer or therapist to select appropriate music for therapeutic use. Likewise, people can also find the songs which make their physiology and psychology relaxed equally and obtain a healthier life.

REFERENCES


Brecheisen, S., Kriegel, H.-P., Kunath, P., & Pryakhin, A. 2006. Hierarchical genre classification for large music collections. Paper presented at the Multimedia and Expo, 2006 IEEE International Conference on.

Cheng, H.-T., Yang, Y.-H., Lin, Y.-C., Liao, I.-B., & Chen, H. H. 2008. Automatic chord recognition for music classification and retrieval. Paper presented at the Multimedia and Expo, 2008 IEEE International Conference on.

Cowie, R., Douglas-Cowie, E., Tsapatsoulis, N., Votsis, G., Kollias, S., Fellenz, W., & Taylor, J. G. 2001. Emotion recognition in human-computer interaction. Signal Processing Magazine, IEEE, 18(1), 32–80.

Dorland, W. A. N. 2011. Illustrated medical dictionary: WB Saunders Company.

Ellis, R. J., & Thayer, J. F. 2010. Music and autonomic nervous system (dys) function. Music perception, 27(4), 317.

Gaston, E. T. 1951. Dynamic music factors in mood change. Music Educators Journal, 37(4), 42–44.

Godleski, J. J., Verrier, R., Koutrakis, P., Catalano, P., Coull, B., Reinisch, U., Wolfson, J. 2000. Mechanisms of morbidity and mortality from exposure to ambient air particles. Research Report (Health Effects Institute) (91), 5.

Han, B.-j., Rho, S., Jun, S., & Hwang, E. 2010. Music emotion classification and context-based music recommendation. Multimedia Tools and Applications, 47(3), 433–460.

Kerr, W. 1942. Psychological effects of music as reported by 162 defense trainees. The Psychological Record.

Kim, K. H., Bang, S., & Kim, S. 2004. Emotion recognition system using short-term monitoring of physiological signals. Medical and biological engineering and computing, 42(3), 419–427.

Levitin, D. J. 2011. This is your brain on music: Understanding a human obsession: Atlantic books.

Lin, C.-R., Liu, N.-H., Wu, Y.-H., & Chen, A. L. 2004. Music classification using significant repeating patterns. Paper presented at the Database Systems for Advanced Applications.

Loh, Q.-J. B., & Emmanuel, S. 2006. ELM for the Classification of Music Genres. Paper presented at the Control, Automation, Robotics and Vision, 2006. ICARCV'06. 9th International Conference on.

Mandel, M. I., & Ellis, D. P. 2008. Multiple-instance learning for music information retrieval. Paper presented at the ISMIR 2008: Proceedings of the 9th International Conference of Music Information Retrieval.

Sachs, C. 1988. Rhythm and Tempo: a study in music history: Columbia University Press.

Scaringella, N., Zoia, G., & Mlynek, D. 2006. Automatic genre classification of music content: a survey. Signal Processing Magazine, IEEE, 23(2), 133–141.

Weihs, C., Ligges, U., Mörchen, F., & Müllensiefen, D. 2007. Classification in music research. Advances in Data Analysis and Classification, 1(3), 255–291.

Witkin, H. A., Lewis, H. B., Hertzman, M., Machover, K., Meissner, P. B., & Wapner, S. 1954. Personality through perception: an experimental and clinical study.

*Bridging Research and Good Practices towards Patient Welfare – Shih & Liang (Eds)*
*© 2015 Taylor & Francis Group, London, ISBN 978-1-138-02716-9*

# A tangible user interface of map for hospital wayfinding

Chun-Wen Chen & Yun-Fong Kao
*Department of Industrial Design, Chang Gung University*

ABSTRACT: Wayfinding in hospitals has been an ongoing problem for both patients and visitors. Interactive wayfinding maps can provide more functions and dynamic content for various users. The aim of this research was to design a simplified, tangible user interface (TUI) for an interactive wayfinding map in a hospital environment. TUI provides a user interface with physical form to represent digital information. We built a simulated hospital map and interfaces with visual marker recognition and three-dimensional environment simulation technologies. The user interface of the system was tested by wayfinding tasks to evaluate user usability and satisfaction. The results showed that spatial knowledge learning, subjective usability, and satisfaction are superior with TUI compared to a graphics user interface.

## 1 INTRODUCTION

Wayfinding in hospitals has been an ongoing problem for users and staff, especially for disabled patients and first-time visitors (Brown et al., 1997; Del Nord, 1999; Allison, 2007; Garip, 2011). Studies on hospital wayfinding mostly focus on building spatial structure and signage design (Wright et al., 1993; Tzeng & Huang, 2009; Rousek & Hallbeck, 2011). Design for emergency evacuation is another important issue in hospital wayfinding (Løvs, 1998; Zheng, 2012). Guidelines developed by experts and governments provide recommendations on designing the built environment and wayfinding system in hospitals (Miller & Lewis, 1999).

Interactive technology provides an opportunity of extending the capability of the available static signage system for wayfinding. Maps are now digitized to be more dynamic and interactive, and interactive wayfinding maps, similar to the "you-are-here" maps in public places, can provide more functions and dynamic content for users.

Graphics user interface (GUI) in information systems has helped users for years. Metaphors are a useful approach utilized in GUI to connect graphics and their corresponding meanings. In general, a user interface consists of input and output or control and representation. Control is for operation of information, and representation is the part a user can perceive (Ishii, 2008). GUI utilizes a physical device, like a keyboard and/or mouse, as the control and a digital screen as the representation (Fig. 1). The physical devices of GUI do not represent information (Ullmer & Ishii, 2000).

Ishii & Ullmer (1997) proposed the concept of tangible user interface (TUI) to connect the digital and physical space. The approach of TUI is to provide a physical form to the user interface to represent digital information. TUI utilizes physical objects as both control and representation as well as a digital screen or projection as representation and outputs physical (tangible device) and digital (sound or image) representations in the same time.

Compared to GUI, TUI can help users more with its better spatial mapping. Sharlin et al. (2004) used Cognitive Map Probe (CMP), which is a TUI system built with interactive physical models, to assess cognitive mapping ability. They found that the TUI system could assess cognitive mapping ability well because of its spatial mapping nature. Kim & Maher (2008) compared the performance of interior space designers with TUI and GUI and found that designers using TUI could focus more on the tasks themselves and the spatial relationship among objects in the tasks.

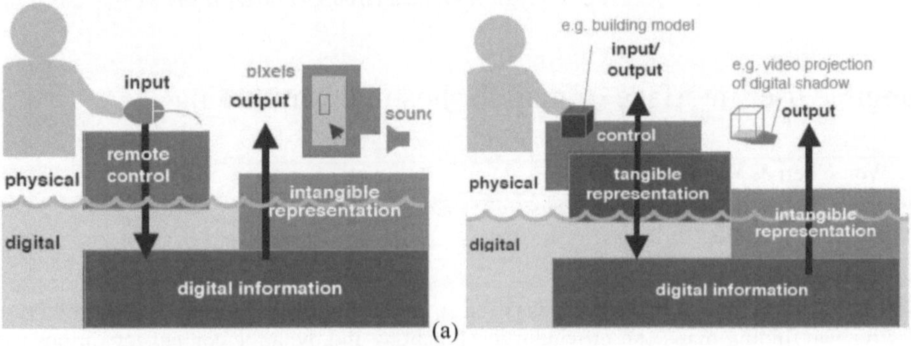


Figure 1. Diagram of a (a) graphic user interface and (b) tangible user interface (Ishii, 2008).

Experimental navigational systems with various TUI technologies have been built and tested. Metadesk is a campus navigational system developed by MIT Media Lab (Ullmer & Ishii, 1997). A building model is used as an interface to control the panning and zooming of the campus map. Göttel (2007) investigated if spatial knowledge can be learned in a virtual environment to the same degree as in a real one, and built the ProBoNO system, using a small human figure as the TUI control, to test this hypothesis. Results showed that spatial knowledge gleaned in a virtual environment can be similar to that in a real environment. Piovesana et al. (2010) built a system combining Google Maps and Street View as its map and virtual environment display. TUI and GUI were compared in spatial knowledge learning, and TUI was found to be superior to GUI.

The aim of this research was to design a simplified and acceptable interface for an interactive wayfinding map for hospital users. We proposed and built a map with TUI for a hospital environment, using visual marker recognition and three-dimensional (3D) environment simulation technologies. The user interface was tested by wayfinding tasks to evaluate usability and satisfaction. Our goals were: (1) to design and build a simulated interactive map with TUI for a hospital; (2) to test and compare spatial (route and survey) knowledge learning and wayfinding performance between TUI and GUI; and (3) to evaluate subjective usability and satisfaction of TUI and GUI.

## 2 METHODS

### 2.1 *Participants*

Thirty-two participants, 16 male and 16 female, were given wayfinding tasks to complete. All participants were college students aged 20–22 years with experience using a 3D simulated environment. All participants tested both the TUI and GUI in different order.

### 2.2 *Materials and design*

Two maze-like simulated hospital plans was used as maps of the test environment (Fig. 2). The two maps were designed differently but with similar complexity to prevent a learning effect between the TUI and GUI.

The simulated 3D environment and interaction programming were integrated in the Unity3D authoring application. All models, including landmarks and walls, were made in the Rhino modeling application based on the plan and then imported into Unity3D (Fig. 3).

The wayfinding map system with TUI consisted of a table surface for back projection, short-focus projector, computer display, webcam, and a computer running Unity3D, reacTIVision, and TUIO applications. A small robot-shaped figure was the main control and representation of the TUI (Fig. 4). The final 3D image, rendered in Unity3D, which simulates the vision from a small

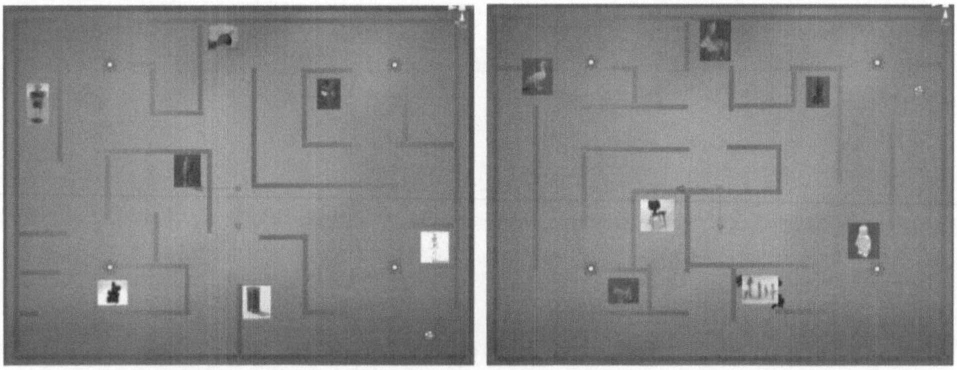

Figure 2. Plans of the simulated environment.

Figure 3. Landmarks of the simulated environment.

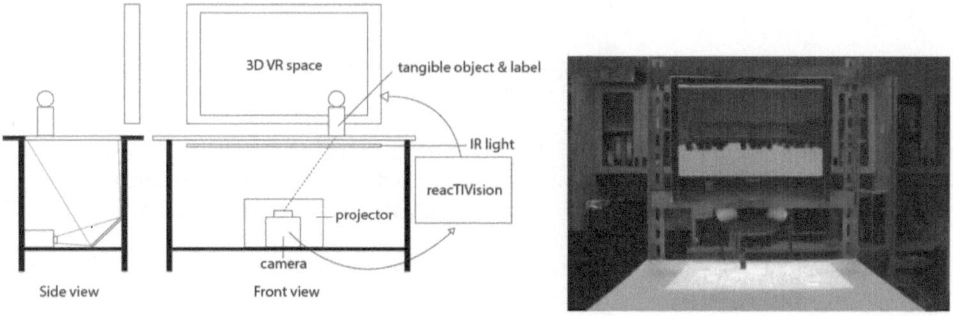


Figure 4. Schematic of the wayfinding map system with TUI.

robot's eyes, was shown on a standing computer display on the table. The map image was projected under a semitransparent table surface.

A fiducial marker attached below the small robot can be detected in the camera image by reacTIVision technology, which determines the position and direction of the fiducial marker in the image (Kaltenbrunner & Bencina, 2007). The data were transmitted to the TUI application (Unity3D) via a TUIO application that transforms the data into a suitable form (Fig. 5).

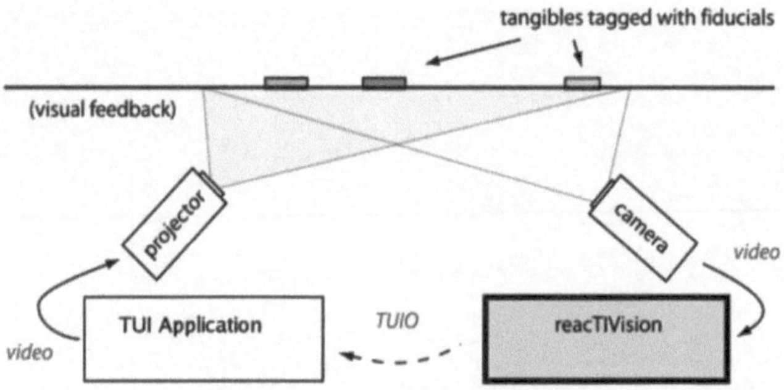


Figure 5. System diagram of reacTIVision (Kaltenbrunner & Bencina, 2007).

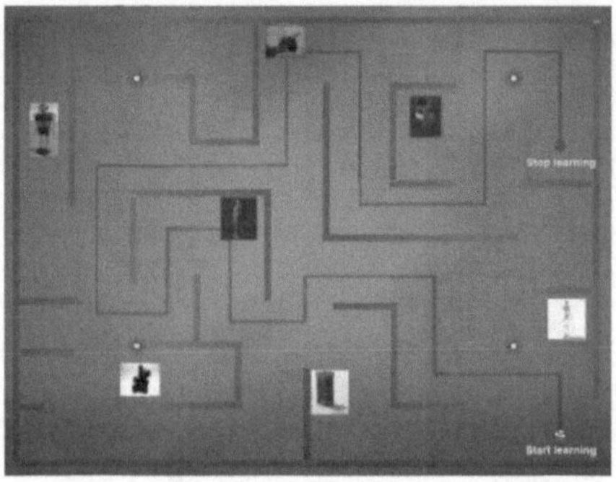

Figure 6. Route knowledge task.

## 2.3 *Tasks*

Three tasks were given to each participant: (1) route knowledge, (2) survey knowledge, and (3) wayfinding performance. Both route knowledge and survey knowledge tasks had two stages: learning and test. In the route knowledge learning stage task, a route line was shown on the map and the participant had to move a small robot to follow the route from start point A to destination point B (Fig. 6). In the test stage, the participant was asked to follow the same route without a line on the map. If the wrong route was selected, the participant had to return to the starting point. The number of errors and complete time were recorded.

In the survey knowledge learning stage task, the system followed a route to show all landmarks and then let the participant browse in the simulated environment by moving the small robot for 10 minutes. In the test stage, two tests were conducted (Fig. 7). The first test was target pointing in which the small robot was put on an assigned position and the participant was asked to turn the small robot to point in the direction of the target, a previously learned landmark, by memory. Three targets were selected, and the error angles were recorded. The second test was landmarks locating in which paper cards were made for seven landmarks in the simulated environment. The participant was asked to put the cards on the correct positions of a paper map. The error distances were recorded.

The wayfinding performance task has three subtasks. Three pairs of start and destination points were assigned, and the participant was asked to navigate the stimulated environment without the map from start to destination three times. Each complete time was recorded.

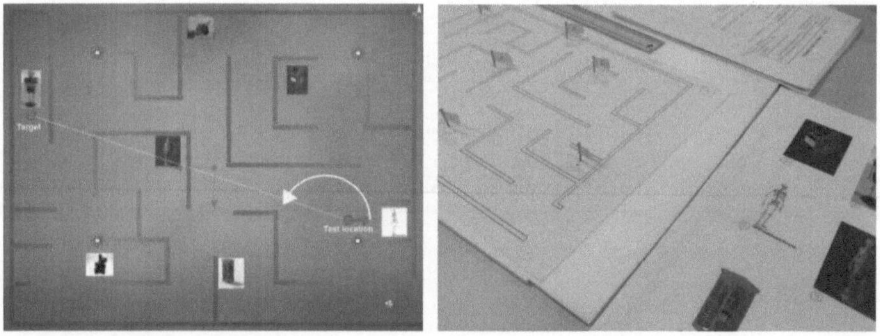


Figure 7. Survey knowledge task.

### 2.4 *Questionnaires*

Subjective evaluation on usability and satisfaction was conducted with the System Usability Scale (SUS) (Brooke, 1996) and Questionnaire for User Interaction Satisfaction (QUIS) (Chin et al., 1988).

SUS is a reliable, low-cost, usability scale that can be applied to global assessments of system usability. It provides broad general measures that can be used to compare usability across a range of contexts. It is a 10-item Likert scale with a score ranging from 0 to 100. It is a composite measure of overall usability, meaning that the score of an individual item is not meaningful.

QUIS is organized into five groups as follows: (1) overall reactions to the system, (2) screen, (3) terminology and system information, (4) learning, and (5) system capabilities. The score of each item ranges from 0 to 9. In this study we omitted terminology and system information for a few corresponding items shown in the system.

### 2.5 *Procedure*

Tests were conducted in a laboratory setting. The wayfinding tasks included operations, such as following paths and locating places in a simulated environment, with the TUI or GUI map to determine wayfinding performance and learning. The test procedure was as follows:

(1) Explanation: The research objective and test contents were described to the participant.
(2) Route knowledge learning: Practice in the simulated environment.
(3) Route knowledge test: Route following.
(4) Survey knowledge learning: Learning the spatial relationship among landmarks and the environment.
(5) Survey knowledge test: Target pointing and landmarks locating.
(6) Wayfinding performance task: Finding destinations in the environment without the map.
(7) Switching interface: Repeating (2)–(6) with another interface (from TUI to GUI or from GUI to TUI).
(8) Questionnaire: Completing the SUS and QUIS for both TUI and GUI.

## 3 RESULTS

### 3.1 *Spatial knowledge*

To compare the route knowledge learned from TUI and GUI, the independent sample *t*-test was used to analyze the error number and complete time (Table 1). In error number analysis, there was a significant difference in the number of errors between TUI and GUI ($p < .001$). In complete time

Table 1. Independent sample *t*-test of route knowledge.

| Data | Interface | N | mean | SD | *t* | *p* |
|---|---|---|---|---|---|---|
| Errors | TUI | 32 | 2.72 | 1.75 | −4.08 | .000*** |
| | GUI | 32 | 4.66 | 2.04 | | |
| Time | TUI | 32 | 63.13 | 25.61 | −3.34 | .002** |
| | GUI | 32 | 95.38 | 48.32 | | |

$^{*}p < .05$, $^{**}p < .01$, $^{***}p < .001$.

Table 2. Independent sample *t*-test of survey knowledge.

| Data | Interface | N | mean | SD | *t* | *p* |
|---|---|---|---|---|---|---|
| Error angle | TUI | 32 | 33.66 | 21.54 | −1.21 | .232 |
| | GUI | 32 | 39.34 | 15.67 | | |
| Error distance | TUI | 32 | 19.47 | 10.53 | −4.47 | .020* |
| | GUI | 32 | 36.03 | 18.10 | | |

$^{*}p < .05$.

Table 3. Independent sample *t*-test of wayfinding complete time.

| Interface | N | mean | SD | *t* | *p* |
|---|---|---|---|---|---|
| TUI | 32 | 26.44 | 9.63 | −1.75 | 0.085 |
| GUI | 32 | 32.28 | 16.25 | | |

analysis, it shows. There was also a significant difference in complete time between TUI and GUI ($p < .01$). The results showed that TUI is superior to GUI in route knowledge tasks.

To compare the survey knowledge learned from TUI and GUI, the independent sample *t*-test was used to analyze the error angle and error distance (Table 2). In error angle analysis, no significant difference was found between TUI and GUI ($p > .05$). However, in error distance analysis, a significant difference was found between TUI and GUI ($p < .05$). The results indicated that TUI is possibly superior to GUI in survey knowledge tasks.

### 3.2 *Wayfinding performance*

To compare wayfinding performance of TUI and GUI, the independent sample *t*-test was used to analyze the complete time (Table 3). No significant difference was found ($p > .05$), indicating similar performance in wayfinding tasks for the two types of interfaces.

### 3.3 *Usability and satisfaction*

To compare the usability of TUI and GUI by SUS, the independent sample *t*-test was utilized to analyze the scores (Table 4). We found a significant difference between TUI and GUI ($p < .001$), demonstrating that TUI is superior to GUI in SUS.

The QUIS scores are listed by group in Table 5. In general, the TUI scores were larger than those of GUI.

To compare the satisfaction of TUI and GUI by QUIS, the independent sample *t*-test was used to analyze the mean scores (Table 6). This showed a significant difference ($p$- $< .001$) between TUI and GUI, with TUI being superior to GUI in QUIS.

Table 4. Independent sample $t$-test of SUS.

| Interface | N | mean | SD | $t$ | $p$ |
|---|---|---|---|---|---|
| TUI | 32 | 73.59 | 13.16 | 7.61 | 0.000*** |
| GUI | 32 | 31.09 | 8.05 | | |

*$p < .05$, **$p < .01$, ***$p < .001$.

Table 5. Scores by groups of QUIS.

| | TUI | | GUI | |
|---|---|---|---|---|
| Interface | mean | SD | mean | SD |
| Overall reactions | 5.89 | 1.13 | 3.24 | 1.31 |
| Screen | 5.70 | 1.31 | 2.99 | 1.36 |
| Learning | 5.81 | 1.26 | 3.21 | 1.50 |
| System capabilities | 5.80 | 1.37 | 2.84 | 1.45 |
| Mean | 5.80 | 1.27 | 3.07 | 1.41 |

N = 32.

Table 6. Independent sample $t$-test of QUIS.

| Interface | N | mean | SD | $t$ | $p$ |
|---|---|---|---|---|---|
| TUI | 32 | 5.80 | 1.27 | 25.74 | 0.000*** |
| GUI | 32 | 3.07 | 1.41 | | |

*$p < .05$, **$p < .01$, ***$p < .001$.

## 4 DISCUSSION

The results of the route knowledge tasks and survey knowledge tasks showed that the interactive wayfinding map with TUI is superior to that with GUI in spatial knowledge learning. This is consistent with the published literature that TUI can provide better spatial knowledge learning. The spatial characteristics of a tangible interface may help users perceive the spatial relationship between interface and environment.

Performance of wayfinding tasks without the map showed no difference between TUI and GUI. Without the map, the function of TUI control (the small robot) is similar to a mouse in GUI. It therefore possibly lacks meaning of representation and its spatial relationship in the physical world, which may explain the similarity.

Unlike GUI, TUI is new to most participants. It can result in ratings of extreme acceptance, very acceptable, or not acceptable. From the results of subjective usability and satisfaction, the wayfinding map with TUI received better scores in both evaluations. Tangible physical objects may arouse more emotion or even pleasure than graphics on display and thereby affect perceived usability. The effect of TUI form design may also be an important issue.

## 5 CONCLUSIONS

In this research, a simulated hospital map and interface were built with visual marker recognition and 3D environment simulation technologies. The system mixed a projected two-dimensional map and 3D virtual environment as display. A tangible robot-shaped figure was used as the control and representation of the user position and direction. The user interface of the system was tested by wayfinding tasks to evaluate user usability and satisfaction. Results showed that spatial knowledge learning and subjective usability and satisfaction of TUI are superior to GUI. Wayfinding maps with TUI could help patients and visitors in hospitals find their way more easily.

## ACKNOWLEDGMENTS

This research was partly sponsored by grants NSC100–2410-H-182–025 and NSC 101-2410-H-182-014-MY2 from the Ministry of Science and Technology, Taiwan.

## REFERENCES

Allison, D. 2007. Hospital as city: Employing urban design strategies for effective wayfinding. *Health Facility Management* 20(6): 61–65.

Brooke, J. 1996. SUS: A "quick and dirty" usability scale. In P.W. Jordan, B. Thomas, B.A. Weerdmeester & I.L. McClelland (eds), *Usability evaluation in industry*: 189-194. London: Taylor & Francis.

Brown, B., Wright, H., & Brown, C. 1997. A post-occupancy evaluation of wayfinding in a pediatric hospital: Research findings and implications for instruction. *Journal of Architectural and Planning Research* 14(1): 35–51.

Chin, J.P., Diehl, V.A. & Norman, K.L. 1988. Development of an instrument measuring user satisfaction of the human-computer interface. In *Proceedings of SIGCHI '88*: 213–218. New York: ACM.

Del Nord, R. 1999. Wayfinding: A quality factor in human design approach to healthcare facilities. *World Hospital Health Service* 35(2): 28–30.

Garip, E. 2011. Environmental cues that affect knowing: A case study in a public hospital building. *Procedia - Social and Behavioral Sciences*: 30, 1770–1776.

Göttel, T. 2007. ProBoNO: Transferring knowledge of virtual environments to real world situations. In *Proceedings of the 6th international conference on Interaction design and children: 81-88*. Aalborg, Denmark: ACM.

Ishii, H. 2008. Tangible bits: Beyond pixels. In *Proceedings of the 2nd international conference on tangible and embedded interaction*: xv–xxv. New York, NY: ACM.

Ishii, H., & Ullmer, B. 1997. Tangible bits: towards seamless interfaces between people, bits and atoms. In *Proceedings of the SIGCHI conference on Human factors in computing systems*: 234–241. Atlanta, GA: ACM.

Kaltenbrunner, M., & Bencina, R. 2007. ReacTIVision: A computer-vision framework for table-based tangible interaction. In *Proceedings of the 1st international conference on Tangible and embedded interaction*: 69–74. Baton Rouge, Louisiana: ACM.

Kim, M.J., & Maher, M.L. 2008. The Impact of Tangible User Interfaces on Designers' Spatial Cognition. *Human–Computer Interaction* 23(2): 101–137.

Løvs, G.G. (1998). Models of wayfinding in emergency evacuations. *European Journal of Operational Research* 105(3): 371–389.

Miller, C., & Lewis, D. 1999. Wayfinding: Guidance for healthcare facilities. London, UK: NHS Estates.

Piovesana, M., Chen, Y.-J., Yu, N.-H., Wu, H.-T., Chan, L.-W., & Hung, Y.-P. 2010. Multi-display map touring with tangible widget. In *Proceedings of the international conference on Multimedia*: 679–682. Firenze, Italy: ACM.

Rousek, J.B., & Hallbeck, M.S. 2011. Improving and analyzing signage within a healthcare setting. *Applied Ergonomics* 42(6): 771–784.

Sharlin, E., Watson, B., Kitamura, Y., Kishino, F., & Itoh, Y. 2004. On tangible user interfaces, humans and spatiality. *Personal and Ubiquitous Computing* 8(5): 338–346.

Tzeng, S.-Y., & Huang, J.-S. 2009. Spatial forms and signage in wayfinding decision points for hospital outpatient services. *Journal of Asian Architecture and Building Engineering* 8(2): 453–460.

Ullmer, B. & Ishii, H. 2000. Emerging Frameworks for Tangible User Interfaces. *IBM Systems Journal* 39(3 & 4): 915–931.

Wright, P., Hull, A.J., & Lickorish, A. 1993. Navigating in a hospital outpatients' department: The merits of maps and wall signs. *Journal of Architectural and Planning Research* 10(1): 76–89.

Zheng, M.-C. 2012. Time constraints in emergencies affecting the use of information signs in wayfinding behavior. *Procedia - Social and Behavioral Sciences* 35: 440–448.

# *Part F: Human factors and product design*

*Bridging Research and Good Practices towards Patient Welfare – Shih & Liang (Eds)*
*© 2015 Taylor & Francis Group, London, ISBN 978-1-138-02716-9*

# A user centered design approach for minimal invasive surgery

A. Albayrak, M. Melles & R.H.M. Goossens
*Delft University of Technology, Faculty of Industrial Design Engineering, Delft, The Netherlands*

ABSTRACT: Most of the products in the operating room are designed without much involvement of the user. It is therefore expected and often observed that many problems arise during the actual use of the product. We studied the influence of user centered design on minimal invasive surgery in a research and design program with seven PhD students in order to generate design requirements and methods for the user centered design of products for minimal invasive surgery that focus on as well the medical professional, the medical environment and even the patient. The aim of this paper is to give an overview of different aims and results of the PhD projects. Various insights in as well users as the interaction with products in the real life environment are reported.

## 1 INTRODUCTION

Healthcare professionals nowadays are part of an ever-lasting cycle of learning to control and using medical devices, and in order to do their daily job, they also have to invest time to learn how to solve usability problems due to unexpected situations.

Although safety in healthcare depends on many different factors, varying from the performance of the medical organization (or the country) as whole, to the availability, quality and usability of the technical equipment, the actual design process of these medical devices can be seen as highly relevant to influence the safety of the healthcare process.

A closer look to the current products shows that a lot of the equipment is designed without much involvement of the user. It is therefore expected and often observed that many problems arise during the actual use of the product in the real context.

To study the influence of user centered design on minimal invasive surgery a research and design program with 7 PhD students was initiated at the faculty Industrial Design Engineering at Delft University of Technology. The program has the aim to generate design requirements and methods for the user centered design of products for minimal invasive surgery that focus on as well the medical professional, the medical environment and even the patient.

The aim of this paper is to provide an overview of the topics and results of the seven PhD projects.

## 2 DESIGN PROJECT

Key factor in all PhD projects was collection and integration of user requirements into the design process. This can be done throughout the design process. Although there are many different ways to define a design cycle, a typical design process consists of the following steps: analysis, idea generation, synthesis, embodiment or prototyping and testing/ evaluation. In all of these steps user involvement can be obtained in different ways, namely interviews, observations, co-design sessions, prototype testing in the lab, in simulated environments and even in real life situations.

By defining multiple design projects focusing on one aim, namely to increase patient safety in the operating room, and by decomposing the tasks during minimal invasive surgery in a set of basic tasks on one side and the user on the other side, we were able to define a large amount of design

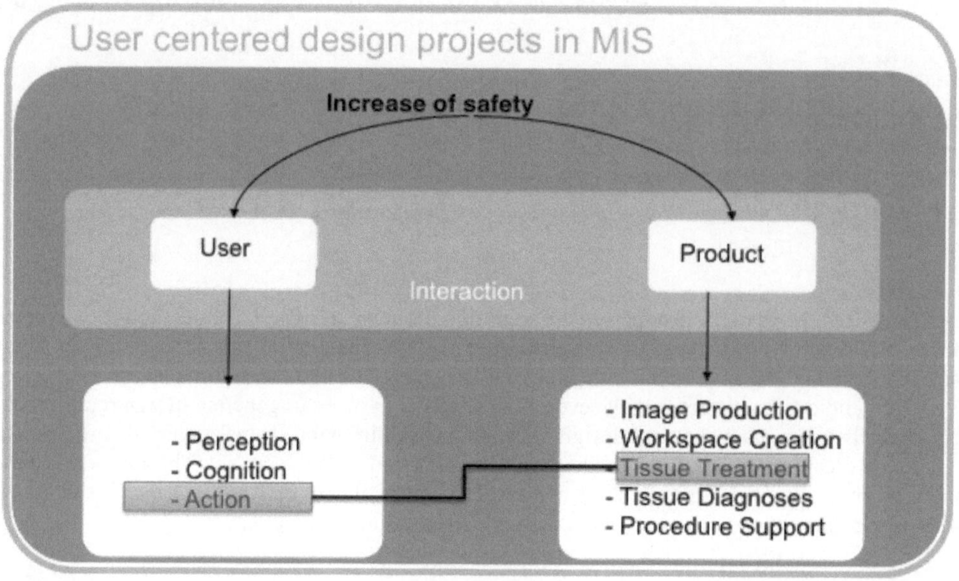


Figure 1. User centered design projects for Minimal Invasive Surgery. By combining aspects of the user and aspects of the product, seven design projects were defined. For example, the combination of action and tissue treatment led to a design project in which a new instrument was designed that is based on vacuum grasping techniques (Vonck, 2012).

projects that were all user centered in some way or the other. The model in figure 1 served as a basic structure for the projects.

## 3 OVERVIEW OF DESIGN CASES

### 3.1 *Case 1. Visualization of the surgeons working area (A. Albayrak)*

PhD-thesis
A. Albayrak
Ergonomics in the operating room: transition from open to image-based surgery (2008)
Open access via www.narcis.nl

This PhD project focuses on the two main factors that influence surgical quality: the human factor and the product factor. The physical aspects of human beings such as posture and vision are studied in relation to the output during surgery. Questions such as: how can the surgeon physically be supported during surgery, are answered. The quality aspects of products that are used during surgery are also investigated: what is the quality of equipment currently used in hospitals? What should be a minimum quality standard for different types of products?

*Result: The quality of the equipment*

Casseres and Albayrak (Cassares and Albayrak 2003) studied the safety of the equipment that is used in Dutch hospitals during minimal invasive surgery. They found in a study in 36 hospitals that 22% of all endoscopic equipment had a defect in electrical isolation and that for 36% of the light guide cables more than 30% of the fibers in the cables were defect.

3.2 *Case 2. The interaction of medical specialists with interfaces for image-guided surgical interventions (S.N. Buzink)*

PhD-thesis
S.N. Buzink
Improving patient safety in image-based procedures: bridging the gap between preferred and actual proficiency (2010) Open access via www.narcis.nl

This PhD project focuses on the human factor, the social factor and on the way these factors have influence on the feed forward in the Surgeons Cockpit (SC) system. The cognitive aspects of the human factor such as skills and experience of the surgery team members are analysed. Also the interaction between the team members and the relevant surgery information that is necessary for each member are analysed. The results of these studies show which information must be available in the feed forward part of the SC system.

*Result: Getting grip on the expertise*

Buzink et al. (Buzink et al. 2009) studied the expertise in manipulation of laparoscopic instruments. And found that the label 'expert', although often used as a single label for experienced surgeons that have performed over 50 procedures, cannot automatically be applied to all skills that are needed for the procedure. Therefore it can be concluded that a more sophisticated approach should be used in order to categorize the skills of the surgeons.

Because surgical techniques that draw from multiple types of image-based procedures (IBP) are increasing, such as Natural Orifice Transluminal Endoscopic Surgery, fusing laparoscopy and flexible endoscopy, and little is known about the relation between psychomotor skills for performing different types of IBP. Buzink et al. (Buzink et al. 2010) studied the relation between psychomotor skills for performing different types of image based procedures. The authors concluded that training and assessment of IBP type specific skills should focus on each type of tasks independently.

3.3 *Case 3. Feedforward in Surgery: How can surgical protocols help to improve surgical quality (L.S.G.L. Wauben)*

PhD-thesis
L.S.G.L. Wauben
Safety in the Operating Theatre | a Multi Factor Approach for Patients and Teams (2010)
Open access via www.narcis.nl

This PhD project focuses on the clinical factor, the political factor, the human factor and the way these factors have influence on the feed forward in the SC. Different protocols that exist for surgical procedures are analyzed and a system is developed in what way these protocols can be implemented in the SC. Also is investigated what administration is necessary of the surgical procedure for applications such as surgery reports, insurance reports and procedure budget. This information is integrated in the SC system.

*Result awareness of ergonomic guidelines*

Wauben et al. (Wauben et al. 2006) studied whether ergonomic guidelines are applied in the operating room by sending 1,292 questionnaires to mainly European surgeons and residents. The response of 22% (284 respondents) showed that 89% of them were unaware of ergonomic guidelines, although 100% stated that they find ergonomics important. The authors concluded that the lack of ergonomic guidelines awareness is a major problem that poses a tough position for ergonomics in the operating room.

*Result practice: The use of protocols*

Wauben et al. (Wauben et al. 2008) contacted fifteen Dutch hospitals for evaluation of their protocols for laparoscopic cholecystectomy. All evaluated protocols were divided into six steps and were compared accordingly. In total 10 protocols were usable for comparison. Concerning the trocar positions, only minor differences were found. The concept of "critical view of safety" was represented in just one protocol. Furthermore, the order of clipping and cutting the cystic artery and duct differed. Descriptions of instruments and apparatus were also inconsistent. It was concluded that the present protocols differ too much to define a universal procedure among surgeons in The Netherlands. The authors propose one (inter)national standardized protocol, including standardized actions.

Wauben et al. (Wauben et al. 2010) assessed the compliance with the Dutch guidelines for writing operative notes in nine hospitals that were asked to send 20 successive LC operative notes. All notes were compared to the Dutch guideline by two reviewers and double-checked by a third reviewer. She concluded that operative notes do not always fully comply with the standards set forth in the guidelines published in the Netherlands. This could influence adjuvant treatment and future patient treatment, and it may make operative notes less suitable background for other purposes. Therefore operative note writing should be taught as part of surgical training, definitions should be provided, and procedure-specific guidelines should be established to improve the quality of the operative notes and their use to improve patient safety.

### 3.4 *Case 4. Augmented reality in Minimally Invasive Surgery (A. Jalote)*

PhD-thesis
A. Jalote-Parmar
Workflow Driven Decision Support Systems: A case of an intra-operative visualization system for surgeons (2009) Open access via www.narcis.nl

The PhD project focuses on the human factor and the clinical factor and the way these factors can be implemented in the interface of the feed-forward and feed-now part of the Surgeons Cockpit. Augmented reality is a recent technology that is similar to the Virtual reality paradigm. It combines 3-dimensional (3D) computer generated objects and text superimposed on real images and video, all in real time. Advancing AR to become user-friendly has rekindled interest in real-time surgical anatomy as a way to maximize the number of safe surgical hands in the next century.

*Result: Framework for designers*

Jalote and Badke-Schaub (Jalote and Badke-Schaub 2008) proposed a framework called the Workflow Integration Matrix (WIM). WIM uses theories of human behavior in problem solving, especially the information processing paradigm.

The proposed framework intends to provide evidence-based decision-making for the development of new surgical technologies. It addresses two needs for designers: first, the analysis of surgical requirements and processes within the surgical workspace, and second, the sharing of the requirements and processes within a multidisciplinary development team.

### 3.5 *Case 5. Feedback in Surgery: How can multisensor measurement help to improve surgical quality (E.P. Westebring)*

PhD-thesis
E.P. Westebring-van der Putten
A Sense of Touch in Laparoscopy: Using Augmented Haptic Feedback to Improve Grasp Control (2011) Open access via www.narcis.nl

This PhD project focuses on the product factor and the implementation of this factor in the feed back and feed now elements of the SC. It is investigated what relevant information of the products that are used during a surgical procedure is necessary. Technological solutions are developed that can measure these information on the products for example by means of sensor technology. Furthermore it is studied how this information should be represented on the interface of the SC in order to develop a user friendly product.

*Result: The use of feedback*

Westebring et al. (Westebring-van der Putten et al. 2009) studied the differences in grasp control during barehanded and laparoscopic lifts. In the article the authors generated a prerequisite of knowledge for the ergonomic design of tools that are used to manipulate objects.

### 3.6 *Case 6. The feasibility of vacuum technique in minimal invasive surgery (D. Vonck)*

PhD-thesis
D. Vonck
The feasibility of vacuum technique in minimal invasive surgery: Improving the patient safety through instrument design (2013) Open access via www.narcis.nl

This PhD project focuses on This thesis focusses on the feasibility of vacuum technique as grasp technique for MIS. This is based on the premise that we strive to a constant performance regardless of the skills and experience of the user. This translates to the research question as: Is vacuum technique feasible as grasping technique for minimal invasive surgery?

This feasibility can be interpreted as the feasibility of a vacuum grasp instrument. The performance is determined by the technical and medical functionality of the instrument, the skills and experience of the surgeon and the comfort of use. This leads to three focus areas. The first to focus points are related to the technical and medical functionality. The third focus point is derived from the skills and experience of the surgeon and comfort of use.

*Result: Safer instruments by design*

Vonck (Vonck et al. 2010, Vonck et al. 2012) used an experimental setting to manipulate healthy pig bowel tissue via two vacuum instruments. During the experiments, two simple manipulations were performed for both prototypes by two experienced surgeons. Each manipulation was repeated 20 times for each prototype at a vacuum level of 60 kPa and 20 times for each prototype at a vacuum level of 20 kPa. All the manipulations were macroscopically assessed by two experienced surgeons in terms of damage to the bowel. In 160 observations, no serosa or seromuscular damages and no perforations were observed. It was concluded that a vacuum grasping technique looks promising for minimal invasive surgery.

### 3.7 *Case 7. Ergonomics of Laparoscopic Surgery Training Environment (D.J. Xiao)*

PhD-thesis
D.J. Xiao
Ergonomic Factors during Laparoscopic Surgery Training (2014) Open access via www.narcis.nl

This study aims to develop an ergonomic simulator setup which integrates the different insights gathered from previous studies and so to help surgeons to improve their skills under ergonomic conditions. This setup can be adjusted to conform to the ergonomic guidelines and can offer a validated laparoscopic training in a realistic OR setting.

*Result: Training in an optimal ergonomic setting*

Xiao (Xiao et al. 2012) evaluated the effect of ergonomic factors on task performance and trainee posture during laparoscopic surgery training. It was concluded that the optimal and non-optimal training setting have different learning effects on trainees' skill learning.

## REFERENCES


Buzink, S. N., Botden, S.M.B.I, Heemskerk, J., Goossens, R.H.M., de Ridder, H., Jakimowicz, J.J. (2009), 'Camera navigation and tissue manipulation; are these laparoscopic skills related?', Surg Endosc, 23 (4), 750–7.

Buzink, S. N., Goossens, R.H.M., Schoon,E.J., de Ridder, H., Jakimowicz, J.J.. (2010), 'Do Basic Psychomotor Skills Transfer Between Different Image-based Procedures?', World J Surg. 34, 933–40.

Cassares, Y.A. and Albayrak, A. (2003), 'Keyhole operations at sight (in Dutch)', (Nederlandse Vereniging voor Endoscopische Chirurgie).

Goossens, R.H.M. (2010) Healthcare. Product Design for Quality. Proceedings AHFE 2010.

Jalote, A. and Badke-Schaub, P. (2008), 'Workflow Integration Matrix: a framework to support the development of surgical information systems', Design Studies, 29, 338–68.

Vonck, D., Goossens, R.H.M., Lopuhaä, H.P., Jakimowicz, J.J., (2012) 'Grasping soft tissue by means of vacuum technique', Medical Engineering Physics', 34(8), 1088–94.

Vonck, D., Goossens, R.H.M., van Eijk, D.J., de Hingh, I.H., Jakimowicz, J.J., (2010) 'Vacuum grasping as a manipulation technique for minimally invasive surgery', Surgical Endoscopy and Other Interventional Techniques, 24(10), 2418–23.

Wauben, L. S.G.L., Goossens, R. H., and Lange, J. F. (2010), 'Evaluation of Operative Notes Concerning Laparoscopic Cholecystectomy: Are Standards Being Met?', World J Surg. 34, 903–9.

Wauben, L. S.G.L., Goossens, R.H.M., van Eijk, D.J., Lange, J.F. (2008), 'Evaluation of protocol uniformity concerning laparoscopic cholecystectomy in the Netherlands', World J Surg, 32 (4), 613–20.

Wauben, L. S.G.L., van Veelen, M.A., Gossot, D., Goossens, R.H.M. (2006), 'Application of ergonomic guidelines during minimally invasive surgery: a questionnaire survey of 284 surgeons', Surg Endosc, 20 (8), 1268–74.

Westebring-van der Putten, E. P., van den Dobbelsteen, J., Goossens, R.H.M., Jakimowicz, J.J., Dankelman, J. (2009), 'Force feedback requirements for efficient laparoscopic grasp control', Ergonomics, 52 (9), 1055–66.

Xiao, D.J., Goossens, R.H.M., Albayrak, A., Jakimowicz, J.J. (2012), 'Ergonomic factors on task performance in laparoscopic surgery training', Applied Ergonomics, 43(3), 548–53.

*Bridging Research and Good Practices towards Patient Welfare – Shih & Liang (Eds)*
*© 2015 Taylor & Francis Group, London, ISBN 978-1-138-02716-9*

# The study of anthropometry of Indonesian infant age 0–2 years in Jakarta

Yanto & Chih-Wei Lu
*Department of Industrial and Systems Engineering, Chung Yuan Christian University, Taiwan, R.O.C*

D.C. Lioenita
*Department of Industrial Engineering, Faculty of Engineering, Atma Jaya Catholic University, Jalan Jendral Sudirman, Jakarta*

ABSTRACT: The aim of this study is to present anthropometric measures of Indonesian Infant age 0–2 years old. A number of 160 subjects was participated in this study and 13 anthropometric data of the subject were measured. The mean and standard deviation (SD) of all anthropometric dimensions were calculated. Results show a gradual increase of subject's stature from age group 0–6 monts to age group 18–24 months for both boys and girls. Gradual increase also exist for weight of the subjects from age group 0–6 monts to age group 18–24 months for both boys and girls. BMI for age based on WHO child growth standard was also calculated for each students. Results on BMI for age showed that 62.5% of infant boys were overweight while girls were 73.7%.

*Keywords*: Anthropometric dimensions, infant, Indonesia.

## 1 INTRODUCTION

Anthropometry is the branch of human sciences that deals with body measurements, particularly with measurements of body size, shape, strength, mobility and flexibility and working capacity (Pheasant and Haslegrave, 2006). Anthropometric measurements are essential as basic descriptive information on body composition and nutritional status (Mokdad and Al-Ansari, 2009). For children, anthropometric data describe health or nutritional status, growth and body development from time to time. Anthropometric data is a common and simplest way to assess children health and nutrion status.

In Indonesia, there was no published anthropometric study carried out for the purpose of describing anthropometric data of infant age 0–2 years. Meanwhile, anthropometric data are necessary to describe nutritional status as well as the growth and development overtime. Hence, this study was carried out to present anthropometric database describing Indonesian infant aged between 0 to 2 years. This data could be used to describe the up-to-date physical characteristics of Indonesian infants as well as for the purpose of the design. By using this anthropometric data, this study also assessed the nutritional status of infants using BMI for age based on WHO child growth standard.

## 2 MATERIALS AND METHODS

### 2.1 *Subject*

There were 160 children participated in this study. Their age varied from age 0 and 2 years old. Since infant has different body composition following age from month to month, the age distribution was stratified based on four group based on Pheasant and Haslegrave (2006). Age group I is the infant with age from 0 to 6 months, age group II with the age from 6 months or more to 12 months, age

Table 1. The age group and gender distribution of students participated in this study.

| | Age group (months) | | | | |
|---|---|---|---|---|---|
| Gender | Age group I | Age group II | Age group III | Age group IV | Total subject |
| Boys | 20 | 20 | 20 | 20 | 80 |
| Girls | 20 | 20 | 20 | 20 | 80 |
| Total | 40 | 40 | 40 | 40 | 160 |

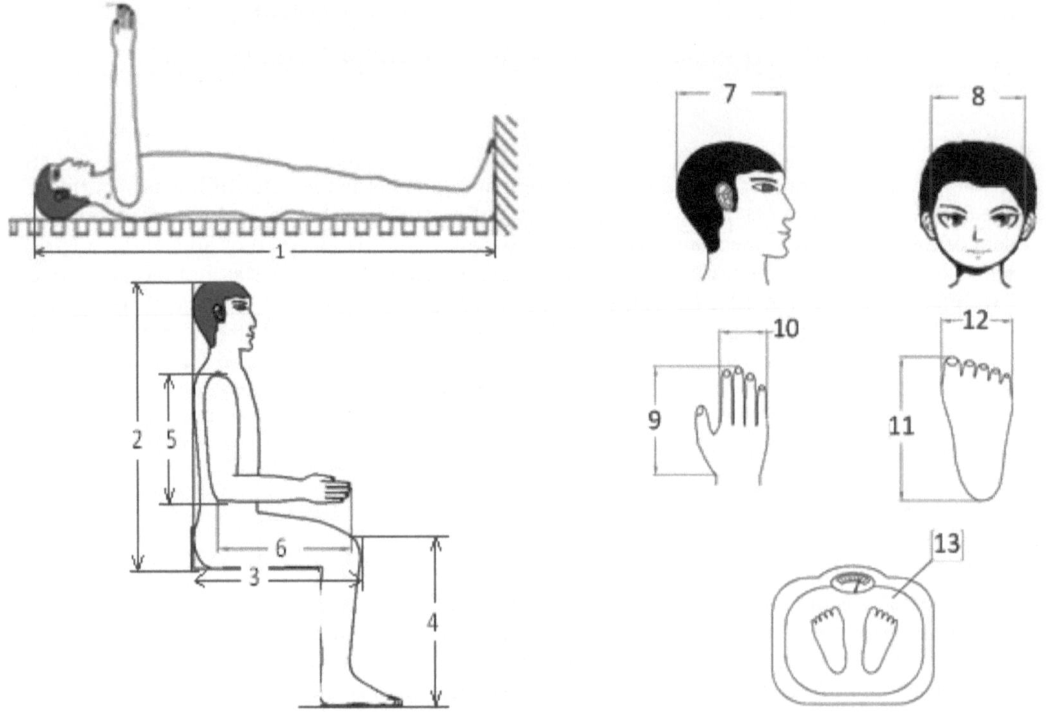


Figure 1. Anthropometric dimensions measured in this study.

group III with the age from 12 months or more to 18 months and age group IV with the age group 18 months or more to 24 months. In particular for subjects within age group I (up to 6 months), the minimum age of children was 1.5 months. Subject's age group stratifications was following the group as published by Pheasant and Haslegrave (2006). Table 1 presents the age group and gender distribution of infants participated in this study.

### 2.2 *Anthropometric dimensions measured*

There were 13 anthropometric dimensions of infant measured this study (illustrated in Figure 1). These dimensions are standard dimensions used by Pheasant and Haslegrave (2006) to describe infant's physical characteristics. Body dimensions, landmarks and the measurement of each body dimension procedures were defined by Pheasant and Haslegrave (2006).

### 2.3 *Measuring equipments*

To measure stature of infants, a self-design equipment called *baby metre* was used (see Figure 2a). An infant weighing scale was used to take the body weight of subjects (Figure 2b). The other dimensions were measured using anthropometer.

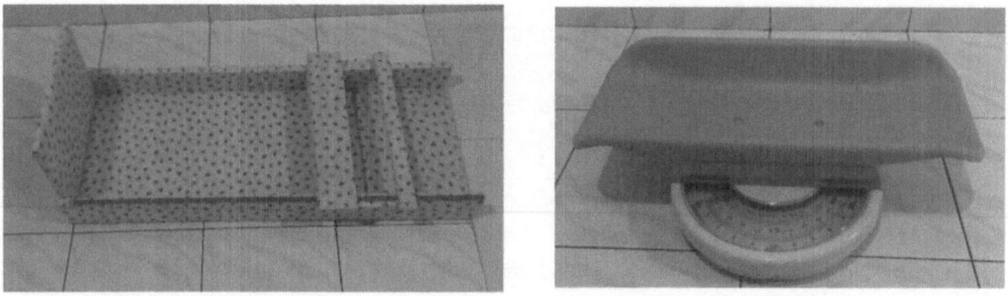

(a) (b)

Figure 2. A self-design baby metre (a) and baby weighing scale (b).

### 2.4 *Measurer*

Two university students were recruited to help in the study. They helped to do anthropometric measurement activity as well as to record measurement output. Prior measurement activity, the students were trained how to measure anthropometric dimensions and to deal with subjects.

### 2.5 *Measuring procedure*

Due to limited physical capability of infant and the extra-need of caution during measurement activity, the study were conducted after their parents (both father and mother) gave permission. During measurement, subjects were always accompanied by their parents (mother or father). They also helped to maintain the subjects in the standard position during measurement. All subjects were wearing light clothing during measurement while shoes, sweater and coats were removed. The anthropometric data collection process were carried out over a period of two months.

Since most of the infant age below 2 years old are still having a great difficulty standing on their own, the stature of subjects were measured while they were lying down on the *baby metre*. Stature of subject taken while they are lying down also called *recumbent length*. During measurement of recumbent length, one experimenter with the help of infant's mother was assigned to hold the subject while the other experimenter did the measurement activity. For the weight of subjects, the assistant together with subject's mother helped to place the subjects on his/her back on the tray of the scale (Figure 3b). The weight of subjects could be seen in display panel. The other dimensions were taken when the subjects being hold by his/her mother or father (see Figure 3c).

### 2.6 *Data analysis*

Mean, standard deviation and percentile values (percentile 5th and 95th) for each anthropometric dimensions were calculated and presented in a standard table following each age group. Normality tests using Anderson-Darling test for each anthropometric dimensions were also conducted to test wether the data were normally distributed. F-test was used to analyze significant difference for stature and weight of both boys and girls among age group I, II, III and IV. All statistical tests were conducted using Minitab 16. Regarding anthropometric data, three indexes were calculated to be used in assessing the nutritional status: weight for age, height (here we use recumbent length) for age, and weight for height (recumbent length). These indexes are then compared to a reference standard of anthropometry to assess the magnitude, distribution and severity of a nutritional problem of subjects (called an *anthropometric indicator*).

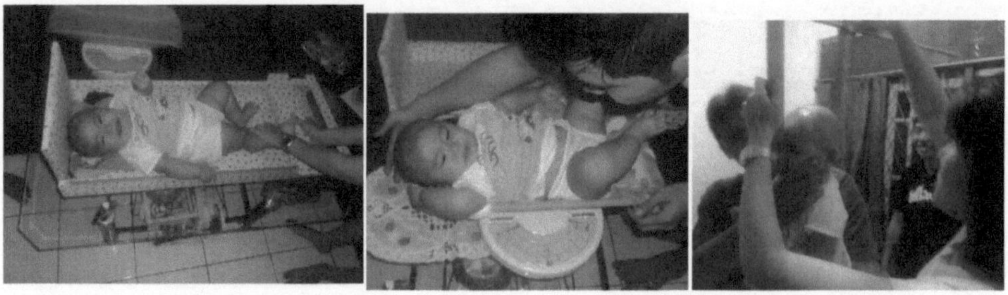

(a). (b) (c)

Figure 3. Measurement Ilustration for stature, weight and head length.

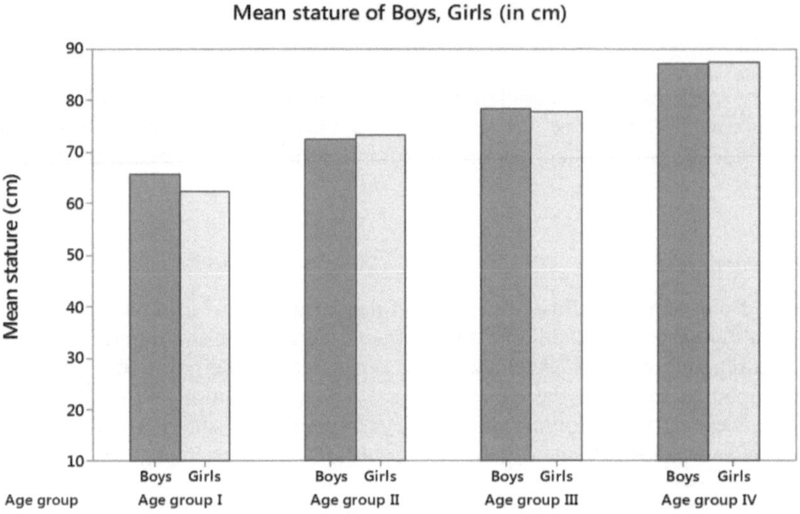


Figure 4. Mean stature (recumbent length) of infant boys and girls, age group I, II, III and IV.

## 3 RESULT AND ANALYSIS

Table 2 shows means and standard deviations anthropometric data of Indonesian infants for age group I (from 0 to 6 months), age group II (6 months or more to 12 months), age group III (12 months or more to 18 months) and age group IV (18 months or more to 24 months) for boys and girls. Figure 4 shows graphical mean stature of both boys and girls for each age group. Figure 5 shows graphical mean weight of boys and girls for each age group years old.

From Table 2, it can be seen that all mean values of anthropometric dimensions increase with increasing age for age group I, II, III and IV either boys and girls. This is a natural phenomenon as infants are in the period of growing and development. Since stature and weight are the most anthropometric dimensions used by researchers as main reference to discuss, we present graphical mean stature and weight of boys and girls in Figure 4 and Figure 5. Figure 4 and Figure 5 show graphically that mean values of both stature and weight increase with increasing age either for boys and girls from age group I, II, III and IV. These findings were supported by results of F-test which showed that there were significant differences of boys's stature among age group I, II, III and IV (p-value = 0.00) as well as weight (p-values = 0.00). The same results were found for stature and

Table 2. Mean and standard deviation of Indonesian infant anthropometric dimensions (in cm, except weight in kg).

| | | | 0–6 months | | <6–12 months | | <12–18 months | | 18–24 months | |
|---|---|---|---|---|---|---|---|---|---|---|
| No. | Dimensions | Sex | Mean | SD | Mean | SD | Mean | SD | Mean | SD |
| 1 | Stature | Boys | 65.69 | 6.87 | 72.45 | 2.84 | 78.38 | 4.41 | 87.24 | 2.18 |
| | (recumbent length) | Girls | 62.36 | 8.80 | 73.46 | 3.37 | 77.73 | 3.27 | 37.57 | 2.39 |
| 2 | Sitting height | Boys | 43.35 | 3.88 | 46.30 | 2.11 | 47.62 | 2.42 | 53.59 | 1.73 |
| | | Girls | 41.85 | 4.54 | 48.03 | 2.09 | 48.39 | 1.43 | 53.89 | 1.65 |
| 3 | Buttock-knee length | Boys | 16.52 | 2.48 | 19.18 | 1.76 | 20.24 | 1.28 | 24.84 | 1.22 |
| | | Girls | 15.07 | 5.60 | 19.59 | 1.85 | 19.73 | 1.10 | 24.81 | 1.25 |
| 4 | Knee height | Boys | 17.04 | 1.92 | 18.73 | 1.58 | 20.70 | 1.53 | 23.73 | 1.24 |
| | | Girls | 16.07 | 2.66 | 19.70 | 1.15 | 20.43 | 1.56 | 23.92 | 1.21 |
| 5 | Shoulder-elbow length | Boys | 12.62 | 1.68 | 14.16 | 0.80 | 14.66 | 1.18 | 16.70 | 1.15 |
| | | Girls | 11.92 | 1.97 | 14.60 | 1.03 | 14.62 | 0.85 | 17.05 | 0.84 |
| 6 | Forearm hand length | Boys | 17.15 | 2.19 | 18.82 | 1.28 | 19.91 | 1.06 | 23.17 | 1.09 |
| | | Girls | 16.04 | 2.72 | 19.40 | 1.10 | 20.04 | 1.12 | 23.37 | 1.12 |
| 7 | Head length | Boys | 14.34 | 1.83 | 15.85 | 0.87 | 16.20 | 0.84 | 17.90 | 0.72 |
| | | Girls | 14.15 | 1.62 | 16.22. | 0.79 | 16.16 | 0.77 | 17.97 | 0.70 |
| 8 | Head breadth | Boys | 11.72 | 1.11 | 12.24 | 0.88 | 12.61 | 1.09 | 13.46 | 0.27 |
| | | Girls | 11.18 | 0.70 | 12.52 | 0.72 | 12.50 | 0.62 | 13.51 | 0.20 |
| 9 | Handlengrh | Boys | 7.64 | 1.06 | 8.58 | 0.41 | 9.28 | 0.69 | 9.99 | 0.45 |
| | | Girls | 7.31 | 1.22 | 8.85 | 0.52 | 8.84 | 0.63 | 10.02 | 0.40 |
| 10 | Hand breadth | Boys | 4.34 | 0.52 | 4.70 | 0.86 | 4.65 | 0.25 | 5.35 | 0.19 |
| | | Girls | 5.35 | 0.54 | 4.69 | 0.51 | 4.53 | 0.63 | 5.32 | 0.22 |
| 11 | Foot length | Boys | 9.66 | 1.38 | 10.83 | 0.63 | 11.98 | 1.25 | 13.41 | 0.72 |
| | | Girls | 9.09 | 1.45 | 11.17 | 0.84 | 11.67 | 0.95 | 15.53 | 0.74 |
| 12 | Foot breadth | Boys | 4.52 | 0.83 | 4.66 | 0.29 | 5.10 | 0.97 | 5.75 | 0.24 |
| | | Girls | 4.19 | 0.56 | 4.86 | 0.56 | 4.96 | 0.62 | 5.72 | 0.28 |
| 13 | Weight | Boys | 8.05 | 1 61 | 9.75 | 0.91 | 11.50 | 1.40 | 14.73 | 0.38 |
| | | Girls | 7.68 | 2.18 | 9.88 | 0.67 | 11.70 | 1.56 | 14.48 | 0.44 |

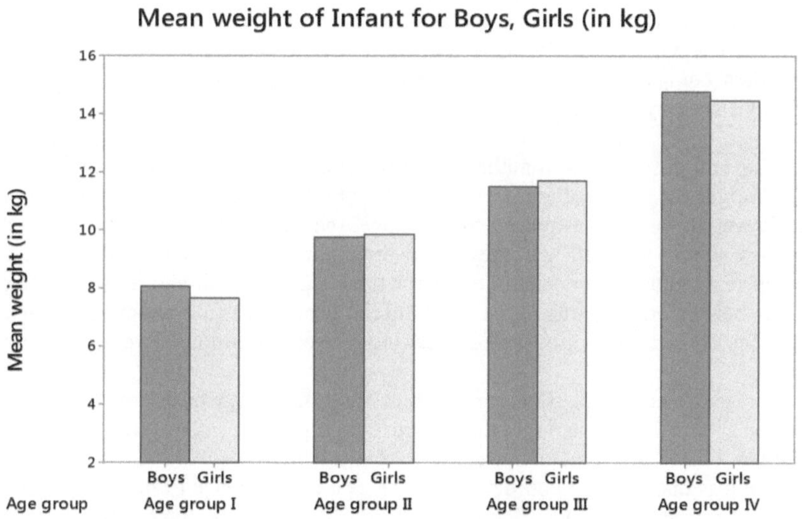


Figure 5. Mean weight of infant boys and girls, age group I, II, III and IV.

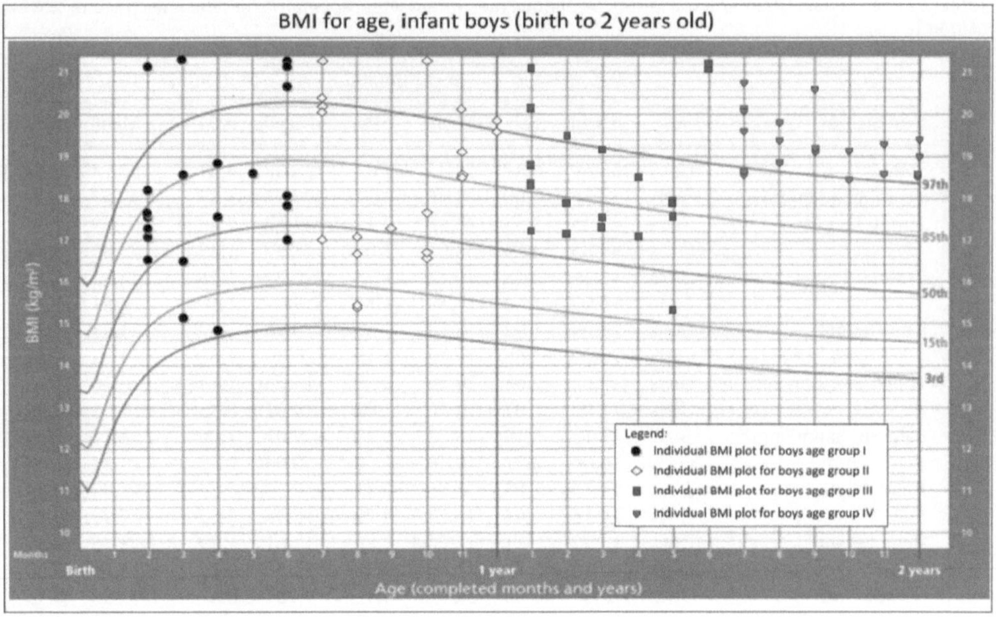


Figure 6. Individual BMI for age plot for infant boys based on WHO child growth standard.

weight of infant girls for age group I, II, III and IV (p-value = 0.00 for stature and p-values = 0.00 for weight).

## 4 DISCUSSION

Regarding anthropometric data obtained in this study (as presented in Table 2), this data were then be used to calculate anthropometric index to determine health or nutrional status of infants. Commonly, there are three indexes that are used in assessing the nutritional status: weight for age, height (here we use recumbent length) for age, and weight for height (recumbent length). These indexes are then compared to a reference standard of anthropometry to assess the magnitude, distribution and severity of a nutritional problem of subjects (here we called an anthropometric indicator).

There are several possibilities of anthropometric classification by comparison with a reference population. Weight for age is often used to tell if a child is normal, overweight or underweight. When a child weighs less than expected for their age, they are underweight, and when they weigh more than they should for their age, they are overweight. Stature (or recumbent length) for age is used to tell if a child is the normal height for their age. A child who is not as long or tall as expected may be stunted, meaning that a child did not grow to its full potential. Weight for height helps to identify children who are wasted, meaning that children do not weight as much as he or she should be.

Figure 6 presents individual BMI for age plot for infant boys based on WHO child growth standard while Figure 7 presents individual BMI for age plot for infant girls. Results on BMI for age showed that 62.5% of infant boys were overweight while girls were 73.7%. An overweight of infants for age less than 1 years could probably be as a consequence of drinking a lot of formula or breast milk. Limiting excessive drinking, limiting snacks and serving to eat prudently will help avoid obesity in infanthood. Some infants will reflexively take whatever being offered by his/her parents so that parents should serve food or drink prudently for their infants.

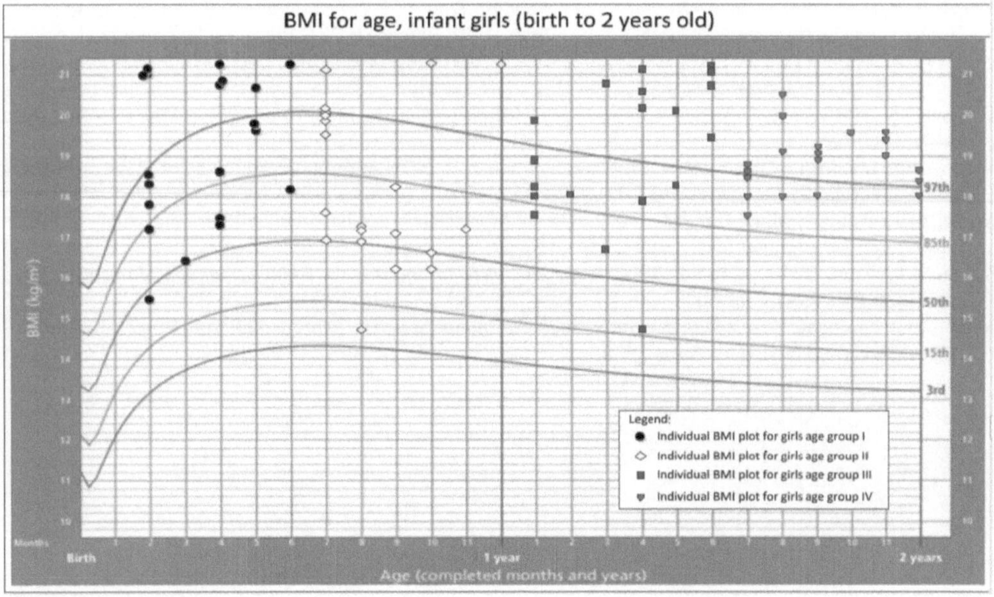


Figure 7. Individual BMI for age plot for infant girls based on WHO child growth standard.

## 5 CONCLUSION

This study aimed to present anthropometric database for Indonesian infant age 0 to 2 years old. As to the aim of this study, 13 anthropometric dimensions were measured and presented. These 13 anthropometric dimensions of infant are defined by Pheasant and Haslegrave (2006). Stature of subject showed that the mean values increased with increasing age group I (from 0 to 6 months), age group II (6 months or more to 12 months), age group III (12 months or more to 18 months) and age group IV ( 18 months or more to 24 months) either boys and girls. This was followed by all other dimensions.

In this study, the authors realize that anthropometric data of infant taken only in the Metropolitan Area of Jakarta while Indonesia consists of many geographical areas. The number of subjects taken in this study was also not representative due to limited subjects. The time, fund and difficulty to get subjects were the limiting factors. However, we can say that this is the first anthropometric database describing Indonesian infant anthropometric data. For further study, the collection of more data and from other geographical is necessary.

## ACKNOWLEDGEMENT

The authors would like to thank to the subject's parents who allowed their children for anthropometric measurement. The authors would also thank to the students who involved in the measurement activity.

## REFERENCES

Department Health and Human Services. 2002. *2000 CDC growth charts for the United States: Methods and development.* National Center for Health Statistics. Vital Health Stat 11(246).

Kroemer, K.H.E. 2003. Engineering Anthropometry. In Karwoski, K., Marras, W.S (eds), *Occupational Ergonomics: Principles of Work Design*: 9-1–9-27. Boca Raton: CRD Press.

Kroemer, K.H.E. 2006. *"Extra-Ordinary" Ergonomics: How to accommodate small dan big persons, the disabled and elderly, expectant mother and children.* Boca Raton, Fl: CRC Press, Taylor and Francis Group.

Mokdad, M. & Al-Ansari, M. 2009. Anthropometrics for the design of Bahraini school furniture. *International Journal of Industrial Ergonomics* 29: 728–735.

Pheasant, S., Haslegrave, C.M. 2006. *Bodyspace: Antropometry, ergonomics and the design of work.* Boca Raton: CRC Press.

Prado-Leon, L.R., Avila-Chaurand, R., Gonzalez-munoz, E.L. 2001. Anthropometric study of Mexican primary school children. *Applied Ergonomics* 32: 339–345.

Wang, M.J., Wang, E.M.Y., Lin, Y.C. 2002. The anthropometric database for children and young adults in Taiwan. *Applied Ergonomics* 33: 583–585.

Wang, M.J., Wang, E.M.Y., Lin, Y.C. 2002. *Anthropometric databook of the Chinese people in Taiwan.* Taiwan: Ergonomics Society of Taiwan.

Yanto. 2010. *Mengolah data statistik parametrik dan nonparametrik untuk penelitian dengan Minitab 15.* Jakarta: Penerbit Universitas Atma Jaya.

*Bridging Research and Good Practices towards Patient Welfare – Shih & Liang (Eds)*
*© 2015 Taylor & Francis Group, London, ISBN 978-1-138-02716-9*

# Effects of virtual scene parameters on shoulder joint motor performance during throwing movement

Chien-Hua Huang
*Rehabilitation Department of Feng Yuan Hospital, Ministry of Health and Welfare, Taiwan, R.O.C.*

Tien-Lung Sun
*Department of Industrial Engineering and Management, Yuan-Ze University, Taiwan, R.O.C.*

ABSTRACT: Throwing movement is an effective modus for upper limb functional training. Repeated practice of throwing movement, however, is boring. Also the patients cannot see their full body motion as a visual feedback. Recent emergence of the Kinect depth sensors allows mark-less and low-cost full body tracking, which makes it possible to display real-time avatar in the game scene so that players can see their body movement. This makes Kinect-based exergame particularly suitable for motor training for people with decreased or incomplete motor ability, e.g. rehabilitation patients, older adults, and so on. This paper develops a Kinect exergame for upper limb throwing training and studies the effect of virtual scene parameters on joint motor performance. The virtual scene parameters studied include: the avatar (yes/no), the viewing angle (0°, 22.5°) and the target distance (3 m, 5 m). The VICON system is employed to collect the subject's upper limb motion data to analyze shoulder joint motion the sagittal, frontal and transvers planes.

## 1 INTRODUCTION

### 1.1 *Background*

New treatment methods are needed to provide a low costs and better motor recovery for stroke. One area in need of improvement is upper extremity rehabilitation because of it tends to be much less recovery of function than the lower extremity following stroke. Some theories suggest that the brain generates movements not by controlling muscles directly, but by controlling the end point trajectory of a moving limb or body. According to these ideas, Emilio Bizzi (1999), try to using a computer generated virtual environment (VE) to training the stroke patient.

Two subjects with hemiplegia were treated using a computer-generated virtual environment to train upper extremity reach in the impaired limb in his study. He used motor recovery and functional ability tests and a real world test for reach task as an evaluation tool pre- and post-VE training. Subjects practiced a virtual task similar to the real task. The results of their study showed reaching errors were reduced by 50% after training. Motor recovery and functional scores showed little to no change. This research suggested that VE training holds future promise for stroke rehabilitation.

Merians et al. (2002) suggested that VR may be useful to augment rehabilitation of the upper limb in patients in the chronic phase following stroke. Their study report three patients, who were in the chronic phase following stroke, participated in a 2-week VR training program, the outcomes showed improvement in the thumb and fingers in range of motion and speed of movement.

Another case report is from Broeren et al. (2004), they investigated whether training in a virtual environment with a haptic device will improve motor function of a stroke subject in 2004. The author mentioned that the advantages of VR training—patients can practice in a controlled environment—individuals can practice simple tasks or more complex tasks and they can measure their success in real time. The participant of their study was a man with right-hemisphere lesion that caused a deficit in the left upper extremity. The subject trained with a 3-dimensional computer game during a 4-week period that consisted of twelve 90-minute sessions. The results showed training with virtual reality and haptic can promote motor rehabilitation.

After nearly 10 years of development, a systematic review of effectiveness for upper limb (UL) motor recovery using VR in stroke rehabilitation was published by Henderson et al. (2007). Two questions about the effectiveness of VR for UL rehabilitation in stroke were posed: (1) Is the use of immersive VR more effective than conventional therapy or no therapy in the rehabilitation of the UL in patients with hemiplegia? (2) Is the use of non-immersive VR more effective than conventional therapy or no therapy in the rehabilitation of the UL in patients with hemiplegia? The result showed an advantage to training in immersive VR environments versus no therapy in UL rehabilitation, and the conclusion is the current evidence on the effectiveness of using VR in the rehabilitation of the UL in patients with stroke is limited but sufficiently encouraging to justify additional clinical trials in this population.

Although recent evidence has demonstrated the efficacy of Virtual Reality (VR) for stroke rehabilitation, but its benefits and limitations in large population of patients have not yet been studied. Turolla et al. (2013) designed a pragmatic clinical trial was conducted among post-stroke patients admitted to their rehabilitation hospital. The objective of their study is to evaluate the effectiveness of non-immersive VR treatment for the restoration of the upper limb motor function and its impact on the activities of daily living capacities in post-stroke patients. 376 subjects were allocated to two treatments groups, receiving combined VR and upper limb conventional (ULC) therapy or ULC therapy alone. The results showed both treatments significantly improved but the improvement obtained with VR rehabilitation was significantly greater than that achieved with ULC therapy alone. The conclusion is the similar as before studies, VR rehabilitation in post-stroke patients seems more effective than conventional interventions in restoring upper limb motor impairments and motor related functional abilities. In summary of the literature review, the above findings, in agreement with previous evidence, suggest that VR treatments represent a valuable therapeutic option and should be more widely considered in rehabilitation programs designed for post-stroke patients.

### 1.2 *Purpose of this study*

Upper limb lift and extend motion such as forward reach and throw are necessary in daily life (Jannink et al. 2008). Joints and muscles need to coordinate for accuracy and stability. The throwing exercise is usually assigned as home program for stroke patients (Li 2007). Previous researches on clinical evaluation of VR for rehabilitation have shown positive evidences for patients suffered from stroke (You et al. 2005) (Hoe et al. 2010), visual vertigo (Keshner et al. 2007) and cerebral palsy (Bryanton et al. 2006) (Chen et al. 2007). Understanding the effect of VR/game parameters on the trainee's motion performance is important for rehabilitation experts to design customized training programs. Hsu et al. (2011) reported that the avatarin the scene will improve the motor performanceof upper limb motion. Ustinova et al. (2010) reported the effect of viewing angles on the user's motor performance.

This paper studies whether the distance between the target and the avatar will influence the upper limb motor performance during throwing movement. The effects of three game scene parameters on the user's shoulder joint motion information during Kinect-based throwing game are studied.

## 2 METHODS AND RESULT

8 volunteers participated in our experiment. They are healthy graduate students from the university. All subjects do not have problems associated with the central nervous system and upper limb muscles. The VICON system (Oxford Metrics Limited, UK, frequency 250 Hz) is employed to capture 3 dimensional kinetics data. The design of this experimental is as follow:

### 2.1 *Independent variables*

Three scene parameters were set as the independent variables, included:

1. With or without an avatar in the scene. (Shown in figure 1.)
2. The different viewing angles. (Shown in figure 2.)
3. The different distance (3 m/5 m) between the avatar and the target. (Shown in figure 3.)

### 2.2 *Dependent variables*

Shoulder joint kinetics data in three planes, including the angle, the largest angular velocity, and the angular acceleration of the shoulder, elbow and wrist1 joint in the sagittal plane (flexion/extension), the frontal plane (abduction/adduction) and the transvers plane(internal rotation/ external rotation).

### 2.3 *Statistics*

Three planes of joint moments about the shoulder, elbow and wrist were calculated using a three-dimensional inverse dynamics approach. Data obtained from the right upper limb during the throw phase were analyzed. Minitab 16.0 was used for calculated mean value for each individual subject ($n = 8$). One-Way ANOVA for data was used to identify statistically significant differences between eight factors (the avatar (yes/no), the viewing angle (0°/22.5°), the real target distance (3 m/5 m) and the virtual target distance (3 m/5 m)) in selected kinematic and kinetic variables of the throwing patterns. All results are presented as means (SD). The level of significance was set at 5%.

### 2.4 *Results*

Table 1 displays the angular acceleration and maximum angular velocity parameters of the different environment design during upper extremity throw movement.

The shoulder joint:

In real environment and target distance is 3 m, the angular acceleration was 2760°/s ± 570.12 and maximal angular velocity was 153.26°/s ± 36.93;

In real environment and target distance is 5 m, the angular acceleration was 4223°/s ± 570.12 and maximal angular velocity was 230.66°/s ± 36.93;

In virtual environment with avatar and the viewing angle is 0° and target distance is 3 m, the angular acceleration was 3116°/s ± 1811 and maximal angular velocity was 192.16°/s ± 94.37;

In virtual environment without avatar and the viewing angle is 0° and target distance is 3 m, the angular acceleration was 1678°/s ± 847 and maximal angular velocity was 171.96°/s ± 49.44;

In virtual environment with avatar and the viewing angle is 0° and target distance is 5 m, the angular acceleration was 3603°/s ± 2845 and maximal angular velocity was 200.42°/s ± 90.55.

In virtual environment without avatar and the viewing angle is 0° and target distance is 5 m, the angular acceleration was 3188°/s ± 2024 and maximal angular velocity was 187.52°/s ± 94.41.

In virtual environment without avatar and the viewing angle is 22.5° and target distance is 3 m, the angular acceleration was 2946°/s ± 2282 and maximal angular velocity was 189.81°/s ± 78.94.

In virtual environment without avatar and the viewing angle is 22.5° and target distance is 5 m, the angular acceleration was 3346°/s ± 1744 and maximal angular velocity was 190.06°/s ± 94.28.

The elbow joint:

In real environment and target distance is 3 m, the angular acceleration was 9621°/s ± 1431 and maximal angular velocity was 655.6°/s ± 48.3;

In real environment and target distance is 5 m, the angular acceleration was 14831°/s ± 4650 and maximal angular velocity was 898.3°/s ± 164.3;

In virtual environment with avatar and the viewing angle is 0° and target distance is 3 m, the angular acceleration was 8038°/s ± 1525 and maximal angular velocity was 659.1°/s ± 88.2;

In virtual environment without avatar and the viewing angle is 0° and target distance is 3 m, the angular acceleration was 8598°/s ± 2016 and maximal angular velocity was 678.1°/s ± 112;

In virtual environment with avatar and the viewing angle is 0° and target distance is 5 m, the angular acceleration was 9927°/s ± 5816 and maximal angular velocity was 717.6°/s ± 177.

In virtual environment without avatar and the viewing angle is 0° and target distance is 5 m, the angular acceleration was 8735°/s ± 1108 and maximal angular velocity was 717.5°/s ± 112.1.

In virtual environment without avatar and the viewing angle is 22.5° and target distance is 3 m, the angular acceleration was 7192°/s ± 1320 and maximal angular velocity was 598.9°/s ± 150.4.

In virtual environment without avatar and the viewing angle is 22.5° and target distance is 5 m, the angular acceleration was 8198°/s ± 1115 and maximal angular velocity was 658.6°/s ± 86.5.

The wrist joint:

In real environment and target distance is 3 m, the angular acceleration was 19857°/s ± 4433 and maximal angular velocity was 1342.5°/s ± 232.2;

In real environment and target distance is 5 m, the angular acceleration was 22472°/s ± 6982 and maximal angular velocity was 1529.6°/s ± 347.3;

In virtual environment with avatar and the viewing angle is 0° and target distance is 3 m, the angular acceleration was 18241°/s ± 4533 and maximal angular velocity was 1393.4°/s ± 276.6;

In virtual environment without avatar and the viewing angle is 0° and target distance is 3 m, the angular acceleration was 17175°/s ± 4056 and maximal angular velocity was 1419.4°/s ± 268.4;

In virtual environment with avatar and the viewing angle is 0° and target distance is 5 m, the angular acceleration was 20265°/s ± 7751 and maximal angular velocity was 1507.3°/s ± 383.4.

In virtual environment without avatar and the viewing angle is 0° and target distance is 5 m, the angular acceleration was 17381°/s ± 2689 and maximal angular velocity was 1449°/s ± 210.2.

In virtual environment without avatar and the viewing angle is 22.5° and target distance is 3 m, the angular acceleration was 14934°/s ± 5582 and maximal angular velocity was 1297.3°/s ± 375.2.

In virtual environment without avatar and the viewing angle is 22.5° and target distance is 5 m, the angular acceleration was 18308°/s ± 5596 and maximal angular velocity was 1395.8°/s ± 261.7.

## 2.5 *Tables*

Table 1. One-Way ANOVA table.

| variables | R3 (N = 8) | R5 (N = 8) | S1 (N = 8) | S2 (N = 8) | S3 (N = 8) | S4 (N = 8) | S5 (N = 8) | S6 (N = 8) | P-value |
|---|---|---|---|---|---|---|---|---|---|
| **shoulder** | | | | | | | | | |
| largest angular velocity | 153.26 (36.93) | 230.66 (56.92) | 192.16 (94.37) | 171.96 (49.44) | 200.42 (90.55) | 187.52 (94.41) | 189.81 (78.94) | 190.06 (94.28) | 0.71 |
| angular acceleration | 2760 (570) | 4223 (648) | 3116 (1811) | 2678 (847) | 3603 (2845) | 3188 (2024) | 2946 (2282) | 3346 (1744) | 0.722 |
| **elbow** | | | | | | | | | |
| largest angular velocity | 655.6 (48.3) | 898.3 (164.3) | 659.1 (88.2) | 678.1 (112.0) | 717.6 (177.0) | 717.5 (112.2) | 598.9 (150.4) | 658.6 (86.5) | **0.001** |
| angular acceleration | 9621 (1431) | 14831 (4650) | 8038 (1525) | 8598 (2016) | 9927 (5816) | 8735 (1108) | 7192 (1320) | 8198 (1115) | **0** |
| **wrist** | | | | | | | | | |
| largest angular velocity | 1342.5 (232.2) | 1529.6 (347.3) | 1393.4 (276.6) | 1419.4 (268.4) | 1507.3 (383.4) | 1449.0 (210.2) | 1297.3 (375.2) | 1395.8 (261.7) | 0.8 |
| angular acceleration | 19857 (4433) | 22472 (6982) | 18241 (4533) | 17175 (4056) | 20265 (7751) | 17381 (2689) | 14934 (5582) | 18308 (5596) | 0.216 |

Values are means (SD) and expressed as degree/second.
R3: Real environment; target distance = 3 m;
R5: Real environment; target distance = 5 m;
S1: Virtual environment with avatar; viewing angle = 0°; target distance = 3 m;
S3: Virtual environment with avatar; viewing angle = 0°; target distance = 5 m;
S2: Virtual environment without avatar; viewing angle = 0°; target distance = 3 m;
S4: Virtual environment without avatar; viewing angle = 0°; target distance = 5 m;
S5: Virtual environment with avatar; viewing angle = 22.5°; target distance = 3 m;
S6: Virtual environment with avatar; viewing angle = 22.5°; target distance = 5 m;

## 2.6 *Figures*

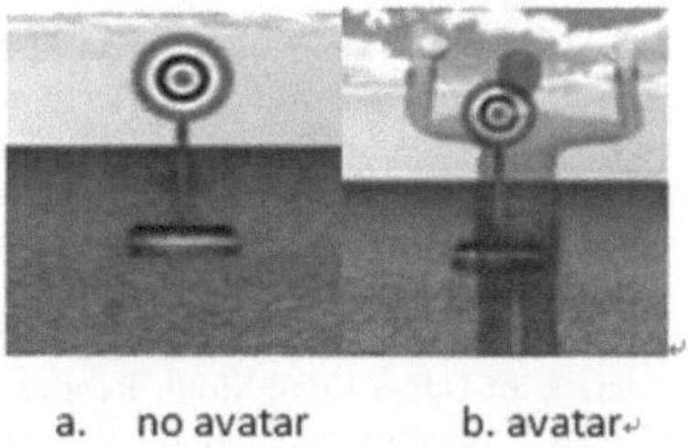


Figure 1. Avatar (yes/no).

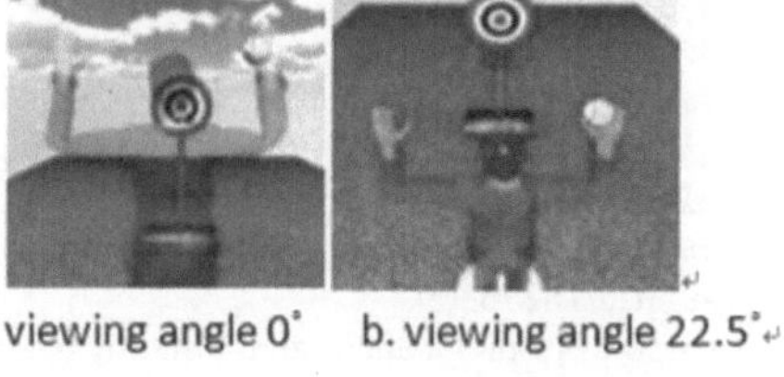


Figure 2. Viewing angle (0/25).

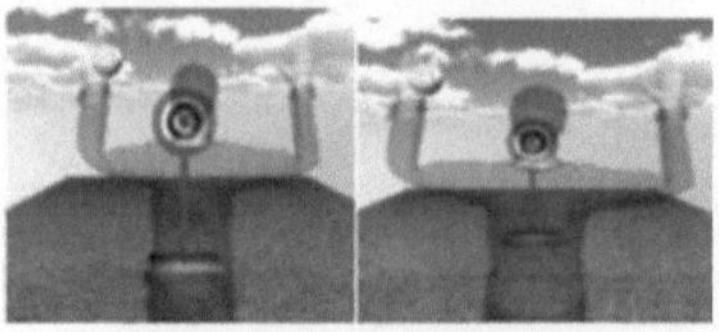


Figure 3. Distance (3/5).

## 3 DISCUSSION AND CONCLUSION

This paper describes the development and evaluation of a Kinect-based exergame for players to practice throwing movement, an important training action for upper limb function. Difference environments design is proposed to influence upper limb kinetic parameters. Only elbow joint kinetic parameters had significant difference between these different environments design. The results indicate that rehabilitation experts can change the environments to design customized motor training program of the elbow joint.

## REFERENCES

Broeren, Jurgen, Martin Rydmark, and Katharina Stibrant Sunnerhagen. 2004. "Virtual Reality and Haptics as a Training Device for Movement Rehabilitation after Stroke: A Single-Case Study." *Archives of Physical Medicine and Rehabilitation* 85 (8) (August): 1247–50. doi:10.1016/j.apmr.2003.09.020. http://www.ncbi.nlm.nih.gov/pubmed/15295748.

Bryanton, C, J Bossé, M Brien, J McLean, a McCormick, and H Sveistrup. 2006. "Feasibility, Motivation, and Selective Motor Control: Virtual Reality Compared to Conventional Home Exercise in Children with Cerebral Palsy." *Cyberpsychology & Behavior: The Impact of the Internet, Multimedia and Virtual Reality on Behavior and Society* 9 (2) (April): 123–8. doi:10.1089/cpb.2006.9.123. http://www.ncbi.nlm.nih.gov/pubmed/16640463.

Chen, Yu-ping, Lin-ju Kang, Tien-yow Chuang, Ji-liang Doong, and Shwn-jan Lee. 2007. "Use of Virtual Reality to Improve Upper-Extremity Control in Children With Cerebral Palsy: A Single-Subject Design." *Physical Therapy* 87 (11): 1441–1457.

Emilio Bizzi, MD Maureen Holden PhD PT Emanuel Todorov PhD Janet Callaban PT. 1999. "Virtual Environment Training Improves Motor Performance in Two Patients with Stroke: Case Report." *NEUROLOGY REPORT* 23 (2): 57–67.

Henderson, Amy, Nicol Korner-Bitensky, and Mindy Levin. 2007. "Virtual Reality in Stroke Rehabilitation: A Systematic Review of Its Effectiveness for Upper Limb Motor Recovery." *Topics in Stroke Rehabilitation* 14 (2): 52–61. doi:10.1310/tsr1402-52. http://www.ncbi.nlm.nih.gov/pubmed/17517575.

Hsu, Jason K, Richard Thibodeau, Stephanie J Wong, Daniel Zukiwsky, Sara Cecile, and David M Walton. 2011. "A 'Wii' Bit of Fun: The Effects of Adding Nintendo Wii(®) Bowling to a Standard Exercise Regimen for Residents of Long-Term Care with Upper Extremity Dysfunction." *Physiotherapy Theory and Practice* 27 (3) (April): 185–93. doi:10.3109/09593985.2010.483267. http://www.ncbi.nlm.nih.gov/pubmed/20698793.

Jannink, Michiel J A, Gelske J van der Wilden, Dorine W Navis, Gerben Visser, Jeanine Gussinklo, and Maarten Ijzerman. 2008. "A Low-Cost Video Game Applied for Training of Upper Extremity Function in Children with Cerebral Palsy: A Pilot Study." *Cyberpsychology & Behavior: The Impact of the Internet, Multimedia and Virtual Reality on Behavior and Society* 11 (1) (March): 27–32. doi:10.1089/cpb.2007.0014. http://www.ncbi.nlm.nih.gov/pubmed/18275309.

Keshner, Emily a, Jefferson Streepey, Yasin Dhaher, and Timothy Hain. 2007. "Pairing Virtual Reality with Dynamic Posturography Serves to Differentiate between Patients Experiencing Visual Vertigo." *Journal of Neuroengineering and Rehabilitation* 4 (January): 24. doi:10.1186/1743-0003-4-24. http://www.pubmedcentral.nih.gov/articlerender.fcgi?artid=1948002&tool=pmcentrez&rendertype=abstract.

Li, Yung-Hsin Yung-Hsin Li1 Jin-Ling Lo. 2007. "Occupational Therapy for a Patient with Traumatic Brain Injury: Comparing the Effect of Remedial Approach and Occupation-Based TreatmentModels." *Journal of Taiwan Occupational Therapy Association* 25 (2): 66–72. http://medcontent.metapress.com/index/A65RM03P4874243N.pdf.

Merians, Alma S, David Jack, Rares Boian, Grigore C Burdea, Sergei V Adamovich, Michael Recce, and Howard Poizner. 2002. "Case Report Virtual Reality – Augmented Rehabilitation for Patients." *Physical Therapy* 82 (9): 898–915.

Turolla, Andrea, Mauro Dam, Laura Ventura, Paolo Tonin, Michela Agostini, Carla Zucconi, Pawel Kiper, Annachiara Cagnin, and Lamberto Piron. 2013. "Virtual Reality for the Rehabilitation of the Upper Limb Motor Function after Stroke: A Prospective Controlled Trial." *Journal of Neuroengineering and Rehabilitation* 10 (1) (January): 85. doi:10.1186/1743-0003-10-85. http://www.pubmedcentral.nih.gov/articlerender.fcgi?artid=3734026&tool=pmcentrez&rendertype=abstract.

Ustinova, K I, J Perkins, L Szostakowski, L S Tamkei, and W a Leonard. 2010. "Effect of Viewing Angle on Arm Reaching While Standing in a Virtual Environment: Potential for Virtual Rehabilitation." *Acta Psychologica* 133 (2) (February): 180–90. doi:10.1016/j.actpsy.2009.11.006. http://www.ncbi.nlm.nih.gov/pubmed/20021998.

You, Sung H, Sung Ho Jang, Yun-Hee Kim, Mark Hallett, Sang Ho Ahn, Yong-Hyun Kwon, Joong Hwi Kim, and Mi Young Lee. 2005. "Virtual Reality-Induced Cortical Reorganization and Associated Locomotor Recovery in Chronic Stroke: An Experimenter-Blind Randomized Study." *Stroke; a Journal of Cerebral Circulation* 36 (6) (June): 1166–71. doi:10.1161/01.STR.0000162715.43417.91. http://www.ncbi.nlm.nih.gov/pubmed/15890990.

Zheng-Yu Hoe, Jue-Long Wang, Yuk-Keung Lo, Shu-Fen Sun, Chao- Hung Chang, Hsin-I Tasi. 2010. "Efficacy of WiiTM-Based Virtual Reality Assisted Rehabilitation for Chronic Stroke Patients." *Taiwan Journal of Physical Medicine and Rehabilitation* 38 (01): 11–18.

*Bridging Research and Good Practices towards Patient Welfare – Shih & Liang (Eds)*
*© 2015 Taylor & Francis Group, London, ISBN 978-1-138-02716-9*

# Fitting pressure garments analysis for burn victim's scars

Pi-Wen Huang & Chih-Wei Lu
*Department of Industrial and Systems Engineering, Chung Yuan Christian University, Taiwan, ROC*

ABSTRACT: After the people suffering from fire, chemical burns, .... etc. has decreased, the only way for the victims to deal with the scars after hospital discharge is to wear pressure garments or wait for the scar healing then go for the plastic reconstructive surgeon. The burn victims usually undergo extreme discomfort in the first three to six months since the scar growth rate accelerates while they also need to endure muscle contracture. Gradually losing the function of individual perspiration system and wearing the pressure garments to combat scar growth lead to the terrible experience. The ultrasound can be used to evaluate scars and determine if the current pressure garments worn is sufficient to prevent scars grow. Pressure garments are made to fit each individual's stature, and the physical therapist also helps to ensure the burn victim's flexibility by having them tried on the garment after it has been made. However, it has been proven that the pressure garments become loose and need to be reworked frequently after patients exercise and lose weight. The loss of one's perspiration system along with the scar growth leads to further discomfort.

According to papers, the optimal pressure each pressure garment supplies is 25 mm Hg. However, none of relevant domestic research studies have been found. In this study, we will first interview the patients and have them experimented, hopefully we will be able to confirm that the optimal pressure each pressure garment supplies is higher than 25 mm Hg.

*Keywords*: Pressure Garment, Keloids, Hypertrophic scar, Contracture, Plastic & Reconstructive Surgeon, Physical Therapist

## 1 INTRODUCTION

According to the National Fire Agency, Taiwan during 2009–2012, the fire accident causes and loss showed that 70.4% of fire accidents in Taiwan were located in the buildings, of which the electrical equipment problems seemed to be the majority of them. Although there has been a significant decrease in the fire accidents due to great efforts on disaster prevention made by our government in this decade, still many people have still suffered from burn injuries owing to fire accidents or other reasons, of which the male's injury and death rate is much higher than female's [2]. It is worth mentioned that not a minority of patients have suffered from serious complications and intubation due to the fire or the other accidents. Reasons for burn injuries can be divided into five categories: (1) Hydrothermal burn, (2) Fire burn, (3) Chemical burns, (4) Electrical burn (5) Others: sunburn, frostbite, and radiation burns.

In Taiwan, burned patients must live in the Intensive Care Unit (ICU) of hospitals. Each adult up 25% of burn area and 20% for a child in the second degree burns are considered as serious burn patients. (http://www.cbf.org.tw/ugC_Pro03- 30.asp).

The skin has multiple layers of ectodermal tissue and dermis, which covers the underlying glands, hair, blood vessels, and nerves. Muscle fibers and collagen bundles of normal skin arrange in order, once burn damage penetrates into dermis, its regular arrangement is destroyed. In the healing process of natural skin or newly-planted skin, the collagen bundles are disorderly piled and abnormal proliferation may occur, it is inevitable to have scars especially on the severe burn areas. The scar tissue, compared to the normal skin tissue, is very different. However, it can still

Table 1. Classification of Burns Based on Depth Characteristics (http://en.wikipedia.org/wiki/Burn).

| Classification | Cause | Appearance | Sensation | Healing time | Scarring |
|---|---|---|---|---|---|
| Superficial burn | Ultraviolet light, very short flash (flame exposure) | Dry and red; blanches with pressure | Painful | 3 to 6 days | None |
| Superficial partial thickness burn | Scald (spill or splash), short flash | Blisters; moist, red and weeping; blanches with pressure | Painful to air and temperature | 7 to 20 days | Unusual; potential pigmentary changes |
| Deep partial thickness burn | Scald (spill), flame, oil, grease | Blisters (easily unroofed); wet or waxy dry; variable color (patchy to cheesy white to red); does not blanch with pressure | Perceptive of pressure only | More than 21 days | Severe (hypertrophic) risk of contracture |
| Full thickness | Scald (immersion), flame, steam, oil, grease, chemical, high-voltage electricity | Waxy white to leathery gray to charred and black; dry and inelastic; does not blanch with pressure | Deep pressure only | Never (if the burn affects more than 2 percent of the total surface area of the body) | Very severe risk of contracture |

reach the result we expect after proper treatments and care. Scar tissues contain many tiny fibers, i.e. collagen fibers, which are piled on one another irregularly (Li Jin et al., 2012).

The healing process of scars can be divided into three stages: (1) Inflammation period: usually the first seven days, (2) Hyperplasia period: from one-and-a-half months to three months, (3) remodeling period: from six months to eight months (P.W.C. Akkermans 1999).

Therefore, it is impossible to predict what level of scab will be formed. It is determined by the depth, the position, area size and the patient's skin type. Obvious scars occur easily on people who have a lighter color of skin. Consequently, the White people have a much lower possibility of getting plump cicatricle, followed by the Oriental, and the Black people. (Li Jin et al., 2012)

Here are several types and characteristics of scars: (1) Superficial scar: mainly happens on abrasions and superficial burns and knife wound, the characteristic of it is slightly rough, pigment changes, flat and soft on some scar area, no malfunction encountered, and usually do not need to be dealt with. (2) Proliferation scar: also called the hypertrophic scar, commonly happen on second and third degree burns and the wound of partial thickness. They are thick, raised, (up to 1–2 centimeters) and red or dark violet, painful and itching, often affect an individual's work and rest. The scar becomes softer and slightly smoother after 6 months approximately, hyperemia decreases and capillary vessels reduce over time, itches and pain gradually improve or disappear.

(3) Atrophic scar: Also called the unstable scar, commonly occurred on the wound of skull after electric shock, large area burn injury, and healing of chronic ulcer. The scar is hard, with few blood vessels, light red or white, deep tissues adhere closely, and cannot bear frictions and heavy burden. It cannot be heal for a long period of time once it's broken and may lead to vicious results. (4) Contracture scar: often happened on deep burn wound of which has not been skin-grafted and merely healed itself. It frequently damages the function of its original seriously, such as ectropion of eyelids and lips, pleural adhesion, joint buckling, contractures, and abnormalities. A long-term scar contracture can affect the underlying muscles, tendons, and nerves, causing joint stiffness, dislocation, and abnormality. (5) Keloid scar: Also called keliods, is an overgrowth of the scar tissues. Keloid scars are commonly happened on burn injury, damages, or minor injuries neglected by individual patients. The edge of scar is obviously higher than the normal skin, and the scar will grow beyond the site of the injury. There scars are pink or purplish red, extremely hard, no elasticity, and bad the blood-supply. Keloid scars occurs frequently on the upper body, such as middle of sternum, shoulders and upper arms, Keloid scars can rarely occur on eyelids, areola, and penis, ...etc. (6) Webbed contractures: The scar is full of creases like ducks' web. Webbed

contractures are generally seen on burn injuries, few of them occur on the skin cut and tears, and inappropriate surgical cut. A large-scale webbed contractures are often found on front necks, armpits, Cubital fossa, and the perineum; while small-scale contractures are commonly found on internal and external eye sockets, nasal fossa and lips, corner of mouths, nostrils, finger and palm sides, web-space, thumb-index web, urethral, vagina, ...etc. (Prof. Dr P.W.C 1999).

Wearing pressure garments made for burn patients helps to produce an appropriate lasting pressure to the scar, prevent an excessive growth of collagen fibers, parallel arrange in order, which decreases the scar hyperplasia, makes it soft and smooth. For the patient with a small-scale of burn area, the earlier you wear pressure garments, the smoother your scar will be. Unfortunately, it can be extremely frustrating for those with a large-scale burn area.

In average, a burn patient must wear the pressure garment for 23 hours daily (taking showers and changing pressure garments excluded). However, it takes much more time for those with a large-scale burn wound in the initial stage since it is quite time-consuming to clear up the wound and change dressings. The patient must wear pressure garments for 9 months to 2 years continuously, or even longer, depending on individual physiques. A considerable number of research papers reveal that many burn patients are unsatisfied with the pressure garments they wear. Their skin becomes relatively fragile because of physical medicine and rehabilitation, and the garments rub against newly-grown skin constantly, making more and more blisters occurred. Undergoing physical medicine and rehabilitation can also hurt the newly-grown skin, making it peeled and a new wound occurred repeatedly, especially around joint areas. In addition, the burn wound (second degree to third degree) is unable to regulate its internal temperature and sweat like normal skin, so as the wound shrink in the healing process. Reasons above bring uncomfortable feelings when they wear.

Serious burn patients cannot wear pressure garments like minor ones at the early beginning, so the scar cannot be pressure-treated and proliferates widely. We can only help with it by taking plastic surgery, planting water polo, skin transplant from the scalp once it turns out keloid or proliferation scars. These are widely popular among burn patients. Not until the scar turns mature and stable can it be adopted. This is definitely helpful for the patients with a small-scale burn wound. However, this seems to be so call "Rob Peter to pay Paul" for the patient with a large-scale burn wound (we could take the skin covered by clothes and have it transplanted on the site we mind).

The medication treatment of pressure garment has been carried out for more than 30 years, well-developed techniques on pressure-garment quality can be found at (Lisa et al., 2005) (Lisa et al., 2006) (Lisa et al., 2007) and (Eric et al., 2005), which brings comfortable feelings when wear it. This is also noted at (F. William et al., 1998). At present, pressure garments are mainly manufactured by Sunshine Social Welfare Foundation and Chen Yang in Taiwan. The former has the garment tailor-made by occupational therapists while the latter mainly focuses on female shape wear and has the garment made by experienced tailors. It is worth mentioning that the pressure garment is usually tailor-made by occupational therapists or physical therapists overseas, such as UK. Pressure treatments are considered effective for minor burn scars: (a) It accelerates the healing of scars, (b) Smooth the scar, (c) Increase the elasticity of scars, (d) Reduce bleeding of scars, (e) Prevent the swelling of the area, (f) Make the collagen rearrange and parallel with epidermis, (g) Slow the synthesizing of collagen, ...etc. (F. William et al., 1998).

## 2 METHODS

This research aims to determine the relation between the depth of scars and other physiological factors of human bodies by the use of relevant ultrasonic wave instruments, and to find out the pressure index which makes patients more comfortable.

### 2.1 *Patient and scar type*

The patient suffered 77% of burn area itself, second to third degrees, and inhalation injury of level 1. This time will only focus and analyze the left hand with third-degree burn injury, and right

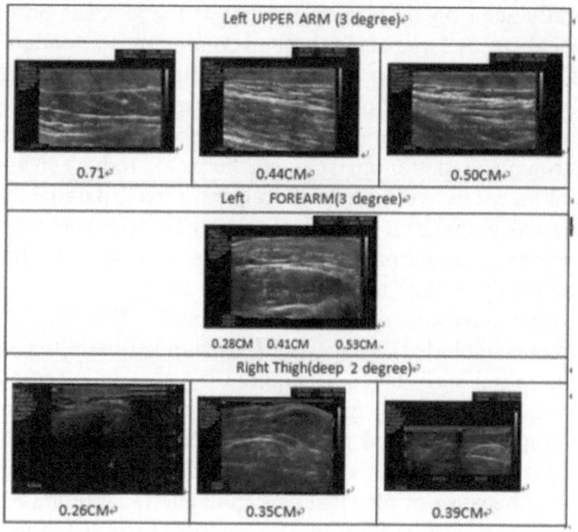


Figure 1. Ultrasound device determines the depth of the scar.

Table 2. ANOVA analysis of scar prominent area.

| One-Way ANOVA: Scar Prominent Versus Area | | | | | |
|---|---|---|---|---|---|
| Source | DF | SS | MS | F | P |
| Area | 3 | 0.2493 | 0.0831 | 8.27 | 0.008 |
| Area Error | 8 | 0.0803 | 0.01 | | |
| Total | 11 | 0.3296 | | | |

S = 0.1002; R-Sq = 75.63%

thigh with second-degree burn injury. This type of scar is called proliferation scar or hypertrophic scar. The thickness refers to Figure 1.

## 2.2 *Experiment*

The experiment has been done on August 7 and December 27, 2013.

The ultrasound Philips HDI 5000 Sonic CT was used to measure the thickness of normal skin and scars, to collect the research data. This experiment took place at Chang Gung Memorial Hospital, Taoyuan Branch; Dr. Wu, working in the Rehabilitation of Chang Gung Memorial Hospital, operated the instrument.

## 2.3 *Data analysis*

Data collected in this experiment were provided by ASUS Intel Core I5-3230M storage device 8G, and were analyzed by statistical software Minitab16.1.0 analysis.

P value <0.05 has showed remarkable standard. According to ANOVA the analysis of Scar prominent & area, even though the same pressure has been adopted on the body, there were still apparently difference among different body areas and depth. If there is some normal skin left on the measured area, its thickness remains the same and will not be affected by the pressure; the normal skin thickness is 0.15 cm for certain. This has revealed that there is a scar depth similarity between the average areas of third-degree burn injury; however, there is nearly half disparity between the average scar depth of second-degree and third-degree.

Table 3. The standard deviation, 5% and 95% average of the burned area (unit:cm).

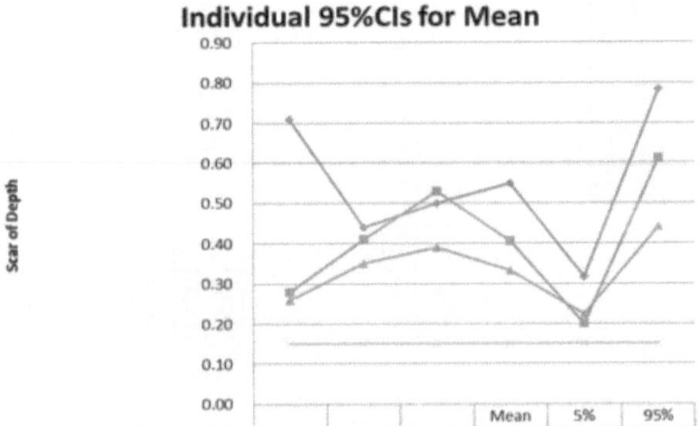


| | | | | Mean | 5% | 95% |
|---|---|---|---|---|---|---|
| Third degree Left upper arm | 0.71 | 0.44 | 0.50 | 0.55 | 0.3168 | 0.7832 |
| Left forarm | 0.28 | 0.4 | 0.53 | 0.41 | 0.2010 | 0.6123 |
| Depth of second degree Right Thigh | 0.26 | 0.35 | 0.39 | 0.33 | 0.2238 | 0.4429 |
| Normal | 0.15 | 0.15 | 0.2 | 0.15 | 0.1500 | 0.1500 |

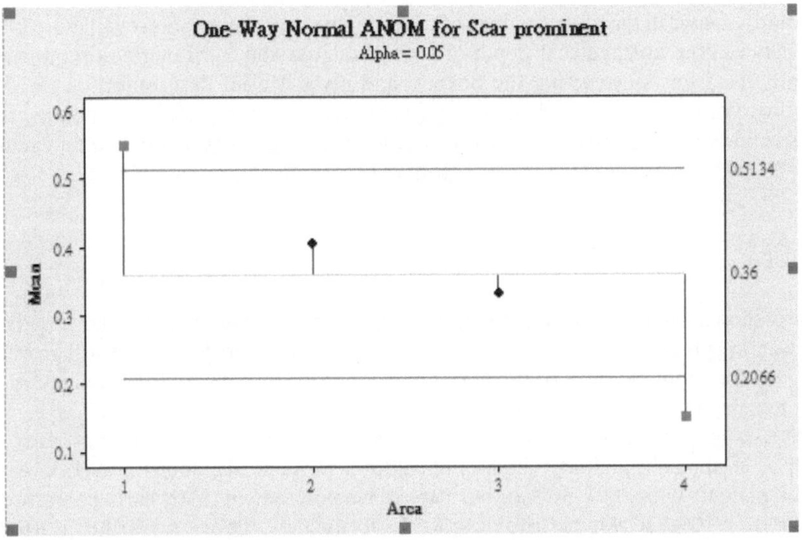


Figure 2. ANOM for Scar prominent.

Figure 2 shows the analysis of means (ANOM) for Scar prominent. 1. Refer to the upper arm, 2. Refer to the forearm, 3. Three samples were adopted from the right thigh, 4. Refer to the thickness of normal skin. Although Level 1 and 2 are all third-degree burn injury, the P value varies due to the patient's upper arm was relatively thicker than the forearm, which contained much fatness before it was burned, this will vary individually.

Depth of second degree to third degree is 0.36 cm on average, however, there is a significant difference. The upper arm is thicker than the forearm thus the average scar prominent between them is completely different.

Based on Figure 3, the average scar depth appears differently among areas of third-degree burn injury. One of the interquartile range (IRQ) Range is 0.27 cm and the other is 0.25 cm, while the IRQ Range = 0.13 cm among areas of second-degree burn. This has revealed that there is a scar

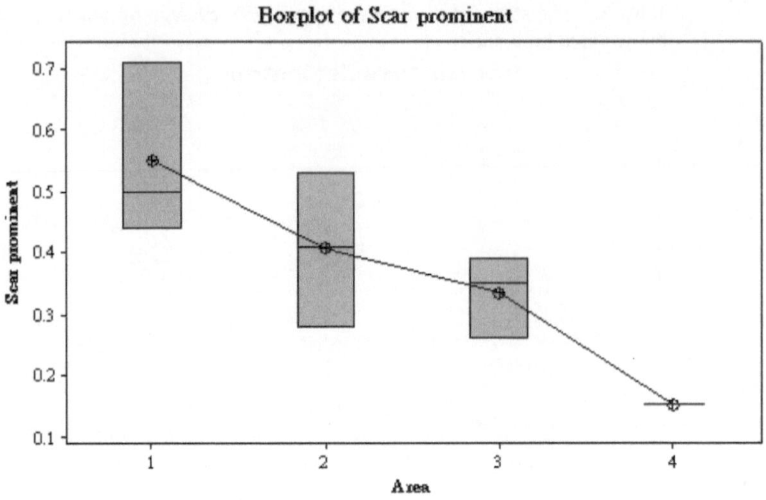


Figure 3. Boxplot of Scar prominent.

depth similarity between the average areas of third-degree burn injury; however, there is nearly half disparity between the average scar depth of second-degree and third-degree (interquartile range, IRQ, usually used for constructing the boxplot and gives a brief description on the distribution of probability. For a symmetrical distribution (where its median equals the midhinge, the average of the third and the first quartiles), half the interquartile range (IRQ) equals the median absolute deviation (MAD). The median is the corresponding measure of central tendency.

## 3 RESULT

It has clearly shown that for the second to the third degree burned patients (both large and extremities injuries), wearing pressure garments with the same pressure applied, their scars still grow differently depending on the depth and size of the burn area, age, gender, race, and family history of genetic keloid, which may all affect the scar thickness.

Patients who have hypertrophic scars grown easily are those with severe burn area 32%–67% (S. Ripper a, et al., 2009). Study (Lisa et al., 2005), (Lisa et al., 2006) and (Lisa et al., 2007) focusing on patients with 20%–30% of burn area have pointed out that wearing pressure garments are able to reduce 10%–20% hypertrophic scars. Unfortunately, there is no further illustration based on the body sites. According to the present study, a better result can be achieved if different pressure can be applied to the burn area based on its depth and size correspondingly.

## ACKNOWLEDGMENT

Having completed the present study, I would like to express my appreciation sincerely to Dr. S.S. Chuang of Plastic Surgery and Dr. P.S. Wu of Rehabilitation, Chang Gung Memorial Hospital.

## REFERENCES

Prof. Dr P.W.C. Akkermans M.A. "Epidermal-dermal crosstalk during burn wound scar maturation" en volgens besluit van het College voor Promoties. De openbare verdediging zal plaatsvinden op woensdag 30 juni 1999.

Li Jin, Liu Hao a,b, Wang Yuxiu, Shi Lingling, He Fengxia "Development of a low cost portable pressure measurement system using for garment design" Measurement 45 (2012) 2114–2120.
Wu-Chien Chien, Lu Paib, Chao-Cheng Lin, Heng-Chang Chen "Epidemiology of hospitalized burns patients in Taiwan" Burns 29 (2003) 582–588.
Lisa Macintyre, Margot Baird "Pressure garments for use in the treatment of hypertrophic scars an evaluation of current construction techniques in NHS hospitals" Burns 31 (2005) 11–14.
Eric Van den Kerckhove, Karel Stappaerts, Steffen Fieuws, Jan Laperre, Patrick Massage e, Mieke Flour, Willy Boeckx "The assessment of erythema and thickness on burn related scars during pressure garment therapy as a preventive measure for hypertrophic scarring" Burns 31 (2005) 696–702.
Lisa Macintyre *, Margot Baird "Review Pressure garments for use in the treatment of hypertrophic scars a review of the problems associated with their use" Burns 32 (2006) 10–15.
Ming-Kun Lu, Yu-San Lin, Pesus Chou, Tao-Hsin Tung "Post- traumatic stress disorder after severe burn in southern Taiwan" Burns 33 (2007) 649–652.
Lisa Macintyre "Designing pressure garments capable of exerting specific pressures on limbs" Burns 33 (2007) 579–586.
S. Ripper a,*, B. Renneberg b, C. Landmann c, G. Weigel a, G. Germann a "Adherence to pressure garment therapy in adult burn patients" Burns 35 (2009) 657–664.
Shih-han Chan a, Yi-Chun Chen b, Tzeng-Ji Chen, Hsu Maae, "Epidemiology of burns in Taiwan: A nationwide report including inpatients and outpatients" burns (2014).
F. Williams, D. Krxapp, M. Wallen "comparison of the characteristics and features of pressure used in the management of burn scars" Burns 24 (1998) 329–335.
Chih-Hsiang Lin, Chia-Fen Chi "The Defect Analysis and Quality Function Deployment Analysis of Requested Alteration for Pressure Garment" National Digital Library These and Dissertation in Taiwan 2008.
P. W.C. Akkermans "Epidermal-dermal crosstalk during burn wound scar maturation" 30 june1999, ISBN 90-73436-48-6
Gerd G Gauglitz,1,2 Hans C Korting,1 Tatiana Pavicic,1 Thomas Ruzicka,1 and Marc G Jeschke2,3,4,5 "Hypertrophic Scarring and Keloids: Pathomechanisms and Current and Emerging Treatment Strategies" MOLMED17(1–2)113–125, JANUARY-FEBRUARY 2011.
Monica C.T. Bloemen, Willem M. van der Veer b, Magda M.W. Ulrich, Paul P.M. van Zuijlen c, Frank B. Niessen,*, Esther Middelkoop "Prevention and curative management of hypertrophic scar formation burns 35(2009) 463–475.
http://en.wikipedia.org/wiki/Burn
http://www.burnsurgery.org/Modules/BurnWound%201/sect_IX.htm (6 of 18) 2008/9/21
http://www.cbf.org.tw/ugC_Pro03- 30.asp

*Bridging Research and Good Practices towards Patient Welfare – Shih & Liang (Eds)*
 *ISBN 978-1-138-02716-9*

# Assessment of range of joint motion using Kinect

J.-L. Lin, H.-C. Chuan & P.-C Kuan
*Department of Information Management, Yuan Ze University, Taiwan*

S.-M. Ni
*Department of Physical Medicine and Rehabilitation, Armed Forced Taoyuan General Hospital, Taiwan*

ABSTRACT: **Objective.** To assess body flexibility, range of joint motion provides a better measurement since it measures the angular motion of a joint and is less dependent on the type of build of the person to be measured. In the literature, many studies used Kinect to track the movement of body, but few used Kinect to track the angular motion of a joint. The objective of this paper is to study whether Kinect can provide a reliable and valid measure for various common joint motions. **Method.** The skeleton tracking function of Microsoft Kinect SDK is used to locate various positions of a subject under test. The range of various joint motions are then defined and calculated using these positions. The joint motions considered include head, arm, wrist, and thigh. Ten graduate students at Yuan Ze University were recruited as subjects for this study, and none of whom reported present joint mobility problem. **Results.** Reliability testing shows that for some joint motions (e.g., arm and thigh), Kinect can provide valid and reliable results. However, for some joint motions, Kinect fails to achieve this goal. **Discussion.** Kinect is often used to track movement of the body's skeleton. However, when using the movement of the body's skeleton to derive the range of the angular motion of a joint, the lengths of the two line segments used to define the angle play a key role on whether the joint motion can be measured properly with Kinect. If the length is small, minor deviation on the endpoints of the two line segments could result in huge deviation on the angle between the two line segments.

## 1 INTRODUCTION

Range of motion (ROM) is important for assessing body mobility (Lea and Gerhardt 1995). It is often measured using a universal goniometer in a clinical setting (Schenker 1956, Rothstein *et al.* 1983, Norkin and White 2009, Yaikwawongs *et al.* 2009). Although the universal goniometer is a convenient instrument, a patient cannot measure ROM using the universal goniometer without the assistance of a trained examiner.

Many new instruments and techniques have been developed for measuring ROM. For example, gravity-dependent goniometers or inclinometers use gravity's effect on fluid levels and pointers to measure ROM (Norkin and White 2009). Digital images analysis has also been used to quantify ROM (Dunlevy *et al.* 2005, Li *et al.* 2008, Bennett *et al.* 2009, Ferriero *et al.* 2011). However, manual intervention is still required to either operate the goniometer or locate the angles to be measured in the digital images.

ROM can be used to reflect a patient's physical rehabilitation progress. With tele-rehabilitation, it is important to measure a patient's ROM not only in a clinic but also at home. However, a trained examiner is often not available in a home setting. An approach to measure ROM without the assistance of others is a key both to make measuring ROM at home feasible, and to support tele-monitoring a patient's physical rehabilitation progress.

Recently, Microsoft Kinect sensor (Kinect for short), originally a new accessory for the Xbox 360 game console, has been used in various applications, e.g., fall detection (Hsieh *et al.* 2012), smoking cessation (Oberoi 2012), and rehabilitation (Chang *et al.* 2011, Lange *et al.* 2011, Smith and Schoene 2012). This paper proposed a method for measuring ROM of active motion using Kinect such that no assistance of others is required. This method uses the skeleton tracking function of Microsoft Kinect SDK to calculate ROM, and thus does not require an examiner to locate the angles to be measured as in previous work. Although the idea of using Kinect to measure joint angle is not entirely new (Bo *et al.* 2011), the reliability and validity of this method have not been studied. Furthermore, Kinect may be applicable to measure ROM of some joints but not for other joints. This paper studies the reliability of using Kinect to measure ROM of various joints.

## 2 MATERIALS AND METHODS

### 2.1 *Subjects*

Ten graduate students (age 24.1 ± 0.47, height 163.9 ± 1.69 cm, weight 67.1 ± 13.42 kg) at Yuan Ze University were recruited as subjects for this study. Five men and five women volunteered, none of whom reported present joint mobility problem. Informed consent was obtained from all subjects prior to their participation, in accordance with the Declaration of Human Rights, Helsinki, 1975.

### 2.2 *Test environment and procedures*

The test environment included two Kinects without tilting placed at two positions of different heights (see Fig. 1). Notably, only one Kinect is required for measuring ROM, but this test used two Kinects at different positions to study whether the relative position between the Kinect and the subject affects the accuracy of the test results. Each Kinect was connected to a PC running an application for collecting the skeletal data from the Kinect, and calculating the joint angles.

Each subject was asked to stand in front of the Kinects (see Fig. 1) and perform various motion at several joint angles (see Fig. 2 for motions of head, Fig. 3 for motions of arm, Fig. 4 for motions of thigh, and Fig. 5 for motions of wrist) (Norkin and White 2009). Although some of these motions involved the movement of either left or right of upper and lower extremity, without sacrificing the test results, we only used the right upper and lower extremity in this test for simplicity. For each motion at each joint angle, we first set a universal goniometer to that angle, and then used the goniometer to adjust the subject's ROM to that angle. Then, the subject was asked to stay in that posture to allow both Kinects to capture the subject's skeletal data. To avoid the interference problem between multiple Kinects (Butler *et al.* 2012, Maimone and Fuchs 2012), one Kinect was blocked while another Kinect was capturing the skeletal data, and vice versa. Finally, the readings from the goniometer and the calculated angles from both Kinects were analyzed, as described later.

Notably, the skeleton tracking function of Microsoft Kinect SDK is used to locate various positions of the body, which are then used to calculate the angles of joints. In this paper,

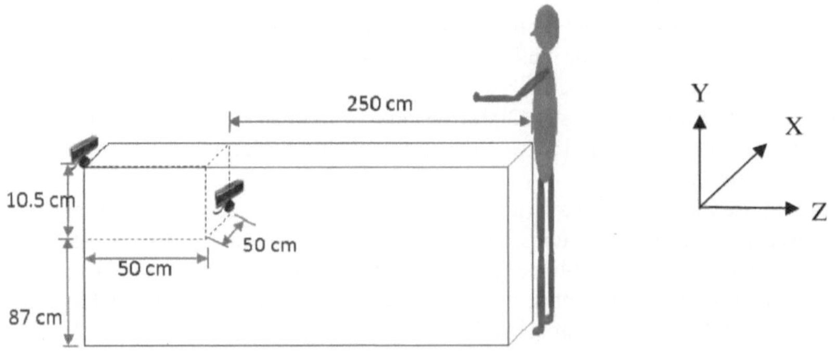


Figure 1. The placement of two Kinects.

the following positions from Kinect SDK are used: HEAD, SHOULDER_CENTER, SHOULDER_RIGHT, ELBOW_RIGHT, WRIST_RIGHT, HAND_RIGHT, SPINE, HIP_CENTER, HIP_RIGHT, HIP_LEFT and KNEE_RIGHT. The coordination system is shown in the right part of Fig. 1. For head motion in Fig. 2, the angle is defined as the included angle between the line segments SHOULDER_CENTER to HEAD and SPINE to SHOULDER_CENTER on the XY plane. For the arm motions in the left and right parts of Fig. 3, the angles are defined as the included angles between the line segments SHOULDER_RIGHT to ELBOW_RIGHT and SHOULDER_CENTER to SPINE on the YZ and XY planes, respectively. For the thigh motions in the left part of Fig. 4,

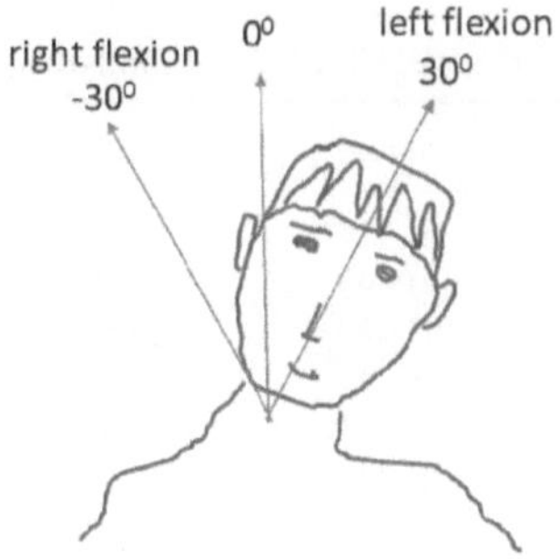


Figure 2. Motion of head. Measures are taken at −30°, 0° and 30°.

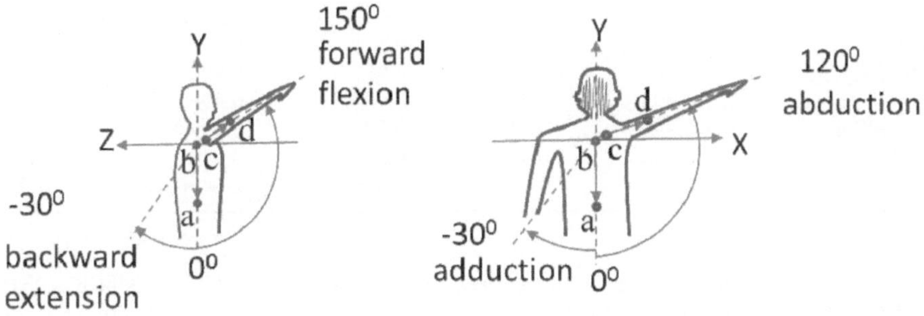


Figure 3. Motion of arm. For forward flexion, measurements are taken from 0° to 150° with an increment of 30°. For backward extension and adduction, measurements are taken from −30° to 0° with an increment of 15°. For abduction, measurements are taken from 0° to 120° with an increment of 30°.

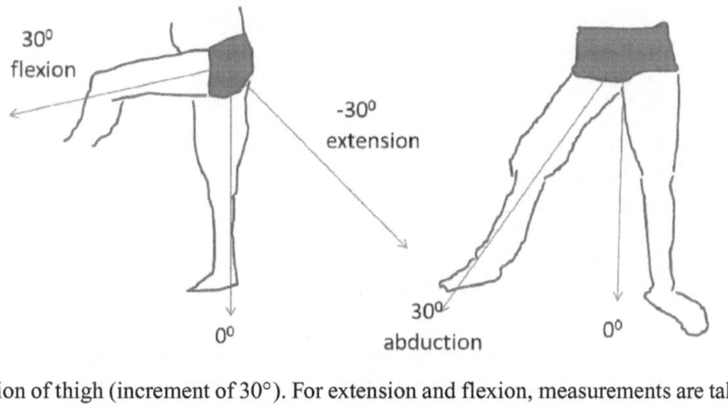


Figure 4. Motion of thigh (increment of 30°). For extension and flexion, measurements are taken from −30° to 30° with an increment of 30°. For abduction, measurements are taken from 0° to 30° with an increment of 15°.

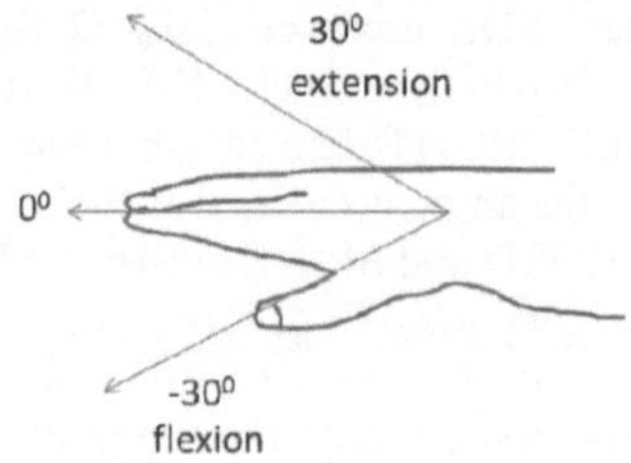


Figure 5. Motion of wrist. Measures are taken at −30°, 0° and 30°.

Table 1. Intra-observer (Kinect 250 and Kinect 300) repeatability testing for various motions.

| Motion | ICC | max | min | Cronbach $\alpha$ | significant |
|---|---|---|---|---|---|
| head (Fig. 2) | 0.973 | 0.944 | 0.987 | 0.973 | V |
| arm (left of Fig. 3) | 0.992 | 0.995 | 0.987 | 0.992 | V |
| arm (right of Fig. 3) | 0.998 | 0.999 | 0.997 | 0.998 | V |
| thigh (left of Fig. 4) | 1 | 1 | 1 | 1 | V |
| thigh (right of Fig. 4) | 0.926 | 0.971 | 0.814 | 0.926 | V |
| wrist (Fig. 5) | 0.521 | 0.772 | −0.007 | 0.521 | X |

Table 2. Intra-observer (Kinect 250 and goniometer) repeatability testing for various motions.

| Motion | ICC | max | min | Cronbach $\alpha$ | significant |
|---|---|---|---|---|---|
| head (Fig. 2) | 0.770 | 0.891 | 0.571 | 0.77 | V |
| arm (left of Fig. 3) | 0.992 | 0.995 | 0.987 | 0.992 | V |
| arm (right of Fig. 3) | 0.995 | 0.994 | 0.985 | 0.995 | V |
| thigh (left of Fig. 4) | 0.999 | 0.999 | 0.998 | 0.999 | V |
| thigh (right of Fig. 4) | 0.940 | 0.976 | 0.849 | 0.940 | V |
| wrist (Fig. 5) | 0 | 0.950 | −0.950 | 2.645E-12 | X |

the angle is defined as the included angle between the line segments HIP_CENTER to SHOULDER_CENTER and HIP_RIGHT to KNEE_RIGHT on the YZ plane. For the thigh motions in the right part of Fig. 4, the angle is defined as the included angle between the line segments HIP_LEFT to HIP_RIGHT and HIP_RIGHT to KNEE_RIGHT on the XY plane. For wrist motion in Fig. 5, the angle is defined as the included angle between the line segments WRIST_RIGHT to HAND_RIGHT and WRIST_RIGHT to ELBOW_RIGHT on the YZ plane.

## 3 RESULTS

For ease of exposition, we denote the Kinect near the sample in Fig. 1 as Kinect 250, and another Kinect as Kinect 300. The two Kinects placed at different positions were used to measure ROMs under two circumstances where the relative position between the Kinect and the sample changed. The results were used to assess intra-observer repeatability. Table 1 shows the intra-observer repeatability between the two Kinects. Both Kinects provide consistent results for all motions except the wrist motion in Fig. 5. Furthermore, ICCs are larger for motions of arm and thigh than for the motion of head.

Next, we compare the measurements of each Kinect against the measurements from the goniometer, and the results are shown in Tables 2 and 3. Here, the measurement from the goniometer is used to check the validity of the measurements from Kinects. Similar to Table 1, only the wrist motion shows low correlation between the measurements from Kinect and the goniometer.

Table 3. Intra-observer (Kinect 300 and goniometer) repeatability testing for various motions.

| Motion | ICC | max | min | Cronbach $\alpha$ | significant |
|---|---|---|---|---|---|
| head (Fig. 2) | 0.795 | 0.902 | 0.569 | 0.795 | V |
| arm (left of Fig. 3) | 0.996 | 0.997 | 0.994 | 0.996 | V |
| arm (right of Fig. 3) | 0.995 | 0.987 | 0.981 | 0.995 | V |
| thigh (left of Fig. 4) | 0.999 | 0.999 | 0.997 | 0.999 | V |
| thigh (right of Fig. 4) | 0.958 | 0.983 | 0.893 | 0.958 | V |
| wrist (Fig. 5) | 0 | 0.950 | −0.950 | 9.999E-15 | X |

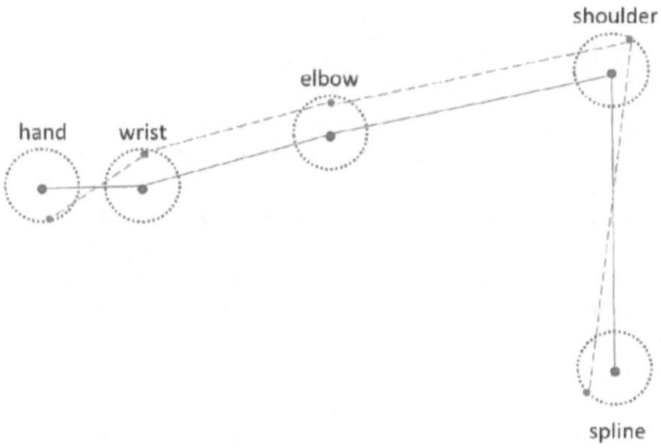


Figure 6. Included angles defined by two longer line segments are more resistant to the errors on the endpoints.

Overall, for all the motions in Figures 2, 3 and 4, Kinect provides valid and reliable results. However, for the wrist motion in Fig. 5, the results of Kinects are neither valid nor reliable. Therefore, although using Kinect to measure ROM has become common and convenient, one should be careful about which ROM is to be measured. The next section discusses the possible causes of the discrepancy of Kinect for these motions.

## 4 DISCUSSION

Previous work on using Kinect to track body motion has achieved very good results. However, this study shows that when using Kinect to track the range of a joint motion, Kinect might not provide reliable measurements for some joint motions. The range of a joint motion is defined as the included angle between two line segments of the body. Kinect can measure the positions of the joints with high accuracy. However, when using these positions to derive the included angle between two line segments of the body, higher accuracy might be required for some joint motions.

Consider Fig. 6 for example. The blue dots indicate the actual positions of the joints, and the red dots indicate the corresponding positions that the Kinect measures. For the wrist motion in Fig. 5, the range of motion is the included angle between the line segment from hand to wrist and the line segment from wrist to elbow. Due to the small length of these two line segments, minor deviation in the positions of these joints could result in a large deviation on the included angle of the two line segments. However, for the arm motion in Fig. 3, the range of motion is the included angle between the line segment from elbow to shoulder and the line segment from shoulder to spine. Since the lengths of these two line segments are longer, their included angle is more resistant to the deviation on the positions of these joints. This observation is consistent to the results in Section 3, where the motions of arm and thigh have larger ICCs than head motion, which in turn has a larger ICC than wrist motion.

## 5 CONCLUSIONS

This paper studies whether Kinect can provide reliable and valid measurements for ranges of joints motions. Our experimental results show that Kinect works well for five out of the six joint motions discussed in Section 2. Furthermore, the lengths of the line segments used to define the range of a joint motion appear to affect whether Kinect is feasible for measuring the included angle between the two line segments. Further study with more variety of joint motions is needed to strengthen this observation.

## REFERENCES


Bennett, D., Hanratty, B., Thompson, N. & Beverland, D., 2009. Measurement of knee joint motion using digital imaging. *Int Orthop,* 33 (6), 1627–31 Available from: http://www.ncbi.nlm.nih.gov/pubmed/19037641.

Bo, A.P.L., Hayashibe, M. & Poignet, P., 2011. Joint angle estimation in rehabilitation with inertial sensors and its integration with kinect. *2011 Annual International Conference of the IEEE Engineering in Medicine and Biology Society, EMBC,.* 3479–83.

Butler, D.A., Izadi, S., Hilliges, O., Molyneaux, D., Hodges, S. & Kim, D., 2012. Shake'n'sense: Reducing interference for overlapping structured light depth cameras. *Proceedings of the 2012 ACM annual conference on Human Factors in Computing Systems.* Austin, Texas, USA: ACM, 1933–36.

Chang, Y.J., Chen, S.F. & Huang, J.D., 2011. A kinect-based system for physical rehabilitation: A pilot study for young adults with motor disabilities. *Research in Developmental Disabilities,* Available from: http://www.ncbi.nlm.nih.gov/pubmed/21784612.

Dunlevy, C., Cooney, M. & Gormely, J., 2005. Procedural considerations for photographic-based joint angle measurements. *Physiother Res Int,* 10 (4), 190–200 Available from: http://www.scopus.com/inward/record.url?eid=2-s2.0-33644985827&partnerID=40&md5=4c26989e34dcfa7507fbf17653e993ff.

Ferriero, G., Sartorio, F., Foti, C., Primavera, D., Brigatti, E. & Vercelli, S., 2011. Reliability of a new application for smartphones (drgoniometer) for elbow angle measurement. *PM and R,* 3 (12), 1153–1154 Available from: http://www.scopus.com/inward/record.url?eid=2-s2.0-84055183510&partnerID=40&md5=2b5e84e281c62c049100d33f64cea814.

Hsieh, J.-C., Chiu, Y.-H., Gu, B.-Y., Chiang, T.-H., Chen, Y.-H. & Lan, H.-T., 2012. A kinect-based fall dector for the elderly homecare. *Telemedicine and e-Health,* 18 (4), A-92.

Lange, B., Chien-Yen, C., Suma, E., Newman, B., Rizzo, A.S. & Bolas, M., Year. Development and evaluation of low cost game-based balance rehabilitation tool using the microsoft kinect sensored. eds. *Engineering in Medicine and Biology Society, EMBC, 2011 Annual International Conference of the IEEE,* 1831–1834.

Lea, R.D. & Gerhardt, J.J., 1995. Range-of-motion measurements. *J Bone Joint Surg Am,* 77 (5), 784–98 Available from: http://www.ncbi.nlm.nih.gov/pubmed/7744906.

Li, G., Van De Velde, S.K. & Bingham, J.T., 2008. Validation of a non-invasive fluoroscopic imaging technique for the measurement of dynamic knee joint motion. *J Biomech,* 41 (7), 1616–22 Available from: http://www.ncbi.nlm.nih.gov/pubmed/18394629.

Maimone, A. & Fuchs, H., Year. Reducing interference between multiple structured light depth sensors using motioned. eds. *Virtual Reality Workshops (VR), 2012 IEEE,* 51–54.

Norkin, C.C. & White, D.J., 2009. *Measurement of joint motion: A guide to goniometry,* 4th ed. Philadelphia: F.A. Davis.

Oberoi, S., 2012. Gaming for health: How virtual environments can heal in the real world. *Telemedicine and e-Health,* 18 (4), A-47.

Rothstein, J.M., Miller, P.J. & Roettger, R.F., 1983. Goniometric reliability in a clinical setting. Elbow and knee measurements. *Physical Therapy,* 63 (10), 1611–15 Available from: http://www.scopus.com/inward/record.url?eid=2-s2.0-0020514697&partnerID=40&md5=4e1372c1931eeeb6e448d2e8f4454371.

Schenker, A.W., 1956. Goniometry; an improved method of joint motion measurement. *N Y State J Med,* 56 (4), 539–45 Available from: http://www.ncbi.nlm.nih.gov/pubmed/13288801.

Smith, S.T. & Schoene, D., 2012. The use of exercise-based videogames for training and rehabilitation of physical function in older adults: Current practice and guidelines for future research. *Aging Health,* 8 (3), 243–52 Available from: http://dx.doi.org/10.2217/ahe.12.30 [Accessed 2012/09/26].

Yaikwawongs, N., Limpaphayom, N. & Wilairatana, V., 2009. Reliability of digital compass goniometer in knee joint range of motion measurement. *J Med Assoc Thai,* 92 (4), 517–22 Available from: http://www.ncbi.nlm.nih.gov/pubmed/19374303.

*Bridging Research and Good Practices towards Patient Welfare – Shih & Liang (Eds)*
 *ISBN 978-1-138-02716-9*

# Synchronizing the arm-speed of toothbrushing with sounds of differing tempos

Tyan Yu Wu & Chi-Tang Lin
*The Department of Industrial Design, Chang Gung University, Taiwan*

ABSTRACT: Toothbrushing, which involves a repetitive back-and-forth arm movement, can be boring. However, a toothbrush integrated with a sound of an appropriate tempo may effectively influence users' emotions and arm movements during the act of brushing. Sounds with differing tempos can cause users to synchronize their arm movements differently, which can further affect their comfort and pleasure. This study attempted to determine an optimal tempo for toothbrushing. The results showed that tempos ranging between 110 bpm and 220 bpm are optimal for toothbrushing. In addition, the participants in the study agreed that an appropriate tempo does enhance comfort, coordination, and pleasure during the act of toothbrushing.

*Keywords*: music toothbrush, tempos, pleasure.

## 1 INTRODUCTION

Oral hygiene defects can be an indicator of early symptoms of systemic disease. Furthermore, oral diseases can have a significant impact on the health of our entire bodies, posing threats to our organs and leading to other diseases such as diabetes, atherosclerosis, pneumonia, and endocarditis (Coleman, Patricia 2002 & Scannapieco, 1999). Nowadays, a variety of methods are available to prevent gingivitis and periodontal disease, such as using chemical toothpastes during regular acts of brushing. Plaque is not easy to detected because of its adhesion to the teeth strongly and the color similar with teeth, therefore the plaque is very difficult to be removed. Based on the research of gingivitis, periodontitis and dental plaque treatment modalities, the results show that a welled Oral hygiene controlled can reduce plaque effectively (Löe, Theilade, & Jensen, 1965), and the most common and effective way to promote oral hygiene is to brush teeth to remove plaque (Wendt, Hallonsten & Koch, 2004; Hidekazu, Mieko, Suguru, Masato, Michiyo & Takashi 2004). Accordingly, selecting the right toothbrush and using the correct movements to brush teeth are critical to maintaining clean teeth.

Traditional toothbrush designs focus on the comfort and softness of the toothbrush handle and the quality of the bristles. A user brushes their teeth by moving their arm back and forth, which in turn moves the toothbrush back and forth in their mouth. However, repetitive arm movements can easily become boring. To enhance the fun of toothbrushing, an interesting brushing tool is needed to improve the task of cleaning teeth. The aim of this study is to explore an optimal sound of tempos to synchronize the arm-speed of toothbrushing.

## 2 LITERATURE REVIEW

Existing musical toothbrushes focus on simple tones and randomly picked pieces of music that are not synchronized with the users' behaviors (e.g., arm movements). Klink, Webber, Smith, and Trippel (2008) indicated that toothbrush designs should not only provide good oral hygiene by protecting teeth, but also look appealing and interesting and provide comfort. In fact, research has demonstrated that musical toothbrushes can clean dental plague effectively and improve gingival quality (Ganesh et al., 2011). However, most musical toothbrushes are designed for children, and their designs do not integrate music into the activity of brushing teeth. Thus, a sophisticated toothbrush that can guide adult users' arms to clean their teeth and yet have fun at the same time is needed. The first part of this study identifies an optimal tempo for synchronizing arm-movement speed of toothbrushing. The result is then integrated into a simple and pleasing rhythm at second part of experiment.

Tempo refers to the speed or pace of a given piece. It is a crucial element of most musical compositions because it can affect the listener's mood (Wikipedia, 2014). Sound is transmitted by air as vibrations, and our ears receive the vibrations and transfer them to nerve impulses in the human auditory system, where these vibrations are processed. The varying pitches, volumes, and timbres of sound waves received by the brain provide the information that allow us to detect changes in sound (Palmer & Pfordresher, 2003). Furthermore, research has found that people can detect physical features in sounds (e.g., sound intensity, sound rolling, and even sound shapes) as well as dynamic states of sounds (Carello, Anderson, Kunkler-Peck, 1998; Hermes, 1998; Grassi, 2005; Giordano & McAdams, 2006). In the auditory psychological study of Özcan and van Egmond (2012), sound attributes were found to be associated with actions, emotions, temporal descriptions, and sources and properties of sounds.

When we perform the act of toothbrushing, we can detect an existing rhythm if we listen carefully. This rhythm is produced by the close relationship between the auditory system and physical movement, wherein the cerebellum is activated in the auditory process to produce movement (Penhune, Virginia, Zatorre, Robert, Evans, & Alan, 1998; Thaut, Michael, Stephan, Klaus, Wunderlich, Gilbert, Schicks, Wilfried, Tellmann, Lutz, Herzog, Hans, McIntosh, Gerald, Seitz, Ruediger, Hömberg, & Volker, 2009). The human nervous system organizes the movement patterns when it perceives messages and enhances action-perception coordination (Kudo, Park, Kay, & Turvey, 2006). For instance, playing a musical instrument requires excellent coordination between music and hand position to integrate auditory and motion loop in the brain (Lu & Tsai, 2009). Moreover, Jerde, Childs, Handy, Nagode, and Pardo (2011) indicated that people can perceive not only the speed but also the rhythm of music to produce physical movements. Based on these findings and on the auditory and body movements coordination theory (Miura, Kudo, & Nakazawa, 2013), we therefore assume that the movement tempo of toothbrushing can be synchronized with the tempo.

People demonstrate emotional responses to musical tone, speed, and rhythm (Bresin & Friberg, 2011). Research has also shown that listening to music can improve performance (e.g., increased work output) and influence psychological (emotional reaction enhancement), psychophysical (decrease perceived exertion) and psychophysiological (improved oxygen consumption) perceptions during exercise (Atkinson, Wilson, & Eubank, 2004; Karageorghis et al., 2010; Szmedra & Bacharach, 1998). Moreover, listening to synchronous music has the effect of improving the physical and psychological effects during exercise (Terry, Karageorghis, Saha, & D'Auria, 2012). For example, jogging in sync with music produces positive emotions. In addition, Trehub and Hannon (2009) showed that the feeling of pleasure produced by listening to music may result not from the complexity of the musical piece but rather from the listener's familiarity with the music. Similar studies indicated a positive correlation between familiarity and preference (Fung, 1996; Krugman, 1943; Zissman & Neimark, 1990). Therefore, we assume that users' arm movements can be synchronized with tempo when brushing teeth. This study focuses primarily on the relationship between tempo sound, pleasure, comfort, and movement coordination during the act of toothbrushing.

## 3 METHOD

### 3.1 *Pretest*

The aim of this pretest is to find a primitive tempo sound through a pilot test.

### 3.1.1 *Participants*
Eight participants (male = 8; female = 8; average ≒ 22 years) took part in the pretest. Participants are college students from Chang Gung University (CGU). Participants who wear teeth braces have oral diseases, or hand limitation/ disease were excluded.

### 3.1.2 *Apparatus and procedure*
The pretest was conducted at pleasurable Products Laboratory of CGU. The experimental environment includes a washbasin and a mirror to simulate the look and atmosphere of a real bathroom. According to Bass Technique (Bass, 1954), small and soft-bristled toothbrushes (i.e., Oral-B Indicator Super Thin Toothbrushes) were used in the experiment. Brushing areas were classified into six blocks, including the outer sides of the upper teeth and insides of the lower teeth. The sequence of toothbrushing started from the upper right to the lower left, changing regions after every 10 seconds. The experiment lasted for a total of 1 minute. A high-fidelity microphone was used to record the sounds of toothbrushing. GoldWave music-editing software was used to analysis the frequency of toothbrushing.

### 3.1.3 *Results*
The statistical results showed the tempo of toothbrushing of the majority of participants to be 218 min-1, which is approximately 220 bmp. The speed of 220 bpm (i.e., quaver; 4/4) was therefore used in the main experiment.

## 3.2 *Main experiment*

In this experiment, we expect to identify an optimal speed of beat through the experiment.

### 3.2.1 *Participants*
A total of 40 participants (male = 20; female = 20; Ave. ≒ 22 years) took part in the main experiment. The participants are design-major students from Chang Gung University. Participants who wear teeth braces, have oral diseases, and hand limitation/disease were excluded.

### 3.2.2 *Stimuli*
In this experiment, we added the speed of 110 bpm to be a stimuli, based on the theory of Trehub and Hannon (2009). The synchronous speed of 220 bpm from the pretest was also used, as well as speeds that were derived from dividing and adding half of each of these two speeds. As a result, four stimuli, 330 bpm, 220 bpm, 110 bpm, and 55 bpm (i.e., quaver; 4/4) were used. In addition, based on the emotion and accent theory of Radocy and Boyle (1997), we added accents to each of the four stimuli, producing an accented task that has 100% intensity on the first half beat and 70% intensity on the second half beat.

In this experiment, a total of 8 stimuli (i.e., 4 accented and 4 unaccented) were used. The tune "Melodic Toms" was used to create the sound of drums from Mixcraft 6 music software. The volume was standardized at 75 dB (ear level), which is safe from an audiological perspective and loud and clear enough to be heard while toothbrushing.

## 3.3 *Apparatus and procedure*

The experimental environment used for the main experiment is the same as the one for the pretest. Following the beat sound from the speakers, participants were asked to brush their upper right teeth using an Oral-B toothbrush. After completing this task, participants were asked to respond to the questionnaire intuitively. The purpose of the questionnaire is to investigate the sense of speed in relation to coordination, comfort, and pleasure when toothbrushing. A 5-item scale (i.e., very fast, fast, appropriate, slow, very slow) was used to check the sense of speed, whereas a 5-point, single-item scale ranging from +1 (very bad) to +5 (very good), with a midpoint of 3 (neutral) was used for coordination, comfort, and pleasure.

Table 1. The speed beats of four stimuli.

| Speed | 55 bpm | 110 bpm | 220 bpm | 330 bpm |
|---|---|---|---|---|
| very fast | 0 (0%) | 2 (2.5%) | 12 (15%) | 51 (63.75%) |
| fast | 1 (1.25%) | 8 (10%) | 35 (43.75%) | 26 (32.5%) |
| appropriate | 13 (16.25%) | 50 (62.5%) | 33 (41.25%) | 3 (3.75%) |
| slow | 40 (50%) | 19 (23.75%) | 0 (0%) | 0 (0%) |
| very slow | 26 (32.5%) | 1 (1.25%) | 0 (0%) | 0 (0%) |

Note: Number of participants (Percentage)

Table 2. The Mean (SD) of Coordinate, Comfort and Pleasure of four stimuli.

| Speed | Coordinate | Comfort | Pleasure |
|---|---|---|---|
| 55bpm | 2.8375 (.9865) | 2.9750 (.9274) | 2.7625 (.7834) |
| 110 bpm | 3.5375 (.8259) | 3.5625 (.6906) | 3.4375 (.7769) |
| 220 bpm | 3.4500 (1.0896) | 3.1875 (.9425) | 3.1500 (.9014) |
| 330 bpm | 2.4500 (.9264) | 2.3875 (.8493) | 2.4000 (.9493) |

## 4 RESULTS AND DISCUSSIONS

Table 1 shows that a tempo at 55 bpm is considered to be too slow and a tempo sound at 330 bpm is considered to be too high for toothbrushing by most participants. Most of the participants selected 220 bpm and 110 bmp as adequate tempos for brushing teeth. However, the majority of participants considered 110 bmp to be the optimal tempo for brushing teeth. Notably, males preferred 220 bpm (50%) over 110 bpm (70%), whereas more females preferred only 110 bmp (55% ). These results indicate that some males prefer a faster speed, whereas females prefer a lower speed at 110 bmp for toothbrushing. Therefore, based on the results, the most effective speed is the tempo between 110 bpm to 220 bpm. Therefore, we suggest that creating a toothbrush can integrate with an adjustable tempo speed, ranging between 110 bmp and 220 bmp.

This study found that most participants prefer the speed 110 bpm, when brush teeth. Oppositely, the most inappropriate speed is 55 bpm and 330 bpm. The reason participants prefer 110 bmp is because of the synchronized tempo. However, the findings of Trehub and Hannon (2009) showed that people prefer to listen to familiar music, which suggests that the familiarity effect may be the reason why participants prefer 110 bmp. Table 2 shows the tempo of 110 bmp to demonstrate a high mean value for coordination, comfort, and pleasure, which indicates that participants preferred a tempo of 110 bmp for coordination, comfort and pleasure while toothbrushing, followed by the mean values for 220 bmp. In table 3, the result of ANOVA test shows no significant differences between 110 bpm and 220 bpm for coordination, pleasure, and comfort. This implies that participants accepted both 110 bmp and 220 bmp because they could not distinguish any differences between these two tempos. Tables 2 and 3 show that 330 bpm is the most uncomfortable speed, whereas 55 bpm showed less comfort when brushing teeth.

The 8 stimuli with both accented and unaccented tempos demonstrated no significant differences for coordination, comfort, and pleasure. These results demonstrate that no significant differences exist between accented and unaccented tempos when brushing teeth. Based on the results of this study, listening to a synchronized tempo sound (e.g., 110–220 bmp) may enhance comfort, coordination, and pleasure during the act of toothbrushing. However, to develop a sophisticated musical toothbrush, not only an ideal tempo sound, but also an elegant rhythm, is required. Future studies can focus on developing an ideal rhythm to be integrated in the musical toothbrush to enhance the experience of toothbrushing.

Table 3. ANOVA test on coordinate, comfort and pleasure to four stimuli.

| Speed | (bpm) | Coordination | Pleasure | Comfort |
|---|---|---|---|---|
| 55 bpm | 110 | .15040 (.000) | .13488 (.000) | .13566 (.000) |
| | 220 | .15040 (.001) | .13488 (.043) | .13566 (.485) |
| | 330 | .15040 (.087) | .13488 (.067) | .13566 (.000) |
| 110 bpm | 55 | .15040 (.000) | .13488 (.000) | .13566 (.000) |
| | 220 | .15040 (.953) | .13488 (.211) | .13566 (.056) |
| | 330 | .15040 (.000) | .13488 (.000) | .13566 (.000) |
| 220 bpm | 55 | .15040 (.001) | .13488 (.043) | .13566 (.485) |
| | 110 | .15040 (.953) | .13488 (.211) | .13566 (.056) |
| | 330 | .15040 (.000) | .13488 (.000) | .13566 (.000) |
| 330 bpm | 55 | .15040 (.087) | .13488 (.067) | .13566 (.000) |
| | 110 | .15040 (.000) | .13488 (.000) | .13566 (.000) |
| | 220 | .15040 (.000) | .13488 (.000) | .13566 (.000) |
| Scheffe's test | | 110, 220 > 55, 330 bpm | 110, 220 > 55, 330 bpm | 110, 220 > 55, 330 bpm |

Notes: ( ) indicates significance.

## REFERENCES


Atkinson, G., Wilson, D., & Eubank, M. 2004. Effects of music on work-rate distribution during a cycling time trial. *International Journal of Sports Medicine, 25*(08), 611–615.

Bass, C. C. 1954. An effective method of personal oral hygiene. *The Journal of the Louisiana State Medical Society: official organ of the Louisiana State Medical Society, 106*(2), 57.

Bresin, R., & Friberg, A. 2011. Emotion rendering in music: range and characteristic values of seven musical variables. *Cortex, 47*(9), 1068–1081.

Carello, C., Anderson, K. L., & Kunkler-Peck, A. J. 1998. Perception of object length by sound. *Psychological science, 9*(3), 211–214.

Coleman, P. 2002. Improving oral health care for the frail elderly: a review of widespread problems and best practices. *Geriatric Nursing, 23*(4), 189–199.

Fung, C. V. 1996. Musicians' and nonmusicians' preferences for world musics: Relation to musical characteristics and familiarity. *Journal of Research in Music Education, 44*(1), 60–83.

Giordano, B. L., & McAdams, S. 2006. Material identification of real impact sounds: Effects of size variation in steel, glass, wood, and plexiglass plates. *The Journal of the Acoustical Society of America, 119*, 1171.

Grassi, M. 2005. Do we hear size or sound? Balls dropped on plates. *Attention, Perception, & Psychophysics, 67*(2), 274–284.

Hermes, D. 1998. Auditory material perception. *IPO Annual Progress Report, 33*, 95.

Karageorghis, C., Priest, D., Williams, L., Hirani, R., Lannon, K., & Bates, B. 2010. Ergogenic and psychological effects of synchronous music during circuit-type exercise. *Psychology of Sport and Exercise, 11*(6), 551–559.

Klink, D., Webber, C., Smith, J., Trippel, M., & Sunshine, T. Children's Musical Electric Toothbrush.

Krugman, H. E. 1943. Affective response to music as a function of familiarity. *The Journal of Abnormal and Social Psychology, 38*(3), 388.

Kudo, K., Park, H., Kay, B. A., & Turvey, M. 2006. Environmental coupling modulates the attractors of rhythmic coordination. *Journal of Experimental Psychology: Human Perception and Performance, 32*(3), 599.

Lu, Yi-hsuan & Tsai, Chen-Gia. 2009. Importance of Motor Imagery for Music Performance: Evidence from Neuroscience. *Guandu Music Journal.* No. 11, 175–190.

Löe, H., Theilade, E., & Jensen, S. B. 1965. Experimental gingivitis in man. *Journal of Periodontology, 36*(3), 177–187.

Palmer, C., & Pfordresher, P. Q. 2003. Incremental planning in sequence production. *Psychological review, 110*(4), 683.

Penhune, V. B., Zatorre, R. J., & Evans, A. C. 1998. Cerebellar contributions to motor timing: a PET study of auditory and visual rhythm reproduction. *Journal of cognitive neuroscience, 10*(6), 752–765.

Sasaki, H., Matsumura, M., Sakashita, S., Tsuji, M., Matsumoto, M., & Ooshima, T. 2004. The effectiveness of a newly designed toothbrush on dental plaque removal in children. *Pediatric dental journal: international journal of Japanese Society of Pediatric Dentistry, 14*(1), 37–40.

Scannapieco, F. A. 1999. Role of oral bacteria in respiratory infection. *Journal of Periodontology, 70*(7), 793–802.

Szmedra, L., & Bacharach, D. 1998. Effect of music on perceived exertion, plasma lactate, norepinephrine and cardiovascular hemodynamics during treadmill running. *International Journal of Sports Medicine, 19*(01), 32–37.

Terry, P. C., Karageorghis, C. I., Saha, A. M., & D'Auria, S. 2012. Effects of synchronous music on treadmill running among elite triathletes. *Journal of Science and Medicine in Sport, 15*(1), 52–57.

Thaut, M. H., Stephan, K. M., Wunderlich, G., Schicks, W., Tellmann, L., Herzog, H., . . . Hömberg, V. 2009. Distinct cortico-cerebellar activations in rhythmic auditory motor synchronization. *Cortex, 45*(1), 44–53.

Trehub, S. E., & Hannon, E. E. 2009. Conventional rhythms enhance infants' and adults' perception of musical patterns. *Cortex, 45*(1), 110–118.

Wendt, L. K., Hallonsten, A. L., Koch, G., & Birkhed, D. 1994. Oral hygiene in relation to caries development and immigrant status in infants and toddlers. *European Journal of Oral Sciences, 102*(5), 269–273.

Zissman, A., & Neimark, E. 1990. The influence of familiarity on evaluations of liking and goodness of several types of music. *The Psychological Record, 40*(4), 481–490.

*Bridging Research and Good Practices towards Patient Welfare – Shih & Liang (Eds)*
*© 2015 Taylor & Francis Group, London, ISBN 978-1-138-02716-9*

# The anthropometric data for Taiwanese female military personnel

Yuh-Chuan Shih & Chien-Chun Pao
*Department of Logistics Management, National Defense University, Taipei City, Taiwan*

Chih-Long Lin
*Department of Craft and Design, National Taiwan University of Arts, New Taipei City, Taiwan*

Li-Wen Liu
*Occupational Hygiene Division, Institute of Occupational Safety and Health, Ministry of Labor, New Taipei City, Taiwan*

ABSTRACT: Anthropometric data are essential and important for product/equipment design, especially for the weapon systems and personal devices. Though several anthropometric databases for general Taiwan populace have been built, there is still a lack of an anthropometric database for the Taiwan military. In addition, with the increasing numbers of female military personnel in Taiwan, the anthropometric data for them becomes more and more important and urgent. Therefore, the aim of this paper is first to define anthropometric dimensions and second to collect the anthropometric data for female military personnel in Taiwan by means of both 3D scanner and manual methods.

Three specified postures including two standing and one sitting were used to collect 105 dimensions from 108 female military personnel. These 105 dimensions included 35 heights, 12 breadths, 21 lengths, 24 circumferences, and 13 dimensions derived from sitting posture. The results showed that there're differences among Taiwanese female military personnel and Taiwanese female civilian and US female military personnel. Taiwanese female military personnel were taller and sturdier than civilian. The heights of Taiwanese and US female military personnel are approximate, but US female military personnel are sturdier. That is, using other researches' anthropometric data to design military clothing and equipment is not appropriate, and to build an anthropometric database for Taiwan military is quite important.

## 1 INTRODUCTION

Anthropometric data are essential and important for product and equipment design for all kinds of occupations. Military anthropometry in the United States has been well-established and applied for about 100 years. They were used for many years in order to provide body size information in military research and development. Knowledge of body size and proportions is indispensable for the design, sizing, and tariffing of military clothing, protective, and personal equipment. Anthropometric data are also required as a basic input in the human factors engineering of military equipment systems such as: aircraft, vehicles, tanks, submarines, and many other types of weapons systems. The anthropometric measurement of the military population provides a standard, statistical description of that population. Researchers and designers then use the information derived from anthropometric data to achieve accommodations, compatibility, integration, safety, improved performance and logistic efficiency in human/equipment systems (US Department of Defense, 1991).

Gender, age, race, geographical region, and occupational groups are the factors to influence the body dimensions. Therefore, misapplying the data could be a serious safety concern (Hsiao *et al.*,

Figure 1. LT body scanner and setup.

2002). So far, there has not yet existed a complete anthropometric database for Taiwanese military, and most of the weapon systems in Taiwan are purchased from USA, and their interior space designs are all based on the US Army's anthropometric data. As a result, these purchased weapon systems could be not suitable for Taiwanese military personnel to operate. In order to support the policy of manufacturing/designing weapon systems by ourselves, to establish the anthropometric database from Taiwanese military personnel becomes important, emergent, and necessary. In addition, recruiting more females to serve in the military is also an important policy in Taiwan. While more females engage in military service their anthropometric data should be helpful for designing/manufacturing the equipment/personal devices for females to use. The percentage of female in the Taiwanese military personnel had increased to 11.59% in 2012. Therefore, the aim of this paper is first to define anthropometric dimensions and second to collect the anthropometric data for Taiwanese female military personnel by means of both 3D scanner and manual methods.

## 2 METHODS

### 2.1 *Equipment and validation*

A LT Body Scanner system made by Logistic technology Cooperation, Ltd. was utilized to scan and collect body anthropometric measurements. This system contains eight scanners, they are divided into four groups, and each group with two scanners is combined and mounted on a metal pole, as shown in Figure 1. These four poles were located to form a square with a diagonal of 1.8 m. that is, the scanned object stands in the center of the square with a distance of 0.9 m away from each pole. The accuracy and the resolution of each scanner is $\pm 1.0$ mm and 2.5 mm $\times$ 2.5 mm, respectively. It only takes 3 seconds to complete scanning a subject.

Prior to the formal measurement, a mannequin and one female subject were repeatedly measured both manually and by scanning to examine the precision and the accuracy of the scan-derived measurements and also to train the measuring skill of assistants. The results showed that the differences between scan-derived measurements and manual measurements in some of the circumferences were larger than the allowable error specified by ISO 20685. Previous studies indicate that circumferences and arcs are much more difficult to obtain the reliable measurements from scanning (Perkins *et al.*, 2000). The manual measurements, which are considered more accurate to be a reference, also contain errors resulting from the measurer's skill, difference in postures, compressed soft tissue, etc. (Gordon & Bradtmiller, 1992, Mckinnon & Istook, 2002). It's hard to define whether

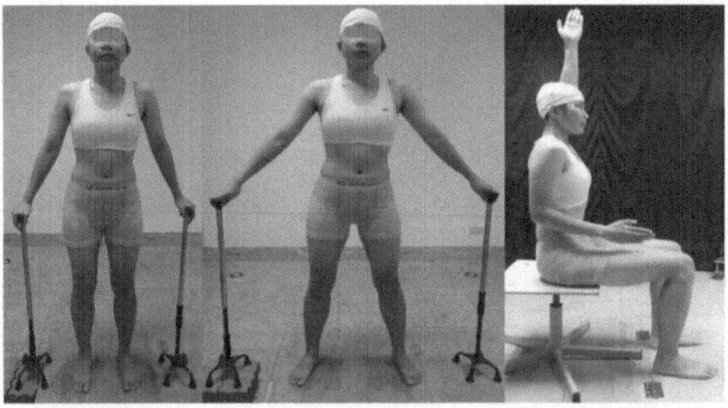


Figure 2. Standardized scanning postures.

the scan-derived measurements or manual measurements of circumferences are much more accurate. Therefore, in our formal anthropometric data collection, both by scanning and manually were presented for circumference data.

### 2.2 *Measuring postures and data collection*

The purpose of a standardized participant posture is important to ensure that measurement differences between individuals are due to body size differences, not due to the variations in body posture (Hotzman *et al.*, 2011). Three specified postures were two standing postures and one sitting posture, as shown in Figure 2. In the first standing posture (Figure 2(a)), participants stood upright, separated their feet for 10 cm, and put their upper extremities slightly away from their trunks and on a stick to balance their bodies. The first one is to derive body dimensions of height, breadth, and length. In the second one (Figure 2(b)), participants stood upright, separated their feet for 30 cm, and put their upper extremities away from the trunks for 45 degrees to collect the circumference measurements. The last is a sitting posture (Figure 2(c)), in which participants were asked to sit on a chair, to stretch their left upper extremities upward, and to bend their right elbows in 90 degrees to gather the sitting-posture data. During scanning all participants should wear fitted not tight and light-color garments.

One hundred and eight female military personnel aged from 18 to 43 years old were both measured manually and scanned with LT Body Scanner to collect 105 dimensions. These 105 dimensions included 35 heights, 12 breadths, 21 lengths, 24 circumferences, and 13 dimensions derived from the sitting posture. Sixty six landmarks were palpated and marked with stickers prior to scanning so that they could be visualized and extracted from the scan images. In the 3D program, users selected landmarks on the scanned images, and the program then performed the measurements automatically.

## 3 RESULTS AND DISCUSSIONS

One hundred and eight female participants containing 38 military personnel (aged from 26 to 43 years old) and 70 cadets (aged from 18 to 24 years old) were recruited. The average age of all participants was 24 years old (standard deviation was 6.5). The age distribution is shown in Figure 3. Thirty eight military personnel contained 27 army officers, 8 navies, 2 air force officers, and 1 military police.

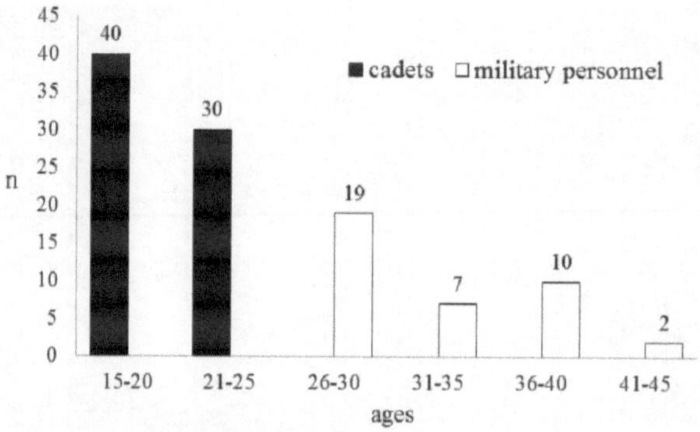


Figure 3. Age distribution for participants.

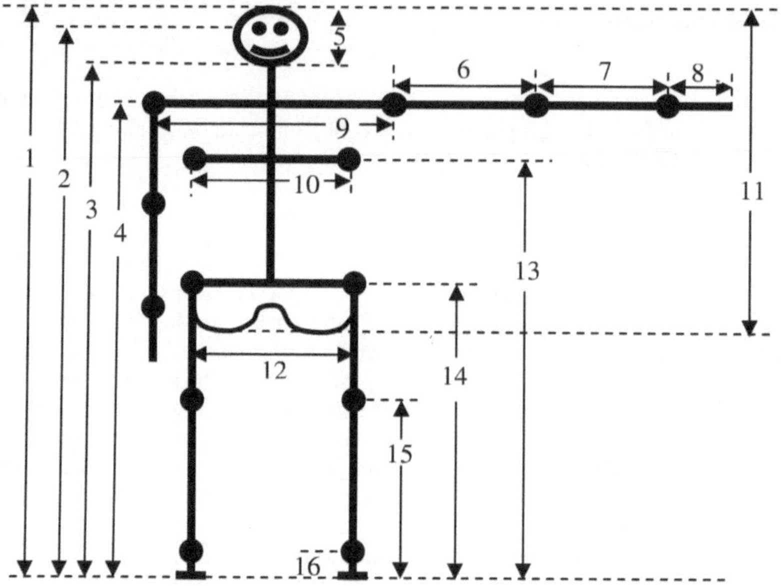


Figure 4. The corresponding selected 15 dimensions.

### 3.1 *Body dimension ratio*

Fifteen practical dimensions (see Figure 4 and Table 1) were selected to execute the Pearson correlation between them and the height (stature), and the coefficients were listed in the Table 1, which indicated that all dimensions existed positive significant correlation with height except chest breadth. Next, the averaged ratios of selected dimensions to the height from 108 participants were calculated and demonstrated in Figure 5.

### 3.2 *Comparison with Taiwanese male cadets*

We compared the corresponding body dimensions with Taiwanese male cadets obtained by Wu (2012) and performed Independent-Samples t-Test ($\alpha = 0.05$). The results shown in Table 2 indicated that there exists a significant difference in weight, BMI, and height. For scan-derived measurements, there exists significant difference in all dimensions except hip breadth (sitting), chest depth, waist circumference, knee circumference, calf circumference, and elbow rest height. For manual measurements of circumference, there exists significant difference in all dimensions

Table 1. Pearson correlation coefficient between height and the other 15 dimensions in height.

| No. | Dimension | *r* | No. | Dimension | *r* |
|---|---|---|---|---|---|
| 1 | Height | – | 9 | Shoulder breadth | 0.406** |
| 2 | Eye height | 0.959** | 10 | Chest breadth | 0.135 |
| 3 | Chin height | 0.952** | 11 | Sitting height | 0.741** |
| 4 | Shoulder height | 0.924** | 12 | Trochanterion breadth | 0.304** |
| 5 | Head length | 0.253** | 13 | Axilla height | 0.913** |
| 6 | Acromion-radiale length | 0.457** | 14 | Trochanterion height | 0.730** |
| 7 | Radiale-stylion length | 0.476** | 15 | Knee height | 0.737** |
| 8 | Hand length | 0.462** | 16 | Lateral malleolus height | 0.328** |

$^{*}p < .05$. $^{**}p < .01$

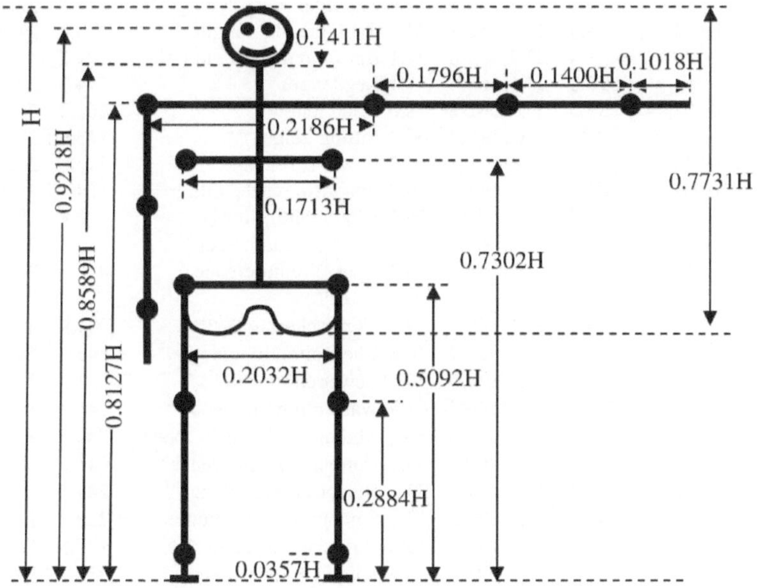


Figure 5. The ratio to height of selected 15 dimensions.

except head circumference, buttock circumference, and thigh circumference. Therefore, the figure between male and female military personnel was obviously different. It implies intuitively that we can not apply the male data to the female population, and it is important to collect the anthropometry data of female military personnel.

### 3.3 *Comparison with Taiwanese female civilian and US female military personnel*

Here we compared our data with both Taiwanese female civilian (Wang, Wang, & Lin, 2002) and US female military personnel (Paquette, Gordon, & Bradtmiller, 2009). First to compare our data with Taiwan female civilian with according to three age group, namely 18~24 (cadets in this study are aged between 18 and 24 years), 25~34, and 35~44 years old (military personnel in this study are aged between 26 and 43 years); second comparing with those 25 manual data of 651 US female military personnel aging upon 17 years old.

1. Height dimensions: Generally speaking, our participants were taller than civilian in many height dimensions, except axilla, lower edge of chest, and lateral malleolus height, but were shorter than US female military personnel, e.g. at cervicale, waist, and crotch height.

Table 2. Comparison with Taiwanese male cadets.

| | This study | | Male cadets | | | This study | | Male cadets | |
|---|---|---|---|---|---|---|---|---|---|
| Dimension | Mean | SD | Mean | SD | Dimension | Mean | SD | Mean | SD |
| Weight(Kg)*** | 55.6 | 5.8 | 67.8 | 7.9 | BMI ($kg/m^2$)** | 21.1 | 2.1 | 22.5 | 2.3 |
| Height*** | 1613.0 | 42.6 | 1740.9 | 56.2 | Eye height*** | 1493.9 | 43.2 | 1604.6 | 60.5 |
| Cervicale height*** | 1366.6 | 42.0 | 1462.8 | 58.1 | Shoulder height*** | 1314.8 | 39.7 | 1397.5 | 56.1 |
| Axilla height*** | 1180.7 | 37.5 | 1308.0 | 54.6 | Nipple height*** | 1161.0 | 38.3 | 1242.8 | 50.2 |
| Waist height*** | 949.9 | 33.2 | 1020.0 | 45.6 | Gluteal furrow height*** | 710.8 | 31.1 | 879.8 | 43.4 |
| Head breadth*** | 169.1 | 4.5 | 174.8 | 9.4 | Chest breadth*** | 277.8 | 17.0 | 345.0 | 18.0 |
| Shoulder breadth*** | 354.5 | 19.5 | 386.6 | 18.5 | Back breadth*** | 286.7 | 18.9 | 349.3 | 19.5 |
| Hip breadth*** | 329.6 | 16.4 | 344.2 | 19.6 | Chest depth | 216.6 | 16.1 | 210.5 | 20.8 |
| Waist depth*** | 189.7 | 20.5 | 202.5 | 19.1 | Abdomen depth** | 214.1 | 23.1 | 223.5 | 18.0 |
| Buttock depth** | 225.8 | 17.6 | 231.5 | 14.4 | Shoulder breadth length*** | 387.7 | 22.7 | 424.6 | 27.7 |
| Chest breadth length*** | 348.9 | 27.3 | 382.4 | 24.1 | Back breadth length*** | 320.4 | 21.9 | 367.5 | 26.9 |
| Back length*** | 431.4 | 20.7 | 479.8 | 21.1 | Acromion-radiale length*** | 291.2 | 14.2 | 314.9 | 17.1 |
| Sleeve length*** | 510.3 | 20.6 | 563.5 | 41.1 | Leg length*** | 774.0 | 34.2 | 1008.1 | 47.9 |
| Navel-knee length | 535.5 | 20.7 | 542.8 | 78.6 | Hand length*** | 165.2 | 7.9 | 184.5 | 8.6 |
| Palm length* | 99.3 | 4.5 | 119.7 | 60.3 | Sitting height*** | 1254.0 | 30.9 | 1320.9 | 47.1 |
| Eye height, sitting*** | 1130.0 | 31.3 | 1196.9 | 45.9 | Elbow rest height | 246.6 | 25.0 | 246.6 | 27.1 |
| Knee height, sitting*** | 467.7 | 18.1 | 516.2 | 26.9 | Fingertip reach, sitting*** | 1249.5 | 38.2 | 1349.1 | 50.3 |
| Elbow height*** | 622.7 | 28.3 | 647.7 | 36.5 | Hip breadth, sitting | 370.8 | 19.9 | 365.0 | 20.6 |
| Head circumference*** | 589.4 | 24.1 | 568.1 | 17.1 | Head circumference$^{M}$ | 582.4 | 23.5 | 577.0 | 11.2 |
| Neck circumference, base** | 445.5 | 22.7 | 456.5 | 20.3 | Neck circumference, base$^{M***}$ | 436.0 | 26.2 | 465.5 | 22.4 |
| Upper chest circ.*** | 889.7 | 51.1 | 1002.4 | 49.8 | Upper chest circ.$^{M***}$ | 841.9 | 49.7 | 932.1 | 128.4 |
| Chest circumference*** | 884.3 | 50.9 | 928.4 | 56.3 | Chest circumference$^{M***}$ | 845.3 | 50.6 | 894.8 | 52.2 |
| Lower chest circ.*** | 763.2 | 39.5 | 873.9 | 50.7 | Lower chest circ.$^{M***}$ | 723.9 | 37.9 | 839.0 | 49.5 |
| Waist circumference | 789.5 | 57.8 | 796.0 | 61.9 | Waist circumference$^{M**}$ | 745.1 | 59.6 | 778.6 | 59.9 |
| Abdomen circumference* | 836.5 | 64.6 | 859.6 | 50.5 | Abdomen circumference$^{M**}$ | 784.5 | 53.9 | 810.9 | 52.3 |
| Buttock circumference | 963.3 | 50.5 | 960.3 | 43.7 | Buttock circumference$^{M}$ | 946.6 | 50.8 | 933.8 | 57.0 |
| Biceps circumference*** | 267.1 | 21.6 | 287.4 | 23.2 | Biceps circumference$^{M***}$ | 244.2 | 19.9 | 273.3 | 21.6 |
| Forearm circumference*** | 246.2 | 12.6 | 270.3 | 17.7 | Forearm circumference$^{M***}$ | 229.2 | 13.1 | 255.9 | 16.1 |
| Wrist circumference** | 163.9 | 14.2 | 171.9 | 15.0 | Wrist circumference$^{M***}$ | 147.5 | 6.7 | 163.3 | 7.1 |
| Thigh circumference* | 588.7 | 42.8 | 603.8 | 35.6 | Thigh circumference$^{M}$ | 556.3 | 37.0 | 563.6 | 36.9 |
| Knee circumference | 375.3 | 23.1 | 369.9 | 19.6 | Knee circumference$^{M**}$ | 355.4 | 19.0 | 364.9 | 17.8 |
| Calf circumference | 367.9 | 27.9 | 372.9 | 22.5 | Calf circumference$^{M***}$ | 357.1 | 27.4 | 381.3 | 22.9 |

$^{M}$for the dimension measured manually *$p < .05$. **$p < .01$. ***$p < .00$

2. Breadth and depth dimensions: Our participants were larger than civilian in head, shoulder breadth, and hip depth, but were smaller than US female military personnel in shoulder breadth.
3. Length dimension: Our participants were longer than civilian in many length dimensions, except back length, but were shorter than US female military personnel in waist back and acromion-radiale length.
4. Circumference dimensions: Our participants were larger than civilian in the circumferences of head, waist, buttock, knee, and calf, but were smaller than US female military personnel in chest circumference, waist, and buttock circumference. On the other hand, the lower chest circumference, abdomen circumference, and wrist circumference of civilian aging between 35 and 44 years were larger than our military personnel.
5. Sitting posture dimensions: The participants in this study were larger than civilian in palm length and hip breadth, and are smaller than US female military personnel in bideltoid breadth, knee height, hand length, and hip breadth.

Table 3. Dimension comparison with other researches.

| Dimension | Subjects | Age group | Mean | −95%CI | +95%CI | 5%ile | 95%ile |
|---|---|---|---|---|---|---|---|
| Height (mm) | Cadets | 18~24 | 1618 | 1540 | 1697 | 1562 | 1680 |
| | Civilian | 18~24 | 1597 | – | – | 1509 | 1684 |
| | Military personnel | 26~43 | 1628 | 1531 | 1726 | 1566 | 1705 |
| | Civilian | 25~34 | 1572 | – | – | 1485 | 1659 |
| | Civilian | 35~44 | 1551 | – | – | 1469 | 1634 |
| | US Military personnel | >17 | 1630 | – | – | 1520 | 1730 |
| Weight (kg) | Cadets | 18~24 | 55.1 | 43.6 | 66.2 | 46.5 | 64.8 |
| | Civilian | 18~24 | 52.0 | – | – | 41.7 | 62.2 |
| | Military personnel | 26~43 | 56.5 | 45.2 | 67.7 | 49.9 | 67.2 |
| | Civilian | 25~34 | 52.1 | – | – | 40.2 | 63.9 |
| | Civilian | 35~44 | 55.0 | – | – | 41.7 | 68.2 |
| | US Military personnel* | >17 | 68.9 | – | – | 51.4 | 93.0 |
| Knee height (mm) | Cadets | 18~24 | 467 | 434 | 501 | 440 | 494 |
| | Civilian | 18~24 | 483 | – | – | 446 | 519 |
| | Military personnel | 26~43 | 468 | 429 | 508 | 441 | 497 |
| | Civilian | 25~34 | 471 | – | – | 431 | 510 |
| | Civilian | 35~44 | 464 | – | – | 433 | 496 |
| | US Military personnel* | >17 | 514 | – | – | 471 | 559 |
| Shoulder breadth (mm) | Cadets | 18~24 | 349 | 313 | 386 | 319 | 378 |
| | Civilian | 18~24 | 338 | – | – | 296 | 379 |
| | Military personnel | 26~43 | 364 | 330 | 399 | 338 | 390 |
| | Civilian* | 25~34 | 324 | – | – | 282 | 366 |
| | Civilian* | 35~44 | 325 | – | – | 280 | 369 |
| | US Military personnel | >17 | 366 | – | – | 333 | 395 |
| Hip breadth (mm) | Cadets | 18~24 | 370 | 329 | 412 | 341 | 406 |
| | Civilian | 18~24 | 323 | – | – | 282 | 364 |
| | Military personnel | 26~43 | 372 | 337 | 406 | 350 | 399 |
| | Civilian | 25~34 | 329 | – | – | 289 | 370 |
| | Civilian | 35~44 | 342 | – | – | 298 | 387 |
| | US Military personnel | >17 | 395 | – | – | 340 | 461 |
| Chest circumference (mm) | Cadets | 18~24 | 879 | 776 | 982 | 809 | 972 |
| | Cadets$^{M}$ | 18~24 | 841 | 739 | 943 | 774 | 924 |
| | Civilian | 18~24 | 822 | – | – | 740 | 904 |
| | Military personnel | 26~43 | 894 | 802 | 986 | 828 | 990 |
| | Military personnel$^{M}$ | 26~43 | 853 | 760 | 945 | 785 | 930 |
| | Civilian | 25~34 | 831 | – | – | 742 | 921 |
| | Civilian | 35~44 | 863 | – | – | 755 | 972 |
| | US Military personnel* | >17 | 955 | – | – | 822 | 1113 |
| Waist circumference (mm) | Cadets | 18~24 | 778 | 670 | 886 | 681 | 861 |
| | Cadets$^{M}$ | 18~24 | 732 | 622 | 841 | 640 | 836 |
| | Civilian | 18~24 | 687 | – | – | 584 | 789 |
| | Military personnel | 26~43 | 812 | 700 | 923 | 742 | 901 |
| | Military personnel$^{M}$ | 26~43 | 770 | 655 | 885 | 690 | 888 |
| | Civilian | 25~34 | 720 | – | – | 607 | 832 |
| | Civilian | 35~44 | 768 | – | – | 630 | 906 |
| | US Military personnel* | >17 | 852 | – | – | 699 | 1029 |
| Hand length (mm) | Cadets | 18~24 | 164 | 149 | 178 | 152 | 174 |
| | Civilian | 18~24 | 168 | – | – | 154 | 183 |
| | Military personnel | 26~43 | 168 | 151 | 184 | 155 | 179 |
| | Civilian | 25~34 | 167 | – | – | 154 | 181 |
| | Civilian | 35~44 | 167 | – | – | 153 | 181 |
| | US Military personnel* | >17 | 183 | – | – | 167 | 200 |

*mean exceeding the 95% confidence interval of our data
$^{M}$dimension measured manually

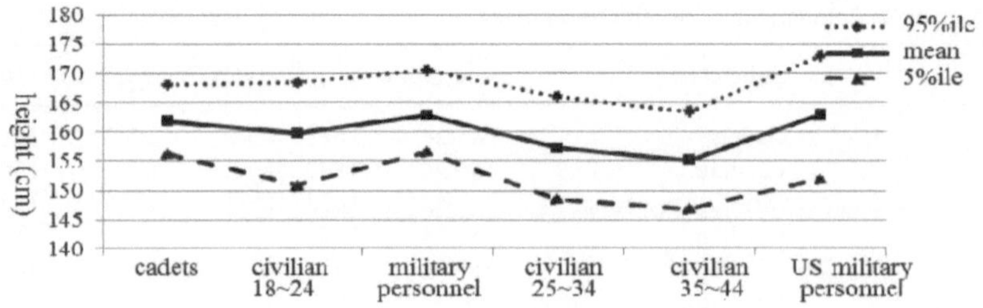


Figure 6. Percentile distribution of height.

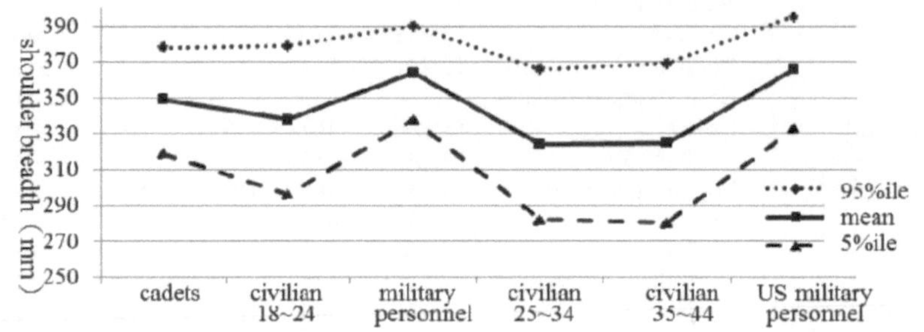


Figure 7. Percentile distribution of shoulder breadth.

### 3.4 *Comparison with Taiwanese female civilian and US female military personnel – mean, the 5th and the 95th percentile*

Mean, the 5th and the 95th percentile are important for products designing. Clearance dimensions should be no less than the 95th percentile user to let the large person fit. The 95th percentile is often being used for maximum value of design limitations; examples include door heights and door widths. Reach distances should be no more than that of the 5th percentile user to let the small person reach. The 5th percentile is often being used for minimum clearances of design limitations; examples include elevator buttons and safety clearances for railings. For postural situations we often have to consider the average or the 50th percentile value to accommodate most of the users. We here chose several important dimensions to compare the mean, the 5th and the 95th percentile of these sample groups, as shown in Table 3.

The percentile distribution of height is shown as Figure 6. The 5th percentile of cadets and military personnel are close to or even higher than the mean of civilian aged between 25 and 44 years. As to the percentile distribution of shoulder breadth shown as Figure 7, it reveals that the shoulder breadth of Taiwanese and US military personnel are larger than the civilian and the distribution are much less dispersal. The 5th percentile of Taiwanese and US military personnel are higher than the means of civilian aged between 25 and 44 years. The mean of Taiwanese military personnel is close to the 95th percentile of civilian aged between 25 and 44 years. Finally, the percentile distribution of hip breadth is shown as Figure 8. The 5th percentile of cadets and military personnel are higher than the means of civilian aged between 18 and 44 years. The 95th percentile of cadets and Taiwanese military personnel are close to the mean of US military personnel. From above comparisons, it demonstrated that we have to notice the difference of the data we will use to design equipment, weapon systems, vehicle workstations, and garment.

## 4 CONCLUSIONS

The literatures have indicated that gender, age, race, geographical region, and occupational groups are the factors to influence the body dimensions. More according to aforementioned comparisons,

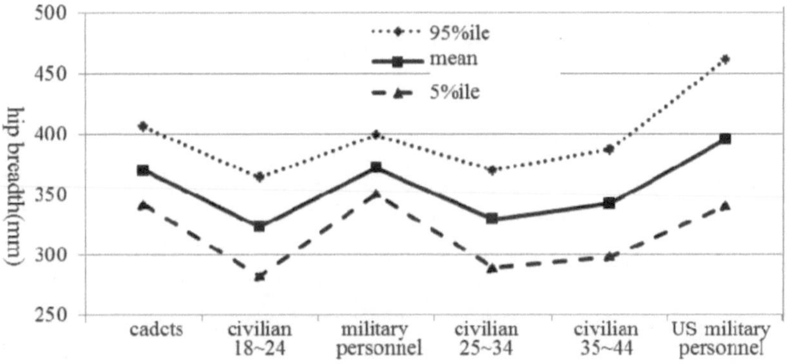


Figure 8. Percentile distribution of hip breadth.

we did find there was significant difference in body dimensions between our sample and Taiwanese female civilian and US female military personnel. This reminds us to carefully apply the anthropometric data from different population on designing the equipment/devices/tools used by female Taiwanese military personnel. Moreover, it had better establish the own anthropometric database of Taiwanese military personnel. The database will be helpful not only to the self-designed equipment/devices/tools but also to requirements specified in purchase contracts. There are several Taiwanese anthropometric databases, but unfortunately, there is still no such data for the military. Therefore, it is the time to build an anthropometric database belonging to Taiwanese military if they really care and want to improve the welfare, safety, and satisfaction of their military personnel and of the users of equipment/devices/tools.

## REFERENCES

Department of Defense 1991. *Military Handbook: Anthropometry of U.S. Military Personnel (ARMY DOD-HDBK-743A).* US: Department of Defense

Gordon, C. C. & Bradtmiller, B. 1992. Interobserver error in a large scale anthropometric survey. *American journal of human biology* 4(2): 253–263.

Hsiao, H., Long, D. & Snyder, K. 2002. Anthropometric differences among occupational groups. *Ergonomics* 45(2): 136–152.

Hotzman, J., Gordon, C. C., Bradtmiller, B., Corner, D. B., Mucher, M., Kristensen, S., Paquette, S. & Blackwell, L. C. 2011. *Measurer's Handbook: US Army and Marine Corps Anthropometric Surveys, 2010-2011(NATICK/TR-11/017).* Retrieved from http://www.dtic.mil/cgi-bin/GetTRDoc?AD=ADA548497

Mckinnon, L. & Istook, C. 2002. Body scanning: The effects of subject respiration and foot positioning on the data integrity of scanned measurements. *Journal of Fashion Marketing and Management* 6(2): 103–121.

Paquette, S., Gordon, C. & Bradtmiller, B. 2009. *Anthropometric survey (ANSUR) II pilot study: methods and summary statistics (Natick/TR-09/014).* Retrieved from http://www.dtic.mil/cgi-bin/GetTRDoc?AD=ADA498172&Location=U2&

Perkins, T., Burnsides, D. B., Robinette, K.M. & Naishadham, D. 2000. Comparative consistency of univariate measures from traditional and 3-D scan anthropometry. *Digital Human Modeling For Design And Engineering Conference And Exposition.*

Wang, M. J., Wang, M. Y. & Lin, Y. C. 2002. *Anthropometric data book of the Chinese people in Taiwan.* Hsinchu: The ergonomics society of Taiwan.

Wu, S.J. 2012. The pilot study of developing anthropometric database for soldiers in Taiwan. Unpublished master thesis, National Defense University.

*Bridging Research and Good Practices towards Patient Welfare – Shih & Liang (Eds)*
*© 2015 Taylor & Francis Group, London, ISBN 978-1-138-02716-9*

# A study of five "F" in product design

Chiu-Wei Chien
*Doctoral Program, Graduate School of Creative Industry Design, College of Design, National Taiwan University of Arts, Taiwan*

Chih-Long Lin
*Craft and Design Department, National Taiwan University of Arts, Taiwan*

Rung-Tai Lin
*Graduate School of Creative Industry Design, College of Design, National Taiwan University of Arts, Taiwan*

ABSTRACT: Via professional product designers' understanding about the two main properties of products, rationality and emotion, we are able to realize the connections between rational features and emotional features of different items. The target sample of this study was the electronic fan which was categorized into different styles based on five characteristics: F1 (Function), F2 (Friendly), F3 (Fun), F4 (Fancy), and F5 (Feeling). Two experiments were performed. Experiment 1 was for the purpose of selecting the representative products of rationality, neutrality, and emotion, and experiment 2 was to reset rational, neutral and emotional of the result from experiment 1. In hopes of obtaining varied design inspirations from the result of this study, these five group products possessed various features and variables respectively. This paper mainly discusses different properties between items of the rational and the emotional. This study also explains that how products of rationality and products of emotion were various whether in shapes or styles according to their properties. Moreover, the confusing quality of F3, F4, and F5 was explicated as well. The goal of this research is to provide suggestions to product designers with regard to the importance of how to design products with both characteristics; rationality and emotion at the same time, and further, to improve their design.

## 1 INTRODUCTION

Lin (2005) said that the five product style-5F could be defined based on features of different eras. In the 1930s, technology determined the trends of design and shape. At that period of time when functionalism was dominant; "form follows function" became the primary guiding principle in design. After World War II, people were getting tired of this kind of design. Therefore, in the 1950s, owing to the rise of ergonomics, "form follows friendly" was the mainstream of designing. Until the 1980s, design of "user friendly" replaced the style of its previous era thanks to the appearance of personal computers. Later, in pursuit of changes and new styles, "form follows fun" came to the stage. The personalized device was the next characteristics of design, so "form follows fancy" emerged. The 21st century is the time of digital technology. The key factor of design at this period is humanity. In other words, the design principle is shifted to "form of feeling." 5F was short for the above descriptions.

Empathic design is a relatively new style and which is also widely reviewed as shown by design research documents. The first principle is balancing rationality and emotion in building understanding of users' experiences. In empathic design this balance is found by combing observations of what people do with interpretations of what people think, feel and dream. (Dandavate et al., 1996; Fulton Suri, 2003) Consequently, the two opposite nouns; rationality and emotions, can be used to explain issues of human personality in human behavior for certain purposes. For example, consider acts of product design itself. Via professional product designers' understanding about the two main properties of products, rationality and emotion, we are able to realize the connections between rational features and emotional features of different items. The two primary purposes for this study are, first, discovering new thoughts on product design by studying the changes between the two properties among 5Fs. Enterprises can take this result as a reference to define their product and products' position when developing new items. Second, by learning changes between these two properties we are capable of knowing what affects designers' understanding about rationality and emotions. The result will help adding harmony to product design; a skill which requires combinations of rationality and emotion, and further perfecting products.

## 2 LITERATURE REVIEW

The definition of quality of rationality and emotions could be traced back to the 1980s when marketing was subjected to market positioning. Aaker and Shansby once made their definition for product properties—all combinations of products' interior and exterior characteristics can be perceived by customers. (Aaker, Shansby, 1982) The fact of how rational a buyer behaved determined the method for Japanese scholars to categorize products. Their explanations a product's of rational properties are: function, quality, price, and company reputation. For products of emotions, the characteristics were: consumers' feeling (if products delight consumers), atmosphere, and trendiness. It is often assumed that designers are highly capable of using their intuition and creative instincts when designing products to convey certain intended product properties, such as trendiness. However, designers and laymen may in some cases differ in their perception of product designs (Blijlevens et al., 2009; Hsuet al., 2000) For this reason, previous research has aimed to provide designers with guidelines on how product properties can be incorporated into product designs (e.g., Dahlgaard, Schütte, Alikalfa, & Dahlgaard Park, 2008; Hsiao & Wang, 1998; Orth & Malkewitz, 2008). Understanding people's experiences of owning and using products, they argued, requires a more holistic approach that includes the emotional domain, i.e., their feelings and experiences. Addressing emotions and rationality in a balanced way will help researchers and designers "to understand those uniquely human traits that are responsible for people's liking, using, and wanting to live with the products [they] design" (Dandavate et al., 1996, p.415). Rationality and emotion are not just nouns that customers are familiar with, but also two terms that a human would use in self-analysis. In addition, buyers tend to take these two qualities as criteria to evaluate images of product. They are also two suitable integrated properties for communications between designers and consumers. Louis Sullivan (1856~1924), an American modern architect, once explained that "all things in nature have a shape, that is to say, a form, an outward semblance, that tells us what they are, that distinguishes them from ourselves and from each other." The appearances of every object represent meanings by themselves. Five styles for 5Fs can be explicated by major events that happened in the history of design from five eras respectively. As shown in table 1.

Table 1. The explanation of 5F style by Design History.

| | | |
|---|---|---|
| F1 | Function | Simple geometric Machine Aesthetic. |
| F2 | Friendly | Human Factors and Ergonomics, focusing on the interaction of people and equipment universal design. |
| F3 | Fun | Design will be centered on the individual and the postmodernism instead modernism. |
| F4 | Fancy | In a globalized environment, product designers need to find out the identify ability for the unique personality of a product. |
| F5 | Feeling | 21st century, smart technology products seemed to be everywhere in our lives, technological advancement affect the product design style which needs more pleasurable emotional design. |

## 3 RESEARCH METHOD

The conceptual framework and study process are shown in figure 1. First of all, the study takes electronic fans as the target sample to control variables and product positioning expecting to obtain new inspirations for design. The design of this experiment is for investigating the impacts on the perceptions of rationality and emotion of five different product design styles. There will be two major experiments with the following purposes: Experiment 1 is for finding out the representative rational, neutral, and emotional samples of five kinds of product design styles, and for determination of impact factors of rationality, neutrality, and emotion. Following experiment 1, experiment 2 asks subjects to reset those samples which will be randomized in the following week in order of the rational, neutral, and emotional features in 5F. Lastly, we utilized SPSS software to do statistical analysis and explain the result in this paper. For the rest of this paper, "5F" is used to signify five styles.

### 3.1 *Experiment samples*

The criteria for sample selection of electric fans and the control variables are as shown in Table 2. F1 (Function) is mainly about table fans with consistent functionality. F2 (Friendly) is mainly about stand fans with consistent functionality. F3 (Fun) is mainly about humorous, cute, and joyful products with style of fun. With the most control variables, F4 (Fancy) is only for the new style of "Tower Fan" with uncontrolled form. F5 (Feeling) is also an uncontrolled diversified style. In this

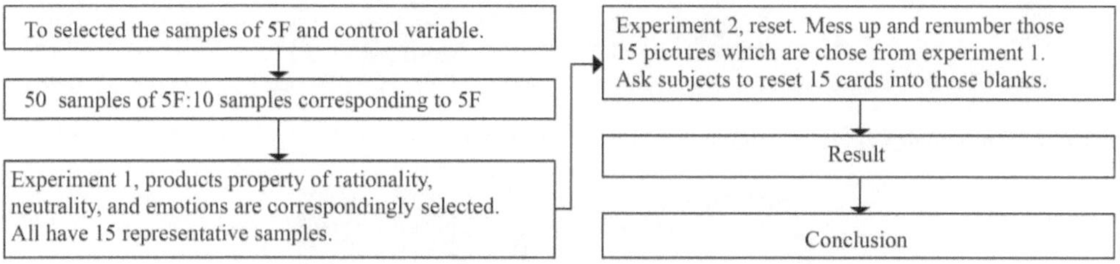


Figure 1. Research process.

Table 2. The samples selected of 5F and control variables.

| 5F | Fan samples selected (all begin with home fans) | Control Variable |
|---|---|---|
| F1:Function | The general table fan appeared in the 1920s to 1930s setting the most important and basic function and form in fan. | Function |
| F2:Friendly | The stand fan appeared In the 1950s. The high stand design because of desk fans needed to rely on a table or platform which is consistent with the desired height. | Function |
| F3:Fun | From the 1970s retro to the 1990s USB fan, as well as modern Japanese cartoon style, etc., funny products are a popular style. | None |
| F4:Fancy | The tower type electric fan was developed from the late 1990s until the 21st century. Because of its fashion type without fan blades and not occupying space it is chosen as the fancy representative. | function, material, color & era |
| F5:Feeling | For either the DYSON bladeless fan (airflow multiplier), or the wooden fan, 21st century fan design displays coexistence and common prosperity with technology and humanity as the sharing philosophy of cultural development and Earth Resources. | None |

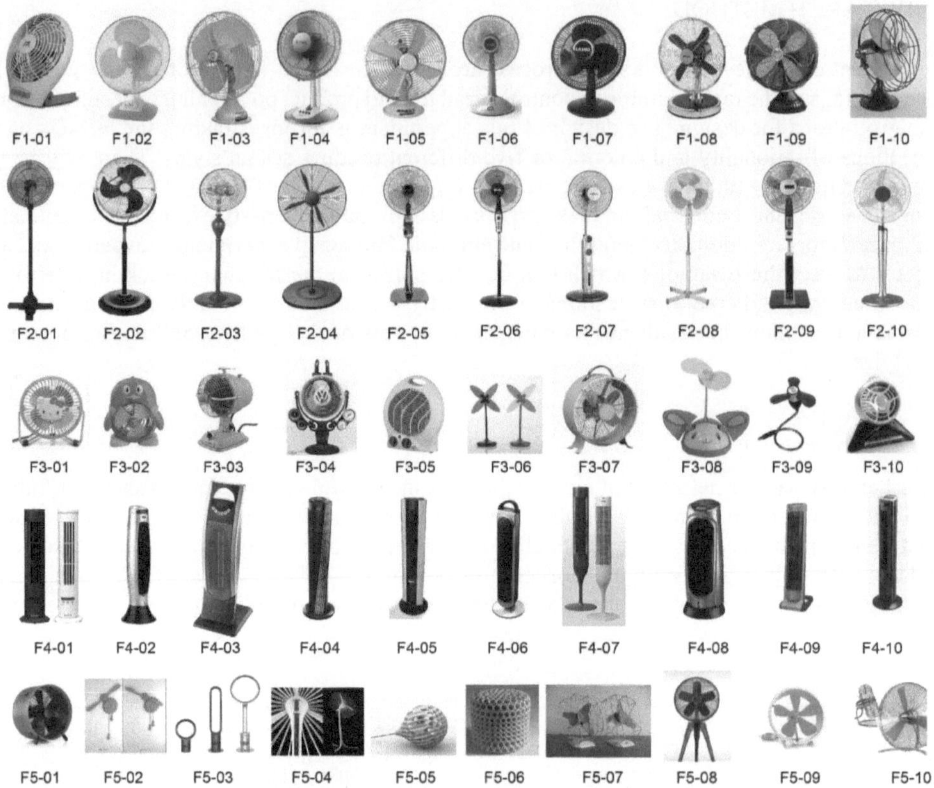


Figure 2. 50 samples of 5F, the bottom is sample no by author.

Figure 3. The situations of experiment 1.

study the five kinds of product styles are also known as "5F". Ten samples will be selected from existing products for each of the five styles, thus there will be a total of 50 experimental samples in Figure 2.

### 3.2 *Experiment I*: indicator samples of rationality/neutrality/emotion and their impact factors

In experiment 1 the representative rational/neutral/emotional products will be selected as indicators. Three fans will be selected from pictures of each of the five product styles representing the most rational/neutral/emotion samples, thus there will be a total of 15 most representative sample pictures. The situation of this experiment is as shown in Figure 3.

3.2.1 *Experimental Subjects*: the experimental subjects in this study are a total of 40 junior and senior students (20 male students and 20 female students) in the Department of Industrial Design of National United University.

3.2.2 *Materials*: the pictures of 50 samples will be printed out in colors as cards of A5 size.

Figure 4. The situations of experiment 2.

3.2.3 *Steps:*

(1) These 50 product picture cards will be divided into 5 groups in accordance with 5 styles such that there will be 10 product picture cards in each group aligned on the table as shown on the left of Figure 2.

(2) The experimental subjects will then select the most representative rational, neutral, and emotional products in accordance with their own design experiences and product perceptions, and fill them in the questionnaires.

### 3.3 *Experiment II: Reset*

Mess up and renumber those 15 pictures which are chose from experiment 1. Ask respondents to place 15 cards into those blanks .The situation of this experiment is as shown in Figure 4.

3.3.1 *Experimental subjects*: the same 40 students in experiment 1 during the first week.
3.3.2 *Materials*: the same color printed product picture cards used in the first week.
3.3.3 *Steps:*

(1) Mess up and renumber those 15 pictures which are chose from experiment 1. If there are more than two items with same numbers, experiment takers have to decide which one is more suitable.

(2) Draw 15 blanks on desk corresponding to three properties of 5F respectively.

(3) Ask respondents to place 15 cards into those blanks, and put numbers of cards in questionnaire. Figure 4 demonstrates how this experiment is conducted.

(4) Employ the mode from SPSS in experiment 2 to reckon which samples are picked by experiment takers most often. The outcome is demonstrated in figure 6. Items that are circled are items that do not match the original one in experiment 1. Original items are put right above those circled pictures and numbers of times are beneath them.

## 4 RESULT AND DISCUSSION

### 4.1 *Experiment I*

Based on responses from the questionnaire for experiment 1, SPSS is used to calculate which samples with the three properties are picked by respondents most often. The result is shown in figure 5. Figures listed below each picture mean the number of times they got selected, and products beneath them are samples with the same number but which are excluded in the end.

At this stage, respondents are analyzed for their recognition about properties of rationality and emotions of the products' 5F for the purpose of comparing communalities and discrepancies in representative products of rationality, neutrality, and emotions about 5F and finding connections between the two qualities: rationality and emotion. The result is explained in the following passages. F1R stands for rational product in the category of Function, F2N represents neutral one in category of Friendly, F3E signifies emotional item in Fun category, and so forth. The results are as follows:

(1) Items that are selected by respondents most often are F3E, F3R, F4N, and F5E. Among these items, F4R and F4N's original code F401 repeats F405 and F406; therefore, F401 is excluded.

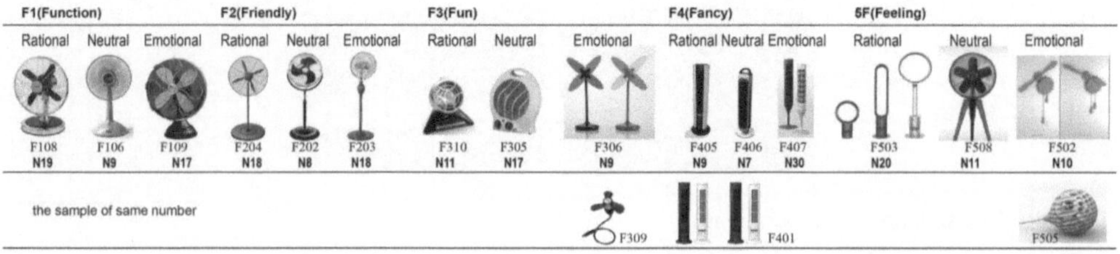


Figure 5. Experiment one elected representative products of rational, neutral, and emotional of 5F and the number (lower row of pictures of the same frequency but weren't elected as representation).

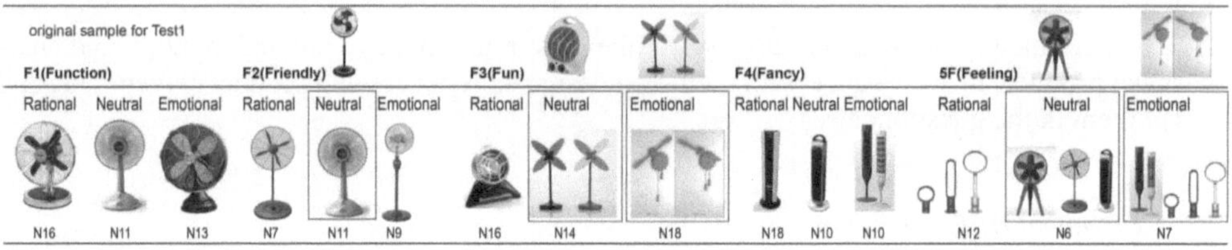


Figure 6. Experiment two: reset and number (The samples in borders are not the same as in experiment one, the above pictures show the original sample in experiment one).

Qualities of F309 are highly similar to that of F3R's representative products F310; both are USB fan, so F309 is ruled out as well. The overall performance of F502, which achieves more than concepts of electronic fans, outmatches F505. In other words, most subjects thought that the quality of feeling is important, consequently, F505 is eliminated.

(2) Products arranged in order of the number of times they are picked out: the top 7 are F4E(N30), F5R(N20), F1R(N19), F2R(N18), F2E(N18), F1E(N17), F3N(N17).

(3) Communalities. The obvious common quality for the 5 rational products is the linear design. Texture and colors of these types of product include silver metal with cool color, achromatic graphite plastic, and silver and blue paint. The curved line is the common feature shared by the 5 emotional products. Textures and colors for this kind are warm golden yellow, bronze metal, plastic with color of wood, plastic with warm high color saturation, and authentic wood. For the neutral products; first, both curved and straight lines are used. Second, colors range from achromatic color, cool, to warm colors. Lastly, for texture, neutral products are primarily designed with plastic then by metal and wood. As a result, common qualities of neutral products are not as high as that of rational or emotional products.

In a preliminary comparison with products that are selected the most from the point (2), there are some differences in results in terms of varied styles and variables in 5F. Sample products for F1 and F2 are a desk fan and a standing fan. Besides, they share the same variable; therefore, samples with high scores are distributed in the rational type F1R(N19), F2R(N18), and the emotional type F2E(N18), F1E(N17). The numbers are very close; 17, 18, and 19 times. F3, F4, and F5 are more diverse than F1, and F2. Among them, F4 owns most variables leaving lines the only condition uncontrolled, and oppositely, F3 and F5 are free from variables. The outcome is that the representative products of F4E, F407, are selected up to 30 times which is the highest number among all 15 samples. We can conclude that it is the variable lines that bring this result. The coke bottle shape and curved lines makes F407 the most typical emotional products. In F3, the highest score goes to the neutral item F305. The result is unusual, for the highest score of every group is normally taken up by emotional or rational products. The reason for this outcome might be F3's fun quality. This quality endows F3 with varied styles regardless of shapes, textures, or colors, and further distracts respondents' attentions from products of rationality and emotions.

In contrast, F305 has no fancy features in shapes, textures, or colors. In other words, it is its average performance that makes it a good option in the group of neutral products. Another style that carries with miscellaneous types but without any variable is F5, feeling. The highest score in this group goes to its rational product F503; probably is because of its popularity and feature of modern technology. Besides, in F5, its two typical emotional items F503 and F505, which are various in shapes, textures, colors, and functions, both get the score of 10. This is an interesting phenomenon. Does it mean F5, feeling, the representative feature of the 21th century, is a varied attribute itself? If combining F5 with the emotional attribute, which is various in styles as well, any amount of variation seems possible for it.

### 4.2 *Experiment II*

The goal of this experiment at this stage is to understand, according to 5F's properties of rationality and emotion, what is the score for products that retain the same positions and how do those products react to variables? After being analyzed, SPSS shows which items are selected most often. The result is shown in figure 5 ,and as follows:

(1) Consistency of positions of 5F after resetting. First, the three products of F1 stay at the same places after resetting. Second, in F2, rational and emotional products retain the same positions while the neutral item is replaced with the neutral product in F1. Next, in F3, rational product remains where it was, neutral product is replaced with emotional product in F3, and emotional product in F5 is substituted for the kind in F3. Then, all three products remain at the same place in F4. Last, the position for rational product is the same as before. F2 and F4 co-exist in neutral group. The emotional group is replaced with emotional product in F4 and rational product in F5.
(2) The top 7 of frequency are F3E(N18), F4R(N18), F1R(N16), F3R(N16), F3N(N14), F1E(N13), F5R(N12).
(3) Consistency. In terms with 5F, products of rationality, neutrality, and emotion in F1 and F4 are put in the original places as in experiment 1 after resetting. Therefore, we can assume that the reason for this result is that F1, function, is has little variation, and F4 has only one variable. So, we can conclude that products with less characteristics are highly consistent in features. Besides, in terms of the three properties, after resetting, rational products in five styles are placed exactly in their original spots. We can make a quick judgment that the high consistency in rational properties makes it easier to define this kind of product. On the contrary, the quality of emotional attribute is just the opposite.
(4) Inconsistency. F3, and F5 which without any control variables are failed to be placed at the same position after this experiment. (only rational products return to their beginning place). F306 in F3 is ranked at the place of neutral, instead of emotion, its original spot. In addition, the place of F3's emotional products is replaced with an emotional product, F502, in F5 which gets the highest score 18. What we achieve at this part of experiment is that the attribute of F502 is equipped with the quality of F3, fun, and the authentic name of F502 is Analog Fan, a wooden fan that has to be pulled by men to generate power—to many experiment takers, the feature of fun plays a more important role than the feature of feeling does. Another interesting result we procure from this experiment is that Dyson's bladeless fan is not only a representative product of rationality in F5, but also a product of emotions which goes together the typical emotional product in F4, F407. It is a particular outcome that one product is selected as a rational and emotional representative simultaneously. It might be owing to that although this product is characterized with modern technology, the delicacy that revealed from its colors, shapes, and textures gives this product the feature of emotion. Whether the line between rationality and emotions becomes blurred in the industry of product design in the 21th century, we cannot be sure. However, in contrast with products that put emphasis on function or friendliness separately in the 20th century, the two properties stand side by side nowadays.

(5) Views on properties of rationality and emotion with three products are listed below:
(a) F407. In experiment 1, the reason why it receives the highest scores of 30 among the 15 items is its single variable, lines. Comparing other products' figures with the coke bottle shaped appearance of F407, its curved lines makes it the best item of emotional products. In experiment 2, it gets the highest score of products of emotional properties in the group of F4 and F5, so its quality of emotions is beyond doubt. However, it is not as clear as its emotional attribute that whether it is F4 or F5 is the group from 5F that it exactly belongs to. (b) F502 is the item that receives the highest score among all 15 products in experiment 2; 18 times in total. Yet, its position in 5F is switched to F3 from F5. The authentic name of F502 is Analog Fan, which is a kind of fan that made of wood and needs to be pulled by men to generate power. For the subjects, feature of fun plays a key role in this item more than feature of feeling does; however, they both belong to emotional attribute. (c) F503 is Dyson's bladeless fan. It is not only the typical rational product in F5's rational, but also an emotional product that co-exists with F407. The feeling of modern technology and delicacy that comes from textures, colors, and figure, make it a representative product of both rationality and emotion.

## 5 CONCLUSIONS

The paper explored and discussed properties of rationality, neutrality, and emotions in five different product styles (5F), including an experiment with two stages. The results are explained as follows:

1. Feature of appearance for product of rationality and emotion: the commonality for rational products is their rather straight lines. Colors and textures for the properties are cool silver metal materials, achromatic black and grey plastic, and silver blue paint. As for emotional products, their common feature is curved lines. Colors and textures are warm gold, bronze metal, plastics in the color of wood, plastics with high color saturation and warm colors, and natural wood color and texture.
2. Feature of consistency in rational and emotional products. Styles with fewer variations are easier to to analyze about what properties they carry. For example, F1, F2, and F4 are three styles less varied in features, so we can say that these kinds of products are highly consistent in qualities. F1 and F2 have little variation in styles and characteristics, and F4 is controlled by its variables. Viewing from their rational and emotional properties, the rational kind has more in common and is easier to define. The emotional properties, on the contrary, do not have many features that are shared between products; a group that is rather hard to define.
3. Exploring properties of rationality and emotion through the following three products, F407, F502, and F503, we can see that people tend to get confused by F3(Fun), F4(Fancy), and F5(Feeling). In other words, they are miscellaneous in features.

## REFERENCES


Aaker, D. A., & Shansby, G. J. (1982). Positioning Your Product. *Business Horizon*, 56–62.

Blijlevens, J., Creusen, M. E. H., & Schoormans, J. P. L. (2009). How consumers perceive product appearance: The identification of three product appearance properties *International Journal of Design, 3*(3), 27–35.

Dahlgaard, J. J., Schütte, S., Alikalfa, E., & Dahlgaard Park, S. M. (2008). Kansei /affective engineering design – A methodology for profound affection and attractive quality creation. *The TQM Journal – International Review of Organizational Improvements, 20*(4), 299–311.

Dandavate, U., Sanders, E. B. -N., & Stuart, S. (1996). Emotions matter: User empathy in the product development process. In *Proceedings of the 40th Annual Meeting* of the *Human Factors and Ergonomics Society* (pp.415–418). Santa Monica, CA: HFES.

Fulton Suri, J. (2003). Empathic design: Informed and inspired by other people's experience. In I. Koskinen, K. Battarbee, & T. Mattelmäki (Eds.), *Empathic design: User experience in product design* (pp. 51–58). Helsinki, Finland: Edita IT Press.

Hsiao, S. -W., & Wang, H. -P. (1998). Applying the semantic transformation method to product form design. *Design Studies, 19*(3), 209–330.

Imagawa A., Hayashi Sumi M., Tada M., & Sunako K. (1989). Emotional consumption, rational consumption – What kind of consumer you are? (Zheng translation). Taipei: keung.

Lin, R.T. (2005). The combination of technology and humanity – The cultural and creative, *Science Development, 396*, 68–75.

Orth, U. R., & Malkewitz, K. (2008). Holistic package design and consumer brand impressions. *Journal of Marketing, 72*(3), 64–71.

# Author index

Albayrak, A. 299, 357
Amemiya, T. 117
Anastasiadis, G. 329
Ang, L.A. 239
Ariyoshi, H. 117

Barraclough, S. 27, 37
Ben-Arieh, D. 251, 259
Bodewes, H. 299

Chang, C.-C. 175
Chang, S.I. 239
Chang, W.-R. 175
Chen, C.-W. 347
Chen, C.-Y. 319
Chen, S.-J. 47
Chen, T.-Y. 319
Chen, Y. 213
Chen, Y.N. 109
Cheng, L.-W. 87
Chien, C.-W. 47, 409
Chiu, M.-C. 337
Chou, C. 319
Chou, F.-H. 281
Chuan, H.-C. 387
Clarkson, P.J. 289

De Capitani, C. 181
Dhamanti, I. 27, 37
Doran, D. 79, 199
Douglas-Mankin, K. 239

Goossens, R.H.M. 299, 329, 357
Guo, Y. 229

Horberry, T. 289
Hsu, J.Y. 213
Hsu, Y.-W. 337
Huang, C.-H. 371
Huang, F.-T. 337
Huang, P.-W. 379
Huang, S.-F. 123
Huang, Y.-Y. 17
Hwan Yun, M. 155
Hwang, S.-L. 337
Hwang, S.L. 109

Iridiastadi, H. 101

Jee, S.-C. 155
Jennifer Tsai, P.-F. 87
Jian, S.-Y. 281
Jiang, B.C. 101
Juan, C.Y. 213, 223

Kajimura, A. 55, 71, 191
Kao, Y.-F. 347
Koike, E. 71
Kuan, P.-C. 387
Kuang, Y. 259
Kwon, S. 155

Lancaster, J. 299, 329
Lee, C.-T. 337
Lee, J.-T. 141
Leggat, S. 27, 37
Lesch, M.F. 175
Leupold, M. 207
Li, A.-H. 309
Liang, F. 229
Liao, S. 37
Lien, C.-W. 141
Lin, C.-L. 47, 399, 409
Lin, C.-T. 393
Lin, J.-L. 387
Lin, R.-T. 409
Lin, R.F. 309
Lioenita, D.C. 363
Liu, B.-S. 141
Liu, L.-W. 399
Lorenzi, F. 181
Lu, C.-W. 363, 379

Margolies, D. 259
Matz, S. 175
Max Liang, S.-F. 87
Melles, M. 329, 357
Menozzi, M. 17
Moes, C.C.M. 329
Mohandoss, M. 163, 167
Montinari, M. 181
Muraki, S. 149

Nagaoka, H. 71
Nakashima, H. 149
Nechols, J. 259
Ni, S.-M. 387
Nishikawa, K. 117
Nurwulan, N.R. 101

Oh, M.-D. 155
Onofrio, R. 181

Pang, R. 309
Pao, C.-C. 399
Patil, V. 289
Pernazza, E. 181
Prieb, M. 269

Rajkumar, J.S. 163, 167
Ranganathan, R. 163, 167
Rau, C.-L. 87
Rys, M. 269
Ryu, T. 155

Schmidt, D. 207
Serranheira, F. 133
Sharan, D. 163, 167
Shih, H.-L. 337
Shih, Y.-C. 399
Shintani, N. 55, 71, 191
Sousa-Uva, A. 133
Sousa-Uva, M. 133
Stock, P. 207
Strasser, H. 3
Su, C.-J. 123
Sui, J. 229
Sun, T.-L. 371
Sun, W. 79, 199
Sunderman, T. 299
Suzaki, Y. 117

Takayama, N. 117
Tatsukawa, S. 55, 71, 191
Tavakkol, B. 239
Taylor, C. 269
Teng, Y.-C. 289
Trucco, P. 181
Tsai, H.-P. 337
Tsai, K.-C. 309

Utarini, A. 37

Vasicheva, V. 61
Vehmeijer, S.B.W. 299, 329

Wang, M.-J. 93
Wang, Q. 229
Ward, J. 289
Wu, J. 259
Wu, J.W. 223
Wu, T.Y. 393

Yang, K.H. 213, 223
Yanto 363
Yeap Loh, P. 149
Yin Wu, C. 93

Zhao, S. 259
Zineldin, J. 61
Zineldin, M. 61
Zülch, G. 207